A First Course in

COMPLEX ANALYSIS

WITH APPLICATIONS

The Jones and Bartlett Publishers Series in Mathematics

Geometry

Geometry with an Introduction to Cosmic Topology
Hitchman (978-0-7637-5457-0) © 2009

Euclidean and Transformational Geometry: A Deductive Inquiry
Libeskind (978-0-7637-4366-6) © 2008

A Gateway to Modern Geometry: The Poincaré Half-Plane, Second Edition
Stahl (978-0-7637-5381-8) © 2008

Understanding Modern Mathematics
Stahl (978-0-7637-3401-5) © 2007

Lebesgue Integration on Euclidean Space, Revised Edition
Jones (978-0-7637-1708-7) © 2001

Precalculus

Precalculus with Calculus Previews (Expanded Volume), Fourth Edition
Zill/Dewar (978-0-7637-6631-3) © 2010

Precalculus: A Functional Approach to Graphing and Problem Solving, Sixth Edition
Smith (978-0-7637-5177-7) © 2010

Precalculus with Calculus Previews (Essentials Version), Fourth Edition
Zill/Dewar (978-0-7637-3779-5) © 2007

Calculus

Calculus of a Single Variable: Early Transcendentals, Fourth Edition
Zill/Wright (978-0-7637-4965-1) © 2010

Multivariable Calculus, Fourth Edition
Zill/Wright (978-0-7637-4966-8) © 2010

Calculus: Early Transcendentals, Fourth Edition
Zill/Wright (978-0-7637-5995-7) © 2010

Exploring Calculus with MATLAB: Topics and Applications
Smaldone (978-0-7637-7002-0) © 2010

Calculus: The Language of Change
Cohen/Henle (978-0-7637-2947-9) © 2005

Applied Calculus for Scientists and Engineers
Blume (978-0-7637-2877-9) © 2005

Calculus: Labs for Mathematica
O'Connor (978-0-7637-3425-1) © 2005

Calculus: Labs for MATLAB
O'Connor (978-0-7637-3426-8) © 2005

Linear Algebra

Linear Algebra: Theory and Applications
Cheney/Kincaid (978-0-7637-5020-6) © 2009

Linear Algebra with Applications, Sixth Edition
Williams (978-0-7637-5753-3) © 2008

Advanced Engineering Mathematics

An Elementary Course in Partial Differential Equations, Second Edition
Amaranath (978-0-7637-6244-5) © 2009

Advanced Engineering Mathematics, Third Edition
Zill/Cullen (978-0-7637-4591-2) © 2006

Complex Analysis

A First Course in Complex Analysis with Applications, Second Edition
Zill/Shanahan (978-0-7637-5772-4) © 2009

Complex Analysis for Mathematics and Engineering, Fifth Edition
Mathews/Howell (978-0-7637-3748-1) © 2006

Classical Complex Analysis
Hahn (978-0-8672-0494-0) © 1996

Real Analysis

Closer and Closer: Introducing Real Analysis
Schumacher (978-0-7637-3593-7) © 2008

The Way of Analysis, Revised Edition
Strichartz (978-0-7637-1497-0) © 2000

Topology

Foundations of Topology, Second Edition
Patty (978-0-7637-4234-8) © 2009

Discrete Math and Logic

Discrete Structures, Logic, and Computability, Third Edition
Hein (978-0-7637-7206-2) © 2010

Essentials of Discrete Mathematics
Hunter (978-0-7637-4892-0) © 2009

Logic, Sets, and Recursion, Second Edition
Causey (978-0-7637-3784-9) © 2006

Numerical Methods

Numerical Mathematics
Grasselli/Pelinovsky (978-0-7637-3767-2) © 2008

Exploring Numerical Methods: An Introduction to Scientific Computing Using MATLAB
Linz (978-0-7637-1499-4) © 2003

Advanced Mathematics

Clinical Statistics: Introducing Clinical Trials, Survival Analysis, and Longitudinal Data Analysis
Korosteleva (978-0-7637-5850-9) © 2009

Harmonic Analysis: A Gentle Introduction
DeVito (978-0-7637-3893-8) © 2007

Beginning Number Theory, Second Edition
Robbins (978-0-7637-3768-9) © 2006

A Gateway to Higher Mathematics
Goodfriend (978-0-7637-2733-8) © 2006

For more information on any of the titles above please visit us online at http://www.jbpub.com/math. Qualified instructors, contact your Publisher's Representative at 1-800-832-0034 or info@jbpub.com to request review copies for course consideration.

SECOND EDITION

A First Course in
COMPLEX
ANALYSIS
WITH APPLICATIONS

Dennis G. Zill
Loyola Marymount University

Patrick D. Shanahan
Loyola Marymount University

JONES AND BARTLETT PUBLISHERS

Sudbury, Massachusetts

BOSTON   TORONTO   LONDON   SINGAPORE

World Headquarters

Jones and Bartlett Publishers
40 Tall Pine Drive
Sudbury, MA 01776
978-443-5000
info@jbpub.com
www.jbpub.com

Jones and Bartlett Publishers
Canada
6339 Ormindale Way
Mississauga, Ontario L5V 1J2
Canada

Jones and Bartlett Publishers
International
Barb House, Barb Mews
London W6 7PA
United Kingdom

Jones and Bartlett's books and products are available through most bookstores and online booksellers. To contact Jones and Bartlett Publishers directly, call 800-832-0034, fax 978-443-8000, or visit our website www.jbpub.com.

Substantial discounts on bulk quantities of Jones and Bartlett's publications are available to corporations, professional associations, and other qualified organizations. For details and specific discount information, contact the special sales department at Jones and Bartlett via the above contact information or send an email to specialsales@jbpub.com.

Production Credits

Acquisitions Editor: Timothy Anderson
Editorial Assistant: Melissa Potter
Senior Marketing Manager: Andrea DeFronzo
Production Director: Amy Rose
Production Manager: Jennifer Bagdigian
Production Assistant: Ashlee Hazeltine
V.P., Manufacturing and Inventory Control: Therese Connell
Composition: Northeast Compositors, Inc.
Cover Design: Kristin E. Parker
Cover and Title Page Image: © Leigh Prather/Dreamstime.com
Printing and Binding: Malloy, Inc.
Cover Printing: Malloy, Inc.

Library of Congress Cataloging-in-Publication Data

Zill, Dennis G., 1940-
   A first course in complex analysis with applications / Dennis G. Zill and Patrick Shanahan. — 2nd ed.
      p. cm.
   Includes index.
   ISBN 978-0-7637-5772-4 (casebound) — ISBN 0763757721(casebound) 1. Functions of complex variables.
I. Shanahan, Patrick, 1968- II. Title. III. Title: Complex analysis.
   QA331.7.Z55 2009
   515'.9—dc22
                                    2008050293

6048
Printed in the United States of America
14 13 12 11 10   10 9 8 7 6 5 4 3 2

For Dana, Kasey, and Cody

Contents

Preface

Philosophy   The first edition of this text grew out the material in Chapters 17–20 of *Advanced Engineering Mathematics, Third Edition* (Jones and Bartlett Publishers, 2006), by Dennis G. Zill and the late Michael R. Cullen. This second edition represents an expansion and revision of that original material and is intended for use either in a one-semester or a one-quarter course. Its aim is to introduce the basic principles and applications of complex analysis to undergraduates who have no prior knowledge of this subject. The writing is straightforward and reflects the no-nonsense style of *Advanced Engineering Mathematics*.

The motivation to adapt the material from *Advanced Engineering Mathematics* into a stand-alone text came from our dissatisfaction with the succession of textbooks that we have used over the years in our departmental undergraduate course offering in complex analysis. It has been our experience that books claiming to be accessible to undergraduates were often written at a level that was too high for our audience. The "audience" for our junior-level course consists of some majors in mathematics, some majors in physics, but mostly majors from electrical engineering and computer science. At our institution, a typical student majoring in science or engineering is not required to take theory-oriented mathematics courses such as methods of proof, linear algebra, abstract algebra, advanced calculus, or introductory real analysis. The only prerequisite for our undergraduate course in complex analysis is the completion of the third semester of the calculus sequence. For the most part, then, calculus is all that we assume by way of preparation for a student to use this text, although some working knowledge of differential equations would be helpful in the sections devoted to applications. We have kept the theory in this text to what we hope is a manageable level, concentrating only on what we feel is necessary in a first course. Many concepts are presented in an informal and conceptual style rather than in the conventional definition/theorem/proof format. We think it would be fair to characterize this text as a continuation of the study of calculus, but this time as the study of the calculus of functions of a complex variable. But do not misinterpret the preceding words; we have not abandoned theory in favor of "cookbook recipes." Proofs of major results are presented and much of the standard terminology is used throughout. Indeed, there are many problems in the exercise sets where a student is asked to prove something. We readily admit that any student—not just majors in mathematics—can gain some mathematical maturity and insight by attempting a proof. However, we also know that most students have no idea how to start a proof. Thus, in some of our "proof" problems the reader is either guided through the starting steps or is provided a strong hint on how to proceed.

Changes in This Edition   First let us say what is not changed in this edition:

- The original underlying philosophy, the number of chapters, and the order of the sections within each chapter are the same as in the first edition. We have purposely kept the number of chapters in this text to seven. This was done for two reasons: to provide an appropriate quantity of material so that most of it can reasonably be covered in a one-term course, and at the same time to minimize the cost.

Our primary goal for this second edition was to enhance the strengths of the original text. As such, we did the following:

- Many new problems, especially conceptual problems, have been added to the exercises. In addition, some existing problems were improved while others were deemed ineffectual and were sent into early retirement.

- Some text and a number of examples have been rewritten in order to either clarify or to expand on the topics under discussion. In some instances we found that clarity of exposition could be improved by adopting the "less is more" attitude.

- Errors and typos in the first edition have been corrected.

- The numbering of the figures, theorems, and definitions has been changed. We have switched to a double-decimal numeration system. For example, the interpretation of "Figure 1.2.3" is

$$\text{Chapter} \rightharpoondown \quad \upharpoonleft \text{Section}$$
$$1.2.3 \longleftarrow \text{Third figure in the section}$$

We feel that this type of numeration will make it easier to find figures, theorems, and definitions when they are referred to in later sections or chapters.

Features of This Text   We have retained many of the features of the previous edition. Each chapter begins with its own opening page that includes a table of contents and a brief introduction describing the material to be covered in the chapter. Moreover, each section in a chapter starts with introductory comments on the specifics covered in that section. Almost every section ends with a feature called Remarks, in which we talk to the students about areas where real and complex calculus differ or discuss additional interesting topics (such as the Riemann sphere and Riemann surfaces), which are related to, but not formally covered in, the section. Several of the longer sections, although unified by subject matter, have been partitioned into subsections; this was done to facilitate covering the material over several class periods. The corresponding exercise sets were divided in the same manner in order to make easier the assignment of homework. Comments, clarifications, and some words of caution are liberally scattered throughout the text by means of marginal annotations.

    We have provided a lot of examples and have tried very hard to supply all pertinent details in their solution. Because applications of complex analysis are often compiled into a single chapter placed at the end of the text, instructors are sometimes hard-pressed to cover any applications in the course. Complex analysis is a powerful tool in applied mathematics. So to facilitate covering this beautiful aspect of the subject, we have chosen to end each chapter with a separate section on applications.

The exercise sets are constructed in a pyramidal fashion, and each set has at least two parts. The first part of an exercise set is a generous supply of routine drill-type problems; the second part consists of conceptual word and geometrical problems. In many exercise sets there is a third part devoted to the use of technology. Since the default operational mode of all computer algebra systems is complex analysis, we have placed an emphasis on that type of software. Although we have discussed the use of Mathematica® in the text proper, the problems are generic in nature.

Each chapter ends with a Chapter Review Quiz. We thought that something more conceptual would be a bit more interesting than the rehashing of the same old problems given in the traditional Chapter Review Exercises.

Answers to selected odd-numbered problems are given in the back of the text. Since the conceptual problems could also be used as topics for classroom discussion, we decided not to include their answers.

For student and instructor resources please contact your Jones and Bartlett Publishers Representative at 1-800-832-0034 or info@jbpub.com.

Acknowledgments   We would like to express our appreciation to our colleagues at Loyola Marymount University who have taught from the text, as well as those instructors who have taken the time to contact us, for their words of encouragement, criticisms, corrections, and thoughtful suggestions. Special recognition is reserved for our colleague Lily Khadjavi who provided numerous helpful suggestions for both the first and second editions of the book. We also wish to acknowledge the valuable input from past LMU students who used this book in preliminary and published versions. Finally, a deeply felt "thank you" goes to the following reviewers who contributed to the first edition of the book:

    Nicolae H. Pavel, Ohio University

    Marcos Jardim, University of Pennsylvania

    Ilia A. Binder, Harvard University

And "thank you" again to those who reviewed this edition for us:

    Joyati Debnath, Winona State University

    Rich Mikula, William Paterson University of New Jersey

    Jim Vance, Wright State University

    Chris Masters, Doane College

    George J. Miel, University of Nevada, Las Vegas

    Jeffrey Lawson, Western Carolina University

    Javad Namazi, Fairleigh Dickinson University

    Irl Bivens, Davidson College

Lastly, we would like to thank our editor, Tim Anderson, for his constant but good-natured prodding, and the staff of the production department for another job well done.

Final Request   Compiling a mathematics text, even one of this modest size, entails juggling thousands of words and symbols. Experience has taught us that errors—typos or just plain mistakes—seem to be an inescapable by-product of the textbook-writing endeavor. We apologize in advance for any errors that you may find and urge you to bring them to our attention. You can email them directly to our editor at tanderson@jbpub.com.

Dennis G. Zill   Patrick D. Shanahan

Complex Numbers and the Complex Plane

Introduction In elementary courses you learned about the existence, and some of the properties, of complex numbers. But in courses in calculus, it is most likely that you did not even see a complex number. In this text we study nothing but complex numbers and the calculus of functions of a complex variable.

We begin with an in-depth examination of the arithmetic and algebra of complex numbers.

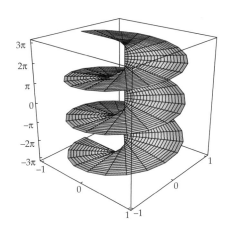

Riemann surface for arg(z). See page 88.

1.1   Complex Numbers and Their Properties

No one person "invented" complex numbers, but controversies surrounding the use of these numbers existed in the sixteenth century. In their quest to solve polynomial equations by formulas involving radicals, early dabblers in mathematics were forced to admit that there were other kinds of numbers besides positive integers. Equations such as $x^2 + 2x + 2 = 0$ and $x^3 = 6x + 4$ that yielded "solutions" $1 + \sqrt{-1}$ and $\sqrt[3]{2 + \sqrt{-2}} + \sqrt[3]{2 - \sqrt{-2}}$ caused particular consternation within the community of fledgling mathematical scholars because *everyone* knew that there are no numbers such as $\sqrt{-1}$ and $\sqrt{-2}$, numbers whose square is negative. Such "numbers" exist only in one's imagination, or as one philosopher opined, "the imaginary, (the) bosom child of complex mysticism." Over time these "imaginary numbers" did not go away, mainly because mathematicians as a group are tenacious and some are even practical. A famous mathematician held that even though "they exist in our imagination . . . nothing prevents us from . . . employing them in calculations." Mathematicians also hate to throw anything away. After all, a memory still lingered that negative numbers at first were branded "fictitious." The concept of *number* evolved over centuries; gradually the set of numbers grew from just positive integers to include rational numbers, negative numbers, and irrational numbers. But in the eighteenth century the number concept took a gigantic evolutionary step forward when the German mathematician Carl Friedrich Gauss put the so-called imaginary numbers—or *complex numbers*, as they were now beginning to be called—on a logical and consistent footing by treating them as an extension of the real number system.

Our goal in this first section is to examine some basic definitions and the arithmetic of complex numbers.

The Imaginary Unit   Even after gaining wide respectability, through the seminal works of Carl Friedrich Gauss and the French mathematician Augustin Louis Cauchy, the unfortunate name "imaginary" has survived down the centuries. The symbol i was originally used as a disguise for the embarrassing symbol $\sqrt{-1}$. We now say that i is the **imaginary unit** and define it by the property $i^2 = -1$. Using the imaginary unit, we build a general complex number out of two real numbers.

Definition 1.1.1   Complex Number

A **complex number** is any number of the form $z = a + ib$ where a and b are real numbers and i is the imaginary unit.

Terminology   The notations $a + ib$ and $a + bi$ are used interchangeably. The real number a in $z = a + ib$ is called the **real part** of z; the real number b is called the **imaginary part** of z. The real and imaginary parts of a complex number z are abbreviated $\mathrm{Re}(z)$ and $\mathrm{Im}(z)$, respectively. For example, if $z = 4 - 9i$, then $\mathrm{Re}(z) = 4$ and $\mathrm{Im}(z) = -9$. A real constant multiple of the imaginary unit is called a **pure imaginary number**. For example, $z = 6i$ is a pure imaginary number. Two complex numbers are **equal** if their corresponding real and imaginary parts are equal. Since this simple concept is sometimes useful, we formalize the last statement in the next definition.

Note: The imaginary part of $z = 4 - 9i$ is -9 **not** $-9i$.

> **Definition 1.1.2 Equality**
>
> Complex numbers $z_1 = a_1 + ib_1$ and $z_2 = a_2 + ib_2$ are **equal**, $z_1 = z_2$, if $a_1 = a_2$ and $b_1 = b_2$.

In terms of the symbols $\text{Re}(z)$ and $\text{Im}(z)$, Definition 1.1.2 states that $z_1 = z_2$ if $\text{Re}(z_1) = \text{Re}(z_2)$ and $\text{Im}(z_1) = \text{Im}(z_2)$.

The totality of complex numbers or the set of complex numbers is usually denoted by the symbol $\mathbf{C}$. Because *any* real number a can be written as $z = a + 0i$, we see that the set $\mathbf{R}$ of real numbers is a subset of $\mathbf{C}$.

Arithmetic Operations Complex numbers can be added, subtracted, multiplied, and divided. If $z_1 = a_1 + ib_1$ and $z_2 = a_2 + ib_2$, these operations are defined as follows.

Addition: $\quad z_1 + z_2 = (a_1 + ib_1) + (a_2 + ib_2) = (a_1 + a_2) + i(b_1 + b_2)$

Subtraction: $\quad z_1 - z_2 = (a_1 + ib_1) - (a_2 + ib_2) = (a_1 - a_2) + i(b_1 - b_2)$

Multiplication: $\quad z_1 \cdot z_2 = (a_1 + ib_1)(a_2 + ib_2)$

$$= a_1 a_2 - b_1 b_2 + i(b_1 a_2 + a_1 b_2)$$

Division: $\quad \dfrac{z_1}{z_2} = \dfrac{a_1 + ib_1}{a_2 + ib_2}, \ a_2 \neq 0, \text{ or } b_2 \neq 0$

$$= \frac{a_1 a_2 + b_1 b_2}{a_2^2 + b_2^2} + i\frac{b_1 a_2 - a_1 b_2}{a_2^2 + b_2^2}$$

The familiar commutative, associative, and distributive laws hold for complex numbers:

Commutative laws: $\quad \begin{cases} z_1 + z_2 = z_2 + z_1 \\ z_1 z_2 = z_2 z_1 \end{cases}$

Associative laws: $\quad \begin{cases} z_1 + (z_2 + z_3) = (z_1 + z_2) + z_3 \\ z_1(z_2 z_3) = (z_1 z_2) z_3 \end{cases}$

Distributive law: $\quad z_1(z_2 + z_3) = z_1 z_2 + z_1 z_3$

In view of these laws, there is no need to memorize the definitions of addition, subtraction, and multiplication.

> *Addition, Subtraction, and Multiplication*
>
> (i) *To add (subtract) two complex numbers, simply add (subtract) the corresponding real and imaginary parts.*
>
> (ii) *To multiply two complex numbers, use the distributive law and the fact that $i^2 = -1$.*

The definition of division deserves further elaboration, and so we will discuss that operation in more detail shortly.

EXAMPLE 1   Addition and Multiplication

If $z_1 = 2 + 4i$ and $z_2 = -3 + 8i$, find (**a**) $z_1 + z_2$ and (**b**) $z_1 z_2$.

Solution (**a**) By adding real and imaginary parts, the sum of the two complex numbers z_1 and z_2 is

$$z_1 + z_2 = (2 + 4i) + (-3 + 8i) = (2 - 3) + (4 + 8)i = -1 + 12i.$$

(**b**) By the distributive law and $i^2 = -1$, the product of z_1 and z_2 is

$$z_1 z_2 = (2 + 4i)(-3 + 8i) = (2 + 4i)(-3) + (2 + 4i)(8i)$$
$$= -6 - 12i + 16i + 32i^2$$
$$= (-6 - 32) + (16 - 12)i = -38 + 4i.$$

☐

Zero and Unity The **zero** in the complex number system is the number $0 + 0i$ and the **unity** is $1 + 0i$. The zero and unity are denoted by 0 and 1, respectively. The zero is the **additive identity** in the complex number system since, for any complex number $z = a + ib$, we have $z + 0 = z$. To see this, we use the definition of addition:

$$z + 0 = (a + ib) + (0 + 0i) = a + 0 + i(b + 0) = a + ib = z.$$

Similarly, the unity is the **multiplicative identity** of the system since, for any complex number z, we have $z \cdot 1 = z \cdot (1 + 0i) = z$.

Conjugate If z is a complex number, the number obtained by changing the sign of its imaginary part is called the **complex conjugate**, or simply **conjugate**, of z and is denoted by the symbol $\bar{z}$. In other words, if $z = a + ib$, then its conjugate is $\bar{z} = a - ib$. For example, if $z = 6 + 3i$, then $\bar{z} = 6 - 3i$; if $z = -5 - i$, then $\bar{z} = -5 + i$. If z is a real number, say, $z = 7$, then $\bar{z} = 7$. From the definitions of addition and subtraction of complex numbers, it is readily shown that the conjugate of a sum and difference of two complex numbers is the sum and difference of the conjugates:

$$\overline{z_1 + z_2} = \bar{z}_1 + \bar{z}_2, \quad \overline{z_1 - z_2} = \bar{z}_1 - \bar{z}_2. \tag{1}$$

Moreover, we have the following three additional properties:

$$\overline{z_1 z_2} = \bar{z}_1 \bar{z}_2, \quad \overline{\left(\frac{z_1}{z_2}\right)} = \frac{\bar{z}_1}{\bar{z}_2}, \quad \bar{\bar{z}} = z. \tag{2}$$

Of course, the conjugate of any finite sum (product) of complex numbers is the sum (product) of the conjugates.

The definitions of addition and multiplication show that the sum and product of a complex number z with its conjugate $\bar{z}$ is a real number:

$$z + \bar{z} = (a + ib) + (a - ib) = 2a \tag{3}$$
$$z\bar{z} = (a + ib)(a - ib) = a^2 - i^2 b^2 = a^2 + b^2. \tag{4}$$

The difference of a complex number z with its conjugate $\bar{z}$ is a pure imaginary number:

$$z - \bar{z} = (a + ib) - (a - ib) = 2ib. \tag{5}$$

Since $a = \text{Re}(z)$ and $b = \text{Im}(z)$, (3) and (5) yield two useful formulas:

$$\text{Re}(z) = \frac{z + \bar{z}}{2} \qquad \text{and} \qquad \text{Im}(z) = \frac{z - \bar{z}}{2i}. \tag{6}$$

However, (4) is *the* important relationship in this discussion because it enables us to approach division in a practical manner.

Division

To divide z_1 by z_2, multiply the numerator and denominator of z_1/z_2 by the conjugate of z_2. That is,

$$\frac{z_1}{z_2} = \frac{z_1}{z_2} \cdot \frac{\bar{z}_2}{\bar{z}_2} = \frac{z_1 \bar{z}_2}{z_2 \bar{z}_2} \tag{7}$$

and then use the fact that $z_2 \bar{z}_2$ is the sum of the squares of the real and imaginary parts of z_2.

The procedure described in (7) is illustrated in the next example.

EXAMPLE 2   Division

If $z_1 = 2 - 3i$ and $z_2 = 4 + 6i$, find z_1/z_2.

Solution We multiply numerator and denominator by the conjugate $\bar{z}_2 = 4 - 6i$ of the denominator $z_2 = 4 + 6i$ and then use (4):

$$\frac{z_1}{z_2} = \frac{2 - 3i}{4 + 6i} = \frac{2 - 3i}{4 + 6i} \frac{4 - 6i}{4 - 6i} = \frac{8 - 12i - 12i + 18i^2}{4^2 + 6^2} = \frac{-10 - 24i}{52}.$$

Because we want an answer in the form $a + bi$, we rewrite the last result by dividing the real and imaginary parts of the numerator $-10 - 24i$ by 52 and reducing to lowest terms:

$$\frac{z_1}{z_2} = -\frac{10}{52} - \frac{24}{52}i = -\frac{5}{26} - \frac{6}{13}i.$$

❑

Inverses In the complex number system, every number z has a unique **additive inverse**. As in the real number system, the additive inverse of $z = a + ib$ is its *negative*, $-z$, where $-z = -a - ib$. For any complex number z, we have $z + (-z) = 0$. Similarly, every *nonzero* complex number z has a **multiplicative inverse**. In symbols, for $z \neq 0$ there exists one and only one nonzero complex number z^{-1} such that $zz^{-1} = 1$. The multiplicative inverse z^{-1} is the same as the **reciprocal** $1/z$.

EXAMPLE 3   Reciprocal

Find the reciprocal of $z = 2 - 3i$.

Solution By the definition of division we obtain

$$\frac{1}{z} = \frac{1}{2-3i} = \frac{1}{2-3i}\,\frac{2+3i}{2+3i} = \frac{2+3i}{4+9} = \frac{2+3i}{13}.$$

Answer should be in the form $a + ib$. ➡ That is,

$$\frac{1}{z} = z^{-1} = \frac{2}{13} + \frac{3}{13}i.$$

You should take a few seconds to verify the multiplication

$$zz^{-1} = (2-3i)\left(\tfrac{2}{13} + \tfrac{3}{13}i\right) = 1.$$

❑

Remarks *Comparison with Real Analysis*

(i) Many of the properties of the real number system **R** hold in the complex number system **C**, but there are some truly remarkable differences as well. For example, the concept of *order* in the real number system does not carry over to the complex number system. In other words, we cannot compare two complex numbers $z_1 = a_1 + ib_1$, $b_1 \neq 0$, and $z_2 = a_2 + ib_2$, $b_2 \neq 0$, by means of inequalities. Statements such as $z_1 < z_2$ or $z_2 \geq z_1$ have no meaning in **C** except in the special case when the two numbers z_1 and z_2 are real. See Problem 55 in Exercises 1.1. Therefore, if you see a statement such as $z_1 = \alpha z_2$, $\alpha > 0$, it is implicit from the use of the inequality $\alpha > 0$ that the symbol α represents a real number.

(ii) Some things that we take for granted as impossible in real analysis, such as $e^x = -2$ and $\sin x = 5$ when x is a real variable, are perfectly correct and ordinary in complex analysis when the symbol x is interpreted as a complex variable. See Example 3 in Section 4.1 and Example 2 in Section 4.3.

We will continue to point out other differences between real analysis and complex analysis throughout the remainder of the text.

EXERCISES 1.1   *Answers to selected odd-numbered problems begin on page ANS-1.*

1. Evaluate the following powers of i.

   (a) i^8 (b) i^{11}

   (c) i^{42} (d) i^{105}

2. Write the given number in the form $a + ib$.

   (a) $2i^3 - 3i^2 + 5i$ (b) $3i^5 - i^4 + 7i^3 - 10i^2 - 9$

   (c) $\dfrac{5}{i} + \dfrac{2}{i^3} - \dfrac{20}{i^{18}}$ (d) $2i^6 + \left(\dfrac{2}{-i}\right)^3 + 5i^{-5} - 12i$

In Problems 3–20, write the given number in the form $a + ib$.

3. $(5 - 9i) + (2 - 4i)$

4. $3(4 - i) - 3(5 + 2i)$

5. $i(5 + 7i)$

6. $i(4 - i) + 4i(1 + 2i)$

7. $(2 - 3i)(4 + i)$

8. $\left(\frac{1}{2} - \frac{1}{4}i\right)\left(\frac{2}{3} + \frac{5}{3}i\right)$

9. $3i + \dfrac{1}{2 - i}$

10. $\dfrac{i}{1 + i}$

11. $\dfrac{2 - 4i}{3 + 5i}$

12. $\dfrac{10 - 5i}{6 + 2i}$

13. $\dfrac{(3 - i)(2 + 3i)}{1 + i}$

14. $\dfrac{(1 + i)(1 - 2i)}{(2 + i)(4 - 3i)}$

15. $\dfrac{(5 - 4i) - (3 + 7i)}{(4 + 2i) + (2 - 3i)}$

16. $\dfrac{(4 + 5i) + 2i^3}{(2 + i)^2}$

17. $i(1 - i)(2 - i)(2 + 6i)$

18. $(1 + i)^2(1 - i)^3$

19. $(3 + 6i) + (4 - i)(3 + 5i) + \dfrac{1}{2 - i}$

20. $(2 + 3i)\left(\dfrac{2 - i}{1 + 2i}\right)^2$

In Problems 21–24, use the **binomial theorem**[*]

$$(A + B)^n = A^n + \frac{n}{1!}A^{n-1}B + \frac{n(n-1)}{2!}A^{n-2}B^2 + \cdots$$
$$+ \frac{n(n-1)(n-2)\cdots(n-k+1)}{k!}A^{n-k}B^k + \cdots + B^n,$$

where $n = 1, 2, 3, \ldots$, to write the given number in the form $a + ib$.

21. $(2 + 3i)^2$

22. $\left(1 - \frac{1}{2}i\right)^3$

23. $(-2 + 2i)^5$

24. $(1 + i)^8$

In Problems 25 and 26, find $\operatorname{Re}(z)$ and $\operatorname{Im}(z)$.

25. $z = \left(\dfrac{i}{3 - i}\right)\left(\dfrac{1}{2 + 3i}\right)$

26. $z = \dfrac{1}{(1 + i)(1 - 2i)(1 + 3i)}$

In Problems 27–30, let $z = x + iy$. Express the given quantity in terms of x and y.

27. $\operatorname{Re}(1/z)$

28. $\operatorname{Re}(z^2)$

29. $\operatorname{Im}(2z + 4\bar{z} - 4i)$

30. $\operatorname{Im}(\bar{z}^2 + z^2)$

In Problems 31–34, let $z = x + iy$. Express the given quantity in terms of the symbols $\operatorname{Re}(z)$ and $\operatorname{Im}(z)$.

31. $\operatorname{Re}(iz)$

32. $\operatorname{Im}(iz)$

33. $\operatorname{Im}((1 + i)z)$

34. $\operatorname{Re}(z^2)$

In Problems 35 and 36, show that the indicated numbers satisfy the given equation. In each case explain why additional solutions can be found.

35. $z^2 + i = 0$, $z_1 = -\dfrac{\sqrt{2}}{2} + \dfrac{\sqrt{2}}{2}i$. Find an additional solution, z_2.

36. $z^4 = -4$; $z_1 = 1 + i$, $z_2 = -1 + i$. Find two additional solutions, z_3 and z_4.

---

[*]Recall that the coefficients in the expansions of $(A + B)^2$, $(A + B)^3$, and so on, can also be obtained using Pascal's triangle.

In Problems 37–42, use Definition 1.1.2 to solve each equation for $z = a + ib$.

37. $2z = i(2 + 9i)$ **38.** $z - 2\bar{z} + 7 - 6i = 0$

39. $z^2 = i$ **40.** $\bar{z}^2 = 4z$

41. $z + 2\bar{z} = \dfrac{2 - i}{1 + 3i}$ **42.** $\dfrac{z}{1 + \bar{z}} = 3 + 4i$

In Problems 43 and 44, solve the given system of equations for z_1 and z_2.

43. $\begin{aligned} iz_1 - \quad\; iz_2 &= 2 + 10i \\ -z_1 + (1 - i)z_2 &= 3 - 5i \end{aligned}$ **44.** $\begin{aligned} iz_1 + (1 + i)z_2 &= 1 + 2i \\ (2 - i)z_1 + \quad 2iz_2 &= 4i \end{aligned}$

Focus on Concepts

45. What can be said about the complex number z if $z = \bar{z}$? If $(z)^2 = (\bar{z})^2$?

46. Think of an alternative solution to Problem 24. Then without doing any significant work, evaluate $(1 + i)^{5404}$.

47. For n a nonnegative integer, i^n can be one of four values: 1, i, -1, and $-i$. In each of the following four cases, express the integer exponent n in terms of the symbol k, where $k = 0, 1, 2, \ldots$.

    (a) $i^n = 1$ **(b)** $i^n = i$

    (c) $i^n = -1$ **(d)** $i^n = -i$

48. There is an alternative to the procedure given in (7). For example, the quotient $(5 + 6i)/(1 + i)$ must be expressible in the form $a + ib$:

$$\frac{5 + 6i}{1 + i} = a + ib.$$

Therefore, $5 + 6i = (1 + i)(a + ib)$. Use this last result and Definition 1.1.2 to find the given quotient. Use this method to find the reciprocal of $3 - 4i$.

49. Assume for the moment that $\sqrt{1 + i}$ makes sense in the complex number system. How would you then demonstrate the validity of the equality

$$\sqrt{1 + i} = \sqrt{\tfrac{1}{2} + \tfrac{1}{2}\sqrt{2}} + i\sqrt{-\tfrac{1}{2} + \tfrac{1}{2}\sqrt{2}}?$$

50. Suppose z_1 and z_2 are complex numbers. What can be said about z_1 or z_2 if $z_1 z_2 = 0$?

51. Suppose the product $z_1 z_2$ of two complex numbers is a nonzero real constant. Show that $z_2 = k\bar{z}_1$, where k is a real number.

52. Without doing any significant work, explain why it follows immediately from (2) and (3) that $z_1\bar{z}_2 + \bar{z}_1 z_2 = 2\text{Re}(z_1\bar{z}_2)$.

53. Mathematicians like to prove that certain "things" within a mathematical system are unique. For example, a proof of a proposition such as "The unity in the complex number system is unique" usually starts out with the assumption that there exist two *different* unities, say, 1_1 and 1_2, and then proceeds to show that this assumption leads to some contradiction. Give one contradiction if it is assumed that two different unities exist.

54. Follow the procedure outlined in Problem 53 to prove the proposition "The zero in the complex number system is unique."

55. A number system is said to be an **ordered system** provided it contains a subset P with the following two properties:

First, for any nonzero number x in the system, either x or $-x$ is (but not both) in P.

Second, if x and y are numbers in P, then both xy and $x + y$ are in P.

In the real number system the set P is the set of *positive* numbers. In the real number system we say x is greater than y, written $x > y$, if and only if $x - y$ is in P. Discuss why the complex number system has no such subset P. [*Hint*: Consider i and $-i$.]

1.2   Complex Plane

A complex number $z = x + iy$ is uniquely determined by an *ordered pair* of real numbers (x, y). The first and second entries of the ordered pairs correspond, in turn, with the real and imaginary parts of the complex number. For example, the ordered pair $(2, -3)$ corresponds to the complex number $z = 2 - 3i$. Conversely, $z = 2 - 3i$ determines the ordered pair $(2, -3)$. The numbers 7, i, and $-5i$ are equivalent to $(7, 0)$, $(0, 1)$, $(0, -5)$, respectively. In this manner we are able to associate a complex number $z = x + iy$ with a *point* (x, y) in a coordinate plane.

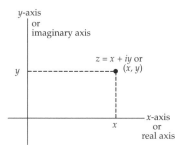

Figure 1.2.1 z-plane

Complex Plane   Because of the correspondence between a complex number $z = x + iy$ and one and only one point (x, y) in a coordinate plane, we shall use the terms *complex number* and *point* interchangeably. The coordinate plane illustrated in Figure 1.2.1 is called the **complex plane** or simply the **z-plane**. The horizontal or x-axis is called the **real axis** because each point on that axis represents a real number. The vertical or y-axis is called the **imaginary axis** because a point on that axis represents a pure imaginary number.

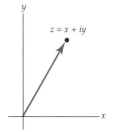

Figure 1.2.2 z as a position vector

Vectors   In other courses you have undoubtedly seen that the numbers in an ordered pair of real numbers can be interpreted as the components of a vector. Thus, a complex number $z = x + iy$ can also be viewed as a two-dimensional position **vector**, that is, a vector whose initial point is the origin and whose terminal point is the point (x, y). See Figure 1.2.2. This vector interpretation prompts us to define the length of the vector z as the distance $\sqrt{x^2 + y^2}$ from the origin to the point (x, y). This length is given a special name.

Definition 1.2.1   Modulus

The **modulus** of a complex number $z = x + iy$ is the real number

$$|z| = \sqrt{x^2 + y^2}. \tag{1}$$

The modulus $|z|$ of a complex number z is also called the **absolute value** of z. We shall use both words *modulus* and *absolute value* throughout this text.

EXAMPLE 1   Modulus of a Complex Number

If $z = 2 - 3i$, then from (1) we find the modulus of the number to be $|z| = \sqrt{2^2 + (-3)^2} = \sqrt{13}$. If $z = -9i$, then (1) gives $|-9i| = \sqrt{(-9)^2} = 9$.

❏

Properties   Recall from (4) of Section 1.1 that for any complex number $z = x + iy$ the product $z\bar{z}$ is a real number; specifically, $z\bar{z}$ is the sum of the squares of the real and imaginary parts of z: $z\bar{z} = x^2 + y^2$. Inspection of (1) then shows that $|z|^2 = x^2 + y^2$. The relations

$$|z|^2 = z\bar{z} \quad \text{and} \quad |z| = \sqrt{z\bar{z}} \tag{2}$$

deserve to be stored in memory. The modulus of a complex number z has the additional properties.

$$|z_1 z_2| = |z_1|\,|z_2| \quad \text{and} \quad \left|\frac{z_1}{z_2}\right| = \frac{|z_1|}{|z_2|}. \tag{3}$$

Note that when $z_1 = z_2 = z$, the first property in (3) shows that

$$\left|z^2\right| = |z|^2. \tag{4}$$

The property $|z_1 z_2| = |z_1|\,|z_2|$ can be proved using (2) and is left as an exercise. See Problem 49 in Exercises 1.2.

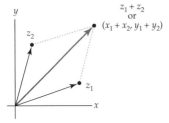

(a) Vector sum

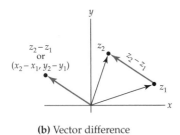

(b) Vector difference

Figure 1.2.3 Sum and difference of vectors

Distance Again   The addition of complex numbers $z_1 = x_1 + iy_1$ and $z_2 = x_2 + iy_2$ given in Section 1.1, when stated in terms of ordered pairs:

$$(x_1,\ y_1) + (x_2,\ y_2) = (x_1 + x_2,\ y_1 + y_2)$$

is simply the component definition of vector addition. The vector interpretation of the sum $z_1 + z_2$ is the vector shown in Figure 1.2.3(a) as the main diagonal of a parallelogram whose initial point is the origin and terminal point is $(x_1 + x_2,\ y_1 + y_2)$. The difference $z_2 - z_1$ can be drawn either starting from the terminal point of z_1 and ending at the terminal point of z_2, or as a position vector whose initial point is the origin and terminal point is $(x_2 - x_1,\ y_2 - y_1)$. See Figure 1.2.3(b). In the case $z = z_2 - z_1$, it follows from (1) and Figure 1.2.3(b) that the **distance between two points** $z_1 = x_1 + iy_1$ and $z_2 = x_2 + iy_2$ in the complex plane is the same as the distance between the origin and the point $(x_2 - x_1,\ y_2 - y_1)$; that is, $|z| = |z_2 - z_1| = |(x_2 - x_1) + i(y_2 - y_1)|$ or

$$|z_2 - z_1| = \sqrt{(x_2 - x_1)^2 + (y_2 - y_1)^2}. \tag{5}$$

When $z_1 = 0$, we see again that the modulus $|z_2|$ represents the distance between the origin and the point z_2.

EXAMPLE 2   Set of Points in the Complex Plane

Describe the set of points z in the complex plane that satisfy $|z| = |z - i|$.

Solution We can interpret the given equation as equality of distances: The distance from a point z to the origin equals the distance from z to the point i. Geometrically, it seems plausible from Figure 1.2.4 that the set of points z lie on a horizontal line. To establish this analytically, we use (1) and (5) to write $|z| = |z - i|$ as:

$$\sqrt{x^2 + y^2} = \sqrt{x^2 + (y-1)^2}$$
$$x^2 + y^2 = x^2 + (y-1)^2$$
$$x^2 + y^2 = x^2 + y^2 - 2y + 1.$$

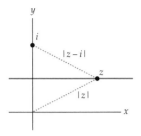

Figure 1.2.4 Horizontal line is the set of points satisfying $|z| = |z - i|$.

The last equation yields $y = \frac{1}{2}$. Since the equality is true for arbitrary x, $y = \frac{1}{2}$ is an equation of the horizontal line shown in color in Figure 1.2.4. Complex numbers satisfying $|z| = |z - i|$ can then be written as $z = x + \frac{1}{2}i$. ❑

Inequalities In the Remarks at the end of the last section we pointed out that no order relation can be defined on the system of complex numbers. However, since $|z|$ is a real number, we can compare the absolute values of two complex numbers. For example, if $z_1 = 3 + 4i$ and $z_2 = 5 - i$, then $|z_1| = \sqrt{25} = 5$ and $|z_2| = \sqrt{26}$ and, consequently, $|z_1| < |z_2|$. In view of (1), a geometric interpretation of the last inequality is simple: *The point* $(3, 4)$ *is closer to the origin than the point* $(5, -1)$.

Now consider the triangle given in Figure 1.2.5 with vertices at the origin, z_1, and $z_1 + z_2$. We know from geometry that the length of the side of the triangle corresponding to the vector $z_1 + z_2$ cannot be longer than the sum of the lengths of the remaining two sides. In symbols we can express this observation by the inequality

This inequality can be derived using the properties of complex numbers in Section 1.1. See Problem 50 in Exercises 1.2. ➡

$$|z_1 + z_2| \leq |z_1| + |z_2|. \tag{6}$$

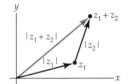

Figure 1.2.5 Triangle with vector sides

The result in (6) is known as the **triangle inequality**. Now from the identity $z_1 = z_1 + z_2 + (-z_2)$, (6) gives

$$|z_1| = |z_1 + z_2 + (-z_2)| \leq |z_1 + z_2| + |-z_2|.$$

Since $|z_2| = |-z_2|$ (see Problem 47 in Exercises 1.2), solving the last result for $|z_1 + z_2|$ yields another important inequality:

$$|z_1| - |z_2| \leq |z_1 + z_2|. \tag{7}$$

But because $z_1 + z_2 = z_2 + z_1$, (7) can be written in the alternative form $|z_1 + z_2| = |z_2 + z_1| \geq |z_2| - |z_1| = -(|z_1| - |z_2|)$ and so combined with the last result implies

$$\big| |z_1| - |z_2| \big| \leq |z_1 + z_2|. \tag{8}$$

It also follows from (6) by replacing z_2 by $-z_2$ that $|z_1 + (-z_2)| \leq |z_1| + |(-z_2)| = |z_1| + |z_2|$. This result is the same as

$$|z_1 - z_2| \leq |z_1| + |z_2|. \tag{9}$$

From (8) with z_2 replaced by $-z_2$, we also find

$$\big| |z_1| - |z_2| \big| \leq |z_1 - z_2|. \tag{10}$$

In conclusion, we note that the triangle inequality (6) extends to any finite sum of complex numbers:

$$|z_1 + z_2 + z_3 + \cdots + z_n| \leq |z_1| + |z_2| + |z_3| + \cdots + |z_n|. \tag{11}$$

The inequalities (6), (8), and (10) will be important when we work with integrals in Chapters 5 and 6.

EXAMPLE 3   An Upper Bound

Find an upper bound for $\left|\dfrac{-1}{z^4 + 3z^2 + 2}\right|$ if $|z| = 2$.

Solution By the second result in (3), the absolute value of a quotient is the quotient of the absolute values. Thus with $|-1| = 1$, we want to find a positive real number M such that

$$\frac{1}{|z^4 + 3z^2 + 2|} \le M.$$

To accomplish this task we want the denominator as small as possible. Since $z^4 + 3z^2 + 2 = (z^2 + 1)(z^2 + 2)$ we can use the first result in (3) and (8) to write

$$\left|z^4 + 3z^2 + 2\right| = \left|z^2 + 1\right| \cdot \left|z^2 + 2\right| \ge \left||z^2| - 1\right| \cdot \left||z^2| - 2\right|. \tag{12}$$

Using $|z| = 2$, (12) becomes

$$\begin{aligned}\left|z^4 + 3z^2 + 2\right| \ge \left||z^2| - 1\right| \cdot \left||z^2| - 2\right| &= \left||z|^2 - 1\right| \cdot \left||z|^2 - 2\right| \\ &= |4 - 1| \cdot |4 - 2| \\ &= 6.\end{aligned}$$

Hence for $|z| = 2$ we have

$$\left|\frac{-1}{z^4 + 3z^2 + 2}\right| = \frac{1}{|z^4 + 3z^2 + 2|} \le \frac{1}{6}.$$

Remarks

We have seen that the triangle inequality $|z_1 + z_2| \le |z_1| + |z_2|$ indicates that the length of the vector $z_1 + z_2$ cannot exceed the sum of the lengths of the individual vectors z_1 and z_2. But the results given in (3) are interesting. The product $z_1 z_2$ and quotient $z_1/z_2, (z_2 \ne 0)$, are complex numbers and so are vectors in the complex plane. The equalities $|z_1 z_2| = |z_1|\,|z_2|$ and $|z_1/z_2| = |z_1|\,/\,|z_2|$ indicate that the lengths of the vectors $z_1 z_2$ and z_1/z_2 are exactly equal to the product of the lengths and to the quotient of the lengths, respectively, of the individual vectors z_1 and z_2.

EXERCISES 1.2   *Answers to selected odd-numbered problems begin on page ANS-2.*

In Problems 1–4, interpret z_1 and z_2 as vectors. Graph z_1, z_2, and the indicated sum and difference as vectors.

1. $z_1 = 4 + 2i$, $z_2 = -2 + 5i$; $z_1 + z_2$, $z_1 - z_2$
2. $z_1 = 1 - i$, $z_2 = 1 + i$; $z_1 + z_2$, $z_1 - z_2$
3. $z_1 = 5 + 4i$, $z_2 = -3i$; $3z_1 + 5z_2$, $z_1 - 2z_2$
4. $z_1 = 4 - 3i$, $z_2 = -2 + 3i$; $2z_1 + 4z_2$, $z_1 - z_2$
5. Given that $z_1 = 5 - 2i$ and $z_2 = -1 - i$, find a vector z_3 in the same direction as $z_1 + z_2$ but four times as long.

6. (a) Plot the points $z_1 = -2 - 8i$, $z_2 = 3i$, $z_3 = -6 - 5i$.

   (b) The points in part (a) determine a triangle with vertices at z_1, z_2, and z_3, respectively. Express each side of the triangle as a difference of vectors.

7. In Problem 6, determine whether the points z_1, z_2, and z_3 are the vertices of a right triangle.

8. The three points $z_1 = 1 + 5i$, $z_2 = -4 - i$, $z_3 = 3 + i$ are vertices of a triangle. Find the length of the median from z_1 to the side $z_3 - z_2$.

In Problems 9–12, find the modulus of the given complex number.

9. $(1 - i)^2$

10. $i(2 - i) - 4\left(1 + \frac{1}{4}i\right)$

11. $\dfrac{2i}{3 - 4i}$

12. $\dfrac{1 - 2i}{1 + i} + \dfrac{2 - i}{1 - i}$

In Problems 13 and 14, let $z = x + iy$. Express the given quantity in terms of x and y.

13. $\left|z - 1 - 3i\right|^2$

14. $\left|z + 5\bar{z}\right|$

In Problems 15 and 16, determine which of the given two complex numbers is closest to the origin. Which is closest to $1 + i$?

15. $10 + 8i,\ \ 11 - 6i$

16. $\frac{1}{2} - \frac{1}{4}i,\ \ \frac{2}{3} + \frac{1}{6}i$

In Problems 17–26, describe the set of points z in the complex plane that satisfy the given equation.

17. $\mathrm{Re}((1 + i)z - 1) = 0$

18. $\left[\mathrm{Im}(i\bar{z})\right]^2 = 2$

19. $|z - i| = |z - 1|$

20. $\bar{z} = z^{-1}$

21. $\mathrm{Im}(z^2) = 2$

22. $\mathrm{Re}(z^2) = \left|\sqrt{3} - i\right|$

23. $|z - 1| = 1$

24. $|z - i| = 2\,|z - 1|$

25. $|z - 2| = \mathrm{Re}(z)$

26. $|z| = \mathrm{Re}(z)$

In Problems 27 and 28, establish the given simultaneous inequality.

27. If $|z| = 2$, then $8 \le |z + 6 + 8i| \le 12$.

28. If $|z| = 1$, then $2 \le \left|z^2 - 3\right| \le 4$.

29. Find an upper bound for the modulus of $3z^2 + 2z + 1$ if $|z| \le 1$.

30. Find an upper and lower bound for the reciprocal of the modulus of $z^4 - 5z^2 + 6$ if $|z| = 2$. [*Hint:* $z^4 - 5z^2 + 6 = \left(z^2 - 3\right)\left(z^2 - 2\right)$.]

In Problems 31 and 32, find a number z that satisfies the given equation.

31. $|z| - z = 2 + i$

32. $|z|^2 + 1 + 12i = 6z$

Focus on Concepts

33. (a) Draw the pair of points $z = a + ib$ and $\bar{z} = a - ib$ in the complex plane if $a > 0$, $b > 0$; $a > 0$, $b < 0$; $a < 0$, $b > 0$; and $a < 0$, $b < 0$.

   (b) In general, how would you describe geometrically the relationship between a complex number $z = a + ib$ and its conjugate $\bar{z} = a - ib$?

   (c) Describe geometrically the relationship between $z = a + ib$ and $z_1 = -a + ib$.

34. How would you describe geometrically the relationship between a nonzero complex number $z = a + ib$ and its

   (a) negative, $-z$?

   (b) inverse, z^{-1}? [*Hint:* Reread Problem 33 and then recall $z^{-1} = \bar{z}/|z|^2$.]

35. Consider the complex numbers $z_1 = 4+i$, $z_2 = -2+i$, $z_3 = -2-2i$, $z_4 = 3-5i$.

(a) Use four different sketches to plot the four pairs of points z_1, iz_1; z_2, iz_2; z_3, iz_3; and z_4, iz_4.

(b) In general, how would you describe geometrically the effect of multiplying a complex number $z = x + iy$ by i? By $-i$?

36. What is the only complex number with modulus 0?

37. Under what circumstances does $|z_1 + z_2| = |z_1| + |z_2|$?

38. Using the complex variable z, find an equation of a circle in the complex plane of radius 5 centered at $3 - 6i$. [*Hint*: Use (5).]

39. Describe the set of points z in the complex plane that satisfy $z = \cos\theta + i\sin\theta$, where θ is measured in radians from the positive x-axis.

40. Using the complex variable z, find an equation of an ellipse in the complex plane with foci $-2 + i$ and $2 + i$ whose major axis is 8 units long.

41. Suppose $z = x + iy$. In (6) of Section 1.1 we saw that x and y could be expressed in terms of z and $\bar{z}$. Use these results to express the following Cartesian equations in complex form.

(a) $x = 3$ (b) $y = 10$

(c) $y = x$ (d) $x + 2y = 8$

42. Using complex notation, find a parametric equation of the line segment between any two distinct complex numbers z_1 and z_2 in the complex plane.

43. Suppose z_1, z_2, and z_3 are three distinct points in the complex plane and k is a real number. Interpret $z_3 - z_2 = k(z_2 - z_1)$ geometrically.

44. Suppose $z_1 \neq z_2$. Interpret $\text{Re}(z_1\bar{z}_2) = 0$ geometrically in terms of vectors z_1 and z_2.

45. Suppose $w = \bar{z}/z$. Without doing any calculations, explain why $|w| = 1$.

46. Without doing any calculations, explain why the inequalities $|\text{Re}(z)| \leq |z|$ and $|\text{Im}(z)| \leq |z|$ hold for all complex numbers z.

47. Show that

(a) $|z| = |-z|$ (b) $|z| = |\bar{z}|$.

48. For any two complex numbers z_1 and z_2, show that

$$|z_1 + z_2|^2 + |z_1 - z_2|^2 = 2\left(|z_1|^2 + |z_2|^2\right).$$

49. In this problem we will start you out in the proof of the first property $|z_1 z_2| = |z_1|\,|z_2|$ in (3). By the first result in (2) we can write $|z_1 z_2|^2 = (z_1 z_2)\overline{(z_1 z_2)}$. Now use the first property in (2) of Section 1.1 to continue the proof.

50. In this problem we guide you through an analytical proof of the triangle inequality (6).

Since $|z_1 + z_2|$ and $|z_1| + |z_2|$ are positive real numbers, we have $|z_1 + z_2| \leq |z_1| + |z_2|$ if and only if $|z_1 + z_2|^2 \leq (|z_1| + |z_2|)^2$. Thus, it suffices to show that $|z_1 + z_2|^2 \leq (|z_1| + |z_2|)^2$.

(a) Explain why $|z_1 + z_2|^2 = |z_1|^2 + 2\text{Re}(z_1\bar{z}_2) + |z_2|^2$.

(b) Explain why $(|z_1| + |z_2|)^2 = |z_1|^2 + 2|z_1\bar{z}_2| + |z_2|^2$.

(c) Use parts (a) and (b) along with the results in Problem 46 to derive (6).

1.3   Polar Form of Complex Numbers

Recall from calculus that a point P in the plane whose rectangular coordinates are (x, y) can also be described in terms of **polar coordinates**. The polar coordinate system, invented by Isaac Newton, consists of point O called the **pole** and the horizontal half-line emanating from the pole called the **polar axis**. If r is a directed distance from the pole to P and θ is an angle of inclination (in radians) measured from the polar axis to the line OP, then the point can be described by the ordered pair (r, θ), called the polar coordinates of P. See Figure 1.3.1.

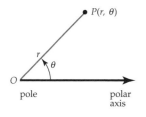

Figure 1.3.1 Polar coordinates

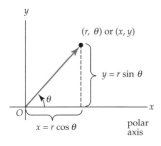

Figure 1.3.2 Polar coordinates in the complex plane

Be careful using $\tan^{-1}(y/x)$ ➡

Polar Form   Suppose, as shown in Figure 1.3.2, that a polar coordinate system is superimposed on the complex plane with the polar axis coinciding with the positive x-axis and the pole O at the origin. Then x, y, r, and θ are related by $x = r \cos \theta$, $y = r \sin \theta$. These equations enable us to express a nonzero complex number $z = x + iy$ as $z = (r \cos \theta) + i(r \sin \theta)$ or

$$z = r\left(\cos \theta + i \sin \theta\right). \tag{1}$$

We say that (1) is the **polar form** or **polar representation** of the complex number z. Again, from Figure 1.3.2 we see that the coordinate r can be interpreted as the distance from the origin to the point (x, y). In other words, we shall adopt the convention that r *is never negative** so that we can take r to be the modulus of z, that is, $r = |z|$. The angle θ of inclination of the vector z, which *will always be measured in radians* from the positive real axis, is positive when measured counterclockwise and negative when measured clockwise. The angle θ is called an **argument** of z and is denoted by $\theta = \arg(z)$. An argument θ of a complex number must satisfy the equations $\cos \theta = x/r$ and $\sin \theta = y/r$. An argument of a complex number z is not unique since $\cos \theta$ and $\sin \theta$ are 2π-periodic; in other words, if θ_0 is an argument of z, then necessarily the angles $\theta_0 \pm 2\pi$, $\theta_0 \pm 4\pi, \ldots$ are also arguments of z. In practice we use $\tan \theta = y/x$ to find θ. However, because $\tan \theta$ is π-periodic, some care must be exercised in using the last equation. A calculator will give only angles satisfying $-\pi/2 < \tan^{-1}(y/x) < \pi/2$, that is, angles in the first and fourth quadrants. We have to choose θ consistent with the quadrant in which z is located; this may require adding or subtracting π to $\tan^{-1}(y/x)$ when appropriate. The following example illustrates how this is done.

EXAMPLE 1   A Complex Number in Polar Form

Express $-\sqrt{3} - i$ in polar form.

Solution   With $x = -\sqrt{3}$ and $y = -1$ we obtain $r = |z| = \sqrt{\left(-\sqrt{3}\right)^2 + (-1)^2}$ $= 2$. Now $y/x = -1/(-\sqrt{3}) = 1/\sqrt{3}$, and so a calculator gives $\tan^{-1}\left(1/\sqrt{3}\right)$ $= \pi/6$, which is an angle whose terminal side is in the first quadrant. But since the point $-\sqrt{3} - i$ lies in the third quadrant, we take the solution of $\tan \theta = -1/(-\sqrt{3}) = 1/\sqrt{3}$ to be $\theta = \arg(z) = \pi/6 + \pi = 7\pi/6$. See Figure 1.3.3. It follows from (1) that a polar form of the number is

$$z = 2\left(\cos \frac{7\pi}{6} + i \sin \frac{7\pi}{6}\right). \tag{2}$$

❑

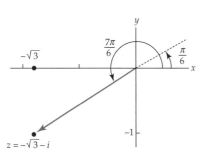

Figure 1.3.3 $\arg\left(-\sqrt{3} - i\right)$

---

*In general, in the polar description (r, θ) of a point P in the Cartesian plane, we can have $r \geq 0$ or $r < 0$.

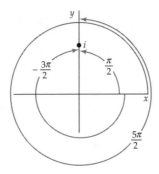

Figure 1.3.4 Some arguments of i

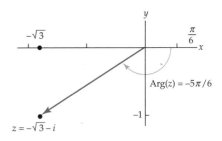

Figure 1.3.5 Principal argument
$z = -\sqrt{3} - i$

Principal Argument The symbol $\arg(z)$ represents a set of values, but the choice of an argument θ of a complex number that lies in the interval $-\pi < \theta \leq \pi$ is called the **principal value** of $\arg(z)$ or the **principal argument** of z. The principal argument of z is *unique* and is represented by the symbol $\mathrm{Arg}(z)$, that is,

$$-\pi < \mathrm{Arg}(z) \leq \pi.$$

For example, if $z = i$, we see in Figure 1.3.4 that some values of $\arg(i)$ are $\pi/2$, $5\pi/2$, $-3\pi/2$, and so on, but $\mathrm{Arg}(i) = \pi/2$. Similarly, we see from Figure 1.3.5 that the argument of $-\sqrt{3} - i$ that lies in the interval $(-\pi, \pi]$, the principal argument of z, is $\mathrm{Arg}(z) = \pi/6 - \pi = -5\pi/6$. Using $\mathrm{Arg}(z)$ we can express the complex number in (2) in the alternative polar form:

$$z = 2\left[\cos\left(-\frac{5\pi}{6}\right) + i\sin\left(-\frac{5\pi}{6}\right)\right].$$

In general, $\arg(z)$ and $\mathrm{Arg}(z)$ are related by

$$\arg(z) = \mathrm{Arg}(z) + 2n\pi, \quad n = 0, \pm 1, \pm 2, \ldots. \tag{3}$$

For example, $\arg(i) = \dfrac{\pi}{2} + 2n\pi$. For the choices $n = 0$ and $n = -1$, (3) gives $\arg(i) = \mathrm{Arg}(i) = \pi/2$ and $\arg(i) = -3\pi/2$, respectively.

Multiplication and Division The polar form of a complex number is especially convenient when multiplying or dividing two complex numbers. Suppose

$$z_1 = r_1(\cos\theta_1 + i\sin\theta_1) \quad \text{and} \quad z_2 = r_2(\cos\theta_2 + i\sin\theta_2),$$

where θ_1 and θ_2 are any arguments of z_1 and z_2, respectively. Then

$$z_1 z_2 = r_1 r_2 \left[\cos\theta_1\cos\theta_2 - \sin\theta_1\sin\theta_2 + i\left(\sin\theta_1\cos\theta_2 + \cos\theta_1\sin\theta_2\right)\right] \tag{4}$$

and, for $z_2 \neq 0$,

$$\frac{z_1}{z_2} = \frac{r_1}{r_2}\left[\cos\theta_1\cos\theta_2 + \sin\theta_1\sin\theta_2 + i\left(\sin\theta_1\cos\theta_2 - \cos\theta_1\sin\theta_2\right)\right]. \tag{5}$$

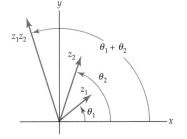

Figure 1.3.6 $\arg(z_1 z_2) = \theta_1 + \theta_2$

From the addition formulas* for the cosine and sine, (4) and (5) can be rewritten as

$$z_1 z_2 = r_1 r_2 \left[\cos(\theta_1 + \theta_2) + i \sin(\theta_1 + \theta_2)\right] \tag{6}$$

and

$$\frac{z_1}{z_2} = \frac{r_1}{r_2} \left[\cos(\theta_1 - \theta_2) + i \sin(\theta_1 - \theta_2)\right]. \tag{7}$$

Inspection of the expressions in (6) and (7) and Figure 1.3.6 shows that the lengths of the two vectors $z_1 z_2$ and z_1/z_2 are the product of the lengths of z_1 and z_2 and the quotient of the lengths of z_1 and z_2, respectively. See (3) of Section 1.2. Moreover, the arguments of $z_1 z_2$ and z_1/z_2 are given by

$$\arg(z_1 z_2) = \arg(z_1) + \arg(z_2) \quad \text{and} \quad \arg\left(\frac{z_1}{z_2}\right) = \arg(z_1) - \arg(z_2). \tag{8}$$

EXAMPLE 2 Argument of a Product and of a Quotient

We have just seen that for $z_1 = i$ and $z_2 = -\sqrt{3} - i$ that $\text{Arg}(z_1) = \pi/2$ and $\text{Arg}(z_2) = -5\pi/6$, respectively. Thus arguments for the product and quotient

$$z_1 z_2 = i(-\sqrt{3} - i) = 1 - \sqrt{3}i \quad \text{and} \quad \frac{z_1}{z_2} = \frac{i}{-\sqrt{3} - i} = -\frac{1}{4} - \frac{\sqrt{3}}{4}i$$

can be obtained from (8):

$$\arg(z_1 z_2) = \frac{\pi}{2} + \left(-\frac{5\pi}{6}\right) = -\frac{\pi}{3} \quad \text{and} \quad \arg\left(\frac{z_1}{z_2}\right) = \frac{\pi}{2} - \left(-\frac{5\pi}{6}\right) = \frac{4\pi}{3}.$$

Additional values of $\arg(z_1 z_2)$ and $\arg(z_1/z_2)$ can be found by adding integer multiples of 2π to $= \pi/3$ or $4\pi/3$, respectively. ❏

Integer Powers of z We can find integer powers of a complex number z from the results in (6) and (7). For example, if $z = r(\cos\theta + i\sin\theta)$, then with $z_1 = z_2 = z$, (6) gives

$$z^2 = r^2 \left[\cos(\theta + \theta) + i\sin(\theta + \theta)\right] = r^2 \left(\cos 2\theta + i \sin 2\theta\right).$$

Since $z^3 = z^2 z$, it then follows that

$$z^3 = r^3 \left(\cos 3\theta + i \sin 3\theta\right),$$

and so on. In addition, if we take $\arg(1) = 0$, then (7) gives

$$z^{-1} = \frac{1}{z} = \frac{1}{r} \left[\cos(0 - \theta) + i\sin(0 - \theta)\right] = r^{-1} \left[\cos(-\theta) + i\sin(-\theta)\right].$$

Continuing in this manner, we obtain a formula for the nth power of z for any integer n:

$$z^n = r^n \left(\cos n\theta + i \sin n\theta\right). \tag{9}$$

When $n = 0$, we get the familiar result $z^0 = 1$.

---

*$\cos(A \pm B) = \cos A \cos B \mp \sin A \sin B$ and $\sin(A \pm B) = \sin A \cos B \pm \cos A \sin B$

EXAMPLE 3   Power of a Complex Number

Compute z^3 for $z = -\sqrt{3} - i$.

Solution   In (2) of Example 1 we saw that a polar form of the given number is $z = 2[\cos(7\pi/6) + i\sin(7\pi/6)]$. Using (9) with $r = 2$, $\theta = 7\pi/6$, and $n = 3$ we get

$$\left(-\sqrt{3} - i\right)^3 = 2^3 \left[\cos\left(3\frac{7\pi}{6}\right) + i\sin\left(3\frac{7\pi}{6}\right)\right] = 8\left[\cos\frac{7\pi}{2} + i\sin\frac{7\pi}{2}\right] = -8i$$

since $\cos(7\pi/2) = 0$ and $\sin(7\pi/2) = -1$. ❏

Note in Example 3, if we also want the value of z^{-3}, then we could proceed in two ways: either find the reciprocal of $z^3 = -8i$ or use (9) with $n = -3$.

de Moivre's Formula   When $z = \cos\theta + i\sin\theta$, we have $|z| = r = 1$, and so (9) yields

$$(\cos\theta + i\sin\theta)^n = \cos n\theta + i\sin n\theta. \tag{10}$$

This last result is known as **de Moivre's formula** and is useful in deriving certain trigonometric identities involving $\cos n\theta$ and $\sin n\theta$. See Problems 33 and 34 in Exercises 1.3.

EXAMPLE 4   de Moivre's Formula

From (10), with $\theta = \pi/6$, $\cos\theta = \sqrt{3}/2$, and $\sin\theta = 1/2$:

$$\left(\frac{\sqrt{3}}{2} + \frac{1}{2}i\right)^3 = \cos 3\theta + i\sin 3\theta = \cos\left(3 \cdot \frac{\pi}{6}\right) + i\sin\left(3 \cdot \frac{\pi}{6}\right)$$

$$= \cos\frac{\pi}{2} + i\sin\frac{\pi}{2} = i.$$

❏

Remarks   *Comparison with Real Analysis*

(*i*) Observe in Example 2 that even though we used the principal arguments of z_1 and z_2 that $\arg(z_1/z_2) = 4\pi/3 \neq \text{Arg}(z_1/z_2)$. Although (8) is true for any arguments of z_1 and z_2, *it is not true*, in general, that $\text{Arg}(z_1 z_2) = \text{Arg}(z_1) + \text{Arg}(z_2)$ and $\text{Arg}(z_1/z_2) = \text{Arg}(z_1) - \text{Arg}(z_2)$. See Problems 37 and 38 in Exercises 1.3.

(*ii*) An argument can be assigned to any *nonzero* complex number z. However, for $z = 0$, $\arg(z)$ cannot be defined in any way that is meaningful.

(iii) If we take arg(z) from the interval $(-\pi, \pi]$, the relationship between a complex number z and its argument is single-valued; that is, every nonzero complex number has precisely *one* angle in $(-\pi, \pi]$. But there is nothing special about the interval $(-\pi, \pi]$; we also establish a single-valued relationship by using the interval $(0, 2\pi]$ to define the principal value of the argument of z. For the interval $(-\pi, \pi]$, the negative real axis is analogous to a barrier that we agree not to cross; the technical name for this barrier is a **branch cut**. If we use $(0, 2\pi]$, the branch cut is the positive real axis. The concept of a branch cut is important and will be examined in greater detail when we study functions in Chapters 2 and 4.

(iv) The "cosine i sine" part of the polar form of a complex number is sometimes abbreviated cis. That is,

$$z = r\left(\cos\theta + i\sin\theta\right) = r\operatorname{cis}\theta.$$

This notation, used mainly in engineering, will not be used in this text.

EXERCISES 1.3   *Answers to selected odd-numbered problems begin on page ANS-2.*

In Problems 1–10, write the given complex number in polar form first using an argument $\theta \neq \operatorname{Arg}(z)$ and then using $\theta = \operatorname{Arg}(z)$.

1. 2

2. -10

3. $-3i$

4. $6i$

5. $1 + i$

6. $5 - 5i$

7. $-\sqrt{3} + i$

8. $-2 - 2\sqrt{3}i$

9. $\dfrac{3}{-1+i}$

10. $\dfrac{12}{\sqrt{3}+i}$

In Problems 11 and 12, use a calculator to write the given complex number in polar form first using an argument $\theta \neq \operatorname{Arg}(z)$ and then using $\theta = \operatorname{Arg}(z)$.

11. $-\sqrt{2} + \sqrt{7}i$

12. $-12 - 5i$

In Problems 13 and 14, write the complex number whose polar coordinates (r, θ) are given in the form $a + ib$. Use a calculator if necessary.

13. $(4, -5\pi/3)$

14. $(2, 2)$

In Problems 15–18, write the complex number whose polar form is given in the form $a + ib$. Use a calculator if necessary.

15. $z = 5\left(\cos\dfrac{7\pi}{6} + i\sin\dfrac{7\pi}{6}\right)$

16. $z = 8\sqrt{2}\left(\cos\dfrac{11\pi}{4} + i\sin\dfrac{11\pi}{4}\right)$

17. $z = 6\left(\cos\dfrac{\pi}{8} + i\sin\dfrac{\pi}{8}\right)$

18. $z = 10\left(\cos\dfrac{\pi}{5} + i\sin\dfrac{\pi}{5}\right)$

In Problems 19 and 20, use (6) and (7) to find $z_1 z_2$ and z_1/z_2. Write the number in the form $a + ib$.

19. $z_1 = 2\left(\cos\dfrac{\pi}{8} + i\sin\dfrac{\pi}{8}\right)$, $z_2 = 4\left(\cos\dfrac{3\pi}{8} + i\sin\dfrac{3\pi}{8}\right)$

20. $z_1 = \sqrt{2}\left(\cos\dfrac{\pi}{4} + i\sin\dfrac{\pi}{4}\right)$, $z_2 = \sqrt{3}\left(\cos\dfrac{\pi}{12} + i\sin\dfrac{\pi}{12}\right)$

In Problems 21–24, write each complex number in polar form. Then use either (6) or (7) to obtain the polar form of the given number. Finally, write the polar form in the form $a + ib$.

21. $(3 - 3i)(5 + 5\sqrt{3}i)$

22. $(4 + 4i)(-1 + i)$

23. $\dfrac{-i}{1 + i}$

24. $\dfrac{\sqrt{2} + \sqrt{6}i}{-1 + \sqrt{3}i}$

In Problems 25–30, use (9) to compute the indicated powers.

25. $\left(1 + \sqrt{3}i\right)^9$

26. $(2 - 2i)^5$

27. $\left(\frac{1}{2} + \frac{1}{2}i\right)^{10}$

28. $(-\sqrt{2} + \sqrt{6}i)^4$

29. $\left[\left(\sqrt{2}\cos\dfrac{\pi}{8} + i\sqrt{2}\sin\dfrac{\pi}{8}\right)\right]^{12}$

30. $\left[\sqrt{3}\left(\cos\dfrac{2\pi}{9} + i\sin\dfrac{2\pi}{9}\right)\right]^6$

In Problems 31 and 32, write the given complex number in polar form and then in the form $a + ib$.

31. $\left(\cos\dfrac{\pi}{9} + i\sin\dfrac{\pi}{9}\right)^{12}\left[2\left(\cos\dfrac{\pi}{6} + i\sin\dfrac{\pi}{6}\right)\right]^5$

32. $\dfrac{\left[8\left(\cos\dfrac{3\pi}{8} + i\sin\dfrac{3\pi}{8}\right)\right]^3}{\left[2\left(\cos\dfrac{\pi}{16} + i\sin\dfrac{\pi}{16}\right)\right]^{10}}$

33. Use de Moivre's formula (10) with $n = 2$ to find trigonometric identities for $\cos 2\theta$ and $\sin 2\theta$.

34. Use de Moivre's formula (10) with $n = 3$ to find trigonometric identities for $\cos 3\theta$ and $\sin 3\theta$.

In Problems 35 and 36, find a positive integer n for which the equality holds.

35. $\left(\dfrac{\sqrt{3}}{2} + \dfrac{1}{2}i\right)^n = -1$

36. $\left(-\dfrac{\sqrt{2}}{2} + \dfrac{\sqrt{2}}{2}i\right)^n = 1$

37. For the complex numbers $z_1 = -1$ and $z_2 = 5i$, verify that:

    (a) $\text{Arg}(z_1 z_2) \neq \text{Arg}(z_1) + \text{Arg}(z_2)$

    (b) $\text{Arg}(-z_2/z_1) \neq \text{Arg}(-z_2) - \text{Arg}(z_1)$.

38. For the complex numbers given in Problem 37, verify that:

    (a) $\arg(z_1 z_2) = \arg(z_1) + \arg(z_2)$

    (b) $\arg(-z_2/z_1) = \arg(-z_2) - \arg(z_1)$.

Focus on Concepts

39. Suppose that $z = r(\cos\theta + i\sin\theta)$. Describe geometrically the effect of multiplying z by a complex number of the form $z_1 = \cos\alpha + i\sin\alpha$, when $\alpha > 0$ and when $\alpha < 0$.

40. Suppose $z = \cos\theta + i\sin\theta$. If n is an integer, evaluate $z^n + \bar{z}^n$ and $z^n - \bar{z}^n$.

41. Write an equation that relates $\arg(z)$ to $\arg(1/z)$, $z \neq 0$.

42. Are there any special cases in which $\text{Arg}(z_1 z_2) = \text{Arg}(z_1) + \text{Arg}(z_2)$? Prove your assertions.

43. How are the complex numbers z_1 and z_2 related if $\arg(z_1) = \arg(z_2)$?

44. Describe the set of points z in the complex plane that satisfy $\arg(z) = \pi/4$.

45. Student A states that, even though she can't find it in the text, she thinks that $\arg(\bar{z}) = -\arg(z)$. For example, she says, if $z = 1 + i$, then $\bar{z} = 1 - i$ and $\arg(z) = \pi/4$ and $\arg(\bar{z}) = -\pi/4$. Student B disagrees because he feels that he has a counterexample: If $z = i$, then $\bar{z} = -i$; we can take $\arg(i) = \pi/2$ and $\arg(-i) = 3\pi/2$ and so $\arg(i) \neq -\arg(-i)$. Take sides and defend your position.

46. Suppose z_1, z_2, and $z_1 z_2$ are complex numbers in the first quadrant and that the points $z = 0$, $z = 1$, z_1, z_2, and $z_1 z_2$ are labeled O, A, B, C, and D, respectively. Study the formula in (6) and then discuss how the triangles OAB and OCD are related.

47. Suppose $z_1 = r_1 (\cos\theta_1 + i\sin\theta_1)$ and $z_2 = r_2 (\cos\theta_2 + i\sin\theta_2)$. If $z_1 = z_2$, then how are r_1 and r_2 related? How are θ_1 and θ_2 related?

48. Suppose z_1 is in the first quadrant. For each z_2, discuss the quadrant in which $z_1 z_2$ could be located.

(**a**) $z_2 = \dfrac{1}{2} + \dfrac{\sqrt{3}}{2} i$ (**b**) $z_2 = -\dfrac{\sqrt{3}}{2} + \dfrac{1}{2} i$

(**c**) $z_2 = -i$ (**d**) $z_2 = -1$

49. (**a**) For $z \neq 1$, verify the identity

$$1 + z + z^2 + \cdots + z^n = \frac{1 - z^{n+1}}{1 - z}.$$

(**b**) Use part (a) and appropriate results from this section to establish that

$$1 + \cos\theta + \cos 2\theta + \cdots + \cos n\theta = \frac{1}{2} + \frac{\sin\left(n + \frac{1}{2}\right)\theta}{2\sin\frac{1}{2}\theta}$$

for $0 < \theta < 2\pi$. The foregoing result is known as **Lagrange's identity** and is useful in the theory of Fourier series.

50. Suppose z_1, z_2, z_3, and z_4 are four distinct complex numbers. Interpret geometrically:

$$\arg\left(\frac{z_1 - z_2}{z_3 - z_4}\right) = \frac{\pi}{2}.$$

1.4 Powers and Roots

Recall from algebra that –2 and 2 are said to be *square roots* of the number 4 because $(-2)^2 = 4$ and $(2)^2 = 4$. In other words, the two square roots of 4 are distinct solutions of the equation $w^2 = 4$. In like manner we say $w = 3$ is a cube root of 27 since $w^3 = 3^3 = 27$. This last equation points us again in the direction of complex variables since any real number has only *one real* cube root and *two complex* roots. In general, we say that a number w is an **nth root** of a nonzero complex number z if $w^n = z$, where n is a positive integer. For example, you are urged to verify that $w_1 = \frac{1}{2}\sqrt{2} + \frac{1}{2}\sqrt{2}i$ and $w_2 = -\frac{1}{2}\sqrt{2} - \frac{1}{2}\sqrt{2}i$ are the two square roots of the complex number $z = i$ because $w_1^2 = i$ and $w_2^2 = i$. See Problem 39 in Exercises 1.1.

We will now demonstrate that there are exactly n solutions of the equation $w^n = z$.

Roots Suppose $z = r(\cos\theta + i\sin\theta)$ and $w = \rho(\cos\phi + i\sin\phi)$ are polar forms of the complex numbers z and w. Then, in view of (9) of Section 1.3, the equation $w^n = z$ becomes

$$\rho^n(\cos n\phi + i \sin n\phi) = r(\cos\theta + i\sin\theta). \tag{1}$$

From (1), we can conclude that

$$\rho^n = r \tag{2}$$

and

$$\cos n\phi + i\sin n\phi = \cos\theta + i\sin\theta. \tag{3}$$

See Problem 47 in Exercises 1.3.

From (2), we define $\rho = \sqrt[n]{r}$ to be the unique positive nth root of the positive real number r. From (3), the definition of equality of two complex numbers implies that

$$\cos n\phi = \cos\theta \quad \text{and} \quad \sin n\phi = \sin\theta.$$

These equalities, in turn, indicate that the arguments θ and ϕ are related by $n\phi = \theta + 2k\pi$, where k is an integer. Thus,

$$\phi = \frac{\theta + 2k\pi}{n}.$$

As k takes on the successive integer values $k = 0,\ 1,\ 2,\ \ldots,\ n-1$ we obtain n *distinct* nth roots of z; these roots have the same modulus $\sqrt[n]{r}$ but different arguments. Notice that for $k \geq n$ we obtain the same roots because the sine and cosine are 2π-periodic. To see why this is so, suppose $k = n + m$, where $m = 0,\ 1,\ 2,\ \ldots$. Then

$$\phi = \frac{\theta + 2(n+m)\pi}{n} = \frac{\theta + 2m\pi}{n} + 2\pi$$

and

$$\sin\phi = \sin\left(\frac{\theta + 2m\pi}{n}\right), \quad \cos\phi = \cos\left(\frac{\theta + 2m\pi}{n}\right).$$

We summarize this result. The n nth roots of a nonzero complex number $z = r(\cos\theta + i\sin\theta)$ are given by

$$w_k = \sqrt[n]{r}\left[\cos\left(\frac{\theta + 2k\pi}{n}\right) + i\sin\left(\frac{\theta + 2k\pi}{n}\right)\right], \tag{4}$$

where $k = 0,\ 1,\ 2,\ \ldots,\ n-1$.

EXAMPLE 1   Cube Roots of a Complex Number
Find the three cube roots of $z = i$.

Solution   Keep in mind that we are basically solving the equation $w^3 = i$. Now with $r = 1$, $\theta = \arg(i) = \pi/2$, a polar form of the given number is given by $z = \cos(\pi/2) + i\sin(\pi/2)$. From (4), with $n = 3$, we then obtain

$$w_k = \sqrt[3]{1}\left[\cos\left(\frac{\pi/2 + 2k\pi}{3}\right) + i\sin\left(\frac{\pi/2 + 2k\pi}{3}\right)\right], \quad k = 0, 1, 2.$$

Hence the three roots are,

$$k = 0, \quad w_0 = \cos\frac{\pi}{6} + i\sin\frac{\pi}{6} = \frac{\sqrt{3}}{2} + \frac{1}{2}i$$

$$k = 1, \quad w_1 = \cos\frac{5\pi}{6} + i\sin\frac{5\pi}{6} = -\frac{\sqrt{3}}{2} + \frac{1}{2}i$$

$$k = 2, \quad w_2 = \cos\frac{3\pi}{2} + i\sin\frac{3\pi}{2} = -i.$$

Principal nth Root On page 15 we pointed out that the symbol $\arg(z)$ really stands for a *set* of arguments for a complex number z. Stated another way, for a given complex number $z \neq 0$, $\arg(z)$ is *infinite-valued*. In like manner, $z^{1/n}$ is n-valued; that is, the symbol $z^{1/n}$ represents the set of n nth roots w_k of z. The *unique* root of a complex number z (obtained by using the principal value of $\arg(z)$ with $k = 0$) is naturally referred to as the **principal nth root** of w. In Example 1, since $\mathrm{Arg}(i) = \pi/2$, we see that $w_0 = \frac{1}{2}\sqrt{3} + \frac{1}{2}i$ is the principal cube root of i. The choice of $\theta = \mathrm{Arg}(z)$ and $k = 0$ guarantees us that when z is a positive real number r, the principal nth root is $\sqrt[n]{r}$.

$\sqrt{4} = 2$ and $\sqrt[3]{27} = 3$ are the principal square root of 4 and the principal cube root of 27, respectively. ➡

Since the roots given by (4) have the same modulus, the n nth roots of a nonzero complex number z lie on a circle of radius $\sqrt[n]{r}$ centered at the origin in the complex plane. Moreover, since the difference between the arguments of any two successive roots w_k and w_{k+1} is $2\pi/n$, the n nth roots of z are equally spaced on this circle, beginning with the root whose argument is θ/n. Figure 1.4.1 shows the three cube roots of i obtained in Example 1 spaced at equal angular intervals of $2\pi/3$ on the circumference of a unit circle beginning with the root w_0 whose argument is $\pi/6$.

As the next example shows, the roots of a complex number do not have to be "nice" numbers as in Example 1.

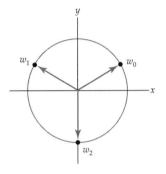

Figure 1.4.1 Three cube roots of i

EXAMPLE 2   **Fourth Roots of a Complex Number**

Find the four fourth roots of $z = 1 + i$.

Solution   In this case, $r = \sqrt{2}$ and $\theta = \arg(z) = \pi/4$. From (4) with $n = 4$, we obtain

$$w_k = \sqrt[8]{2}\left[\cos\left(\frac{\pi/4 + 2k\pi}{4}\right) + i\sin\left(\frac{\pi/4 + 2k\pi}{4}\right)\right], \quad k = 0, 1, 2, 3.$$

With the aid of a calculator we find

$$k = 0, \quad w_0 = \sqrt[8]{2}\left[\cos\frac{\pi}{16} + i\sin\frac{\pi}{16}\right] \approx 1.0696 + 0.2127i$$

$$k = 1, \quad w_1 = \sqrt[8]{2}\left[\cos\frac{9\pi}{16} + i\sin\frac{9\pi}{16}\right] \approx -0.2127 + 1.0696i$$

$$k = 2, \quad w_2 = \sqrt[8]{2}\left[\cos\frac{17\pi}{16} + i\sin\frac{17\pi}{16}\right] \approx -1.0696 - 0.2127i$$

$$k = 3, \quad w_3 = \sqrt[8]{2}\left[\cos\frac{25\pi}{16} + i\sin\frac{25\pi}{16}\right] \approx 0.2127 - 1.0696i.$$

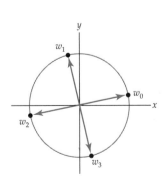

Figure 1.4.2 Four fourth roots of $1 + i$

As shown in Figure 1.4.2, the four roots lie on a circle centered at the origin of radius $r = \sqrt[8]{2} \approx 1.09$ and are spaced at equal angular intervals of $2\pi/4 = \pi/2$ radians, beginning with the root whose argument is $\pi/16$. Since $\theta = \mathrm{Arg}(z) = \pi/4$, the principal fourth root is w_0. ❑

> **Remarks** *Comparison with Real Analysis*
>
> (*i*) As a consequence of (4), we can say that the complex number system is *closed* under the operation of extracting roots. This means that for any z in **C**, $z^{1/n}$ is also in **C**. The real number system does not possess a similar closure property since, if x is in **R**, $x^{1/n}$ is not necessarily in **R**.
>
> (*ii*) Geometrically, the n nth roots of a complex number z can also be interpreted as the vertices of a regular polygon with n sides that is inscribed within a circle of radius $\sqrt[n]{r}$ centered at the origin. You can see the plausibility of this fact by reinspecting Figures 1.4.1 and 1.4.2. See Problem 19 in Exercises 1.4.
>
> (*iii*) When m and n are positive integers with no common factors, then (4) enables us to define a **rational power** of z, that is, $z^{m/n}$. It can be shown that the set of values $(z^{1/n})^m$ is the same as the set of values $(z^m)^{1/n}$. This set of n common values is defined to be $z^{m/n}$. See Problems 25 and 26 in Exercises 1.4.

EXERCISES 1.4   *Answers to selected odd-numbered problems begin on page ANS-2.*

In Problems 1–14, use (4) to compute all roots. Give the principal nth root in each case. Sketch the roots w_0, w_1, ..., w_{n-1} on an appropriate circle centered at the origin.

1. $(8)^{1/3}$ **2.** $(-1)^{1/4}$

3. $(-9)^{1/2}$ **4.** $(-125)^{1/3}$

5. $(i)^{1/2}$ **6.** $(-i)^{1/3}$

7. $(-1+i)^{1/3}$ **8.** $(1+i)^{1/5}$

9. $(-1+\sqrt{3}i)^{1/2}$ **10.** $(-1-\sqrt{3}i)^{1/4}$

11. $(3+4i)^{1/2}$ **12.** $(5+12i)^{1/2}$

13. $\left(\dfrac{16i}{1+i}\right)^{1/8}$ **14.** $\left(\dfrac{1+i}{\sqrt{3}+i}\right)^{1/6}$

15. (a) Verify that $(4+3i)^2 = 7+24i$.

   (b) Use part (a) to find the two values of $(7+24i)^{1/2}$.

16. Rework Problem 15 using (4).

17. Find all solutions of the equation $z^4 + 1 = 0$.

18. Use the fact that $8i = (2+2i)^2$ to find all solutions of the equation

$$z^2 - 8z + 16 = 8i.$$

The n distinct **nth roots of unity** are the solutions of the equation $w^n = 1$. Problems 19–24 deal with roots of unity.

19. (a) Show that the n nth roots of unity are given by

$$(1)^{1/n} = \cos\frac{2k\pi}{n} + i\sin\frac{2k\pi}{n}, \quad k = 0,1,2,\ldots,n-1.$$

(b) Find n nth roots of unity for $n = 3$, $n = 4$, and $n = 5$.

(c) Carefully plot the roots of unity found in part (b). Sketch the regular polygons formed with the roots as vertices. [*Hint*: See (*ii*) in the Remarks.]

20. Suppose w is a cube root of unity corresponding to $k = 1$. See Problem 19(a).

 (a) How are w and w^2 related?

 (b) Verify by direct computation that

$$1 + w + w^2 = 0.$$

 (c) Explain how the result in part (b) follows from the basic definition that w is a cube root of 1, that is, $w^3 = 1$. [*Hint*: Factor.]

21. For a fixed n, if we take $k = 1$ in Problem 19(a), we obtain the root

$$w_n = \cos\frac{2\pi}{n} + i\sin\frac{2\pi}{n}.$$

Explain why the n nth roots of unity can then be written

$$1, w_n, w_n^2, w_n^3, \ldots, w_n^{n-1}.$$

22. Consider the equation $(z + 2)^n + z^n = 0$, where n is a positive integer. By any means, solve the equation for z when $n = 1$. When $n = 2$.

23. Consider the equation in Problem 22.

 (a) In the complex plane, determine the location of all solutions z when $n = 5$. [*Hint*: Write the equation in the form $[(z + 2)/(-z)]^5 = 1$ and use part (a) of Problem 19.]

 (b) Reexamine the solutions of the equation in Problem 22 for $n = 1$ and $n = 2$. Form a conjecture as to the location of all solutions of $(z + 2)^n + z^n = 0$.

24. For the n nth roots of unity given in Problem 21, show that

$$1 + w_n + w_n^2 + w_n^3 + \cdots + w_n^{n-1} = 0.$$

[*Hint*: Multiply the sum $1 + w_n + w_n^2 + w_n^3 + \cdots + w_n^{n-1}$ by $w_n - 1$.]

Before working Problems 25 and 26, read (*iii*) in the Remarks. If m and n are positive integers with no common factors, then the n values of the rational power $z^{m/n}$ are

$$w_k = \sqrt[n]{r^m}\left[\cos\frac{m}{n}(\theta + 2k\pi) + i\sin\frac{m}{n}(\theta + 2k\pi)\right], \tag{5}$$

$k = 0, 1, 2, \ldots, n - 1$. The w_k are the n distinct solutions of $w^n = z^m$.

25. (a) First compute the set of values $i^{1/2}$ using (4). Then compute $(i^{1/2})^3$ using (9) of Section 1.3.

 (b) Now compute i^3. Then compute $(i^3)^{1/2}$ using (4). Compare these values with the results of part (b).

 (c) Lastly, compute $i^{3/2}$ using formula (5).

26. Use (5) to find all solutions of the equation $w^2 = (-1 + i)^5$.

Focus on Concepts

27. The vector given in Figure 1.4.3 represents one value of $z^{1/n}$. Using only the figure and trigonometry—that is, do not use formula (4)—find the remaining values of $z^{1/n}$ when $n = 3$. Repeat for $n = 4$ and $n = 5$.

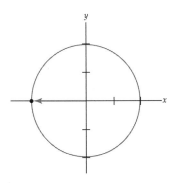

Figure 1.4.3 Figure for Problem 27

28. Suppose n denotes a nonnegative integer. Determine the values of n such that $z^n = 1$ possesses only real solutions. Defend your answer with sound mathematics.

29. (a) Proceed as in Example 2 to find the approximate values of the two square roots w_0 and w_1 of $1 + i$.

    (b) Show that the exact values of the roots in part (a) are

$$w_0 = \sqrt{\frac{1+\sqrt{2}}{2}} + i\sqrt{\frac{\sqrt{2}-1}{2}}, \quad w_1 = -\sqrt{\frac{1+\sqrt{2}}{2}} - i\sqrt{\frac{\sqrt{2}-1}{2}}.$$

30. Discuss: What geometric significance does the result in Problem 24 have?

31. Discuss: A real number can have a complex nth root. Can a nonreal complex number have a real nth root?

32. Suppose w is located in the first quadrant and is a cube root of a complex number z. Can there exist a second cube root of z located in the first quadrant? Defend your answer with sound mathematics.

33. Suppose z is a complex number that possesses a fourth root w that is neither real nor pure imaginary. Explain why the remaining fourth roots are neither real nor pure imaginary.

34. Discuss: The vectors in Figure 1.4.4 represent the four fourth roots of a complex number z. Can you determine from this figure alone which root is the principal fourth root of z?

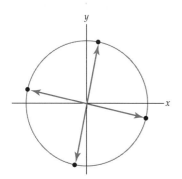

Figure 1.4.4 Figure for Problem 34

Computer Lab Assignments

In Problems 35–40, use a CAS* to first find $z^n = w$ for the given complex number and the indicated value of n. Then, using the output and the same value of n, determine whether $w^{1/n} = (z^n)^{1/n} = z$. If not, explain why not.

35. $z = 2.5 - i$; $n = 10$ 36. $z = -0.5 + 0.3i$; $n = 5$

37. $z = 1 + 3i$; $n = 8$ 38. $z = 2 + 2i$; $n = 12$

39. $z = i$; $n = 21$ 40. $z = -1 + \sqrt{3}\,i$; $n = 11$

1.5 Sets of Points in the Complex Plane

In the preceding sections we examined some rudiments of the algebra and geometry of complex numbers. But we have barely scratched the surface of the subject known as complex analysis; the main thrust of our study lies ahead. Our goal in the chapters that follow is to examine functions of a single complex variable $z = x + iy$ and the calculus of these functions.

Before introducing the notion of a function in Chapter 2, we need to state some essential definitions and terminology about sets in the complex plane.

Figure 1.5.1 Circle of radius ρ

Circles Suppose $z_0 = x_0 + iy_0$. Since $|z - z_0| = \sqrt{(x - x_0)^2 + (y - y_0)^2}$ is the distance between the points $z = x + iy$ and $z_0 = x_0 + iy_0$, the points $z = x + iy$ that satisfy the equation

$$|z - z_0| = \rho, \quad \rho > 0, \tag{1}$$

lie on a **circle** of radius ρ centered at the point z_0. See Figure 1.5.1.

---

*Throughout this text we shall use the abbreviation CAS for "computer algebra system."

EXAMPLE 1   **Two Circles**

(a) $|z| = 1$ is an equation of a unit circle centered at the origin.

(b) By rewriting $|z - 1 + 3i| = 5$ as $|z - (1 - 3i)| = 5$, we see from (1) that the equation describes a circle of radius 5 centered at the point $z_0 = 1 - 3i$.

❏

Disks and Neighborhoods   The points z that satisfy the inequality $|z - z_0| \leq \rho$ can be either *on* the circle $|z - z_0| = \rho$ or *within* the circle. We say that the set of points defined by $|z - z_0| \leq \rho$ is a **disk** of radius ρ centered at z_0. But the points z that satisfy the strict inequality $|z - z_0| < \rho$ lie within, and not on, a circle of radius ρ centered at the point z_0. This set is called a **neighborhood** of z_0. Occasionally, we will need to use a neighborhood of z_0 that also *excludes* z_0. Such a neighborhood is defined by the simultaneous inequality $0 < |z - z_0| < \rho$ and is called a **deleted neighborhood** of z_0. For example, $|z| < 1$ defines a neighborhood of the origin, whereas $0 < |z| < 1$ defines a deleted neighborhood of the origin; $|z - 3 + 4i| < 0.01$ defines a neighborhood of $3 - 4i$, whereas the inequality $0 < |z - 3 + 4i| < 0.01$ defines a deleted neighborhood of $3 - 4i$.

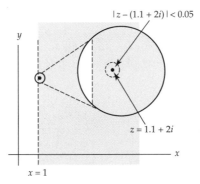

Figure 1.5.2 Open set

Open Sets   A point z_0 is said to be an **interior point** of a set S of the complex plane if there exists some neighborhood of z_0 that lies entirely within S. If every point z of a set S is an interior point, then S is said to be an **open set**. See Figure 1.5.2. For example, the inequality $\text{Re}(z) > 1$ defines a *right half-plane*, which is an open set. All complex numbers $z = x + iy$ for which $x > 1$ are in this set. If we choose, for example, $z_0 = 1.1 + 2i$, then a neighborhood of z_0 lying entirely in the set is defined by $|z - (1.1 + 2i)| < 0.05$. See Figure 1.5.3. On the other hand, the set S of points in the complex plane defined by $\text{Re}(z) \geq 1$ is *not* open because every neighborhood of a point lying on the line $x = 1$ must contain points in S and points not in S. See Figure 1.5.4.

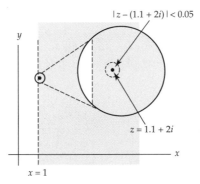

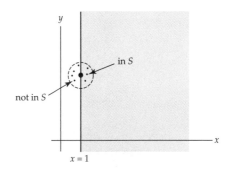

Figure 1.5.3 Open set with magnified view of point near $x = 1$

Figure 1.5.4 Set S not open

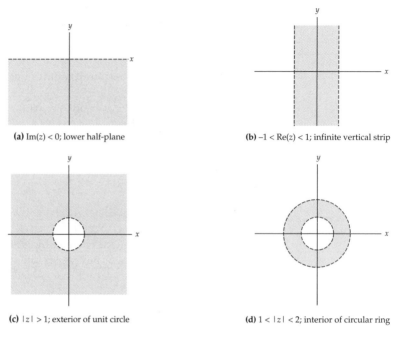

(a) Im(z) < 0; lower half-plane

(b) −1 < Re(z) < 1; infinite vertical strip

(c) |z| > 1; exterior of unit circle

(d) 1 < |z| < 2; interior of circular ring

Figure 1.5.5 Four examples of open sets

EXAMPLE 2   Some Open Sets

Figure 1.5.5 illustrates some additional open sets.   ❏

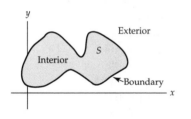

Figure 1.5.6 Interior, boundary, and exterior of set S

If *every* neighborhood of a point z_0 of a set S contains at least one point of S and at least one point not in S, then z_0 is said to be a **boundary point** of S. For the set of points defined by $\text{Re}(z) \geq 1$, the points on the vertical line $x = 1$ are boundary points. The points that lie on the circle $|z - i| = 2$ are boundary points for the disk $|z - i| \leq 2$ as well as for the neighborhood $|z - i| < 2$ of $z = i$. The collection of boundary points of a set S is called the **boundary** of S. The circle $|z - i| = 2$ is the boundary for both the disk $|z - i| \leq 2$ and the neighborhood $|z - i| < 2$ of $z = i$. A point z that is neither an interior point nor a boundary point of a set S is said to be an **exterior point** of S; in other words, z_0 is an exterior point of a set S if there exists *some* neighborhood of z_0 that contains no points of S. Figure 1.5.6 shows a typical set S with interior, boundary, and exterior.

An open set S can be as simple as the complex plane with a single point z_0 deleted. The boundary of this "punctured plane" is z_0, and the only candidate for an exterior point is z_0. However, S has no exterior points since no neighborhood of z_0 can be free of points of the plane.

Annulus   The set S_1 of points satisfying the inequality $\rho_1 < |z - z_0|$ lie exterior to the circle of radius ρ_1 centered at z_0, whereas the set S_2 of points satisfying $|z - z_0| < \rho_2$ lie interior to the circle of radius ρ_2 centered at z_0. Thus, if $0 < \rho_1 < \rho_2$, then the set of points satisfying the simultaneous inequality

$$\rho_1 < |z - z_0| < \rho_2, \tag{2}$$

is the intersection of the sets S_1 and S_2. This intersection is an open circular ring centered at z_0. Figure 1.5.5(d) illustrates such a ring centered at the origin. The set defined by (2) is called an open **circular annulus**. By allowing $\rho_1 = 0$, we obtain a deleted neighborhood of z_0.

Figure 1.5.7 Connected set

Domain If any pair of points z_1 and z_2 in a set S can be connected by a polygonal line that consists of a finite number of line segments joined end to end that lies entirely in the set, then the set S is said to be **connected**. See Figure 1.5.7. An open connected set is called a **domain**. Each of the open sets in Figure 1.5.5 is connected and so each is a domain. The set of numbers z satisfying $\text{Re}(z) \neq 4$ is an open set but is not connected since it is not possible to join points on either side of the vertical line $x = 4$ by a polygonal line without leaving the set (bear in mind that the points on the line $x = 4$ are not in the set). A neighborhood of a point z_0 is a connected set.

Note: "Closed" does not mean "not open." ➡

Regions A **region** is a domain in the complex plane together with all, some, or none of its boundary points. Since an open set does not contain any boundary points, it is automatically a region. A set that contains all of its boundary points is said to be **closed**. The disk defined by $|z - z_0| \leq \rho$ is an example of a closed region and is referred to as a **closed disk**. A neighborhood of a point z_0 defined by $|z - z_0| < \rho$ is an open set or an open region and is said to be an **open disk**. If the center z_0 is deleted from either a closed disk or an open disk, the regions defined by $0 < |z - z_0| \leq \rho$ or $0 < |z - z_0| < \rho$ are called **punctured disks**. *A punctured open disk is the same as a deleted neighborhood of z_0.* A region can be neither open nor closed; the annular region defined by the inequality $1 \leq |z - 5| < 3$ contains only some of its boundary points (the points lying on the circle $|z - 5| = 1$), and so it is *neither open nor closed.* In (2) we defined a circular annular region; in a more general interpretation, an annulus or **annular region** may have the appearance shown in Figure 1.5.8.

Figure 1.5.8 Annular region

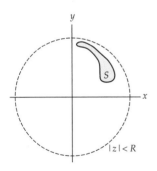

Figure 1.5.9 The set S is bounded since some neighborhood of the origin encloses S entirely.

Bounded Sets Finally, we say that a set S in the complex plane is **bounded** if there exists a real number $R > 0$ such that $|z| < R$ for every z in S. That is, S is bounded if it can be completely enclosed within *some* neighborhood of the origin. In Figure 1.5.9, the set S shown in color is bounded because it is contained entirely within the dashed circular neighborhood of the origin. A set is **unbounded** if it is not bounded. For example, the set in Figure 1.5.5(d) is bounded, whereas the sets in Figures 1.5.5(a), 1.5.5(b), and 1.5.5(c) are unbounded.

Remarks *Comparison with Real Analysis*

In your study of mathematics you undoubtedly found that you had to deal with the concept of *infinity*. For example, in a course in calculus you dealt with *limits at infinity*, where the behavior of functions was examined as x either increased or decreased without bound. Since we have exactly two directions on a number line, it is convenient to represent the notions of "increasing without bound" and "decreasing without bound" symbolically by $x \to +\infty$ and $x \to -\infty$, respectively. It turns out that we can get along just fine without the designation $\pm\infty$ by dealing with an "ideal point" called the **point at infinity**, which is denoted simply by ∞. To do this we identify any real number a with a point $(x_0,\ y_0)$ on a unit circle $x^2 + y^2 = 1$ by drawing a straight line from the point $(a,\ 0)$ on the x-axis or horizontal number line to the point $(0,\ 1)$ on the circle. The point $(x_0,\ y_0)$ is the point of intersection of the line with the circle. See Problem 47 in Exercises 1.5. It should be clear from Figure 1.5.10(a) that the farther the point $(a,\ 0)$ is from the origin, the nearer $(x_0,\ y_0)$ is to $(0,\ 1)$. The only point on the circle that does not correspond to a real number a is $(0,\ 1)$. To complete the correspondence with *all* the points on the circle, we identify $(0,\ 1)$ with ∞. The set consisting of the real numbers $\mathbf{R}$ adjoined with ∞ is called the **extended real-number system**.

In our current study the analogue of the number line is the complex plane. Recall that since $\mathbf{C}$ is not ordered, the notions of z either "increasing" or "decreasing" have no meaning. However, we know that by increasing the modulus $|z|$ of a complex number z, the number moves farther from the origin. If we allow $|z|$ to become unbounded, say, along the real and imaginary axes, we do not have to distinguish between "directions" on these axes by notations such as $z \to +\infty$, $z \to -\infty$, $z \to +i\infty$, and $z \to -i\infty$. In complex analysis, only a single notion of ∞ is used because we can extend the complex number system $\mathbf{C}$ in a manner analogous to that just described for the real number system $\mathbf{R}$. This time, however, we associate a complex number with a point on a unit sphere called the **Riemann sphere**. By drawing a line from the number $z = a + ib$, written as $(a,\ b,\ 0)$, in the complex plane to the north pole $(0,\ 0,\ 1)$ of the sphere $x^2 + y^2 + u^2 = 1$, we determine a unique point $(x_0,\ y_0,\ u_0)$ on a unit sphere. As can be visualized from Figure 1.5.10(b), a complex number with a very large modulus is far from the origin $(0,\ 0,\ 0)$ and, correspondingly, the point $(x_0,\ y_0,\ u_0)$ is close to $(0,\ 0,\ 1)$. In this manner each complex number is identified with a single point on the sphere. See Problems 48–50 in Exercises 1.5. Because the point $(0,\ 0,\ 1)$ corresponds to no number z in the plane, we correspond it with ∞. Of course, the system consisting of $\mathbf{C}$ adjoined with the "ideal point" ∞ is called the **extended complex-number system**.

This way of corresponding or mapping the complex numbers onto a sphere—north pole $(0,\ 0,\ 1)$ excluded—is called a **stereographic projection**.

For a finite number z, we have $z + \infty = \infty + z = \infty$, and for $z \neq 0$, $z \cdot \infty = \infty \cdot z = \infty$. Moreover, for $z \neq 0$ we write $z/0 = \infty$ and for $z \neq \infty$, $z/\infty = 0$. Expressions such as $\infty - \infty$, ∞/∞, ∞^0, and 1^∞ cannot be given a meaningful definition and are called indeterminate.

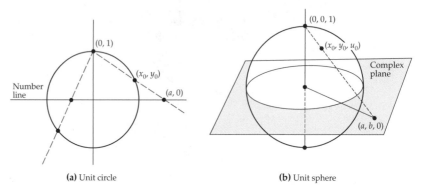

(a) Unit circle   **(b)** Unit sphere

Figure 1.5.10 The method of correspondence in (b) is a stereographic projection.

EXERCISES 1.5 *Answers to selected odd-numbered problems begin on page ANS-3.*

In Problems 1–12, sketch the graph of the given equation in the complex plane.

1. $|z - 4 + 3i| = 5$

2. $|z + 2 + 2i| = 2$

3. $|z + 3i| = 2$

4. $|2z - 1| = 4$

5. $\mathrm{Re}(z) = 5$

6. $\mathrm{Im}(z) = -2$

7. $\mathrm{Im}(\bar{z} + 3i) = 6$

8. $\mathrm{Im}(z - i) = \mathrm{Re}(z + 4 - 3i)$

9. $|\mathrm{Re}\,(1 + i\bar{z})| = 3$

10. $z^2 + \bar{z}^2 = 2$

11. $\mathrm{Re}(z^2) = 1$

12. $\arg(z) = \pi/4$

In Problems 13–24, sketch the set S of points in the complex plane satisfying the given inequality. Determine whether the set is (**a**) open, (**b**) closed, (**c**) a domain, (**d**) bounded, or (**e**) connected.

13. $\mathrm{Re}(z) < -1$

14. $|\mathrm{Re}\,(z)| > 2$

15. $\mathrm{Im}(z) > 3$

16. $\mathrm{Re}\,((2 + i)z + 1) > 0$

17. $2 < \mathrm{Re}(z - 1) < 4$

18. $-1 \leq \mathrm{Im}(z) < 4$

19. $\mathrm{Re}(z^2) > 0$

20. $\mathrm{Im}(z) < \mathrm{Re}(z)$

21. $|z - i| > 1$

22. $2 < |z - i| < 3$

23. $1 \leq |z - 1 - i| < 2$

24. $2 \leq |z - 3 + 4i| \leq 5$

25. Give the boundary points of the sets in Problems 13–24.

26. Consider the set S consisting of the complex plane with the circle $|z| = 5$ deleted. Give the boundary points of S. Is S connected?

In Problems 27 and 28, sketch the set of points in the complex plane satisfying the given inequality.

27. $0 \leq \arg(z) \leq \pi/6$

28. $-\pi < \arg(z) < \pi/2$

In Problems 29 and 30, describe the shaded set in the given figure using $\arg(z)$ and an inequality.

29.

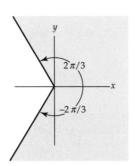

Figure 1.5.11 Figure for Problem 29

30.

Figure 1.5.12 Figure for Problem 30

In Problems 31 and 32, solve the given pair of simultaneous equations.

31. $|z| = 2$, $|z - 2| = 2$

32. $|z - i| = 5$, $\arg(z) = \pi/4$

Focus on Concepts

33. On page 28 we stated that if $\rho_1 > 0$, then the set of points satisfying $\rho_1 < |z - z_0|$ is the exterior to the circle of radius ρ_1 centered at z_0. In general, describe the set if $\rho_1 = 0$. In particular, describe the set defined by $|z + 2 - 5i| > 0$.

34. **(a)** What are the boundary points of a deleted neighborhood of z_0?

    (b) What are the boundary points of the complex plane?

    (c) Give several examples, not including the one given on page 29, of a set S in the complex plane that is neither open nor closed.

35. Use complex notation and inequalities in parts (a) and (b).

    (a) Make up a list of five sets in the complex plane that are connected.

    (b) Make up a list of five sets in the complex plane that are not connected.

36. Consider the disk centered at z_0 defined by $|z - z_0| \leq \rho$. Demonstrate that this set is bounded by finding an $R > 0$ so that all points z in the disk satisfy $|z| < R$. [*Hint*: See the discussion on inequalities in Section 1.2.]

37. Suppose z_0 and z_1 are distinct points. Using only the concept of distance, describe in words the set of points z in the complex plane that satisfy $|z - z_0| = |z - z_1|$.

38. Using only the concept of distance, describe in words the set of points z in the complex plane that satisfies $|z - i| + |z + i| = 1$.

In Problems 39 and 40, describe the shaded set in the given figure by filling in the two blanks in the set notation

$$\{ \, z: \, \underline{\hspace{2cm}} \; \text{and/or} \; \underline{\hspace{2cm}} \}$$

using complex notation for equations or inequalities and one of the words *and* or *or*.

39.

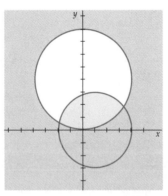

40.

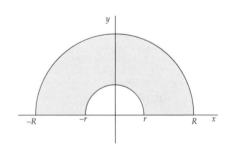

Figure 1.5.13 Figure for Problem 39   Figure 1.5.14 Figure for Problem 40

41. Consider the set S of points in the complex plane defined by $\{i/n\}$, $n = 1, 2, 3, \ldots$. Discuss which of the following terms apply to S: *boundary, open, closed, connected, bounded*.

42. Consider a finite set S of complex numbers $\{z_1, z_2, z_3, \ldots, z_n\}$. Discuss whether S is necessarily bounded. Defend your answer with sound mathematics.

43. A set S is said to be **convex** if each pair of points P and Q in S can be joined by a line segment $\overline{PQ}$ such that every point on the line segment also lies in S.

Determine which of the sets S in the complex plane defined by the following conditions are convex.

(a) $|z - 2 + i| < 3$ (b) $1 < |z| < 2$

(c) $x > 2$, $y \le -1$ (d) $y < x^2$

(e) $\mathrm{Re}(z) \le 5$ (f) $\mathrm{Re}(z) \ne 0$

44. Discuss: Is a convex set, defined in Problem 43, necessarily connected?

45. Discuss and decide whether the empty set $\varnothing$ is an open set.

46. Suppose S_1 and S_2 are open sets in the complex plane.

   (a) Discuss: Is the union $S_1 \cup S_2$ an open set? If you think the statement is true, try to prove it. If you think the statement is false, find a counterexample.

   (b) Repeat part (a) for the intersection $S_1 \cap S_2$.

Before answering Problems 47–50, reread the Remarks on the end of this section.

47. Find the point (x_0, y_0) on the unit circle that corresponds to each of the real numbers: $-\frac{1}{4}$, $\frac{1}{2}$, -3, 1, 10. See Figure 1.5.10(a).

48. Find a point (x_0, y_0, u_0) on the unit sphere that corresponds to the complex number $2 + 5i$. See Figure 1.5.10(b).

49. Describe the set of points on the unit sphere that correspond to each of the following sets in the complex plane.

   (a) the numbers on unit circle $|z| = 1$

   (b) the numbers within the open disk $|z| < 1$

   (c) the numbers that are exterior to unit circle, that is, $|z| > 1$

50. Express the coordinates of the point (x_0, y_0, u_0) on the unit sphere in Figure 1.5.10(b) in terms of the coordinates of the point $(a, b, 0)$ in the complex plane. Use these formulas to verify your answer to Problem 48. [*Hint*: First show that all points on the line containing $(0, 0, 1)$ and $(a, b, 0)$ are of the form $(ta, tb, 1 - t)$.]

1.6   Applications

In this section we are going to examine a few simple applications of complex numbers. It will be assumed in the discussion that the reader has some familiarity with methods for solving elementary ordinary differential equations.

    We saw how to find roots of complex numbers in Section 1.4. With that background we are now in a position to examine how to solve a quadratic equation with complex coefficients using the complex version of the quadratic formula. We then examine how complex numbers and the complex exponential are used in differential equations. This last discussion leads us to Euler's formula and a new compact way of writing the polar form of a complex number. Lastly, we explore some ways complex numbers are used in electrical engineering.

Algebra You probably encountered complex numbers for the first time in a beginning course in algebra where you learned that roots of polynomial equations can be complex as well as real. For example, any second degree, or quadratic, polynomial equation can be solved by completing the square. In the general case, $ax^2 + bx + c = 0$, where the coefficients $a \ne 0$, b, and c are real, completion of the square in x yields the quadratic formula:

$$x = \frac{-b \pm \sqrt{b^2 - 4ac}}{2a}. \tag{1}$$

When the discriminant $b^2 - 4ac$ is negative, the roots of the equation are complex. For example, by (1) the two roots of $x^2 - 2x + 10 = 0$ are

$$x = \frac{-(-2) \pm \sqrt{(-2)^2 - 4(1)(10)}}{2(1)} = \frac{2 \pm \sqrt{-36}}{2}. \tag{2}$$

In beginning courses the imaginary unit i is written $i = \sqrt{-1}$ and the assumption is made that the laws of exponents hold so that a number such as $\sqrt{-36}$ can be written $\sqrt{-36} = \sqrt{36}\sqrt{-1} = 6i$. Let us denote the two complex roots in (2) as $z_1 = 1 + 3i$ and $z_2 = 1 - 3i$.

Note: The roots z_1 and z_2 are conjugates. See Problem 22 in Exercises 1.6.

Quadratic Formula The quadratic formula is perfectly valid when the coefficients $a \neq 0$, b, and c of a quadratic polynomial equation $az^2 + bz + c = 0$ are complex numbers. Although the formula can be obtained in exactly the same manner as (1), we choose to write the result as

$$z = \frac{-b + (b^2 - 4ac)^{1/2}}{2a}. \tag{3}$$

Notice that the numerator of the right-hand side of (3) looks a little different than the traditional $-b \pm \sqrt{b^2 - 4ac}$ given in (1). Bear in mind that when $b^2 - 4ac \neq 0$, the symbol $\left(b^2 - 4ac\right)^{1/2}$ represents the set of two square roots of the complex number $b^2 - 4ac$. Thus, (3) gives *two* complex solutions. From this point on we reserve the use of the symbol $\sqrt{}$ to real numbers where $\sqrt{a}$ denotes the nonnegative root of the real number $a \geq 0$. The next example illustrates the use of (3).

Interpretation of $\sqrt{}$ in this text

EXAMPLE 1   Using the Quadratic Formula

Solve the quadratic equation $z^2 + (1 - i)z - 3i = 0$.

Solution   From (3), with $a = 1$, $b = 1 - i$, and $c = -3i$, we have

$$z = \frac{-(1-i) + [(1-i)^2 - 4(-3i)]^{1/2}}{2} = \frac{1}{2}\left[-1 + i + (10i)^{1/2}\right]. \tag{4}$$

To compute $(10i)^{1/2}$ we use (4) of Section 1.4 with $r = \sqrt{10}$, $\theta = \pi/2$, and $n = 2$. The two square roots of $10i$ are:

$$w_0 = \sqrt{10}\left(\cos\frac{\pi}{4} + i\sin\frac{\pi}{4}\right) = \sqrt{10}\left(\frac{1}{\sqrt{2}} + \frac{1}{\sqrt{2}}i\right) = \sqrt{5} + \sqrt{5}i$$

and

$$w_1 = \sqrt{10}\left(\cos\frac{5\pi}{4} + i\sin\frac{5\pi}{4}\right) = \sqrt{10}\left(-\frac{1}{\sqrt{2}} - \frac{1}{\sqrt{2}}i\right) = -\sqrt{5} - \sqrt{5}i.$$

Therefore, (4) gives two values:

$$z_1 = \frac{1}{2}\left[-1 + i + \left(\sqrt{5} + \sqrt{5}i\right)\right] \quad \text{and} \quad z_2 = \frac{1}{2}\left[-1 + i + \left(-\sqrt{5} - \sqrt{5}i\right)\right].$$

These solutions of the original equation, written in the form $z = a + ib$, are

Note: The roots z_1 and z_2 are **not** conjugates. See Problem 26 in Exercises 1.6.

$$z_1 = \frac{1}{2}\left(\sqrt{5} - 1\right) + \frac{1}{2}\left(\sqrt{5} + 1\right)i \quad \text{and} \quad z_2 = -\frac{1}{2}\left(\sqrt{5} + 1\right) - \frac{1}{2}\left(\sqrt{5} - 1\right)i.$$

❑

Factoring a Quadratic Polynomial By finding all the roots of a polynomial equation we can factor the polynomial completely. This statement follows as a corollary to an important theorem that will be proved in Section 5.5. For the present, note that if z_1 and z_2 are the roots defined by (3), then a quadratic polynomial $az^2 + bz + c$ factors as

$$az^2 + bz + c = a(z - z_1)(z - z_2). \tag{5}$$

For example, we have already used (2) to show that the quadratic equation $x^2 - 2x + 10 = 0$ has roots $z_1 = 1 + 3i$ and $z_2 = 1 - 3i$. With $a = 1$, (5) enables us to factor the polynomial $x^2 - 2x + 10$ using complex numbers:

$$x^2 - 2x + 10 = [x - (1 + 3i)][x - (1 - 3i)] = (x - 1 - 3i)(x - 1 + 3i).$$

Similarly, the factorization of the quadratic polynomial $z^2 + (1 - i)z - 3i$ in Example 1 is

$$z^2 + (1-i)z - 3i = \left[z - \frac{1}{2}(\sqrt{5}-1) - \frac{1}{2}(\sqrt{5}+1)i\right]\left[z + \frac{1}{2}(\sqrt{5}+1) + \frac{1}{2}(\sqrt{5}-1)i\right]$$

Because a first course in calculus deals principally with real quantities, you probably did not see any complex numbers until you took a course in differential equations or in electrical engineering.

Differential Equations The first step in solving a linear second-order ordinary differential equation $ay'' + by' + cy = f(x)$ with real coefficients a, b, and c is to solve the associated homogeneous equation $ay'' + by' + cy = 0$. The latter equation possesses solutions of the form $y = e^{mx}$. To see this, we substitute $y = e^{mx}$, $y' = me^{mx}$, and $y'' = m^2 e^{mx}$ into $ay'' + by' + cy = 0$:

$$ay'' + by' + cy = am^2 e^{mx} + bme^{mx} + ce^{mx} = e^{mx}\left(am^2 + bm + c\right) = 0.$$

From $e^{mx}\left(am^2 + bm + c\right) = 0$, we see that $y = e^{mx}$ is a solution of the homogeneous equation whenever m is root of the polynomial equation $am^2 + bm + c = 0$. The latter quadratic equation is known as the **auxiliary equation**. Now when the coefficients of a polynomial equation are *real*, the equation cannot have just *one* complex root; that is, complex roots must always appear in conjugate pairs. Thus, if the auxiliary equation possesses complex roots $\alpha + i\beta, \alpha - i\beta, \beta > 0$, then two solutions of $ay'' + by' + cy = 0$ are complex exponential functions $y = e^{(\alpha+i\beta)x}$ and $y = e^{(\alpha-i\beta)x}$. In order to obtain real solutions of the differential equation we use **Euler's formula**

$$e^{i\theta} = \cos\theta + i\sin\theta, \tag{6}$$

where θ is real. With θ replaced, in turn, by β and $-\beta$, we use (6) to write

$$e^{(\alpha+i\beta)x} = e^{\alpha x}e^{i\beta x} = e^{\alpha x}(\cos\beta x + i\sin\beta x) \text{ and } e^{(\alpha-i\beta)x} = e^{\alpha x}e^{-i\beta x} = e^{\alpha x}(\cos\beta x - i\sin\beta x). \tag{7}$$

Now since the differential equation is homogeneous, the linear combinations

$$y_1 = \frac{1}{2}\left[e^{(\alpha+i\beta)x} + e^{(\alpha-i\beta)x}\right] \quad \text{and} \quad y_2 = \frac{1}{2i}\left[e^{(\alpha+i\beta)x} - e^{(\alpha-i\beta)x}\right]$$

are also solutions. But in view of (7), both of the foregoing expressions are real functions

$$y_1 = e^{\alpha x} \cos \beta x \quad \text{and} \quad y_2 = e^{\alpha x} \sin \beta x. \tag{8}$$

EXAMPLE 2   **Solving a Differential Equation**

Solve the differential equation $y'' + 2y' + 2y = 0$.

Solution   We apply the quadratic formula to the auxiliary equation $m^2 + 2m + 2 = 0$ and obtain the complex roots $m_1 = -1 + i$ and $m_2 = \overline{m}_1 = -1 - i$. With the identifications $\alpha = -1$ and $\beta = 1$, (8) gives us two solutions $y_1 = e^{-x} \cos x$ and $y_2 = e^{-x} \sin x$.   ❏

You may recall that the so-called general solution of a homogeneous linear nth-order differential equation consists of a linear combination of n linearly independent solutions. Thus in Example 2, the general solution of the given second-order differential equation is $y_1 = c_1 y_1 + c_2 y_2 = c_1 e^{-x} \cos x + c_2 e^{-x} \sin x$, where c_1 and c_2 are arbitrary constants.

Exponential Form of a Complex Number   We hasten to point out that the results given in (6) and (7) were assumptions at that point because the complex exponential function has not been defined as yet. As a brief preview of the material in Sections 2.1 and 4.1, the complex exponential e^z is the complex number *defined* by

$$e^z = e^{x+iy} = e^x (\cos y + i \sin y). \tag{9}$$

Although the proof is postponed until Section 4.1, (9) can be used to show that the familiar law of exponents holds for complex numbers z_1 and z_2:

$$e^{z_1} e^{z_2} = e^{z_1 + z_2}. \tag{10}$$

Because of (10), the results in (7) are valid. Moreover, note that Euler's formula (4) is a special case of (9) when z is a pure imaginary number, that is, with $x = 0$ and y replaced by θ. Euler's formula provides a notational convenience for several concepts considered earlier in this chapter. The polar form of a complex number z, $z = r(\cos \theta + i \sin \theta)$, can now be written compactly as

$$z = re^{i\theta}. \tag{11}$$

This convenient form is called the **exponential form** of a complex number z. For example, $i = e^{\pi i/2}$ and $1 + i = \sqrt{2} e^{\pi i/4}$. Also, the formula for the n nth roots of a complex number, (4) of Section 1.4, becomes

$$z^{1/n} = \sqrt[n]{r} e^{i(\theta + 2k\pi)/n}, k = 0, 1, 2, \ldots, n - 1. \tag{12}$$

Electrical Engineering   In applying mathematics to physical situations, engineers and mathematicians often approach the same problem in completely different ways. For example, consider the problem of finding the steady-state current $i_p(t)$ in an LRC-series circuit in which the charge $q(t)$ on the capacitor for time $t > 0$ is described by the differential equation

$$L\frac{d^2q}{dt^2} + R\frac{dq}{dt} + \frac{1}{C}q = E_0 \sin \gamma t \tag{13}$$

where the positive constants L, R, and C are, in turn, the inductance, resistance, and capacitance. Now to find the **steady-state current** $i_p(t)$, we first find the **steady-state charge** on the capacitor by finding a particular solution $q_p(t)$ of (13). Proceeding as we would in a course in differential equations, we will use the method of undetermined coefficients to find $q_p(t)$. This entails assuming a particular solution of the form $q_p(t) = A\sin\gamma t + B\cos\gamma t$, substituting this expression into the differential equation, simplifying, equating coefficients, and solving for the unknown coefficients A and B. It is left as an exercise to show that $A = E_0 X/(-\gamma Z^2)$ and $B = E_0 R/(-\gamma Z^2)$, where the quantities

$$X = L\gamma - 1/C\gamma \quad \text{and} \quad Z = \sqrt{X^2 + R^2} \tag{14}$$

are called, respectively, the **reactance** and **impedance** of the circuit. Thus, the steady-state solution or steady-state charge in the circuit is

$$q_p(t) = -\frac{E_0 X}{\gamma Z^2}\sin\gamma t - \frac{E_0 R}{\gamma Z^2}\cos\gamma t.$$

From this solution and $i_p(t) = q_p'(t)$ we obtain the steady-state current:

$$i_p(t) = \frac{E_0}{Z}\left(\frac{R}{Z}\sin\gamma t - \frac{X}{Z}\cos\gamma t\right). \tag{15}$$

Electrical engineers often solve circuit problems such as this by using complex analysis. First of all, to avoid confusion with the current i, an electrical engineer will denote the imaginary unit i by the symbol j; in other words, $j^2 = -1$. Since current i is related to charge q by $i = dq/dt$, the differential equation (13) is the same as

$$L\frac{di}{dt} + Ri + \frac{1}{C}q = E_0 \sin \gamma t. \tag{16}$$

Now in view of Euler's formula (6), if θ is replaced by the symbol γ, then the impressed voltage $E_0 \sin\gamma t$ is the same as $\text{Im}(E_0 e^{j\gamma t})$. Because of this last form, the method of undetermined coefficients suggests that we try a solution of (16) in the form of a constant multiple of a complex exponential, that is, $i_p(t) = \text{Im}(Ae^{j\gamma t})$. We substitute this last expression into equation (16), assume that the complex exponential satisfies the "usual" differentiation rules, use the fact that charge q is an antiderivative of the current i, and equate coefficients of $e^{j\gamma t}$. The result is $(jL\gamma + R + 1/jC\gamma)A = E_0$ and from this we obtain

$$A = \frac{E_0}{R + j\left(L\gamma - \dfrac{1}{C\gamma}\right)} = \frac{E_0}{R + jX}, \tag{17}$$

where X is the reactance given in (14). The denominator of the last expression, $Z_c = R + j(L\gamma - 1/C\gamma) = R + jX$, is called the **complex impedance** of the circuit. Since the modulus of the complex impedance is $|Z_c| = \sqrt{R^2 + (L\gamma - 1/C\gamma)^2}$, we see from (14) that the impedance Z and the complex impedance Z_c are related by $Z = |Z_c|$.

Now from the exponential form of a complex number given in (11), we can write the complex impedance as

$$Z_c = |Z_c| \, e^{j\theta} = Z e^{j\theta} \quad \text{where} \quad \tan\theta = \frac{L\gamma - \dfrac{1}{C\gamma}}{R}.$$

Hence (17) becomes $A = E_0/Z_c = E_0/(Ze^{j\theta})$, and so the steady-state current is given by

$$i_p(t) = \text{Im}\left(\frac{E_0}{Z} e^{-j\theta} e^{j\gamma t}\right). \tag{18}$$

You are encouraged to verify that the expression in (18) is the same as that given in (15).

The problem of finding the steady-state current in an LRC-series circuit described by the differential equation

$$L\frac{di}{dt} + Ri + \frac{1}{C}q = E_0 \cos\gamma t \tag{19}$$

can be solved in a similar manner. In this case, the impressed voltage $E_0 \cos\gamma t$ suggests that we try a solution of the form $i_p(t) = \text{Re}(Ae^{j\gamma t})$. With the same analysis as above, we obtain the solution of (19) given by

$$i_p(t) = \text{Re}\left(\frac{E_0}{Z} e^{-j\theta} e^{j\gamma t}\right). \tag{20}$$

Remarks *Comparison with Real Analysis*

We have seen in this section that if z_1 is a complex root of a polynomial equation, then $z_2 = \bar{z}_1$ is another root whenever all the coefficients of the polynomial are real, but that $\bar{z}_1$ is not necessarily a root of the equation when at least one coefficient is not real. In the latter case, we can obtain another factor of the polynomial by dividing it by $z - z_1$. We note that synthetic division is valid in complex analysis. See Problems 27 and 28 in Exercises 1.6.

EXERCISES 1.6 *Answers to selected odd-numbered problems begin on page ANS-5.*

In Problems 1–6, solve the given quadratic equation using the quadratic formula (3). Then use (5) to factor the polynomial.

1. $z^2 + iz - 2 = 0$ **2.** $iz^2 - z + i = 0$

3. $z^2 - (1+i)z + 6 - 17i = 0$ **4.** $z^2 - (1+9i)z - 20 + 5i = 0$

5. $z^2 + 2z - \sqrt{3}i = 0$

6. $3z^2 + (2-3i)z - 1 - 3i = 0$ [*Hint:* See Problem 15 in Exercises 1.4.]

In Problems 7–12, express the given complex number in the exponential form $z = re^{i\theta}$.

7. -10 **8.** $-2\pi i$

9. $-4 - 4i$ **10.** $\dfrac{2}{1+i}$

11. $(3-i)^2$ **12.** $(1+i)^{20}$

In Problems 13–16, find solutions of the given homogeneous differential equation.

13. $y'' - 4y' + 13y = 0$ **14.** $3y'' + 2y' + y = 0$

15. $y'' + y' + y = 0$ **16.** $y'' + 2y' + 4y = 0$

In Problems 17–20, find the steady-state charge $q_p(t)$ and steady-state current $i_p(t)$ for the LRC-series circuit described by the given differential equation. Find the complex impedance and impedance of the circuit. Use the complex method discussed on pages 37 and 38.

17. $\dfrac{di}{dt} + 6i + 25q = 10\cos 5t$ **18.** $\dfrac{di}{dt} + 6i + 25q = 10\sin 5t$

19. $\dfrac{di}{dt} + i + 2q = 100\sin t$ **20.** $\dfrac{di}{dt} + i + 2q = 100\cos t$

Focus on Concepts

21. Discuss how (3) can be used to find the four roots of $z^4 - 2z^2 + 1 - 2i = 0$. Carry out your ideas.

22. If z_1 is a root of a polynomial equation with *real* coefficients, then its conjugate $z_2 = \bar{z}_1$ is also a root. Prove this result in the case of a quadratic equation $az^2 + bz + c = 0$, where $a \neq 0$, b, and c are real. Start with the properties of conjugates given in (1) and (2) of Section 1.1.

In Problems 23 and 24, use Problem 22 and (5) of this section to factor the given quadratic polynomial if the indicated complex number is one root.

23. $4z^2 + 12z + 34 = 0;\quad z_1 = -\dfrac{3}{2} + \dfrac{5}{2}i$

24. $5z^2 - 2z + 4 = 0;\quad z_1 = \dfrac{1}{5} + \dfrac{\sqrt{19}}{5}i$

25. (a) Find a quadratic polynomial equation for which $2 - i$ is one root.

   (b) Is your answer to part (a) unique? Elaborate in detail.

26. If z_1 is a root of a quadratic equation with leading coefficient 1 and at least one *nonreal coefficient*, then $\bar{z}_1$ is *not* a root. Prove this result in the case of a quadratic equation $z^2 + bz + c = 0$, where at least one of b or c is a nonzero complex number.

In Problems 27 and 28, factor the given quadratic polynomial if the indicated complex number is one root. [*Hint*: Consider long division or synthetic division.]

27. $3iz^2 + (9 - 16i)z - 17 - i;\; z_1 = 5 + 2i$

28. $4z^2 + (-13 + 18i)z - 5 - 10i;\; z_1 = 3 - 4i$

In Problems 29 and 30, establish the plausibility of Euler's formula (6) in the two ways that are specified.

29. The Maclaurin series $e^x = \sum_{n=0}^{\infty} \dfrac{1}{n!}x^n = 1 + x + \dfrac{1}{2!}x^2 + \dfrac{1}{3!}x^3 + \cdots$ is known to converge for all values of x. Taking the last statement at face value, substitute $x = i\theta$, θ real, into the series and simplify the powers of i^n. See what results when you separate the series into real and imaginary parts.

30. (a) Verify that $y_1 = \cos\theta$ and $y_2 = \sin\theta$ satisfy the homogeneous linear second-order differential equation $\dfrac{d^2y}{d\theta^2} + y = 0$. Since the set of solutions consisting of y_1 and y_2 is linearly independent, the general solution of the differential equation is $y = c_1\cos\theta + c_2\sin\theta$.

   (b) Verify that $y = e^{i\theta}$, where i is the imaginary unit and θ is a real variable, also satisfies the differential equation given in part (a).

(c) Since $y = e^{i\theta}$ is a solution of the differential equation, it must be obtainable from the general solution given in part (a); in other words, there must exist *specific* coefficients c_1 and c_2 such that $e^{i\theta} = c_1 \cos\theta + c_2 \sin\theta$. Verify from $y = e^{i\theta}$ that $y(0) = 1$ and $y'(0) = i$. Use these two conditions to determine c_1 and c_2.

31. Find a homogeneous linear second-order differential equation for which $y = e^{-5x} \cos 2x$ is a solution.

32. (a) By differentiating Equation (13) with respect to t, show that the current in the *LRC*-series is described by

$$L\frac{d^2 i}{dt^2} + R\frac{di}{dt} + \frac{1}{C}i = E_0\, \gamma \cos\gamma t.$$

(b) Use the method of undetermined coefficients to find a particular solution $i_{p_1}(t) = Ae^{j\gamma t}$ of the differential equation

$$L\frac{d^2 i}{dt^2} + R\frac{di}{dt} + \frac{1}{C}i = E_0\gamma e^{j\gamma t}.$$

(c) How can the result of part (b) be used to find a particular solution $i_p(t)$ of the differential equation in part (a). Carry out your thoughts and verify that $i_p(t)$ is the same as (15).

Computer Lab Assignments

In Problems 33–36, use a CAS as an aid in factoring the given quadratic polynomial.

33. $z^2 - 3iz - 2$

34. $z^2 - \sqrt{3}z - i$

35. $iz^2 - (2 + 3i)z + 1 + 5i$

36. $(3 + i)z^2 + (1 + 7i)z - 10$

In Problems 37 and 38, use a CAS to solve the given polynomial equation. In *Mathematica* the command **Solve** will find all roots of polynomial equations up to degree four by means of a formula.

37. $z^3 - 4z^2 + 10 = 0$

38. $z^4 + 4iz^2 + 10i = 0$

In Problems 39 and 40, use a CAS to solve the given polynomial equation. The command **NSolve** in *Mathematica* will approximate all roots of polynomial equations of degree five or higher.

39. $z^5 - z - 12 = 0$

40. $z^6 - z^4 + 3iz^3 - 1 = 0$

Projects

41. Cubic Formula In this project you are asked to investigate the solution of a cubic polynomial equation by means of a formula using radicals, that is, a combination of square roots and cube roots of expressions involving the coefficients.

(a) To solve a general cubic equation $z^3 + az^2 + bz + c = 0$ it is sufficient to solve a **depressed cubic equation** $x^3 = mx + n$ since the general cubic equation can be reduced to this special case by eliminating the term az^2. Verify this by means of the substitution $z = x - a/3$ and identify m and n.

(b) Use the procedure outlined in part (a) to find the depressed cubic equation for $z^3 + 3z^2 - 3z - 9 = 0$.

(c) A solution of $x^3 = mx + n$ is given by

$$x = \left[\frac{n}{2} + \left(\frac{n^2}{4} - \frac{m^3}{27}\right)^{1/2}\right]^{1/3} + \left[\frac{n}{2} - \left(\frac{n^2}{4} - \frac{m^3}{27}\right)^{1/2}\right]^{1/3}.$$

Use this formula to solve the depressed cubic equation found in part (b).

(**d**) Graph the polynomial $z^3 + 3z^2 - 3z - 9$ and the polynomial from the depressed cubic equation in part (b); then estimate the x-intercepts from the graphs.

(**e**) Compare your results from part (d) with the solutions found in part (c). Resolve any apparent differences. Find the three solutions of $z^3 + 3z^2 - 3z - 9 = 0$.

(**f**) Do some additional reading to find geometrically motivated proofs (using a square and a cube) to derive the quadratic formula and the formula given in part (c) for the solution of the depressed cubic equation. Why is the name *quadratic* formula used when the prefix *quad* stems from the Latin word for the number four?

42. **Complex Matrices** In this project we assume that you have either had some experience with matrices or are willing to learn something about them.

Certain complex matrices, that is, matrices whose entries are complex numbers, are important in applied mathematics. An $n \times n$ complex matrix A is said to be:

$$\textit{Hermitian} \quad \text{if } \bar{A}^T = A,$$
$$\textit{Skew-Hermitian} \quad \text{if } \bar{A}^T = -A,$$
$$\textit{Unitary} \quad \text{if } \bar{A}^T = A^{-1}.$$

Here the symbol $\bar{A}$ means the conjugate of the matrix A, which is the matrix obtained by taking the conjugate of each entry of A. $\bar{A}^T$ is then the transpose of $\bar{A}$, which is the matrix obtained by interchanging the rows with the columns. The negative $-A$ is the matrix formed by negating all the entries of A; the matrix A^{-1} is the multiplicative inverse of A.

(**a**) Which of the following matrices are Hermitian, skew-Hermitian, or unitary?

$$A = \begin{pmatrix} 3i & 10 & -10-2i \\ -10 & 0 & 4+i \\ 10-2i & -4+i & -5i \end{pmatrix}$$

$$B = \begin{pmatrix} 1 & 0 & 0 \\ 0 & \frac{2+i}{\sqrt{10}} & \frac{-2+i}{\sqrt{10}} \\ 0 & \frac{2+i}{\sqrt{10}} & \frac{2-i}{\sqrt{10}} \end{pmatrix}$$

$$C = \begin{pmatrix} 2 & 1+7i & -6+2i \\ 1-7i & 4 & 1+i \\ -6-2i & 1-i & 0 \end{pmatrix}$$

(**b**) What can be said about the entries on the main diagonal of a Hermitian matrix? Prove your assertion.

(c) What can be said about the entries on the main diagonal of a skew-Hermitian matrix? Prove your assertion.

(d) Prove that the eigenvalues of a Hermitian matrix are real.

(e) Prove that the eigenvalues of a skew-Hermitian matrix are either pure imaginary or zero.

(f) Prove that the eigenvalues of unitary matrix are unimodular; that is, $|\lambda| = 1$. Describe where these eigenvalues are located in the complex plane.

(g) Prove that the modulus of a unitary matrix is one, that is, $|\det A| = 1$.

(h) Do some additional reading and find an application of each of these types of matrices.

(i) What are the real analogues of these three matrices?

CHAPTER 1 REVIEW QUIZ   *Answers to selected odd-numbered problems begin on page ANS-5.*

In Problems 1–22, answer true or false. If the statement is false, justify your answer by either explaining why it is false or giving a counterexample; if the statement is true, justify your answer by either proving the statement or citing an appropriate result in this chapter.

1. $\operatorname{Re}(z_1 z_2) = \operatorname{Re}(z_1)\operatorname{Re}(z_2)$

2. $\operatorname{Im}(4 + 7i) = 7i$

3. $|z - 1| = |\bar{z} - 1|$

4. If $\operatorname{Im}(z) > 0$, then $\operatorname{Re}(1/z) > 0$.

5. $i < 10i$

6. If $z \neq 0$, then $\operatorname{Arg}(z + \bar{z}) = 0$.

7. $|x + iy| \leq |x| + |y|$

8. $\arg(\bar{z}) = \arg\left(\dfrac{1}{z}\right)$

9. If $\bar{z} = -z$, then z is pure imaginary.

10. $\arg(-2 + 10i) = \pi - \tan^{-1}(5) + 2n\pi$ for n an integer.

11. If z is a root of a polynomial equation $a_n z^n + a_{n-1} z^{n-1} + \cdots + a_1 z + a_0 = 0$, then $\bar{z}$ is also a root.

12. For any nonzero complex number z, there are an infinite number of values for $\arg(z)$.

13. If $|z - 2| < 2$, then $|\operatorname{Arg}(z)| < \pi/2$.

14. The set S of complex numbers $z = x + iy$ whose real and imaginary parts are related by $y = \sin x$ is a bounded set.

15. The set S of complex numbers z satisfying $|z| < 1$ or $|z - 3i| < 1$ is a domain.

16. Consider a set of S of complex numbers. If the set A of all real parts of the numbers in S is bounded and the set B of all imaginary parts of the numbers in S is bounded, then necessarily the set S is bounded.

17. The sector defined by $-\pi/6 < \arg(z) \leq \pi/6$ is neither open nor closed.

18. For $z \neq 0$, there are exactly five values of $z^{3/5} = (z^3)^{1/5}$.

19. A boundary point of a set S is a point in S.

20. The complex plane with the real and imaginary axes deleted has no boundary points.

21. $\text{Im}\left(e^{i\theta}\right) = \sin\theta$.

22. The equation $z^n = 1$, n a positive integer, will have only real solutions for $n = 1$ and $n = 2$.

In Problems 23–50, try to fill in the blanks without referring back to the text.

23. If $a + ib = \dfrac{3-i}{2+3i} + \dfrac{2-2i}{1-5i}$, then $a = $ _____ and $b = $ _____.

24. If $z = \dfrac{4i}{-3-4i}$, then $|z| = $ _____ .

25. If $|z| = \text{Re}(z)$, then z is _____ .

26. If $z = 3 + 4i$, then $\text{Re}\left(\dfrac{z}{\bar{z}}\right) = $ _____ .

27. The principal argument of $z = -1 - i$ is _____ .

28. $\overline{z_1^2 + \bar{z}_2^2} = $ _____ .

29. $\arg\left((1+i)^5\right) = $ _____ , $\left|(1+i)^6\right| = $ _____ , $\text{Im}\left((1+i)^7\right) = $ _____ , and $\text{Re}\left((1+i)^8\right) = $ _____ .

30. $\left(\frac{1}{2} + \frac{\sqrt{3}}{2}i\right)^{483} = $ _____ .

31. If z is a point in the second quadrant, then $i\bar{z}$ is in the _____ quadrant.

32. $i^{127} - 5i^9 + 2i^{-1} = $ _____ .

33. Of the three points $z_1 = 2.5 + 1.9i$, $z_2 = 1.5 - 2.9i$, and $z_3 = -2.4 + 2.2i$, _____ is the farthest from the origin.

34. If $3i\bar{z} - 2z = 6$, then $z = $ _____ .

35. If $2x - 3yi + 9 = -x + 2yi + 5i$, then $z = $ _____ .

36. If $z = \dfrac{5}{-\sqrt{3}+i}$, then $\text{Arg}(z) = $ _____ .

37. If $z \neq 0$ is a real number, then $z + z^{-1}$ is real. Other complex numbers $z = x + iy$ for which $z + z^{-1}$ is real are defined by $|z| = $ _____ .

38. The position vector of length 10 passing through $(1, -1)$ is the same as the complex number $z = $ _____ .

39. The vector $z = (2 + 2i)(\sqrt{3} + i)$ lies in the _____ quadrant.

40. The boundary of the set S of complex numbers z satisfying both $\text{Im}(y) > 0$ and $|z - 3i| > 1$ is _____ .

41. In words, the region in the complex plane for which $\text{Re}(z) < \text{Im}(z)$ is _____ .

42. The region in the complex plane consisting of the two disks $|z + i| \leq 1$ and $|z - i| \leq 1$ is _____ (connected/not connected).

43. Suppose that z_0 is not a real number. The circles $|z - z_0| = |\bar{z}_0 - z_0|$ and $|z - \bar{z}_0| = |z_0 - \bar{z}_0|$ intersect on the _____ (real axis/imaginary axis).

44. In complex notation, an equation of the circle with center -1 that passes through $2 - i$ is _____ .

45. A positive integer n for which $(1 + i)^n = 4096$ is $n = $ _____ .

46. $\left|\dfrac{(4 - 5i)^{658}}{(5 + 4i)^{658}}\right| = $ _____ .

47. From $(\cos\theta + i\sin\theta)^4 = \cos 4\theta + i\sin 4\theta$ we get the real trigonometric identities $\cos 4\theta = $ _____ and $\sin 4\theta = $ _____ .

48. When z is a point within the open disk defined by $|z| < 4$, an upper bound for $\left| z^3 - 2z^2 + 6z + 2 \right|$ is given by _____ .

49. In Problem 22 in Exercises 1.6 we saw that if z_1 is a root of a polynomial equation with real coefficients, then its conjugate $z_2 = \bar{z}_1$ is also a root. Assume that the cubic polynomial equation $az^3 + bz^2 + cz + d = 0$, where a, b, c, and d are real, has exactly three roots. One of the roots must be real because _____ .

50. **(a)** Interpret the circular mnemonic for positive integer powers of i given in Figure 1.R.1. Use this mnemonic to find i^n for the following values of n:

$$5, \ \ 9, \ 13, \ 17, \ 21, \ \dots \ ; \ i^n = \text{_____}$$
$$6, \ 10, \ 14, \ 18, \ 22, \ \dots \ ; \ i^n = \text{_____}$$
$$7, \ 11, \ 15, \ 19, \ 23, \ \dots \ ; \ i^n = \text{_____}$$
$$8, \ 12, \ 16, \ 20, \ 24, \ \dots \ ; \ i^n = \text{_____}$$

(b) Reinspect the powers n in the four rows in part (a) and then divide each these powers by 4. Based on your discovery, discern an easy rule for determining i^n for any positive integer n. Use your rule to compute

$$i^{33} = \text{\_\_\_\_} \, , \, i^{68} = \text{\_\_\_\_} \, , \, i^{87} = \text{\_\_\_\_} \, , \, i^{102} = \text{\_\_\_\_} \, , \, i^{624} = \text{\_\_\_\_} \, .$$

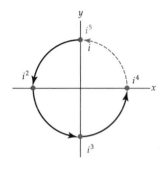

Figure 1.R.1 Figure for Problem 50

Complex Functions and Mappings

Chapter Outline

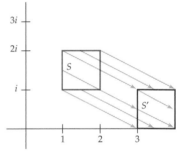

Image of a square under translation. See page 63.

Introduction In the last chapter we introduced complex numbers and examined some of their algebraic and geometric properties. In this chapter we turn our attention to the study of functions from a set of complex numbers to another set of complex numbers. Unlike the functions studied in elementary calculus, we shall see that we cannot draw the graph of a complex function. Therefore, we introduce the notion of a mapping as an alternative way of graphically representing a complex function. The concepts of a limit of a complex function and continuity of a complex function are also introduced in this chapter.

2.1   Complex Functions

One of the most important concepts in mathematics is that of a function. You may recall from previous courses that a function is a certain kind of correspondence between two sets; more specifically:

> A *function* f from a set A to a set B is a rule of correspondence that assigns to each element in A one and only one element in B.

We often think of a function as a rule or a machine that accepts *inputs* from the set A and returns *outputs* in the set B. In elementary calculus we studied functions whose inputs and outputs were real numbers. Such functions are called **real-valued functions of a real variable**. In this section we begin our study of functions whose inputs and outputs are complex numbers. Naturally, we call these functions **complex functions of a complex variable**, or **complex functions** for short. As we will see, many interesting and useful complex functions are simply generalizations of well-known functions from calculus.

Function   Suppose that f is a function from the set A to the set B. If f assigns to the element a in A the element b in B, then we say that b is the **image** of a under f, or the **value** of f at a, and we write $b = f(a)$. The set A—the set of inputs—is called the **domain** of f and the set of images in B—the set of outputs—is called the **range** of f. We denote the domain and range of a function f by $\mathrm{Dom}(f)$ and $\mathrm{Range}(f)$, respectively. As an example, consider the "squaring" function $f(x) = x^2$ defined for the real variable x. Since any real number can be squared, the domain of f is the set $\mathbf{R}$ of all real numbers. That is, $\mathrm{Dom}(f) = A = \mathbf{R}$. The range of f consists of all real numbers x^2 where x is a real number. Of course, $x^2 \geq 0$ for all real x, and it is easy to see from the graph of f that the range of f is the set of all *nonnegative* real numbers. Thus, $\mathrm{Range}(f)$ is the interval $[0, \infty)$. The range of f need not be the same as the set B. For instance, f can be viewed as a function from $A = \mathbf{R}$ to $B = \mathbf{R}$, in which case the range is contained in but not equal to B.

As the following definition indicates, a *complex* function is a function whose inputs and outputs are complex numbers.

Definition 2.1.1   Complex Function

A **complex function** is a function f whose domain and range are subsets of the set $\mathbf{C}$ of complex numbers.

Notation used throughout this text.   ➡   A complex function is also called a **complex-valued function of a complex variable**. For the most part we will use the usual symbols f, g, and h to denote complex functions. In addition, inputs to a complex function f will typically be denoted by the variable z and outputs by the variable $w = f(z)$. When referring to a complex function we will use three notations interchangeably, for example, $f(z) = z - i$, $w = z - i$, or, simply, the function $z - i$. Throughout this text the notation $w = f(z)$ will always denote a complex function, whereas the notation $y = f(x)$ will be reserved to represent a real-valued function of a real variable x.

EXAMPLE 1 Complex Function

(a) The expression $z^2 - (2+i)z$ can be evaluated at any complex number z and always yields a single complex number, and so $f(z) = z^2 - (2+i)z$ defines a complex function. Values of f are found by using the arithmetic operations for complex numbers given in Section 1.1. For instance, at the points $z = i$ and $z = 1 + i$ we have:

$$f(i) = (i)^2 - (2+i)(i) = -1 - 2i + 1 = -2i$$

and $\quad f(1+i) = (1+i)^2 - (2+i)(1+i) = 2i - 1 - 3i = -1 - i.$

(b) The expression $g(z) = z + 2\,\mathrm{Re}(z)$ also defines a complex function. Some values of g are:

$$g(i) = i + 2\mathrm{Re}\,(i) = i + 2(0) = i$$

and $\quad g(2-3i) = 2 - 3i + 2\mathrm{Re}\,(2-3i) = 2 - 3i + 2(2) = 6 - 3i.$ ❏

When the domain of a complex function is not explicitly stated, we assume the domain to be the set of all complex numbers z for which $f(z)$ is defined. This set is sometimes referred to as the *natural domain* of f. For example, the functions $f(z) = z^2 - (2+i)z$ and $g(z) = z + 2\,\mathrm{Re}(z)$ in Example 1 are defined for all complex numbers z, and so, $\mathrm{Dom}(f) = \mathbf{C}$ and $\mathrm{Dom}(g) = \mathbf{C}$. The complex function $h(z) = z/\left(z^2+1\right)$ is not defined at $z = i$ and $z = -i$ because the denominator $z^2 + 1$ is equal to 0 when $z = \pm i$. Therefore, $\mathrm{Dom}(h)$ is the set of all complex numbers except i and $-i$.

In the section introduction, we defined a real-valued function of a real variable to be a function whose domain and range are subsets of the set $\mathbf{R}$ of real numbers. Because $\mathbf{R}$ is a subset of the set $\mathbf{C}$ of the complex numbers, *every* real-valued function of a real variable is also a complex function. We will soon see that real-valued functions of two real variables x and y are also special types of complex functions. These functions will play an important role in the study of complex analysis. In order to avoid repeating the cumbersome terminology *real-valued function of a real variable* and *real-valued function of two real variables,* we use the term **real function** from this point on to refer to any type of function studied in a single or multivariable calculus course.

Real and Imaginary Parts of a Complex Function It is often helpful to express the inputs and the outputs of a complex function in terms of their real and imaginary parts. If $w = f(z)$ is a complex function, then the image of a complex number $z = x + iy$ under f is a complex number $w = u + iv$. By simplifying the expression $f(x+iy)$, we can write the real variables u and v in terms of the real variables x and y. For example, by replacing the symbol z with $x+iy$ in the complex function $w = z^2$, we obtain:

$$w = u + iv = (x+iy)^2 = x^2 - y^2 + 2xyi. \tag{1}$$

From (1) the real variables u and v are given by $u = x^2 - y^2$ and $v = 2xy$, respectively. This example illustrates that, if $w = u + iv = f(x+iy)$ is a complex function, then both u and v are real functions of the two real variables x and y. That is, by setting $z = x+iy$, we can express any complex function $w = f(z)$ in terms of two real functions as:

$$f(z) = u(x,y) + iv(x,y). \tag{2}$$

The functions $u(x, y)$ and $v(x, y)$ in (2) are called the **real** and **imaginary parts of f**, respectively.

EXAMPLE 2   Real and Imaginary Parts of a Function

Find the real and imaginary parts of the functions: (a) $f(z) = z^2 - (2+i)z$ and (b) $g(z) = z + 2\operatorname{Re}(z)$.

Solution In each case, we replace the symbol z by $x + iy$, then simplify.

(a) $f(z) = (x+iy)^2 - (2+i)(x+iy) = x^2 - 2x + y - y^2 + (2xy - x - 2y)i$.
So, $u(x, y) = x^2 - 2x + y - y^2$ and $v(x, y) = 2xy - x - 2y$.

(b) Since $g(z) = x + iy + 2\operatorname{Re}(x+iy) = 3x + iy$, we have $u(x, y) = 3x$ and $v(x, y) = y$. ❏

A function f can be defined without using the symbol z. ➡

Every complex function is completely determined by the real functions $u(x, y)$ and $v(x, y)$ in (2). Thus, a complex function $w = f(z)$ can be defined by arbitrarily specifying two real functions $u(x, y)$ and $v(x, y)$, even though $w = u + iv$ may not be obtainable through familiar operations performed solely on the symbol z. For example, if we take, say, $u(x, y) = xy^2$ and $v(x, y) = x^2 - 4y^3$, then $f(z) = xy^2 + i(x^2 - 4y^3)$ defines a complex function. In order to find the value of f at the point $z = 3 + 2i$, we substitute $x = 3$ and $y = 2$ into the expression for f to obtain $f(3 + 2i) = 3 \cdot 2^2 + i(3^2 - 4 \cdot 2^3) = 12 - 23i$.

We note in passing that complex functions defined in terms of $u(x, y)$ and $v(x, y)$ can always be expressed, if desired, in terms of operations on the symbols z and $\bar{z}$. See Problem 32 in Exercises 2.1.

Exponential Function   In Section 1.6 we informally introduced the complex exponential function e^z. This complex function is an example of one that is defined by specifying its real and imaginary parts.

Definition 2.1.2   Complex Exponential Function

The function e^z defined by:

$$e^z = e^x \cos y + ie^x \sin y \tag{3}$$

is called the **complex exponential function**.

By Definition 2.1.2, the real and imaginary parts of the complex exponential function are $u(x, y) = e^x \cos y$ and $v(x, y) = e^x \sin y$, respectively. Values of the complex exponential function $w = e^z$ are found by expressing the point z as $z = x + iy$ and then substituting the values of x and y in (3). The following example illustrates this procedure.

EXAMPLE 3   Values of the Complex Exponential Function

Find the values of the complex exponential function e^z at the following points.

(a) $z = 0$ (b) $z = i$ (c) $z = 2 + \pi i$

Solution In each part we substitute $x = \text{Re}(z)$ and $y = \text{Im}(z)$ into (3) and then simplify.

(a) For $z = 0$, we have $x = 0$ and $y = 0$, and so $e^0 = e^0 \cos 0 + ie^0 \sin 0$. Since $e^0 = 1$ (for the real exponential function), $\cos 0 = 1$, and $\sin 0 = 0$, $e^0 = e^0 \cos 0 + i \sin 0$ simplifies to $e^0 = 1$.

(b) For $z = i$, we have $x = 0$ and $y = 1$, and so:

$$e^i = e^0 \cos 1 + ie^0 \sin 1 = \cos 1 + i \sin 1 \approx 0.5403 + 0.8415i.$$

(c) For $z = 2 + \pi i$, we have $x = 2$ and $y = \pi$, and so $e^{2+\pi i} = e^2 \cos \pi + ie^2 \sin \pi$. Since $\cos \pi = -1$ and $\sin \pi = 0$, this simplifies to $e^{2+\pi i} = -e^2$.   ❏

Exponential Form of a Complex Number The complex exponential function enables us to express the polar form of a nonzero complex number $z = r(\cos \theta + i \sin \theta)$ in a particularly convenient and compact form:

$$z = re^{i\theta}. \tag{4}$$

We call (4) the **exponential form** of the complex number z. For example, a polar form of the complex number $3i$ is $3 \left[\cos (\pi/2) + i \sin (\pi/2) \right]$, whereas an exponential form of $3i$ is $3e^{i\pi/2}$. Bear in mind that in the exponential form (4) of a complex number, the value of $\theta = \arg(z)$ is not unique. This is similar to the situation with the polar form of a complex number. You are encouraged to verify that $\sqrt{2}e^{i\pi/4}$, $\sqrt{2}e^{i9\pi/4}$, and $\sqrt{2}e^{i17\pi/4}$ are all valid exponential forms of the complex number $1 + i$.

If z is a real number, that is, if $z = x + 0i$, then (3) gives $e^z = e^x \cos 0 + ie^x \sin 0 = e^x$. In other words, the complex exponential function agrees with the usual real exponential function for real z. Many well-known properties of the real exponential function are also satisfied by the complex exponential function. For instance, if z_1 and z_2 are complex numbers, then (3) can be used to show that:

$$e^0 = 1, \tag{5}$$

$$e^{z_1} e^{z_2} = e^{z_1 + z_2}, \tag{6}$$

$$\frac{e^{z_1}}{e^{z_2}} = e^{z_1 - z_2}, \tag{7}$$

$$\left(e^{z_1} \right)^n = e^{nz_1} \text{ for } n = 0, \pm 1, \pm 2, \dots . \tag{8}$$

Proofs of properties (5)–(8) will be given in Section 4.1 where the complex exponential function is discussed in greater detail.

While the real and complex exponential functions have many similarities, they also have some surprising and important differences. Perhaps the most unexpected difference is:

The complex exponential function is periodic.

In Problem 33 in Exercises 2.1 you are asked to show that that $e^{z+2\pi i} = e^z$ for all complex numbers z. This result implies that the complex exponential function has a pure imaginary period $2\pi i$.

Polar Coordinates   Up to this point, the real and imaginary parts of a complex function were determined using the Cartesian description $x + iy$ of the complex variable z. It is equally valid, and, oftentimes, more convenient to express the complex variable z using either the polar form $z = r(\cos\theta + i\sin\theta)$ or, equivalently, the exponential form $z = re^{i\theta}$. Given a complex function $w = f(z)$, if we replace the symbol z with $r(\cos\theta + i\sin\theta)$, then we can write this function as:

$$f(z) = u(r, \theta) + iv(r, \theta). \tag{9}$$

We still call the real functions $u(r, \theta)$ and $v(r, \theta)$ in (9) the real and imaginary parts of f, respectively. For example, replacing z with $r(\cos\theta + i\sin\theta)$ in the function $f(z) = z^2$, yields, by de Moivre's formula,

$$f(z) = (r(\cos\theta + i\sin\theta))^2 = r^2\cos 2\theta + ir^2\sin 2\theta.$$

Thus, using the polar form of z we have shown that the real and imaginary parts of $f(z) = z^2$ are

The functions $u(r, \theta)$ and $v(r, \theta)$ are not the same as the functions $u(x, y)$ and $v(x, y)$.

$$u(r, \theta) = r^2\cos 2\theta \quad \text{and} \quad v(r, \theta) = r^2\sin 2\theta, \tag{10}$$

respectively. Because we used a polar rather than a Cartesian description of the variable z, the functions u and v in (10) are not the same as the functions u and v in (1) previously computed for the function z^2.

As with Cartesian coordinates, a complex function can be defined by specifying its real and imaginary parts in polar coordinates. The expression $f(z) = r^3\cos\theta + (2r\sin\theta)i$, therefore, defines a complex function. To find the value of this function at, say, the point $z = 2i$, we first express $2i$ in polar form:

$$2i = 2\left(\cos\frac{\pi}{2} + i\sin\frac{\pi}{2}\right).$$

We then set $r = 2$ and $\theta = \pi/2$ in the expression for f to obtain:

$$f(2i) = (2)^3\cos\frac{\pi}{2} + \left(2\cdot 2\sin\frac{\pi}{2}\right)i = 8\cdot 0 + (4\cdot 1)i = 4i.$$

Remarks   *Comparison with Real Analysis*

(i) The complex exponential function provides a good example of how complex functions can be similar to and, at the same time, different from their real counterparts. Both the complex and the real exponential function satisfy properties (5)–(8). On the other hand, the complex exponential function is periodic and, from part (c) of Example 3, a value of the complex exponential function can be a negative real number. Neither of these properties are shared by the real exponential function.

(ii) In this section we made the important observation that every complex function can be defined in terms of two real functions $u(x, y)$ and $v(x, y)$ as $f(z) = u(x, y) + iv(x, y)$. This implies that the study of complex functions is closely related to the study of real multivariable functions of two real variables. The notions of limit, continuity, derivative, and integral of such real functions will be used to develop and aid our understanding of the analogous concepts for complex functions.

(*iii*) On page 47 we discussed that real-valued functions of a real variable and real-valued functions of two real variables can be viewed as special types of complex functions. Other special types of complex functions that we encounter in the study of complex analysis include the following:

Real-valued functions of a complex variable are functions $y = f(z)$ where z is a complex number and y is a real number. The functions $x = \mathrm{Re}(z)$ and $r = |z|$ are both examples of this type of function.

Complex-valued functions of a real variable are functions $w = f(t)$ where t is a real number and w is a complex number. It is customary to express such functions in terms of two real-valued functions of the real variable t, $w(t) = x(t) + iy(t)$. For example, $w(t) = 3t + i\cos t$ is a complex-valued function of a real variable t.

These special types of complex functions will appear in various places throughout the text.

EXERCISES 2.1   *Answers to selected odd-numbered problems begin on page ANS-6.*

In Problems 1–8, evaluate the given complex function f at the indicated points.

1. $f(z) = z^2 \bar{z} - 2i$ **(a)** $2i$ **(b)** $1 + i$ **(c)** $3 - 2i$

2. $f(z) = -z^3 + 2z + \bar{z}$ **(a)** i **(b)** $2 - i$ **(c)** $1 + 2i$

3. $f(z) = \log_e |z| + i\mathrm{Arg}(z)$ **(a)** 1 **(b)** $4i$ **(c)** $1 + i$

4. $f(z) = |z|^2 - 2\mathrm{Re}(iz) + z$ **(a)** $3 - 4i$ **(b)** $2 - i$ **(c)** $1 + 2i$

5. $f(z) = (xy - x^2) + i(3x + y)$ **(a)** $3i$ **(b)** $4 + i$ **(c)** $3 - 5i$

6. $f(z) = e^z$ **(a)** $2 - \pi i$ **(b)** $\dfrac{\pi}{3}i$ **(c)** $\log_e 2 - \dfrac{5\pi}{6}i$

7. $f(z) = r + i\cos^2\theta$ **(a)** 3 **(b)** $-2i$ **(c)** $2 - i$

8. $f(z) = r\sin 3\theta + i\cos 2\theta$ **(a)** -2 **(b)** $1 + i$ **(c)** $-5i$

In Problems 9–16, find the real and imaginary parts u and v of the given complex function f as functions of x and y.

9. $f(z) = 6z - 5 + 9i$ 10. $f(z) = -3z + 2\bar{z} - i$

11. $f(z) = z^3 - 2z + 6$ 12. $f(z) = z^2 + \bar{z}^2$

13. $f(z) = \dfrac{\bar{z}}{z + 1}$ 14. $f(z) = z + \dfrac{1}{z}$

15. $f(z) = e^{2z+i}$ 16. $f(z) = e^{z^2}$

In Problems 17–22, find the real and imaginary parts u and v of the given complex function f as functions of r and θ.

17. $f(z) = \bar{z}$ 18. $f(z) = |z|$

19. $f(z) = z^4$ 20. $f(z) = z + \dfrac{1}{z}$

21. $f(z) = e^z$ 22. $f(z) = x^2 + y^2 - yi$

In Problems 23–26, find the natural domain of the given complex function f.

23. $f(z) = 2\text{Re}(z) - iz^2$

24. $f(z) = \dfrac{3z + 2i}{z^3 + 4z^2 + z}$

25. $f(z) = \dfrac{iz}{|z - 1|}$

26. $f(z) = \dfrac{iz}{|z| - 1}$

Focus on Concepts

27. Discuss: Do the following expressions define complex functions $f(z)$? Defend your answer.

(a) $\arg(z)$ (b) $\text{Arg}(z)$ (c) $\cos(\arg(z)) + i\sin(\arg(z))$

(d) $z^{1/2}$ (e) $|z|$ (f) $\text{Re}(z)$

28. Find the range of each of the following complex functions.

(a) $f(z) = \text{Im}(z)$ defined on the closed disk $|z| \le 2$

(b) $f(z) = (1 + i)|z|$ defined on the square $0 \le \text{Re}(z) \le 1, 0 \le \text{Im}(z) \le 1$

(c) $f(z) = \bar{z}$ defined on the upper half-plane $\text{Im}(z) > 0$

29. Find the natural domain and the range of each of the following complex functions.

(a) $f(z) = \dfrac{z}{|z|}$. [*Hint*: In order to determine the range, consider $|f(z)|$.]

(b) $f(z) = 3 + 4i + \dfrac{5z}{|z|}$.

(c) $f(z) = \dfrac{z + \bar{z}}{z - \bar{z}}$.

30. Give an example of a complex function whose natural domain consists of all complex numbers except 0, $1 + i$, and $1 - i$.

31. Determine the natural domain and range of the complex function $f(z) = \cos(x - y) + i\sin(x - y)$.

32. Suppose that $z = x + iy$. Reread Section 1.1 and determine how to express x and y in terms of z and $\bar{z}$. Then write the following functions in terms of the symbols z and $\bar{z}$.

(a) $f(z) = x^2 + y^2$ (b) $f(z) = x - 2y + 2 + (6x + y)i$

(c) $f(z) = x^2 - y^2 - (5xy)i$ (d) $f(z) = 3y^2 + (3x^2)i$

33. In this problem we examine some properties of the complex exponential function.

(a) If $z = x + iy$, then show that $|e^z| = e^x$.

(b) Are there any complex numbers z with the property that $e^z = 0$? [*Hint*: Use part (a).]

(c) Show that $f(z) = e^z$ is a function that is periodic with pure imaginary period $2\pi i$. That is, show that $e^{z + 2\pi i} = e^z$ for all complex numbers z.

34. Use (3) to prove that $e^{\bar{z}} = \overline{e^z}$ for all complex z.

35. What can be said about z if $|e^{-z}| < 1$?

36. Let $f(z) = \dfrac{e^{iz} + e^{-iz}}{2}$.

(a) Show that f is periodic with real period 2π.

(b) Suppose that z is real. That is, $z = x + 0i$. Use (3) to rewrite $f(x + 0i)$. What well-known real function do you get?

37. What is the period of each of the following complex functions?

(a) $f(z) = e^{z+\pi}$ (b) $f(z) = e^{\pi z}$

(c) $f(z) = e^{2iz}$ (d) $f(z) = e^{3z+i}$

38. If $f(z)$ is a complex function with pure imaginary period i, then what is the period of the function $g(z) = f(iz - 2)$?

2.2   Complex Functions As Mappings

Recall that if f is a real-valued function of a real variable, then the graph of f is a curve in the Cartesian plane. Graphs are used extensively to investigate properties of *real* functions in elementary courses. However, we'll see that the graph of a *complex* function lies in four-dimensional space, and so we cannot use graphs to study complex functions. In this section we discuss the concept of a *complex mapping*, which was developed by the German mathematician Bernhard Riemann to give a geometric representation of a complex function. The basic idea is this. Every complex function describes a correspondence between points in two copies of the complex plane. Specifically, the point z in the z-plane is associated with the unique point $w = f(z)$ in the w-plane. We use the alternative term **complex mapping** in place of "complex function" when considering the function as this correspondence between points in the z-plane and points in the w-plane. The geometric representation of a complex mapping $w = f(z)$ consists of two figures: the first, a subset S of points in the z-plane, and the second, the set S' of the images of points in S under $w = f(z)$ in the w-plane.

Mappings A useful tool for the study of real functions in elementary calculus is the graph of the function. Recall that if $y = f(x)$ is a real-valued function of a real variable x, then the graph of f is defined to be the set of all points $(x, f(x))$ in the two-dimensional Cartesian plane. An analogous definition can be made for a complex function. However, if $w = f(z)$ is a complex function, then both z and w lie in a complex plane. It follows that the set of all points $(z, f(z))$ lies in four-dimensional space (two dimensions from the input z and two dimensions from the output w). Of course, a subset of four-dimensional space cannot be easily illustrated. Therefore:

We cannot draw the graph of a complex function.

The concept of a complex mapping provides an alternative way of giving a geometric representation of a complex function. As described in the section introduction, we use the term **complex mapping** to refer to the correspondence determined by a complex function $w = f(z)$ between points in a z-plane and images in a w-plane. If the point z_0 in the z-plane corresponds to the point w_0 in the w-plane, that is, if $w_0 = f(z_0)$, then we say that f **maps** z_0 onto w_0 or, equivalently, that z_0 is **mapped** onto w_0 by f.

As an example of this type of geometric thinking, consider the real function $f(x) = x + 2$. Rather than representing this function with a line of slope 1 and y-intercept $(0, 2)$, consider how one copy of the real line (the x-line) is mapped onto another copy of the real line (the y-line) by f. Each point on the x-line is mapped onto a point two units to the right on the y-line (0 is mapped onto 2, while 3 is mapped onto 5, and so on). Therefore, the real function $f(x) = x + 2$ can be thought of as a mapping that *translates* each point in the real line two units to the right. You can visualize the action of this mapping by imagining the real line as an infinite rigid rod that is physically moved two units to the right.

In order to create a geometric representation of a complex mapping, we begin with two copies of the complex plane, the z-plane and the w-plane, drawn either side-by-side or one above the other. A complex mapping is represented by drawing a set S of points in the z-plane and the corresponding set of images of the points in S under f in the w-plane. This idea is illustrated in Figure 2.2.1 where a set S in the z-plane is shown in color in Figure 2.2.1(a) and a set labeled S', which represents the set of the images of points in S under $w = f(z)$, is shown in gray in Figure 2.2.1(b). From this point on we will use notation similar to that in Figure 2.2.1 when discussing mappings.

Notation: S'

*If $w = f(z)$ is a complex mapping and if S is a set of points in the z-plane, then we call the set of images of the points in S under f the **image of S under f**, and we denote this set by the symbol S'.***

If the set S has additional properties, such as S is a domain or a curve, then we also use symbols such as D and D' or C and C', respectively, to denote the set and its image under a complex mapping. The notation $f(C)$ is also sometimes used to denote the image of a curve C under $w = f(z)$.

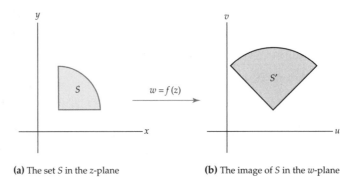

(a) The set S in the z-plane   (b) The image of S in the w-plane

Figure 2.2.1 The image of a set S under a mapping $w = f(z)$

An illustration like Figure 2.2.1 is meant to convey information about the general relationship between an arbitrary point z and its image $w = f(z)$. As such, the set S needs to be chosen with some care. For example, if f is a function whose domain and range are the set of complex numbers $\mathbf{C}$, then choosing $S = \mathbf{C}$ will result in a figure consisting solely of two complex planes. Clearly, such an illustration would give no insight into *how* points in the z-plane are mapped onto points in the w-plane by f.

EXAMPLE 1   **Image of a Half-Plane under $w = iz$**

Find the image of the half-plane $\mathrm{Re}(z) \geq 2$ under the complex mapping $w = iz$ and represent the mapping graphically.

Solution Let S be the half-plane consisting of all complex points z with $\mathrm{Re}(z) \geq 2$. We proceed as illustrated in Figure 2.2.1. Consider first the vertical boundary line $x = 2$ of S shown in color in Figure 2.2.2(a). For any

---

*The set S is sometimes called the **pre-image** of S' under f.

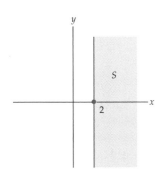

(a) The half-plane S

$w = iz$

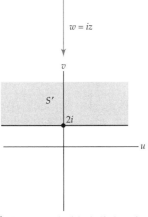

(b) The image, S', of the half-plane S

Figure 2.2.2 The mapping $w = iz$

point z on this line we have $z = 2 + iy$ where $-\infty < y < \infty$. The value of $f(z) = iz$ at a point on this line is $w = f(2 + iy) = i(2 + iy) = -y + 2i$. Because the set of points $w = -y + 2i$, $-\infty < y < \infty$, is the line $v = 2$ in the w-plane, we conclude that the vertical line $x = 2$ in the z-plane is mapped onto the horizontal line $v = 2$ in the w-plane by the mapping $w = iz$. Therefore, the vertical line shown in color in Figure 2.2.2(a) is mapped onto the horizontal line shown in black in Figure 2.2.2(b) by this mapping.

Now consider the entire half-plane S shown in color in Figure 2.2.2(a). This set can be described by the two simultaneous inequalities,

$$x \geq 2 \quad \text{and} \quad -\infty < y < \infty. \tag{1}$$

In order to describe the image of S, we express the mapping $w = iz$ in terms of its real and imaginary parts u and v; then we use the bounds given by (1) on x and y in the z-plane to determine bounds on u and v in the w-plane. By replacing the symbol z by $x + iy$ in $w = iz$, we obtain $w = i(x+iy) = -y + ix$, and so the real and imaginary parts of $w = iz$ are:

$$u = -y \quad \text{and} \quad v = x. \tag{2}$$

After replacing the variables x and y in (1) using the equations $x = v$ and $y = -u$ from (2), we obtain $v \geq 2$ and $-\infty < u < \infty$. That is, the set S', the image of S under $w = iz$, consists of all points $w = u + iv$ in the w-plane that satisfy the simultaneous inequalities $v \geq 2$ and $-\infty < u < \infty$. In words, the set S' consists of all points in the half-plane lying on or above the horizontal line $v = 2$. This image can also be described by the single inequality $\text{Im}(w) \geq 2$. In summary, the half-plane $\text{Re}(z) \geq 2$ shown in color in Figure 2.2.2(a) is mapped onto the half-plane $\text{Im}(w) \geq 2$ shown in gray in Figure 2.2.2(b) by the complex mapping $w = iz$. ❏

In Example 1, the set S and its image S' are both half-planes. This might lead you to believe that there is some simple geometric way to visualize the image of other sets in the complex plane under the mapping $w = iz$. (We will see that this is the case in Section 2.3.) For most mappings, however, the relationship between S and S' is more complicated. This is illustrated in the following example.

EXAMPLE 2   **Image of a Line under $w = z^2$**

Find the image of the vertical line $x = 1$ under the complex mapping $w = z^2$ and represent the mapping graphically.

Solution Let C be the set of points on the vertical line $x = 1$ or, equivalently, the set of points $z = 1 + iy$ with $-\infty < y < \infty$. We proceed as in Example 1. From (1) of Section 2.1, the real and imaginary parts of $w = z^2$ are $u(x, y) = x^2 - y^2$ and $v(x, y) = 2xy$, respectively. For a point $z = 1 + iy$ in C, we have $u(1, y) = 1 - y^2$ and $v(1, y) = 2y$. This implies that the image of S is the set of points $w = u + iv$ satisfying the simultaneous equations:

$$u = 1 - y^2 \tag{3}$$

and

$$v = 2y \tag{4}$$

for $-\infty < y < \infty$. Equations (3) and (4) are *parametric equations* in the real parameter y, and they define a curve in the w-plane. We can find a Cartesian

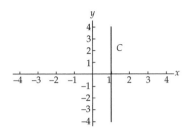

(a) The vertical line Re(z) = 1

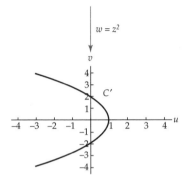

(b) The image of C is the parabola
$u = 1 - \frac{1}{4}v^2$

Figure 2.2.3 The mapping $w = z^2$

In a parametrization ➡
$z(t) = x(t) + iy(t)$,
t is a **real** variable.

equation in u and v for this curve by eliminating the parameter y. In order to do so, we solve (4) for y and then substitute this expression into (3):

$$u = 1 - \left(\frac{v}{2}\right)^2 = 1 - \frac{v^2}{4}. \tag{5}$$

Since y can take on any real value and since $v = 2y$, it follows that v can take on any real value in (5). Consequently, C'—the image of C—is a parabola in the w-plane with vertex at $(1,0)$ and u-intercepts at $(0, \pm 2)$. See Figure 2.2.3(b). In conclusion, we have shown that the vertical line $x = 1$ shown in color in Figure 2.2.3(a) is mapped onto the parabola $u = 1 - \frac{1}{4}v^2$ shown in black in Figure 2.2.3(b) by the complex mapping $w = z^2$. ❑

In contrast to Figure 2.2.2, the representation of the mapping $w = z^2$ shown in Figure 2.2.3 gives little insight into what the images of other sets in the plane might be. Mapping by this "complex squaring function" will be examined in greater detail in Section 2.4.

Parametric Curves in the Complex Plane

For a simple complex function, the manner in which the complex plane is mapped might be evident after analyzing the image of a single set, but for most functions an understanding of the mapping is obtained only after looking at the images of a variety of sets. We can often gain a good understanding of a complex mapping by analyzing the images of *curves* (one-dimensional subsets of the complex plane) and this process is facilitated by the use of parametric equations.

If $x = x(t)$ and $y = y(t)$ are real-valued functions of a real variable t, then the set C of all points $(x(t), y(t))$, where $a \leq t \leq b$, is called a **parametric curve**. The equations $x = x(t)$, $y = y(t)$, for $a \leq t \leq b$ are called **parametric equations** of C. A parametric curve can be regarded as lying in the complex plane by letting x and y represent the real and imaginary parts of a point in the complex plane. In other words, if $x = x(t)$, $y = y(t)$, and $a \leq t \leq b$ are parametric equations of a curve C in the Cartesian plane, then the equation $z(t) = x(t) + iy(t)$, $a \leq t \leq b$, is a description of the curve C in the complex plane. For example, consider the parametric equations $x = \cos t$, $y = \sin t$, $0 \leq t \leq 2\pi$, of a curve C in the xy-plane (the curve C is a circle centered at $(0,0)$ with radius 1). The equation $z(t) = \cos t + i \sin t$, $0 \leq t \leq 2\pi$, describes the same curve C in the complex plane. If, say, $t = \pi/2$, then the point $(\cos \frac{\pi}{2}, \sin \frac{\pi}{2}) = (0, 1)$ is on the curve C in the Cartesian plane, while the point $z(\pi/2) = \cos \frac{\pi}{2} + i \sin \frac{\pi}{2} = i$ represents this point on C in the complex plane. This discussion is summarized in the following definition.

Definition 2.2.1   Parametric Curves in the Complex Plane

If $x(t)$ and $y(t)$ are real-valued functions of a real variable t, then the set C consisting of all points $z(t) = x(t) + iy(t)$, $a \leq t \leq b$, is called a **parametric curve** or a **complex parametric curve**. The complex-valued function of the real variable t, $z(t) = x(t) + iy(t)$, is called a **parametrization** of C.

Properties of curves in the Cartesian plane such as continuous, differentiable, smooth, simple, and closed can all be reformulated to be properties of curves in the complex plane. These properties are important in the study of the complex integral and will be discussed in Chapter 5.

Two of the most elementary curves in the plane are lines and circles. Parametrizations of these curves in the complex plane can be derived from parametrizations in the Cartesian plane. It is also relatively easy to find these parametrizations directly by using the geometry of the complex plane. For example, suppose that we wish to find a parametrization of the line in the complex plane containing the points z_0 and z_1. We know from Chapter 1 that $z_1 - z_0$ represents the vector originating at z_0 and terminating at z_1, shown in color in Figure 2.2.4. If z is any point on the line containing z_0 and z_1, then inspection of Figure 2.2.4 indicates that the vector $z - z_0$ is a real multiple of the vector $z_1 - z_0$. Therefore, if z is on the line containing z_0 and z_1, then there is a real number t such that $z - z_0 = t(z_1 - z_0)$. Solving this equation for z gives a parametrization $z(t) = z_0 + t(z_1 - z_0) = z_0(1 - t) + z_1 t$, $-\infty < t < \infty$, for the line. Note that if we restrict the parameter t to the interval $[0, 1]$, then the points $z(t)$ range from z_0 to z_1, and this gives a parametrization of the line segment from z_0 to z_1. On the other hand, if we restrict t to the interval $[0, \infty]$, then we obtain a parametrization of the ray emanating from z_0 and containing z_1. These parametrizations are included in the following summary.

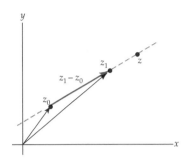

Figure 2.2.4 Parametrization of a line

These are not the only parametrizations possible! See Problem 28 in Exercises 2.2. ➡

Common Parametric Curves in the Complex Plane

Line
A parametrization of the line containing the points z_0 and z_1 is:

$$z(t) = z_0(1 - t) + z_1 t, \quad -\infty < t < \infty. \tag{6}$$

Line Segment
A parametrization of the line segment from z_0 to z_1 is:

$$z(t) = z_0(1 - t) + z_1 t, \quad 0 \le t \le 1. \tag{7}$$

Ray
A parametrization of the ray emanating from z_0 and containing z_1 is:

$$z(t) = z_0(1 - t) + z_1 t, \quad 0 \le t < \infty. \tag{8}$$

Circle
A parametrization of the circle centered at z_0 with radius r is:

$$z(t) = z_0 + r(\cos t + i \sin t), \quad 0 \le t \le 2\pi. \tag{9}$$

In exponential notation, this parametrization is:

$$z(t) = z_0 + re^{it}, \quad 0 \le t \le 2\pi. \tag{10}$$

Restricting the values of the parameter t in (9) or (10) gives parametrizations of circular arcs. For example, by setting $z_0 = 0$ and $r = 1$ in (10) we see that $z(t) = e^{it}$, $0 \le t \le \pi$, is a parametrization of the semicircular arc of the unit circle centered at the origin and lying in the upper half-plane $\text{Im}(z) \ge 0$.

Parametric curves are important in the study of complex mappings because it is easy to determine a parametrization of the image of a parametric curve. For example, if $w = iz$ and C is the line $x = 2$ given by $z(t) = 2 + it$, $-\infty < t < \infty$, then the value of $f(z) = iz$ at a point on this line is $w = f(2 + it) = i(2 + it) = -t + 2i$, and so the image of $z(t)$ is $w(t) = -t + 2i$. Put another way, $w(t) = -t + 2i$, $-\infty < t < \infty$, is a parametrization of the image C'. Thus, C' is the line $v = 2$. In summary, we have the following procedure for finding the images of curves under a complex mapping.

Image of a Parametric Curve under a Complex Mapping

If $w = f(z)$ is a complex mapping and if C is a curve parametrized by $z(t)$, $a \le t \le b$, then

$$w(t) = f(z(t)), \quad a \le t \le b \tag{11}$$

is a parametrization of the image, C' of C under $w = f(z)$.

In some instances it is convenient to represent a complex mapping using a single copy of the complex plane. We do so by superimposing the w-plane on top of the z-plane, so that the real and imaginary axes in each copy of the plane coincide. Because such a figure simultaneously represents both the z- and the w-planes, we omit all labels x, y, u, and v from the axes. For example, if we plot the half-plane S and its image S' from Example 1 in the same copy of the complex plane, then we see that the half-plane S' may be obtained by rotating the half-plane S through an angle $\pi/2$ radians counter-clockwise about the origin. This observation about the mapping $w = iz$ will be verified in Section 2.3. In the following examples, we represent a complex mapping using a single copy of the complex plane.

EXAMPLE 3   Image of a Parametric Curve

Use (11) to find the image of the line segment from 1 to i under the complex mapping $w = \overline{iz}$.

Solution Let C denote the line segment from 1 to i and let C' denote its image under $f(z) = \overline{iz}$. By identifying $z_0 = 1$ and $z_1 = i$ in (7), we obtain a parametrization $z(t) = 1 - t + it$, $0 \le t \le 1$, of C. The image C' is then given by (11):

$$w(t) = f(z(t)) = \overline{i(1 - t + it)} = -i(1 - t) - t, \quad 0 \le t \le 1.$$

With the identifications $z_0 = -i$ and $z_1 = -1$ in (7), we see that $w(t)$ is a parametrization of the line segment from $-i$ to -1. Therefore, C' is the line segment from $-i$ to -1. This mapping is depicted in Figure 2.2.5 using a single copy of the complex plane. In Figure 2.2.5, the line segment shown in color is mapped onto the line segment shown in black by $w = \overline{iz}$.   ❑

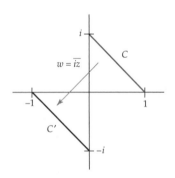

Figure 2.2.5 The mapping $w = \overline{iz}$

EXAMPLE 4   Image of a Parametric Curve

Find the image of the semicircle shown in color in Figure 2.2.6 under the complex mapping $w = z^2$.

Solution Let C denote the semicircle shown in Figure 2.2.6 and let C' denote its image under $f(z) = z^2$. We proceed as in Example 3. By setting $z_0 = 0$

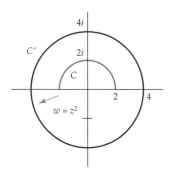

Figure 2.2.6 The mapping $w = z^2$

and $r = 2$ in (10) we obtain the following parametrization of C:

$$z(t) = 2e^{it}, \quad 0 \le t \le \pi.$$

Thus, from (11) we have that:

$$w(t) = f(z(t)) = \left(2e^{it}\right)^2 = 4e^{2it}, \quad 0 \le t \le \pi, \tag{12}$$

is a parametrization of C'. If we set $t = \frac{1}{2}s$ in (12), then we obtain a new parametrization of C':

$$W(s) = 4e^{is}, \ 0 \le s \le 2\pi. \tag{13}$$

From (10) with $z_0 = 0$ and $r = 4$, we find that (13) defines a circle centered at 0 with radius 4. Therefore, the image C' is the circle $|w| = 4$. We represent this mapping in Figure 2.2.6 where the semicircle shown in color is mapped onto the circle shown in black by $w = z^2$. ❑

Use of Computers   Computer algebra systems such as *Maple* and *Mathematica* perform standard algebraic operations with complex numbers. This capability combined with the ability to graph a parametric curve makes these systems excellent tools for exploring properties of complex mappings. In *Mathematica*, for example, a complex function can be defined using the command

$$\mathbf{f[z\_]} := \textit{an expression in } z.$$

A complex parametrization can be defined similarly using the command

$$\mathbf{g[t\_]} := \textit{an expression in } t.$$

From (11), it follows that $\mathbf{w[t\_]} := \mathbf{f[g[t]]}$ is a parametrization of the image of the curve. This image can be graphed using the parametric plot command:

$$\mathbf{ParametricPlot[\ \{Re[w[t]], Im[w[t]]\},\ \{t,\ a,\ b\}]}$$

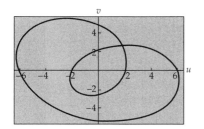

Figure 2.2.7 The image of a circle under $w = z^2 + iz - \text{Re}(z)$

where $\mathbf{a}$ and $\mathbf{b}$ are the upper and lower bounds on $\mathbf{t}$ respectively. For example, *Mathematica* was used to produce Figure 2.2.7, which shows the image of the circle $|z| = 2$ under the complex mapping $w = z^2 + iz - \text{Re}(z)$.

Remarks   *Comparison with Real Analysis*

(*i*) In this section we introduced an important difference between real and complex analysis, namely, that we cannot graph a complex function. Instead, we represent a complex function with two figures: the first a subset S in the complex plane, and the second, the image S' of the set S under a complex mapping. A complete understanding of a complex mapping is obtained when we understand the relationship between *any* set S and its image S'.

(*ii*) Complex mappings are closely related to parametric curves in the plane. In later sections, we use this relationship to help visualize the notions of limit, continuity, and differentiability of complex functions. Parametric curves will also be of central importance in the study of complex integrals much as they were in the study of real line integrals.

EXERCISES 2.2   *Answers to selected odd-numbered problems begin on page ANS-6.*

In Problems 1–8, proceed as in Example 1 or Example 2 to find the image S' of the set S under the given complex mapping $w = f(z)$.

  1. $f(z) = i\bar{z}$; S is the horizontal line $y = 3$

  2. $f(z) = \bar{z}$; S is the line $y = x$

  3. $f(z) = 3z$; S is the half-plane $\mathrm{Im}(z) > 2$

  4. $f(z) = 3iz$; S is the infinite vertical strip $2 \le \mathrm{Re}(z) < 3$

  5. $f(z) = (1 + i)z$; S is the vertical line $x = 2$

  6. $f(z) = (1 - i)z$; S is the line $y = 2x + 1$

  7. $f(z) = iz + 4$; S is the half-plane $\mathrm{Im}(z) \le 1$

  8. $f(z) = iz + \bar{z}$; S is the vertical line $x = 3$

In Problems 9–14, find the image of the given line under the complex mapping $w = z^2$.

  9. $y = 1$ **10.** $x = -3$

 11. $x = 0$ **12.** $y = 0$

 13. $y = x$ **14.** $y = -x$

In Problems 15–20, **(a)** plot the parametric curve C given by $z(t)$ and describe the curve in words, **(b)** find a parametrization of the image, C', of C under the given complex mapping $w = f(z)$, and **(c)** plot C' and describe this curve in words.

 15. $z(t) = 2(1 - t) + it, 0 \le t \le 1$; $f(z) = 3z$

 16. $z(t) = i(1 - t) + (1 + i)t$, $0 \le t < \infty$; $f(z) = -z$

 17. $z(t) = 1 + 2e^{it}$, $0 \le t \le 2\pi$; $f(z) = z + 1 - i$

 18. $z(t) = i + e^{it}$, $0 \le t \le \pi$; $f(z) = (z - i)^2$

 19. $z(t) = t$, $0 \le t \le 2$; $f(z) = e^{i\pi z}$

 20. $z(t) = 4e^{it}$, $0 \le t \le \pi$, $f(z) = \mathrm{Re}(z)$

In Problems 21–26, use parametrizations to find the image, C', of the curve C under the given complex mapping $w = f(z)$.

 21. $f(z) = z^3$; C is the positive imaginary axis

 22. $f(z) = iz$; C is the circle $|z - 1| = 2$

 23. $f(z) = 1/z$; C is the circle $|z| = 2$

 24. $f(z) = 1/z$; C is the line segment from $1 - i$ to $2 - 2i$

 25. $f(z) = z + \bar{z}$; C is the semicircle of the unit circle $|z| = 1$ in the upper half-plane $\mathrm{Im}(z) \ge 0$

 26. $f(z) = e^z$; C is the ray emanating from the origin and containing $2 + \sqrt{3}i$

Focus on Concepts

 27. In this problem we will find the image of the line $x = 1$ under the complex mapping $w = 1/z$.

    (a) The line $x = 1$ consists of all points $z = 1 + iy$ where $-\infty < y < \infty$. Find the real and imaginary parts u and v of $f(z) = 1/z$ at a point $z = 1 + iy$ on this line.

(b) Show that $\left(u - \frac{1}{2}\right)^2 + v^2 = \frac{1}{4}$ for the functions u and v from part (a).

(c) Based on part (b), describe the image of the line $x = 1$ under the complex mapping $w = 1/z$.

(d) Is there a point on the line $x = 1$ that maps onto 0? Do you want to alter your description of the image in part (c)?

28. Consider the parametrization $z(t) = i(1 - t) + 3t$, $0 \le t \le 1$.

    (a) Describe in words this parametric curve.

    (b) What is the difference between the curve in part (a) and the curve defined by the parametrization $z(t) = 3(1 - t) + it$, $0 \le t \le 1$?

    (c) What is the difference between the curve in part (a) and the curve defined by the parametrization $z(t) = \frac{3}{2}t + i\left(1 - \frac{1}{2}t\right)$, $0 \le t \le 2$?

    (d) Find a parametrization of the line segment from $1 + 2i$ to $2 + i$ where the parameter satisfies $0 \le t \le 3$.

29. Use parametrizations to find the image of the circle $|z - z_0| = R$ under the mapping $f(z) = iz - 2$.

30. Consider the line $y = mx + b$ in the complex plane.

    (a) Give a parametrization $z(t)$ for the line.

    (b) Describe in words the image of the line under the complex mapping $w = z + 2 - 3i$.

    (c) Describe in words the image of the line under the complex mapping $w = 3z$.

31. The complex mapping $w = \bar{z}$ is called **reflection** about the real axis. Explain why.

32. Let $f(z) = az$ where a is a complex constant and $|a| = 1$.

    (a) Show that $|f(z_1) - f(z_2)| = |z_1 - z_2|$ for all complex numbers z_1 and z_2.

    (b) Give a geometric interpretation of the result in (a).

    (c) What does your answer to (b) tell you about the image of a circle under the complex mapping $w = az$.

33. In this problem we investigate the effect of the mapping $w = az$, where a is a complex constant and $a \ne 0$, on angles between rays emanating from the origin.

    (a) Let C be a ray in the complex plane emanating from the origin. Use parametrizations to show that the image C' of C under $w = az$ is also a ray emanating from the origin.

    (b) Consider two rays C_1 and C_2 emanating from the origin such that C_1 contains the point $z_1 = a_1 + ib_1$ and C_2 contains the point $z_2 = a_2 + ib_2$. In multivariable calculus, you saw that the angle θ between the rays C_1 and C_2 (which is the same as the angle between the position vectors $(a_1,\ b_1)$ and $(a_2,\ b_2)$) is given by:

$$\theta = \arccos\left(\frac{a_1 a_2 + b_1 b_2}{\sqrt{a_1^2 + b_1^2}\,\sqrt{a_2^2 + b_2^2}}\right) = \arccos\left(\frac{z_1 \bar{z}_2 + \bar{z}_1 z_2}{2\,|z_1|\,|z_2|}\right). \qquad (14)$$

    Let C_1' and C_2' be the images of C_1 and C_2 under $w = az$. Use part (a) and (14) to show that the angle between C_1' and C_2' is the same as the angle between C_1 and C_2.

34. Consider the complex mapping $w = z^2$.

    (a) Repeat Problem 33(a) for the mapping $w = z^2$.

    (b) Experiment with different rays. What effect does the complex mapping $w = z^2$ appear to have on angles between rays emanating from the origin?

In Problems 35–38, use a CAS to (**a**) plot the image of the unit circle under the given complex mapping $w = f(z)$, and (**b**) plot the image of the line segment from 1 to $1 + i$ under the given complex mapping $w = f(z)$.

35. $f(z) = z^2 + (1 + i)z - 3$ **36.** $f(z) = iz^3 + z - i$

37. $f(z) = z^4 - z$ **38.** $f(z) = z^3 - \bar{z}$

2.3   Linear Mappings

Recall that a real function of the form $f(x) = ax + b$ where a and b are any *real* constants is called a linear function. In keeping with the similarities between real and complex analysis, we define a **complex linear function** to be a function of the form $f(z) = az + b$ where a and b are any *complex* constants. Just as real linear functions are the easiest types of real functions to graph, complex linear functions are the easiest types of complex functions to visualize as mappings of the complex plane. In this section, we will show that every nonconstant complex linear mapping can be described as a composition of three basic types of motions: a translation, a rotation, and a magnification.

Before looking at a general complex linear mapping $f(z) = az + b$, we investigate three special types of linear mappings called translations, rotations, and magnifications. Throughout this section we use the symbols T, R, and M to represent mapping by translation, rotation, and magnification, respectively.

Translations   A complex linear function

$$T(z) = z + b, \quad b \neq 0, \tag{1}$$

is called a **translation**. If we set $z = x + iy$ and $b = x_0 + iy_0$ in (1), then we obtain:

$$T(z) = (x + iy) + (x_0 + iy_0) = x + x_0 + i(y + y_0).$$

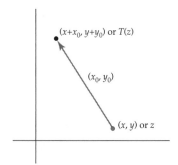

Figure 2.3.1 Translation

Thus, the image of the point (x, y) under T is the point $(x + x_0, y + y_0)$. From Figure 2.3.1 we see that if we plot (x, y) and $(x + x_0, y + y_0)$ in the same copy of the complex plane, then the vector originating at (x, y) and terminating at $(x + x_0, y + y_0)$ is (x_0, y_0); equivalently, if we plot z and $T(z)$ in the same copy of the complex plane, then the vector originating at z and terminating at $T(z)$ is (x_0, y_0). Therefore, the linear mapping $T(z) = z + b$ can be visualized in a single copy of the complex plane as the process of *translating* the point z along the vector (x_0, y_0) to the point $T(z)$. Since (x_0, y_0) is the vector representation of the complex number b, the mapping $T(z) = z + b$ is also called a *translation by* b.

EXAMPLE 1   Image of a Square under Translation

Find the image S' of the square S with vertices at $1 + i$, $2 + i$, $2 + 2i$, and $1 + 2i$ under the linear mapping $T(z) = z + 2 - i$.

Solution We will represent S and S' in the same copy of the complex plane. The mapping T is a translation, and so S' can be determined as follows.

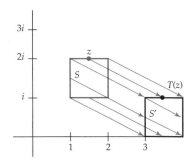

Figure 2.3.2 Image of a square under translation

Identifying $b = 2 + i(-1)$ in (1), we plot the vector $(2, -1)$ originating at each point in S. See Figure 2.3.2. The set of terminal points of these vectors is S', the image of S under T. Inspection of Figure 2.3.2 indicates that S' is a square with vertices at:

$$T(1+i) = (1+i) + (2-i) = 3 \qquad T(2+i) = (2+i) + (2-i) = 4$$

$$T(2+2i) = (2+2i) + (2-i) = 4+i \quad T(1+2i) = (1+2i) + (2-i) = 3+i.$$

Therefore, the square S shown in color in Figure 2.3.2 is mapped onto the square S' shown in black by the translation $T(z) = z + 2 - i$. ❏

From our geometric description, we see that a translation does not change the shape or size of a figure in the complex plane. That is, the image of a line, circle, or triangle under a translation will also be a line, circle, or triangle, respectively. See Problems 23 and 24 in Exercises 2.3. A mapping with this property is sometimes called a *rigid motion*.

Rotations   A complex linear function

$$R(z) = az, \quad |a| = 1, \tag{2}$$

is called a **rotation**. Although it may seem that the requirement $|a| = 1$ is a major restriction in (2), it is not. Keep in mind that the constant a in (2) is a complex constant. If α is any nonzero complex number, then $a = \alpha/|\alpha|$ is a complex number for which $|a| = 1$. So, for any nonzero complex number α, we have that $R(z) = \dfrac{\alpha}{|\alpha|} z$ is a rotation.

Consider the rotation R given by (2) and, for the moment, assume that $\text{Arg}(a) > 0$. Since $|a| = 1$ and $\text{Arg}(a) > 0$, we can write a in exponential form as $a = e^{i\theta}$ with $0 < \theta \le \pi$. If we set $a = e^{i\theta}$ and $z = re^{i\phi}$ in (2), then by property (6) of Section 2.1 we obtain the following description of R:

$$R(z) = e^{i\theta} r e^{i\phi} = r e^{i(\theta + \phi)}. \tag{3}$$

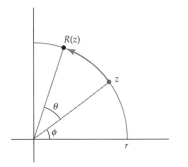

Figure 2.3.3 Rotation

From (3), we see that the modulus of $R(z)$ is r, which is the same as the modulus of z. Therefore, if z and $R(z)$ are plotted in the same copy of the complex plane, then both points lie on a circle centered at 0 with radius r. See Figure 2.3.3. Observe also from (3) that an argument of $R(z)$ is $\theta + \phi$, which is θ radians greater than an argument of z. Therefore, the linear mapping $R(z) = az$ can be visualized in a single copy of the complex plane as the process of *rotating* the point z counterclockwise through an angle of θ radians about the origin to the point $R(z)$. See Figure 2.3.3. In a similar manner, if $\text{Arg}(a) < 0$, then the linear mapping $R(z) = az$ can be visualized in a single copy of the complex plane as the process of rotating points *clockwise* through an angle of θ radians about the origin. For this reason the angle $\theta = \text{Arg}(a)$ is called an **angle of rotation** of R.

EXAMPLE 2   Image of a Line under Rotation

Find the image of the real axis $y = 0$ under the linear mapping

$$R(z) = \left(\tfrac{1}{2}\sqrt{2} + \tfrac{1}{2}\sqrt{2}\, i \right) z.$$

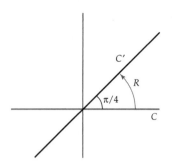

Figure 2.3.4 Image of a line under rotation

Solution Let C denote the real axis $y = 0$ and let C' denote the image of C under R. Since $\left|\frac{1}{2}\sqrt{2} + \frac{1}{2}\sqrt{2}\,i\right| = 1$, the complex mapping $R(z)$ is a rotation. In order to determine the angle of rotation, we write the complex number $\frac{1}{2}\sqrt{2} + \frac{1}{2}\sqrt{2}\,i$ in exponential form $\frac{1}{2}\sqrt{2} + \frac{1}{2}\sqrt{2}\,i = e^{i\pi/4}$. If z and $R(z)$ are plotted in the same copy of the complex plane, then the point z is rotated counterclockwise through $\pi/4$ radians about the origin to the point $R(z)$. The image C' is, therefore, the line $v = u$, which contains the origin and makes an angle of $\pi/4$ radians with the real axis. This mapping is depicted in a single copy of the complex plane in Figure 2.3.4 where the real axis shown in color is mapped onto the line shown in black by $R(z) = \left(\frac{1}{2}\sqrt{2} + \frac{1}{2}\sqrt{2}\,i\right)z$. ❏

As with translations, rotations are rigid motions and will not change the shape or size of a figure in the complex plane. Thus, the image of a line, circle, or triangle under a rotation will also be a line, circle, or triangle, respectively.

Magnifications   The final type of special linear function we consider is a magnification. A complex linear function

$$M(z) = az, \ a > 0, \tag{4}$$

is called a **magnification**. Recall from the Remarks at the end of Section 1.1 that since there is no concept of order in the complex number system, it is implicit in the inequality $a > 0$ that the symbol a represents a real number. Therefore, if $z = x + iy$, then $M(z) = az = ax + iay$, and so the image of the point $(x,\ y)$ is the point $(ax,\ ay)$. Using the exponential form $z = re^{i\theta}$ of z, we can also express the function in (4) as:

$$M(z) = a\left(re^{i\theta}\right) = (ar)\,e^{i\theta}. \tag{5}$$

The product ar in (5) is a positive real number since both a and r are positive real numbers, and from this it follows that the magnitude of $M(z)$ is ar. Assume that $a > 1$. Then from (5) we have that the complex points z and $M(z)$ have the same argument θ but different moduli $r < ar$. If we plot both z and $M(z)$ in the same copy of the complex plane, then $M(z)$ is the unique point on the ray emanating from 0 and containing z whose distance from 0 is ar. Since $a > 1$, $M(z)$ is a times farther from the origin than z. Thus, the linear mapping $M(z) = az$ can be visualized in a single copy of the complex plane as the process of *magnifying* the modulus of the point z by a factor of a to obtain the point $M(z)$. See Figure 2.3.5. The real number a is called the **magnification factor** of M. If $0 < a < 1$, then the point $M(z)$ is a times closer to the origin than the point z. This special case of a magnification is called a **contraction**.

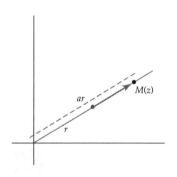

Figure 2.3.5 Magnification

EXAMPLE 3   Image of a Circle under Magnification

Find the image of the circle C given by $|z| = 2$ under the linear mapping $M(z) = 3z$.

Solution Since M is a magnification with magnification factor of 3, each point on the circle $|z| = 2$ will be mapped onto a point with the same argument but with modulus magnified by 3. Thus, each point in the image will have modulus $3 \cdot 2 = 6$. The image points can have any argument since the points z in the circle $|z| = 2$ can have any argument. Therefore, the image C' is the

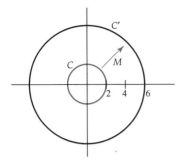

Figure 2.3.6 Image of a circle under magnification

circle $|w| = 6$ that is centered at the origin and has radius 6. In Figure 2.3.6 we illustrate this mapping in a single copy of the complex plane. Under the mapping $M(z) = 3z$, the circle C shown in color in Figure 2.3.6 is mapped onto the circle C' shown in black in Figure 2.3.6. ❑

Although a magnification mapping will change the size of a figure in the complex plane, it will not change its basic shape. For example, the image of a triangle S under a magnification $M(z) = az$ is also a triangle S'. Since the lengths of the sides of S' are all a times longer than the lengths of the sides S, it follows that S and S' are similar triangles.

Linear Mappings   We are now ready to show that a general linear mapping $f(z) = az + b$ is a composition of a rotation, a magnification, and a translation. Recall that if f and g are two functions, then the *composition* of f and g is the function $f \circ g$ defined by $f \circ g(z) = f(g(z))$. The value $w = f \circ g(z)$ is determined by first evaluating the function g at z, then evaluating the function f at $g(z)$. In a similar manner, the image, S'', of set S under a composition $w = f \circ g(z)$ is determined by first finding the image, S', of S under g, and then finding the image S'' of S' under f.

Now suppose that $f(z) = az + b$ is a complex linear function. We assume that $a \neq 0$; otherwise, our mapping would be the **constant map** $f(z) = b$, which maps every point in the complex plane onto the single point b. Observe that we can express f as:

$$f(z) = az + b = |a| \left(\frac{a}{|a|} z \right) + b. \tag{6}$$

Now, step by step, we investigate what happens to a point z_0 under the composition in (6). First z_0 is multiplied by the complex number $a/|a|$. Since $\left| \dfrac{a}{|a|} \right| = \dfrac{|a|}{|a|} = 1$, the complex mapping $w = \dfrac{a}{|a|} z$ is a rotation that rotates the point z_0 through an angle of $\theta = \text{Arg} \left(\dfrac{a}{|a|} \right)$ radians about the origin. The angle of rotation can also be written as $\theta = \text{Arg}(a)$ since $1/|a|$ is a real number. Let z_1 be the image of z_0 under this rotation by $\text{Arg}(a)$. The next step in (6) is to multiply z_1 by $|a|$. Because $|a| > 0$ is a real number, the complex mapping $w = |a| z$ is a magnification with a magnification factor $|a|$. Now let z_2 be the image of z_1 under magnification by $|a|$. The last step in our linear mapping in (6) is to add b to z_2. The complex mapping $w = z + b$ translates z_2 by b onto the point $w_0 = f(z_0)$. We now summarize this description of a linear mapping.

> ## Image of a Point under a Linear Mapping
>
> Let $f(z) = az + b$ be a linear mapping with $a \neq 0$ and let z_0 be a point in the complex plane. If the point $w_0 = f(z_0)$ is plotted in the same copy of the complex plane as z_0, then w_0 is the point obtained by
>
> (i)   rotating z_0 through an angle of $\text{Arg}(a)$ about the origin,
>
> (ii) magnifying the result by $|a|$, and
>
> (iii) translating the result by b.

This description of the image of a point z_0 under a linear mapping also describes the image of any set of points S. In particular, the image, S', of a set S under $f(z) = az + b$ is the set of points obtained by rotating S through $\text{Arg}(a)$, magnifying by $|a|$, and then translating by b.

From (6) we see that every nonconstant complex linear mapping is a composition of *at most* one rotation, one magnification, and one translation. We emphasize the phrase "at most" in order to stress the fact that one or more of the maps involved may be the **identity mapping** $f(z) = z$ (which maps every complex number onto itself). For instance, the linear mapping $f(z) = 3z + i$ involves a magnification by 3 and a translation by i, but no rotation. It is also evident from (6) that if $a \neq 0$ is a complex number, $R(z)$ is a rotation through $\text{Arg}(a)$, $M(z)$ is a magnification by $|a|$, and $T(z)$ is a translation by b, then the composition $f(z) = T \circ M \circ R(z) = T(M(R(z)))$ is a complex linear function. In addition, since the composition of any finite number of linear functions is again a linear function, it follows that the composition of finitely many rotations, magnifications, and translations is a linear mapping.

We have seen that translations, rotations, and magnifications all preserve the basic shape of a figure in the complex plane. A linear mapping, therefore, will also preserve the basic shape of a figure in the complex plane. This observation is an important property of complex linear mappings and is worth repeating.

> *A complex linear mapping $w = az + b$ with $a \neq 0$ can distort the size of a figure in the complex plane, but it cannot alter the basic shape of the figure.*

Note: The order in which you perform the steps in a linear mapping is important! ➡

When describing a linear function as a composition of a rotation, a magnification, and a translation, keep in mind that the order of composition is important. In order to see that this is so, consider the mapping $f(z) = 2z + i$, which magnifies by 2, then translates by i; so, 0 maps onto i under f. If we reverse the order of composition—that is, if we translate by i, then magnify by 2—the effect is 0 maps onto $2i$. Therefore, reversing the order of composition can give a different mapping. In some special cases, however, changing the order of composition does not change the mapping. See Problems 27 and 28 in Exercises 2.3.

A complex linear mapping can always be represented as a composition in more than one way. The complex mapping $f(z) = 2z + i$, for example, can also be expressed as $f(z) = 2(z + i/2)$. Therefore, magnification by 2 followed by translation by i is the same mapping as translation by $i/2$ followed by magnification by 2.

EXAMPLE 4   Image of a Rectangle under a Linear Mapping

Find the image of the rectangle with vertices $-1 + i$, $1 + i$, $1 + 2i$, and $-1 + 2i$ under the linear mapping $f(z) = 4iz + 2 + 3i$.

Solution Let S be the rectangle with the given vertices and let S' denote the image of S under f. Because f is a linear mapping, our foregoing discussion implies that S' has the same shape as S. That is, S' is also a rectangle. Thus,

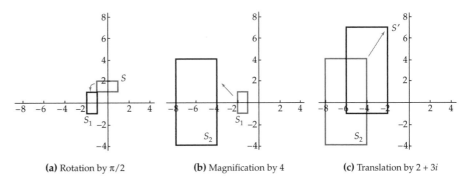

(a) Rotation by $\pi/2$ **(b)** Magnification by 4   **(c)** Translation by $2 + 3i$

Figure 2.3.7 Linear mapping of a rectangle

in order to determine S', we need only find its vertices, which are the images of the vertices of S under f:

$$f(-1 + i) = -2 - i \qquad\qquad f(1 + i) = -2 + 7i$$
$$f(1 + 2i) = -6 + 7i \qquad\qquad f(-1 + 2i) = -6 - i.$$

Therefore, S' is the rectangle with vertices $-2-i$, $-2+7i$, $-6+7i$, and $-6-i$. ❏

The linear mapping $f(z) = 4iz + 2 + 3i$ in Example 4 can also be viewed as a composition of a rotation, a magnification, and a translation. Because $\text{Arg}(4i) = \pi/2$ and $|4i| = 4$, f acts by rotating through an angle of $\pi/2$ radians about the origin, magnifying by 4, then translating by $2 + 3i$. This sequence of mappings is depicted in Figure 2.3.7. In Figure 2.3.7(a), the rectangle S shown in color is rotated through $\pi/2$ onto the rectangle S_1 shown in black; in Figure 2.3.7(b), the rectangle S_1 shown in color is magnified by 4 onto the rectangle S_2 shown in black; and finally, in Figure 2.3.7(c), the rectangle S_2 shown in color is translated by $2 + 3i$ onto the rectangle S' shown in black.

EXAMPLE 5   A Linear Mapping of a Triangle

Find a complex linear function that maps the equilateral triangle with vertices $1 + i$, $2 + i$, and $\frac{3}{2} + \left(1 + \frac{1}{2}\sqrt{3}\right) i$ onto the equilateral triangle with vertices i, $\sqrt{3} + 2i$, and $3i$.

Solution Let S_1 denote the triangle with vertices $1 + i$, $2 + i$, and $\frac{3}{2} + \left(1 + \frac{1}{2}\sqrt{3}\right) i$ shown in color in Figure 2.3.8(a), and let S' represent the triangle with vertices i, $3i$, and $\sqrt{3} + 2i$ shown in black in Figure 2.3.8(d). There are many ways to find a linear mapping that maps S_1 onto S'. One approach is the following: We first translate S_1 to have one of its vertices at the origin. If we decide that the vertex $1 + i$ should be mapped onto 0, then this is accomplished by the translation $T_1(z) = z - (1 + i)$. Let S_2 be the image of S_1 under T_1. Then S_2 is the triangle with vertices 0, 1, and $\frac{1}{2} + \frac{1}{2}\sqrt{3}i$ shown in black in Figure 2.3.8(a). From Figure 2.3.8(a), we see that the angle between the imaginary axis and the edge of S_2 containing the vertices 0 and $\frac{1}{2} + \frac{1}{2}\sqrt{3}i$ is $\pi/6$. Thus, a rotation through an angle of $\pi/6$ radians counterclockwise about the origin will map S_2 onto a triangle with two vertices on the imaginary axis. This rotation is given by $R(z) = \left(e^{i\pi/6}\right) z = \left(\frac{1}{2}\sqrt{3} + \frac{1}{2}i\right) z$, and the image of S_2 under R is the triangle S_3 with vertices at 0, $\frac{1}{2}\sqrt{3} + \frac{1}{2}i$, and i shown in black in Figure 2.3.8(b). It is easy to verify that each side of the triangle S_3 has length 1. Because each side of the desired triangle S' has length 2, we next magnify S_3 by a factor of 2. The magnification $M(z) = 2z$ maps the triangle S_3 shown in color in Figure 2.3.8(c) onto the triangle S_4 with vertices 0, $\sqrt{3}+i$, and $2i$ shown in black in Figure 2.3.8(c). Finally, we translate S_4 by i using the mapping $T_2(z) = z+i$. This translation maps the triangle S_4 shown in color in Figure 2.3.8(d) onto the triangle S' with vertices i, $\sqrt{3}+2i$, and $3i$ shown in black in Figure 2.3.8(d).

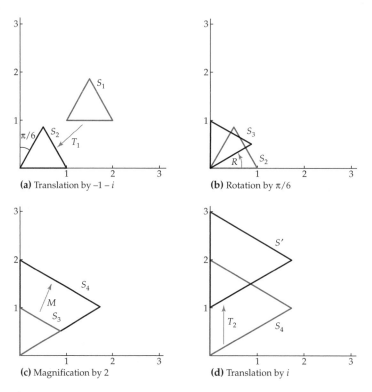

Figure 2.3.8 Linear mapping of a triangle

In conclusion, we have found that the linear mapping:

$$f(z) = T_2 \circ M \circ R \circ T_1(z) = \left(\sqrt{3}+i\right)z + 1 - \sqrt{3} + \sqrt{3}\,i$$

maps the triangle S_1 onto the triangle S'. ❏

Remarks *Comparison with Real Analysis*

The study of differential calculus is based on the principle that real linear functions are the easiest types of functions to understand (whether it be from an algebraic, numerical, or graphical point of view). One of the many uses of the derivative of a real function f is to find a linear function that approximates f in a neighborhood of a point x_0. In particular, recall that the linear approximation of a differentiable function $f(x)$ at $x = x_0$ is the linear function $l(x) = f(x_0) + f'(x_0)(x - x_0)$. Geometrically, the graph of the linear approximation is the tangent line to the graph of f at the point $(x_0,\ f(x_0))$. Although there is no analogous geometric interpretation for

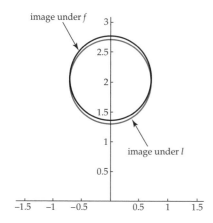

image under f

image under l

Figure 2.3.9 Linear approximation of $w = z^2$

complex functions, the linear approximation formula can be applied to complex functions once an appropriate definition of the derivative of complex function is given. That is, if $f'(z_0)$ represents the derivative of the complex function $f(z)$ at z_0 (this will be defined in Section 3.1), then the **linear approximation** of f in a neighborhood of z_0 is the complex linear function $l(z) = f(z_0) + f'(z_0)(z - z_0)$. Geometrically, $l(z)$ approximates how $f(z)$ acts as a complex mapping near the point z_0. For example, we will see in Chapter 3 that the derivative of the complex function $f(z) = z^2$ is $f'(z) = 2z$. Therefore, the linear approximation of $f(z) = z^2$ at $z_0 = 1+i$ is $l(z) = 2i + 2(1+i)(z-1-i) = 2\sqrt{2}\left(e^{i\pi/4}z\right) - 2i$. Near the point $z = 1 + i$ the mapping $w = z^2$ can be approximated by the linear mapping consisting of the composition of rotation through $\pi/4$, magnification by $2\sqrt{2}$, and translation by $-2i$. In Figure 2.3.9, the image of the circle $|z - (1 + i)| = 0.25$ under f is shown in black and the image of this circle under l is shown in color. Figure 2.3.9 indicates that, for the circle $|z - (1 + i)| = 0.25$, the linear mapping l gives an accurate approximation of the complex mapping f.

EXERCISES 2.3   *Answers to selected odd-numbered problems begin on page ANS-7.*

In Problems 1–6, (**a**) find the image of the closed disk $|z| \leq 1$ under the given linear mapping $w = f(z)$ and (**b**) represent the linear mapping with a sequence of plots as in Figure 2.3.7.

1. $f(z) = z + 3i$

2. $f(z) = z + 2 - i$

3. $f(z) = 3iz$

4. $f(z) = (1 + i)z$

5. $f(z) = 2z - i$

6. $f(z) = (6 - 5i)z + 1 - 3i$

In Problems 7–12, (**a**) find the image of the triangle with vertices 0, 1, and i under the given linear mapping $w = f(z)$ and (**b**) represent the linear mapping with a sequence of plots as in Figure 2.3.7.

7. $f(z) = z + 2i$

8. $f(z) = 3z$

9. $f(z) = e^{i\pi/4}z$

10. $f(z) = \frac{1}{2}iz$

11. $f(z) = -3z + i$

12. $f(z) = (1 - i)z - 2$

In Problems 13–16, express the given linear mapping $w = f(z)$ as a composition of a rotation, magnification, and a translation as in (6). Then describe the action of the linear mapping in words.

13. $f(z) = 3iz + 4$

14. $f(z) = 5\left(\cos\dfrac{\pi}{5} + i\sin\dfrac{\pi}{5}\right)z + 7i$

15. $f(z) = -\dfrac{1}{2}z + 1 - \sqrt{3}i$

16. $f(z) = (3 - 2i)z + 12$

In Problems 17–20, find a linear mapping that maps the set S onto the set S'. (Note: there may be more than one linear mapping that works.)

17. S is the triangle with vertices 0, 1, and $1 + i$. S' is the triangle with vertices $2i$, $3i$, and $-1 + 3i$.

18. S is the circle $|z - 1| = 3$. S' is the circle $|w + i| = 5$.

19. S is the imaginary axis. S' is the line through the points i and $1 + 2i$.

20. S is the square with vertices $1 + i$, $-1 + i$, $-1 - i$, and $1 - i$. S' is the square with vertices 1, $2 + i$, $1 + 2i$, and i.

21. Find two different linear mappings that map the square with vertices 0, 1, $1+i$, and i, onto the square with vertices -1, 0, i, $-1+i$.

22. Find two different linear mappings that map the half-plane $\text{Re}(z) \geq 2$ onto the half-plane $\text{Re}(w) \geq 5$.

23. Consider the line segment parametrized by $z(t) = z_0 (1-t) + z_1 t$, $0 \leq t \leq 1$.

    (a) Find a parametrization of the image of the line segment under the translation $T(z) = z + b$, $b \neq 0$. Describe the image in words.

    (b) Find a parametrization of the image of the line segment under the rotation $R(z) = az$, $|a| = 1$. Describe the image in words.

    (c) Find a parametrization of the image of the line segment under the magnification $M(z) = az$, $a > 0$. Describe the image in words.

24. Repeat Problem 23 for the circle parametrized by $z(t) = z_0 + re^{it}$, $0 \leq t \leq 2\pi$.

25. In parts (a)–(c), express the given composition of mappings as a linear mapping $f(z) = az + b$.

    (a) rotation through $\pi/4$, magnification by 2, and translation by $1+i$

    (b) magnification by 2, translation by $\sqrt{2}$, and rotation through $\pi/4$

    (c) translation by $\frac{1}{2}\sqrt{2}$, rotation through $\pi/4$, then magnification by 2

    (d) What do you notice about the linear mappings in (a)–(c)?

26. Consider the complex linear mapping $f(z) = \left(1 + \sqrt{3}i\right) z + i$. In each part, find the translation T, rotation R, and magnification M that satisfy the given equation and then describe the mapping f in words using T, R, and M.

    (a) $f(z) = T \circ M \circ R(z)$ (b) $f(z) = M \circ T \circ R(z)$

    (c) $f(z) = R \circ M \circ T(z)$

Focus on Concepts

27. (a) Prove that the composition of two translations $T_1(z) = z + b_1$, $b_1 \neq 0$, and $T_2(z) = z + b_2$, $b_2 \neq 0$, is a translation or the identity mapping. Does the order of composition matter?

    (b) Prove that the composition of two rotations $R_1(z) = a_1 z$, $|a_1| = 1$, and $R_2(z) = a_2 z$, $|a_2| = 1$, is a rotation or the identity mapping. Does the order of composition matter?

    (c) Prove that the composition of two magnifications $M_1(z) = a_1 z$, $a_1 > 0$, and $M_2(z) = a_2 z$, $a_2 > 0$, is a magnification or the identity mapping. Does the order of composition matter?

28. We say that two mappings f and g **commute** if $f \circ g(z) = g \circ f(z)$ for all z. That is, two mappings commute if the order in which you compose them does not change the mapping.

    (a) Can a translation and a nonidentity rotation commute?

    (b) Can a translation and a nonidentity magnification commute?

    (c) Can a nonidentity rotation and a nonidentity magnification commute?

29. Recall from Problem 31 in Exercises 2.2 that the mapping $f(z) = \bar{z}$ is called **reflection** about the real axis. Using the mapping $f(z) = \bar{z}$ and any linear mappings, find a mapping g that reflects about the imaginary axis. That is, express the mapping $g(x + iy) = -x + iy$ in terms of complex constants and the symbol $\bar{z}$.

30. Describe how to obtain the image $w_0 = f(z_0)$ of a point z_0 under the mapping $f(z) = a\bar{z} + b$ in terms of translation, rotation, magnification, and reflection.

31. What can you say about a linear mapping f if you know that $|z| = |f(z)|$ for all complex numbers z?

32. What can you say about a linear mapping f if you know that $|z_2 - z_1| = |f(z_2) - f(z_1)|$ for all complex numbers z_1 and z_2?

33. A **fixed point** of a mapping f is a point z_0 with the property that $f(z_0) = z_0$.

    (a) Does the linear mapping $f(z) = az + b$ have a fixed point z_0? If so, then find z_0 in terms of a and b.

    (b) Give an example of a complex linear mapping that has no fixed points.

    (c) Give an example of a complex linear mapping that has more than one fixed point. [*Hint*: There is only one such mapping.]

    (d) Prove that if z_0 is a fixed point of the complex linear mapping f and if f commutes with the complex linear mapping g (see Problem 28), then z_0 is a fixed point of g.

34. Suppose that the set S is mapped onto the set S' by the complex mapping $w = f(z)$. If $S = S'$ as subsets of a single copy of the complex plane, then S is said to be **invariant** under f. Notice that it is not necessary that $f(z) = z$ for all z in S in order for S to be invariant under f.

    (a) Explain why the closed disk $|z| \leq 2$ is invariant under the rotation $R(z) = az$, $|a| = 1$.

    (b) What are the invariant sets under a translation $T(z) = z + b$, $b \neq 0$?

    (c) What are the invariant sets under a magnification $M(z) = az$, $a > 0$?

35. In this problem we show that a linear mapping is uniquely determined by the images of two points.

    (a) Let $f(z) = az + b$ be a complex linear function with $a \neq 0$ and assume that $f(z_1) = w_1$ and $f(z_2) = w_2$. Find two formulas that express a and b in terms of z_1, z_2, w_1, and w_2. Explain why these formulas imply that the linear mapping f is uniquely determined by the images of two points.

    (b) Show that a linear function is *not* uniquely determined by the image of one point. That is, find two different linear functions f_1 and f_2 that agree at one point.

36. Find a complex linear function $f(z) = az + b$ that rotates the point z counterclockwise through an angle of θ radians about the point z_0 in the complex plane. [*Hint*: By Problem 35, this mapping is uniquely determined by the images of two points.]

37. **(a)** Given two complex numbers w_1 and w_2, does there always exist a linear function that maps 0 onto w_1 and 1 onto w_2? Explain.

    (b) Given three complex numbers w_1, w_2, and w_3, does there always exist a linear function that maps 0 onto w_1, 1 onto w_2, and i onto w_3? Explain.

38. In Chapter 1 we used the triangle inequality to obtain bounds on the modulus of certain expressions in z given a bound on the modulus of z. For example, if $|z| \leq 1$, then $|(1 + i)z - 2| \leq 2 + \sqrt{2}$. Justify this inequality using linear mappings, then determine a value of z for which $|(1 + i)z - 2| = 2 + \sqrt{2}$.

39. Consider the complex function $f(z) = 2iz + 1 - i$ defined on the closed annulus $2 \leq |z| \leq 3$.

    (a) Use linear mappings to determine upper and lower bounds on the modulus of $f(z) = 2iz + 1 - i$. That is, find real values L and M such that $L \leq |2iz + 1 - i| \leq M$.

(b) Find values of z that attain your bounds in (a). In other words, find z_0 and z_1 such that $|f(z_0)| = L$ and $|f(z_1)| = M$.

(c) Determine upper and lower bounds on the modulus of the function $g(z) = 1/f(z)$ defined on the closed annulus $2 \leq |z| \leq 3$.

Projects

40. Groups of Isometries   In this project we investigate the relationship between complex analysis and the Euclidean geometry of the Cartesian plane.

The **Euclidean distance** between two points $(x_1, \ y_1)$ and $(x_2, \ y_2)$ in the Cartiesian plane is

$$d\left((x_1, y_1),(x_2, y_2)\right) = \sqrt{(x_2 - x_1)^2 + (y_2 - y_1)^2}.$$

Of course, if we consider the complex representations $z_1 = x_1 + iy_1$ and $z_2 = x_2 + iy_2$ of these points, then the Euclidean distance is given by the modulus

$$d(z_1, z_2) = |z_2 - z_1|.$$

A function from the plane to the plane that preserves the Euclidean distance between every pair of points is called a **Euclidean isometry** of the plane. In particular, a complex mapping $w = f(z)$ is a Euclidean isometry of the plane if

$$|z_2 - z_1| = |f(z_1) - f(z_2)|$$

for every pair of complex numbers z_1 and z_2.

(a) Prove that every linear mapping of the form $f(z) = az + b$ where $|a| = 1$ is a Euclidean isometry.

A **group** is an algebraic structure that occurs in many areas of mathematics. A group is a set G together with a special type of function $*$ from $G \times G$ to G. The function $*$ is called a **binary operation** on G, and it is customary to use the notation $a * b$ instead of $*(a, b)$ to represent a value of $*$. We now give the formal definition of a group. A **group** is a set G together with a binary operation $*$ on G, which satisfies the following three properties:

   (i) for all elements a, b, and c in G, $a * (b * c) = (a * b) * c$,

   (ii) there exists an element e in G such that $e * a = a * e = a$ for all a in G, and

   (iii) for every element a in G there exists an element b in G such that $a * b = b * a = e$. (The element b is called the **inverse** of a in G and is denoted by a^{-1}.)

Let $\text{Isom}_+(\mathbf{E})$ denote the set of all complex functions of the form $f(z) = az + b$ where $|a| = 1$. In the remaining part of this project you are asked to demonstrate that $\text{Isom}_+(\mathbf{E})$ is a group with composition of functions as the binary operation. This group is called the **group of orientation-preserving isometries of the Euclidean plane**.

(b) Prove that composition of functions is a binary operation on $\text{Isom}_+(\mathbf{E})$. That is, prove that if f and g are functions in $\text{Isom}_+(\mathbf{E})$, then the function $f \circ g$ defined by $f \circ g(z) = f\left(g(z)\right)$ is an element in $\text{Isom}_+(\mathbf{E})$.

(c) Prove that the set $\text{Isom}_+(\mathbf{E})$ with composition satisfies property (i) of a group.

(d) Prove that the set Isom$_+(\mathbf{E})$ with composition satisfies property (*ii*) of a group. That is, show that there exists a function e in Isom$_+(\mathbf{E})$ such that $e \circ f = f \circ e = f$ for all functions f in Isom$_+(\mathbf{E})$.

(e) Prove that the set Isom$_+(\mathbf{E})$ with composition satisfies property (*iii*) of a group.

2.4 Special Power Functions

A **complex polynomial function** is a function of the form $p(z) = a_n z^n + a_{n-1} z^{n-1} + \ldots + a_1 z + a_0$ where n is a positive integer and $a_n, a_{n-1}, \ldots, a_1, a_0$ are complex constants. In general, a complex polynomial mapping can be quite complicated, but in many special cases the action of the mapping is easily understood. For instance, the complex linear functions studied in Section 2.3 are complex polynomials of degree $n = 1$.

In this section we study complex polynomials of the form $f(z) = z^n$, $n \geq 2$. Unlike the linear mappings studied in the previous section, the mappings $w = z^n$, $n \geq 2$, do not preserve the basic shape of every figure in the complex plane. Associated to the function z^n, $n \geq 2$, we also have the *principal nth root function* $z^{1/n}$. The principal nth root functions are inverse functions of the functions z^n defined on a sufficiently restricted domain. Consequently, complex mappings associated to z^n and $z^{1/n}$ are closely related.

Power Functions Recall that a real function of the form $f(x) = x^a$, where a is a real constant, is called a power function. We form a complex power function by allowing the input or the exponent a to be a complex number. In other words, a **complex power function** is a function of the form $f(z) = z^\alpha$ where α is a complex constant. If α is an integer, then the power function z^α can be evaluated using the algebraic operations on complex numbers from Chapter 1. For example, $z^2 = z \cdot z$ and $z^{-3} = \dfrac{1}{z \cdot z \cdot z}$. We can also use the formulas for taking roots of complex numbers from Section 1.4 to define power functions with fractional exponents of the form $1/n$. For instance, we *can* define $z^{1/4}$ to be the function that gives the principal fourth root of z. In this section we will restrict our attention to special complex power functions of the form z^n and $z^{1/n}$ where $n \geq 2$ and n is an integer. More complicated complex power functions such as $z^{\sqrt{2}-i}$ will be discussed in Section 4.2 following the introduction of the complex logarithmic function.

2.4.1 The Power Function z^n

In this subsection we consider complex power functions of the form z^n, $n \geq 2$. It is natural to begin our investigation with the simplest of these functions, the **complex squaring function** z^2.

The Function z^2 Values of the complex power function $f(z) = z^2$ are easily found using complex multiplication. For example, at $z = 2 - i$, we have $f(2 - i) = (2 - i)^2 = (2 - i) \cdot (2 - i) = 3 - 4i$. Understanding the complex mapping $w = z^2$, however, requires a little more work. We begin by expressing this mapping in exponential notation by replacing the symbol z with $re^{i\theta}$:

$$w = z^2 = \left(re^{i\theta}\right)^2 = r^2 e^{i2\theta}. \tag{1}$$

From (1) we see that the modulus r^2 of the point w is the square of the modulus r of the point z, and that the argument 2θ of w is twice the argument

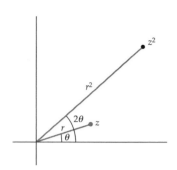

Figure 2.4.1 The mapping $w = z^2$

θ of z. If we plot both z and w in the same copy of the complex plane, then w is obtained by magnifying z by a factor of r and then by rotating the result through the angle θ about the origin. In Figure 2.4.1 we depict the relationship between z and $w = z^2$ when $r > 1$ and $\theta > 0$. If $0 < r < 1$, then z is contracted by a factor of r, and if $\theta < 0$, then the rotation is clockwise.

It is important to note that the magnification or contraction factor and the rotation angle associated to $w = f(z) = z^2$ depend on where the point z is located in the complex plane. For example, since $f(2) = 4$ and $f(i/2) = -\frac{1}{4}$, the point $z = 2$ is magnified by 2 but not rotated, whereas the point $z = i/2$ is contracted by $\frac{1}{2}$ and rotated through $\pi/2$. In general, the squaring function z^2 does not magnify the modulus of points on the unit circle $|z| = 1$ and it does not rotate points on the positive real axis.

The description of the mapping $w = z^2$ in terms of a magnification and rotation can be used to visualize the image of some special sets. For example, consider a ray emanating from the origin and making an angle of ϕ with the positive real axis. All points on this ray have an argument of ϕ and so the images of these points under $w = z^2$ have an argument of 2ϕ. Thus, the images lie on a ray emanating from the origin and making an angle of 2ϕ with the positive real axis. Moreover, since the modulus ρ of a point on the ray can be any value in the interval $[0, \infty)$, the modulus ρ^2 of a point in the image can also be any value in the interval $[0, \infty)$. This implies that a ray emanating from the origin making an angle of ϕ with the positive real axis is mapped onto a ray emanating from the origin making an angle 2ϕ with the positive real axis by $w = z^2$. We can also justify this mapping property of z^2 by parametrizing the ray and its image using (8) and (11) of Section 2.2.

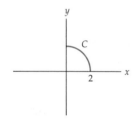

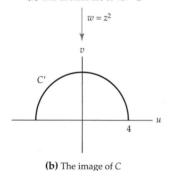

(a) The circular arc of $|z| = 2$

(b) The image of C

Figure 2.4.2 The mapping $w = z^2$

EXAMPLE 1   Image of a Circular Arc under $w = z^2$

Find the image of the circular arc defined by $|z| = 2$, $0 \le \arg(z) \le \pi/2$, under the mapping $w = z^2$.

Solution   Let C be the circular arc defined by $|z| = 2$, $0 \le \arg(z) \le \pi/2$, shown in color in Figure 2.4.2(a), and let C' denote the image of C under $w = z^2$. Since each point in C has modulus 2 and since the mapping $w = z^2$ squares the modulus of a point, it follows that each point in C' has modulus $2^2 = 4$. This implies that the image C' must be contained in the circle $|w| = 4$ centered at the origin with radius 4. Since the arguments of the points in C take on every value in the interval $[0, \pi/2]$ and since the mapping $w = z^2$ doubles the argument of a point, it follows that the points in C' have arguments that take on every value in the interval $[2 \cdot 0, 2 \cdot (\pi/2)] = [0, \pi]$. That is, the set C' is the semicircle defined by $|w| = 4$, $0 \le \arg(w) \le \pi$. In conclusion, we have shown that $w = z^2$ maps the circular arc C shown in color in Figure 2.4.2(a) onto the semicircle C' shown in black in Figure 2.4.2(b).   ❏

An alternative way to find the image in Example 1 would be to use a parametrization. From (10) of Section 2.2, the circular arc C can be parametrized by $z(t) = 2e^{it}$, $0 \le t \le \pi/2$, and by (11) of Section 2.2 its image C' is given by $w(t) = f(z(t)) = 4e^{i2t}$, $0 \le t \le \pi/2$. By replacing the parameter

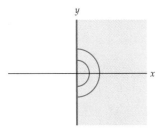

(a) The half-plane Re(z) > 0

$w = z^2$

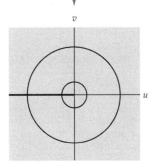

(b) Image of the half-plane in (a)

Figure 2.4.3 The mapping $w = z^2$

t in $w(t)$ with the new parameter $s = 2t$, we obtain $W(s) = 4e^{is}$, $0 \leq s \leq \pi$, which is a parametrization of the semicircle $|w| = 4$, $0 \leq \arg(w) \leq \pi$.

In a similar manner, we find that the squaring function maps a semicircle $|z| = r$, $-\pi/2 \leq \arg(z) \leq \pi/2$, onto a circle $|w| = r^2$. Since the right half-plane $\mathrm{Re}\,(z) \geq 0$ consists of the collection of semicircles $|z| = r$, $-\pi/2 \leq \arg(z) \leq \pi/2$, where r takes on every value in the interval $[0, \infty)$, we have that the image of this half-plane consists of the collection of circles $|w| = r^2$ where r takes on any value in $[0, \infty)$. This implies that $w = z^2$ maps the right half-plane $\mathrm{Re}\,(z) \geq 0$ onto the entire complex plane. We illustrate this property in Figure 2.4.3. Observe that the images of the two semicircles centered at 0 shown in color in Figure 2.4.3(a) are the two circles shown in black in Figure 2.4.3(b). Since $w = z^2$ squares the modulus of a point, the semicircle with smaller radius in Figure 2.4.3(a) is mapped onto the circle with smaller radius in Figure 2.4.3(b), while the semicircle with larger radius in Figure 2.4.3(a) is mapped onto the circle with larger radius in Figure 2.4.3(b). We also see in Figure 2.4.3 that the ray emanating from the origin and containing the point i and the ray emanating from the origin and containing the point $-i$ are both mapped onto the nonpositive real axis. Thus, the imaginary axis shown in color in Figure 2.4.3(a) is mapped onto the set consisting of the point $w = 0$ together with the negative u-axis shown in black in Figure 2.4.3(b).

In order to gain a deeper understanding of the mapping $w = z^2$ we next consider the images of vertical and horizontal lines in the complex plane.

EXAMPLE 2   **Image of a Vertical Line under $w = z^2$**

Find the image of the vertical line $x = k$ under the mapping $w = z^2$.

Solution In this example it is convenient to work with real and imaginary parts of $w = z^2$ which, from (1) in Section 2.1, are $u(x, y) = x^2 - y^2$ and $v(x, y) = 2xy$, respectively. Since the vertical line $x = k$ consists of the points $z = k + iy$, $-\infty < y < \infty$, it follows that the image of this line consists of all points $w = u + iv$ where

$$u = k^2 - y^2, \quad v = 2ky, \quad -\infty < y < \infty. \tag{2}$$

If $k \neq 0$, then we can eliminate the variable y from (2) by solving the second equation for $y = v/2k$ and then substituting this expression into the remaining equation and inequality. After simplification, this yields:

$$u = k^2 - \frac{v^2}{4k^2}, \quad -\infty < v < \infty. \tag{3}$$

Thus, the image of the line $x = k$ (with $k \neq 0$) under $w = z^2$ is the set of points in the w-plane satisfying (3). This image is a parabola that opens in the direction of the negative u-axis, has its vertex at $(k^2, 0)$, and has v-intercepts at $(0, \pm 2k^2)$. Notice that the image given by (3) is unchanged if k is replaced

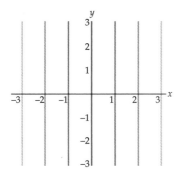

(a) Vertical lines in the z-plane

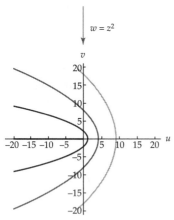

(b) The images of the lines in (a)

Figure 2.4.4 The mapping $w = z^2$

by $-k$. This implies that if $k \neq 0$, then the pair of vertical lines $x = k$ and $x = -k$ are both mapped onto the parabola $u = k^2 - v^2 / \left(4k^2\right)$ by $w = z^2$.

The action of the mapping $w = z^2$ on vertical lines is depicted in Figure 2.4.4. The vertical lines $x = k$, $k \neq 0$, shown in color in Figure 2.4.4(a) are mapped onto the parabolas shown in black in Figure 2.4.4(b). In particular, from (3) we have that the lines $x = 3$ and $x = -3$ shown in color in Figure 2.4.4(a) are mapped onto the parabola with vertex at $(9, 0)$ shown in black in Figure 2.4.4(b). In a similar manner, the lines $x = \pm 2$ are mapped onto the parabola with vertex at $(4, 0)$, and the lines $x = \pm 1$ are mapped onto the parabola with vertex at $(1, 0)$. In the case when $k = 0$, it follows from (2) that the image of the line $x = 0$ (which is the imaginary axis) is given by:

$$u = -y^2, \quad v = 0, \quad -\infty < y < \infty.$$

Therefore, we also have that the imaginary axis is mapped onto the negative real axis by $w = z^2$. See Figure 2.4.4. ❑

With minor modifications, the method of Example 2 can be used to show that a horizontal line $y = k$, $k \neq 0$, is mapped onto the parabola

$$u = \frac{v^2}{4k^2} - k^2 \tag{4}$$

by $w = z^2$. Again we see that the image in (4) is unchanged if k is replaced by $-k$, and so the pair of horizontal lines $y = k$ and $y = -k$, $k \neq 0$, are both mapped by $w = z^2$ onto the parabola given by (4). If $k = 0$, then the horizontal line $y = 0$ (which is the real axis) is mapped onto the positive real axis. Therefore, the horizontal lines $y = k$, $k \neq 0$, shown in color in Figure 2.4.5(a) are mapped by $w = z^2$ onto the parabolas shown in black in Figure 2.4.5(b). Specifically, the lines $y = \pm 3$ are mapped onto the parabola with vertex at $(-9, 0)$, the lines $y = \pm 2$ are mapped onto the parabola with vertex at $(-4, 0)$, and the lines $y = \pm 1$ are mapped onto the parabola with vertex at $(-1, 0)$.

EXAMPLE 3   **Image of a Triangle under $w = z^2$**

Find the image of the triangle with vertices 0, $1 + i$, and $1 - i$ under the mapping $w = z^2$.

Solution Let S denote the triangle with vertices at 0, $1 + i$, and $1 - i$, and let S' denote its image under $w = z^2$. Each of the three sides of S will be treated separately. The side of S containing the vertices 0 and $1 + i$ lies on a ray emanating from the origin and making an angle of $\pi/4$ radians with the positive x-axis. By our previous discussion, the image of this segment must lie on a ray making an angle of $2 \left(\pi/4\right) = \pi/2$ radians with the positive u-axis. Furthermore, since the moduli of the points on the edge containing 0 and $1+i$ vary from 0 to $\sqrt{2}$, the moduli of the images of these points vary from $0^2 = 0$ to $\left(\sqrt{2}\right)^2 = 2$. Thus, the image of this side is a vertical line segment from 0 to $2i$ contained in the v-axis and shown in black in Figure 2.4.6(b). In a similar manner, we find that the image of the side of S containing the vertices 0 and

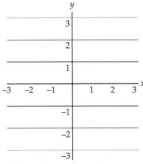

(a) Horizontal lines in the z-plane

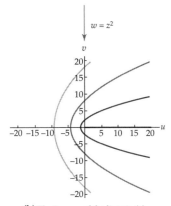

(b) The images of the lines in (a)

Figure 2.4.5 The mapping $w = z^2$

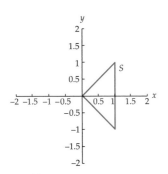

(a) A triangle in the z-plane

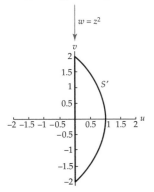

(b) The image of the triangle in (a)

Figure 2.4.6 The mapping $w = z^2$

$1 - i$ is a vertical line segment from 0 to $-2i$ contained in the v-axis. See Figure 2.4.6. The remaining side of S contains the vertices $1 - i$ and $1 + i$. This side consists of the set of points $z = 1 + iy$, $-1 \leq y \leq 1$. Because this side is contained in the vertical line $x = 1$, it follows from (2) and (3) of Example 2 that its image is a parabolic segment given by:

$$u = 1 - \frac{v^2}{4}, \quad -2 \leq v \leq 2.$$

Thus, we have shown that the image of triangle S shown in color in Figure 2.4.6(a) is the figure S' shown in black in Figure 2.4.6(b). ◻

The Function z^n, $n > 2$　An analysis similar to that used for the mapping $w = z^2$ can be applied to the mapping $w = z^n$, $n > 2$. By replacing the symbol z with $re^{i\theta}$ we obtain:

$$w = z^n = r^n e^{in\theta}. \tag{5}$$

Consequently, if z and $w = z^n$ are plotted in the same copy of the complex plane, then this mapping can be visualized as the process of magnifying or contracting the modulus r of z to the modulus r^n of w, and by rotating z about the origin to increase an argument θ of z to an argument $n\theta$ of w.

We can use this description of $w = z^n$ to show that a ray emanating from the origin and making an angle of ϕ radians with the positive x-axis is mapped onto a ray emanating from the origin and making an angle of $n\phi$ radians with the positive u-axis. This property is illustrated for the mapping $w = z^3$ in Figure 2.4.7. Each ray shown in color in Figure 2.4.7(a) is mapped onto a ray shown in black in Figure 2.4.7(b). Since the mapping $w = z^3$ increases the argument of a point by a factor of 3, the ray nearest the x-axis in the first quadrant in Figure 2.4.7(a) is mapped onto the ray in the first quadrant in Figure 2.4.7(b), and the remaining ray in the first quadrant in Figure 2.4.7(a) is mapped onto the ray in the second quadrant in Figure 2.4.7(b). Similarly, the ray nearest the x-axis in the fourth quadrant in Figure 2.4.7(a) is mapped onto the ray in the fourth quadrant in Figure 2.4.7(b), and the remaining ray in the fourth quadrant of Figure 2.4.7(a) is mapped onto the ray in the third quadrant in Figure 2.4.7(b).

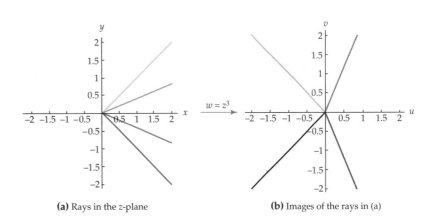

(a) Rays in the z-plane　　　　**(b)** Images of the rays in (a)

Figure 2.4.7 The mapping $w = z^3$

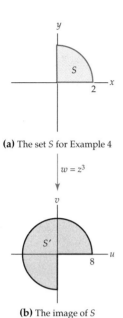

(a) The set S for Example 4

$w = z^3$

(b) The image of S

Figure 2.4.8 The mapping $w = z^3$

EXAMPLE 4   **Image of a Circular Wedge under $w = z^3$**

Determine the image of the quarter disk defined by the inequalities $|z| \leq 2$, $0 \leq \arg(z) \leq \pi/2$, under the mapping $w = z^3$.

Solution Let S denote the quarter disk and let S' denote its image under $w = z^3$. Since the moduli of the points in S vary from 0 to 2 and since the mapping $w = z^3$ cubes the modulus of a point, it follows that the moduli of the points in S' vary from $0^3 = 0$ to $2^3 = 8$. In addition, because the arguments of the points in S vary from 0 to $\pi/2$ and because the mapping $w = z^3$ triples the argument of a point, we also have that the arguments of the points in S' vary from 0 to $3\pi/2$. Therefore, S' is the set given by the inequalities $|w| \leq 8$, $0 \leq \arg(w) \leq 3\pi/2$, shown in gray in Figure 2.4.8(b). In summary, the set S shown in color in Figure 2.4.8(a) is mapped onto the set S' shown in gray in Figure 2.4.8(b) by $w = z^3$.   ❑

2.4.2 The Power Function $z^{1/n}$

We now investigate complex power functions of the form $z^{1/n}$ where n is an integer and $n \geq 2$. We begin with the case $n = 2$.

Principal Square Root Function $z^{1/2}$ In (4) of Section 1.4 we saw that the n nth roots of a nonzero complex number $z = r\left(\cos\theta + i\sin\theta\right) = re^{i\theta}$ are given by:

$$\sqrt[n]{r}\left[\cos\left(\frac{\theta + 2k\pi}{n}\right) + i\sin\left(\frac{\theta + 2k\pi}{n}\right)\right] = \sqrt[n]{r}e^{i(\theta + 2k\pi)/n}$$

where $k = 0, 1, 2, \ldots, n - 1$. In particular, for $n = 2$, we have that the two square roots of z are:

$$\sqrt{r}\left[\cos\left(\frac{\theta + 2k\pi}{2}\right) + i\sin\left(\frac{\theta + 2k\pi}{2}\right)\right] = \sqrt{r}e^{i(\theta + 2k\pi)/2} \qquad (6)$$

for $k = 0, 1$. The formula in (6) does not define a function because it assigns *two* complex numbers (one for $k = 0$ and one for $k = 1$) to the complex number z. However, by setting $\theta = \text{Arg}(z)$ and $k = 0$ in (6) we can define a function that assigns to z the unique principal square root. Naturally, this function is called the **principal square root function**.

Definition 2.4.1   Principal Square Root Function

The function $z^{1/2}$ defined by:

$$z^{1/2} = \sqrt{|z|}e^{i\text{Arg}(z)/2} \qquad (7)$$

is called the **principal square root function**.

If we set $\theta = \text{Arg}(z)$ and replace z with $re^{i\theta}$ in (7), then we obtain an alternative description of the principal square root function for $|z| > 0$:

$$z^{1/2} = \sqrt{r}\, e^{i\theta/2}, \quad \theta = \text{Arg}(z). \qquad (8)$$

Note ➡ You should take note that the symbol $z^{1/2}$ used in Definition 2.4.1 represents something different from the same symbol used in Section 1.4. In (7) we use $z^{1/2}$ to represent the *value* of the principal square root of the complex number z, whereas in Section 1.4 the symbol $z^{1/2}$ was used to represent the *set* of two square roots of the complex number z. This repetition of notation is unfortunate, but widely used. For the most part, the context in which you see the symbol $z^{1/2}$ should make it clear whether we are referring to the principal square root or the set of square roots. In order to avoid confusion we will, at times, also explicitly state "the principal square root function $z^{1/2}$" or "the function $f(z) = z^{1/2}$ given by (7)."

EXAMPLE 5   Values of $z^{1/2}$

Find the values of the principal square root function $z^{1/2}$ at the following points:

(a) $z = 4$ (b) $z = -2i$ (c) $z = -1 + i$

Solution In each part we use (7) to determine the value of $z^{1/2}$.

(a) For $z = 4$, we have $|z| = |4| = 4$ and $\text{Arg}(z) = \text{Arg}(4) = 0$, and so from (7) we obtain:

$$4^{1/2} = \sqrt{4}\, e^{i(0/2)} = 2e^{i(0)} = 2.$$

(b) For $z = -2i$, we have $|z| = |-2i| = 2$ and $\text{Arg}(z) = \text{Arg}(-2i) = -\pi/2$, and so from (7) we obtain:

$$(-2i)^{1/2} = \sqrt{2}\, e^{i(-\pi/2)/2} = \sqrt{2}\, e^{-i\pi/4} = 1 - i.$$

(c) For $z = -1 + i$, we have $|z| = |-1 + i| = \sqrt{2}$ and $\text{Arg}(z) = \text{Arg}(-1 + i) = 3\pi/4$, and so from (7) we obtain:

$$(-1 + i)^{1/2} = \sqrt{(\sqrt{2})}\, e^{i(3\pi/4)/2} = \sqrt[4]{2}\, e^{i(3\pi/8)} \approx 0.4551 + 1.0987i.$$

❏

It is important that we use the *principal argument* when we evaluate the principal square root function of Definition 2.4.1. Using a different choice for the argument of z can give a different function. For example, in Section 1.4 we saw that the two square roots of i are $\frac{1}{2}\sqrt{2} + \frac{1}{2}\sqrt{2}\,i$ and $-\frac{1}{2}\sqrt{2} - \frac{1}{2}\sqrt{2}\,i$. For $z = i$ we have that $|z| = 1$ and $\text{Arg}(z) = \pi/2$. It follows from (7) that:

$$i^{1/2} = \sqrt{1}\, e^{i(\pi/2)/2} = 1 \cdot e^{i\pi/4} = \frac{\sqrt{2}}{2} + \frac{\sqrt{2}}{2}\,i.$$

Therefore, only the first of these two roots of i is the value of the principal square root function. Of course, we can define other "square root" functions. For example, suppose we let θ be the unique value of the argument of z in the interval $\pi < \theta \leq 3\pi$. Then $f(z) = \sqrt{|z|}e^{i\theta/2}$ defines a function for which $f(i) = -\frac{1}{2}\sqrt{2} - \frac{1}{2}\sqrt{2}i$. The function f is not the principal square root function but it is closely related. Since $\pi < \mathrm{Arg}(z) + 2\pi \leq 3\pi$, it follows that $\theta = \mathrm{Arg}(z) + 2\pi$, and so

$$f(z) = \sqrt{|z|}e^{i\theta/2} = \sqrt{|z|}e^{i(\mathrm{Arg}(z)+2\pi)/2} = \sqrt{|z|}e^{i\mathrm{Arg}(z)/2}e^{i\pi}.$$

Because $e^{i\pi} = -1$, the foregoing expression simplifies to $f(z) = -\sqrt{|z|}e^{i\mathrm{Arg}(z)/2}$. That is, we have shown that the function $f(z) = \sqrt{|z|}e^{i\theta/2}$, $\pi < \theta \leq 3\pi$, is the negative of the principal square root function $z^{1/2}$.

Inverse Functions   The principal square root function $z^{1/2}$ given by (7) is *an* inverse function of the squaring function z^2 examined in the first part of this section. Before elaborating on this statement, we need to review some general terminology regarding inverse functions.

A real function must be one-to-one in order to have an inverse function. The same is true for a complex function. The definition of a one-to-one complex function is analogous to that for a real function. Namely, a complex function f is **one-to-one** if each point w in the range of f is the image of a unique point z, called the *pre-image* of w, in the domain of f. That is, f is one-to-one if whenever $f(z_1) = f(z_2)$, then $z_1 = z_2$. Put another way, if $z_1 \neq z_2$, then $f(z_1) \neq f(z_2)$. This says that a one-to-one complex function will not map distinct points in the z-plane onto the same point in the w-plane. For example, the function $f(z) = z^2$ is not one-to-one because $f(i) = f(-i) = -1$. If f is a one-to-one complex function, then for any point w in the range of f there is a unique pre-image in the z-plane, which we denote by $f^{-1}(w)$. This correspondence between a point w and its pre-image $f^{-1}(w)$ defines the inverse function of a one-to-one complex function.

Definition 2.4.2   **Inverse Function**

If f is a one-to-one complex function with domain A and range B, then the **inverse function of f**, denoted by f^{-1}, is the function with domain B and range A defined by $f^{-1}(z) = w$ if $f(w) = z$.

It follows immediately from Definition 2.4.2 that if a set S is mapped onto a set S' by a one-to-one function f, then f^{-1} maps S' onto S. In other words, the complex mappings f and f^{-1} "undo" each other. It also follows from Definition 2.4.2 that if f has an inverse function, then $f\left(f^{-1}(z)\right) = z$ and $f^{-1}\left(f(z)\right) = z$. That is, the two compositions $f \circ f^{-1}$ and $f^{-1} \circ f$ are the identity function.

EXAMPLE 6   **Inverse Function of $f(z) = z + 3i$**

Show that the complex function $f(z) = z + 3i$ is one-to-one on the entire complex plane and find a formula for its inverse function.

Solution One way of showing that f is one-to-one is to show that the equality $f(z_1) = f(z_2)$ implies the equality $z_1 = z_2$. For the function $f(z) = z + 3i$, this follows immediately since $z_1 + 3i = z_2 + 3i$ implies that $z_1 = z_2$. As with real functions, the inverse function of f can often be found algebraically by solving the equation $z = f(w)$ for the symbol w. Solving $z = w + 3i$ for w, we obtain $w = z - 3i$ and so $f^{-1}(z) = z - 3i$. ❑

Solve the equation $z = f(w)$ for w to find a formula for $w = f^{-1}(z)$. ➡

From Section 2.3 we know that the mapping $f(z) = z + 3i$ in Example 6 is a translation by $3i$ and that $f^{-1}(z) = z - 3i$ is a translation by $-3i$. Suppose we represent these mappings in a single copy of the complex plane. Since translating a point by $3i$ and then translating the image by $-3i$ moves the original point back onto itself, we see that the mappings f and f^{-1} "undo" each other as expected.

Inverse Functions of z^n, $n > 2$

We now describe how to obtain an inverse function of the power function z^n, $n \geq 2$. This requires some explanation since the function $f(z) = z^n$, $n \geq 2$, is *not* one-to-one. In order to see that this is so, consider the points $z_1 = re^{i\theta}$ and $z_2 = re^{i(\theta + 2\pi/n)}$ with $r \neq 0$. Because $n \geq 2$, the points z_1 and z_2 are distinct. That is, $z_1 \neq z_2$. From (5) we have that $f(z_1) = r^n e^{in\theta}$ and $f(z_2) = r^n e^{i(n\theta + 2\pi)} = r^n e^{in\theta} e^{i2\pi} = r^n e^{in\theta}$. Therefore, f is not one-to-one since $f(z_1) = f(z_2)$ but $z_1 \neq z_2$. In fact, the n distinct points $z_1 = re^{i\theta}$, $z_2 = re^{i(\theta + 2\pi/n)}$, $z_3 = re^{i(\theta + 4\pi/n)}$, ..., $z_n = re^{i(\theta + 2(n-1)\pi/n)}$ are all mapped onto the single point $w = r^n e^{in\theta}$ by $f(z) = z^n$. This fact is illustrated for $n = 6$ in Figure 2.4.9. The six points $z_1, z_2, \ldots, z_6$ with equal modulus and arguments that differ by $2\pi/6 = \pi/3$ shown in Figure 2.4.9 are all mapped onto the same point by $f(z) = z^6$.

The preceding discussion appears to imply that the function $f(z) = z^n$, $n \geq 2$, does not have an inverse function because it is not one-to-one. You have encountered this problem before, when defining inverse functions for certain real functions in elementary calculus. For example, the real functions $f(x) = x^2$ and $g(x) = \sin x$ are not one-to-one, yet we still have the inverse functions $f^{-1}(x) = \sqrt{x}$ and $g^{-1}(x) = \arcsin x$. The key to defining these inverse functions is to appropriately restrict the domains of the functions to sets on which the functions are one-to-one. For example, whereas the real function $f(x) = x^2$ defined on the interval $(-\infty, \infty)$ is not one-to-one, the same function defined on the interval $[0, \infty)$ is one-to-one. Similarly, $g(x) = \sin x$ is not one-to-one on the interval $(-\infty, \infty)$, but it is one-to-one on the interval $[-\pi/2, \pi/2]$. The function $f^{-1}(x) = \sqrt{x}$ is the inverse function of the function $f(x) = x^2$ defined on the interval $[0, \infty)$. Since $\text{Dom}(f) = [0, \infty)$ and $\text{Range}(f) = [0, \infty)$, the domain and range of $f^{-1}(x) = \sqrt{x}$ are both $[0, \infty)$ as well. See Figure 2.4.10. In a similar manner, $g^{-1}(x) = \arcsin x$ is the inverse function of the function $g(x) = \sin x$ defined on $[-\pi/2, \pi/2]$. The domain and range of g^{-1} are $[-1, 1]$ and $[-\pi/2, \pi/2]$, respectively. See Figure 2.4.11. We use this same idea for the complex power function z^n, $n \geq 2$. That is, in order to define an inverse function for $f(z) = z^n$, $n \geq 2$, we must restrict the domain of f to a set on which f is a one-to-one function. One such choice of this "restricted domain" when $n = 2$ is found in Example 7.

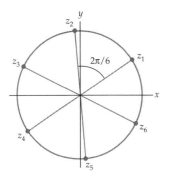

Figure 2.4.9 Points mapped onto the same point by $f(z) = z^6$

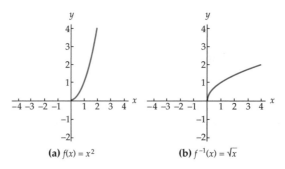

Figure 2.4.10 $f(x) = x^2$ defined on $[0, \infty)$ and its inverse function.

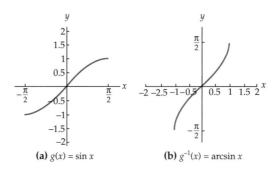

Figure 2.4.11 $g(x) = \sin x$ defined on $[-\pi/2, \pi/2]$ and its inverse function.

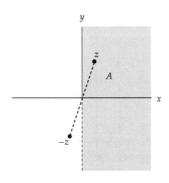

Figure 2.4.12 A domain on which $f(z) = z^2$ is one-to-one

EXAMPLE 7   A Restricted Domain for $f(z) = z^2$

Show that $f(z) = z^2$ is a one-to-one function on the set A defined by $-\pi/2 < \arg(z) \le \pi/2$ and shown in color in Figure 2.4.12.

Solution As in Example 6, we show that f is one-to-one by demonstrating that if z_1 and z_2 are in A and if $f(z_1) = f(z_2)$, then $z_1 = z_2$. Suppose that z_1 and z_2 are in A, then we may write $z_1 = r_1 e^{i\theta_1}$ and $z_2 = r_2 e^{i\theta_2}$ with $-\pi/2 < \theta_1 \le \pi/2$ and $-\pi/2 < \theta_2 \le \pi/2$. If $f(z_1) = f(z_2)$, then it follows from (1) that:

$$r_1^2 e^{i2\theta_1} = r_2^2 e^{i2\theta_2}. \tag{9}$$

From (9) we conclude that the complex numbers $r_1^2 e^{i2\theta_1}$ and $r_2^2 e^{i2\theta_2}$ have the same modulus and principal argument:

$$r_1^2 = r_2^2 \quad \text{and} \quad \text{Arg}\left(r_1^2 e^{i2\theta_1}\right) = \text{Arg}\left(r_2^2 e^{i2\theta_2}\right). \tag{10}$$

Remember: $\text{Arg}(z)$ is in the interval $(-\pi, \pi]$.

Because both r_1 and r_2 are positive, the first equation in (10) implies that $r_1 = r_2$. Moreover, since $-\pi/2 < \theta_1 \le \pi/2$ and $-\pi/2 < \theta_2 \le \pi/2$, it follows that $-\pi < 2\theta_1 \le \pi$, and $-\pi < 2\theta_2 \le \pi$. This means that $\text{Arg}(r_1^2 e^{i2\theta_1}) = 2\theta_1$ and $\text{Arg}(r_2^2 e^{i2\theta_2}) = 2\theta_2$. This fact combined with the second equation in (10) implies that $2\theta_1 = 2\theta_2$, or $\theta_1 = \theta_2$. Therefore, we conclude that $z_1 = z_2$, and this proves that f is a one-to-one function on A. ❑

An Inverse of $f(z) = z^2$

In Example 7 we saw that the squaring function z^2 is one-to-one on the set A defined by $-\pi/2 < \arg(z) \le \pi/2$. It follows from Definition 2.4.2 that this function has a well-defined inverse function f^{-1}. We now proceed to show that this inverse function is the principal square root function $z^{1/2}$ from Definition 2.4.1. In order to do so, let $z = re^{i\theta}$ and $w = \rho e^{i\phi}$ where θ and ϕ are the principal arguments of z and w respectively. Suppose that $w = f^{-1}(z)$. Since the range of f^{-1} is the domain of f, the principal argument ϕ of w must satisfy:

$$-\frac{\pi}{2} < \phi \le \frac{\pi}{2}. \tag{11}$$

On the other hand, by Definition 2.4.2, $f(w) = w^2 = z$. Hence, w is one of the two square roots of z given by (6). That is, either $w = \sqrt{r}e^{i\theta/2}$ or $w = \sqrt{r}e^{i(\theta+2\pi)/2}$. Assume that w is the latter. That is, assume that

$$w = \sqrt{r}e^{i(\theta+2\pi)/2}. \tag{12}$$

Because $\theta = \mathrm{Arg}(z)$, we have $-\pi < \theta \le \pi$, and so, $\pi/2 < (\theta + 2\pi)/2 \le 3\pi/2$. From this and (12) we conclude that the principal argument ϕ of w must satisfy either $-\pi < \phi \le -\pi/2$ or $\pi/2 < \phi \le \pi$. However, this cannot be true since $-\pi/2 < \phi \le \pi/2$ by (11), and so our assumption in (12) must be incorrect. Therefore, we conclude that $w = \sqrt{r}e^{i\theta/2}$, which is the value of the principal square root function $z^{1/2}$ given by (8).

Since $z^{1/2}$ is an inverse function of $f(z) = z^2$ defined on the set $-\pi/2 < \arg(z) \le \pi/2$, it follows that the domain and range of $z^{1/2}$ are the range and domain of f, respectively. In particular, $\mathrm{Range}(z^{1/2}) = A$, that is, the range of $z^{1/2}$ is the set of complex w satisfying $-\pi/2 < \arg(w) \le \pi/2$. In order to find $\mathrm{Dom}(z^{1/2})$ we need to find the range of f. On page 75 we saw that $w = z^2$ maps the right half-plane $\mathrm{Re}(z) \ge 0$ onto the entire complex plane. See Figure 2.4.3. The set A is equal to the right half-plane $\mathrm{Re}(z) \ge 0$ excluding the set of points on the ray emanating from the origin and containing the point $-i$. That is, A does not include the point $z = 0$ or the points satisfying $\arg(z) = -\pi/2$. However, we have seen that the image of the set $\arg(z) = \pi/2$—the positive imaginary axis—is the same as the image of the set $\arg(z) = -\pi/2$. Both sets are mapped onto the negative real axis. Since the set $\arg(z) = \pi/2$ is contained in A, it follows that the only difference between the image of the set A and the image of the right half-plane $\mathrm{Re}(z) \ge 0$ is the image of the point $z = 0$, which is the point $w = 0$. That is, the set A defined by $-\pi/2 < \arg(z) \le \pi/2$ is mapped onto the entire complex plane excluding the point $w = 0$ by $w = z^2$, and so the domain of $f^{-1}(z) = z^{1/2}$ is the entire complex plane $\mathbf{C}$ excluding 0. In summary, we have shown that the principal square root function $w = z^{1/2}$ maps the complex plane $\mathbf{C}$ excluding 0 onto the set defined by $-\pi/2 < \arg(w) \le \pi/2$. This mapping is depicted in Figure 2.4.13. The circle $|z| = r$ shown in color in Figure 2.4.13(a) is mapped onto the semicircle $|w| = \sqrt{r}$, $-\pi/2 < \arg(w) \le \pi/2$, shown in black in Figure 2.4.13(b) by $w = z^{1/2}$. Furthermore, the negative real axis shown in color in Figure 2.4.13(a) is mapped by $w = z^{1/2}$ onto the positive imaginary axis shown in black in Figure 2.4.13(b). Of course, if needed, the principal square root function can be extended to include the point 0 in its domain.

The Mapping $w = z^{1/2}$

As a mapping, the function z^2 squares the modulus of a point and doubles its argument. Because the principal square

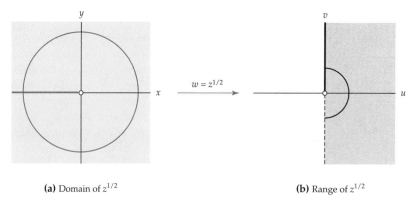

(a) Domain of $z^{1/2}$ (b) Range of $z^{1/2}$

Figure 2.4.13 The principal square root function $w = z^{1/2}$

root function $z^{1/2}$ is an inverse function of z^2, it follows that the mapping $w = z^{1/2}$ takes the square root of the modulus of a point and halves its principal argument. That is, if $w = z^{1/2}$, then we have $|w| = \sqrt{|z|}$ and $\text{Arg}(w) = \frac{1}{2}\text{Arg}(z)$. These relations follow directly from (7) and are helpful in determining the images of sets under $w = z^{1/2}$.

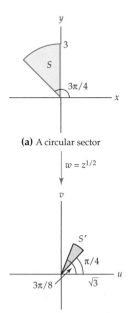

(a) A circular sector

$w = z^{1/2}$

(b) The image of the set in (a)

Figure 2.4.14 The mapping $w = z^{1/2}$

EXAMPLE 8   Image of a Circular Sector under $w = z^{1/2}$

Find the image of the set S defined by $|z| \leq 3$, $\pi/2 \leq \arg(z) \leq 3\pi/4$, under the principal square root function.

Solution Let S' denote the image of S under $w = z^{1/2}$. Since $|z| \leq 3$ for points in S and since $z^{1/2}$ takes the square root of the modulus of a point, we must have that $|w| \leq \sqrt{3}$ for points w in S'. In addition, since $\pi/2 \leq \arg(z) \leq 3\pi/4$ for points in S and since $z^{1/2}$ halves the argument of a point, it follows that $\pi/4 \leq \arg(w) \leq 3\pi/8$ for points w in S'. Therefore, we have shown that the set S shown in color in Figure 2.4.14(a) is mapped onto the set S' shown in gray in Figure 2.4.14(b) by $w = z^{1/2}$. ❑

Principal nth Root Function By modifying the argument given in Example 7 that the function $f(z) = z^2$ is one-to-one on the set defined by $-\pi/2 < \arg(z) \leq \pi/2$, we can show that the complex power function $f(z) = z^n$, $n > 2$, is one-to-one on the set defined by

$$-\frac{\pi}{n} < \arg(z) \leq \frac{\pi}{n}. \tag{13}$$

It is also relatively easy to see that the image of the set defined by (13) under the mapping $w = z^n$ is the entire complex plane **C** excluding $w = 0$. Therefore, there is a well-defined inverse function for f. Analogous to the case $n = 2$, this inverse function of z^n is called the **principal nth root function** $z^{1/n}$. The domain of $z^{1/n}$ is the set of all nonzero complex numbers, and the range of $z^{1/n}$ is the set of complex numbers w satisfying $-\pi/n < \arg(w) \leq \pi/n$. A purely algebraic description of the principal nth root function is given by the following formula, which is analogous to (7).

Definition 2.4.3   Principal nth Root Functions

For $n \geq 2$, the function $z^{1/n}$ defined by:

$$z^{1/n} = \sqrt[n]{|z|}e^{i\mathrm{Arg}(z)/n} \tag{14}$$

is called the **principal nth root function**.

Notice that the principal square root function $z^{1/2}$ from Definition 2.4.1 is simply a special case of (14) with $n = 2$. Notice also that in Definition 2.4.3 we use the symbol $z^{1/n}$ to represent something different than the same symbol used in Section 1.4. As with the symbol $z^{1/2}$, whether $z^{1/n}$ represents the principal nth root or the set of principal nth roots will be clear from the context or stated explicitly.

By setting $z = re^{i\theta}$ with $\theta = \mathrm{Arg}(z)$ we can also express the principal nth root function as:

$$z^{1/n} = \sqrt[n]{r}e^{i\theta/n}, \quad \theta = \mathrm{Arg}(z). \tag{15}$$

EXAMPLE 9   Values of $z^{1/n}$

Find the value of the given principal nth root function $z^{1/n}$ at the given point z.

(a) $z^{1/3}$; $z = i$ (b) $z^{1/5}$; $z = 1 - \sqrt{3}i$

Solution In each part we use (14).

(a) For $z = i$, we have $|z| = 1$ and $\mathrm{Arg}(z) = \pi/2$. Substituting these values in (14) with $n = 3$ we obtain:

$$i^{1/3} = \sqrt[3]{1}e^{i(\pi/2)/3} = e^{i\pi/6} = \frac{\sqrt{3}}{2} + \frac{1}{2}i.$$

(b) For $z = 1 - \sqrt{3}i$, we have $|z| = 2$ and $\mathrm{Arg}(z) = -\pi/3$. Substituting these values in (14) with $n = 5$ we obtain:

$$\left(1 - \sqrt{3}i\right)^{1/5} = \sqrt[5]{2}e^{i(-\pi/3)/5} = \sqrt[5]{2}e^{-i(\pi/15)} \approx 1.1236 - 0.2388i.$$

❏

Multiple-Valued Functions In Section 1.4 we saw that a nonzero complex number z has n distinct nth roots in the complex plane. This means that the process of "taking the nth root" of a complex number z *does not* define a complex function because it assigns a set of n complex numbers to the complex number z. We introduced the symbol $z^{1/n}$ in Section 1.4 to represent the set consisting of the n nth roots of z. A similar type of process is that of finding the argument of a complex number z. Because the symbol $\arg(z)$ represents an infinite set of values, it also does not represent a complex function. These types of operations on complex numbers are examples of **multiple-valued functions**. This term often leads to confusion since a multiple-valued function is *not* a function; a function, by definition, must be *single-valued*. However unfortunate, the term *multiple-valued function* is a standard one in complex analysis and so we shall use it from this point on. We will adopt the following functional notation for multiple-valued functions.

> ### Notation: Multiple-Valued Functions
>
> *When representing multiple-valued functions with functional notation, we will use uppercase letters such as $F(z) = z^{1/2}$ or $G(z) = \arg(z)$. Lowercase letters such as f and g will be reserved to represent functions.*

This notation will help avoid confusion associated to the symbols like $z^{1/n}$. For example, we should assume that $g(z) = z^{1/3}$ refers to the principal cube root function defined by (14) with $n = 3$, whereas, $G(z) = z^{1/3}$ represents the multiple-valued function that assigns the three cube roots of z to the value of z. Thus, we have $g(i) = \frac{1}{2}\sqrt{3} + \frac{1}{2}i$ from Example 9(a) and $G(i) = \left\{\frac{1}{2}\sqrt{3} + \frac{1}{2}i, -\frac{1}{2}\sqrt{3} + \frac{1}{2}i, -i\right\}$ from Example 1 of Section 1.4.

Remarks

(i) In (5) in Exercises 1.4 we defined a rational power of z. One way to define a power function $z^{m/n}$ where m/n is a rational number is as a composition of the principal nth root function and the power function z^m. That is, we can define $z^{m/n} = \left(z^{1/n}\right)^m$. Thus, for $m/n = 2/3$ and $z = 8i$ you should verify that $(8i)^{2/3} = 2 + 2\sqrt{3}i$. Of course, using a root other than the principal nth root gives a possibly different function.

(ii) You have undoubtedly noticed that the complex linear mappings studied in Section 2.3 are much easier to visualize than the mappings by complex power functions studied in this section. In part, mappings by complex power functions are more intricate because these functions are not one-to-one. The visualization of a complex mapping that is **multiple-to-one** is difficult and it follows that the multiple-valued functions, which are "inverses" to the multiple-to-one functions, are also hard to visualize. A technique attributed to the mathematician Bernhard Riemann (1826–1866) for visualizing multiple-to-one and multiple-valued functions is to construct a **Riemann surface** for the mapping. Since a rigorous description of Riemann surfaces is beyond the scope of this text, our discussion of these surfaces will be informal.

We begin with a description of a Riemann surface for the complex squaring function $f(z) = z^2$ defined on the closed unit disk $|z| \leq 1$. On page 81 we saw that $f(z) = z^2$ is not one-to-one. It follows from Example 7 that $f(z) = z^2$ is one-to-one on the set A defined by $|z| \leq 1$, $-\pi/2 < \arg(z) \leq \pi/2$. Under the complex mapping $w = z^2$, the set A shown in color in Figure 2.4.15(a) is mapped onto the closed unit disk $|w| \leq 1$ shown in gray in Figure 2.4.15(a). In a similar manner, we can show that $w = z^2$ is a one-to-one mapping of the set B defined by $|z| \leq 1$, $\pi/2 < \arg(z) \leq 3\pi/2$, onto the closed unit disk $|w| \leq 1$. See Figure 2.4.15(b). Since the unit disk $|z| \leq 1$ is

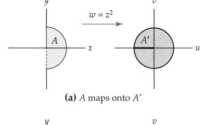

(a) *A maps onto A′*

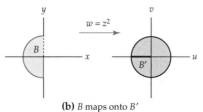

(b) *B maps onto B′*

Figure 2.4.15 The mapping $w = z^2$

Figure 2.4.16 The cut disks A' and B'

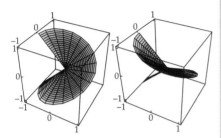

(a) The cut disks A' and B' in xyz-space

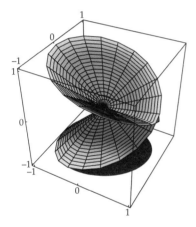

(b) Riemann surface in xyz-space

Figure 2.4.17 A Riemann surface for $f(z) = z^2$

the union of the sets A and B, the image of the disk $|z| \leq 1$ under $w = z^2$ covers the disk $|w| \leq 1$ twice (once by A and once by B). We visualize this "covering" by considering *two* image disks for $w = z^2$. Let A' denote the image of A under f and let B' denote the image of B under f. See Figure 2.4.15. Now imagine that the disks A' and B' have been cut open along the negative real axis as shown in Figure 2.4.16. The edges shown in black in Figure 2.4.16 of the cut disks A' and B' are the images of edges shown in color of A and B, respectively. Similarly, the dashed edges of A' and B' shown in Figure 2.4.16 are the images of the dashed edges of A and B, respectively. We construct a Riemannn surface for $f(z) = z^2$ by stacking the cut disks A' and B' one atop the other in xyz-space and attaching them by gluing together their edges. The black edge of A' shown in Figure 2.4.16 is glued to the dashed edge of B' shown in Figure 2.4.16, and the dashed edge of A' is glued to the black edge of B'. After attaching in this manner we obtain the Riemann surface shown in Figure 2.4.17. We assume that this surface is lying directly above the closed unit disk $|w| \leq 1$. Although $w = z^2$ is not a one-to-one mapping of the closed unit disk $|z| \leq 1$ onto the closed unit disk $|w| \leq 1$, it is a one-to-one mapping of the closed unit disk $|z| \leq 1$ onto the Riemann surface that we have constructed. Moreover, the two-to-one covering nature of the mapping $w = z^2$ can be visualized by mapping the disk $|z| \leq 1$ onto the Riemann surface, then projecting the points of the Riemann surface vertically down onto the disk $|w| \leq 1$. In addition, by reversing the order in this procedure, we can also use this Riemann surface to help visualize the multiple-valued function $F(z) = z^{1/2}$.

Another interesting Riemann surface is one for the multiple-valued function $G(z) = \arg(z)$ defined on the punctured disk $0 < |z| \leq 1$. To construct this surface, we take a copy of the punctured disk $0 < |z| \leq 1$ and cut it open along the negative real axis. See Figure 2.4.18(a). Call this cut disk A_0 and let it represent the points in the domain written in exponential notation as $re^{i\theta}$ with $-\pi < \theta \leq \pi$. Take another copy of the cut disk, call it A_1, and let it represent the points in the domain written as $re^{i\theta}$ with $\pi < \theta \leq 3\pi$. Similarly, let A_{-1} be a cut disk that represents the points in the domain written as $re^{i\theta}$ with $-3\pi < \theta \leq -\pi$. Continue in this manner to produce an infinite set of cut disks $\ldots A_{-2}$, A_{-1}, A_0, A_1, A_2 $\ldots$. In general, the cut disk A_n represents points in the domain of G expressed as $re^{i\theta}$ with $(2n-1)\pi < \theta \leq (2n+1)\pi$. Now place each disk A_n in xyz-space so that the point $re^{i\theta}$ with $(2n-1)\pi < \theta \leq (2n+1)\pi$ lies at height θ directly above the point $re^{i\theta}$ in the xy-plane. The cut disk A_0 placed in xyz-space is shown in Figure 2.4.18(b). The collection of all the cut disks in xyz-space forms the Riemann surface for the multiple-valued function $G(z) = \arg(z)$ shown in Figure 2.4.19. The Riemann surface indicates how this multiple-valued function maps the punctured disk $0 < |z| \leq 1$ onto the real line. Namely, a vertical line passing through any point z in $0 < |z| \leq 1$ intersects this Riemann surface at infinitely many points. The heights of the points of intersection represent different choices for the argument of z. Therefore, by horizontally projecting the points of intersection onto the vertical axis, we see the infinitely many images of $G(z) = \arg(z)$.

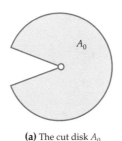

(a) The cut disk A_0

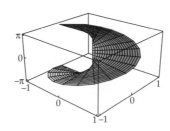

(b) A_0 placed in xyz-space

Figure 2.4.18 The cut disk A_0

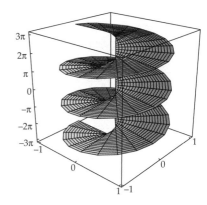

Figure 2.4.19 The Riemann surface for
$G(z) = \arg(z)$

EXERCISES 2.4   *Answers to selected odd-numbered problems begin on page ANS-8.*

2.4.1 The Power Function z^n

In Problems 1–14, find the image of the given set under the mapping $w = z^2$.
Represent the mapping by drawing the set and its image.

1. the ray $\arg(z) = \dfrac{\pi}{3}$

2. the ray $\arg(z) = -\dfrac{3\pi}{4}$

3. the line $x = 3$

4. the line $y = -5$

5. the line $y = -\dfrac{1}{4}$

6. the line $x = \dfrac{3}{2}$

7. the positive imaginary axis

8. the line $y = x$

9. the circular arc $|z| = \dfrac{1}{2}$, $0 \leq \arg(z) \leq \pi$

10. the circular arc $|z| = \dfrac{4}{3}$, $-\dfrac{\pi}{2} \leq \arg(z) \leq \dfrac{\pi}{6}$

11. the triangle with vertices 0, 1, and $1 + i$

12. the triangle with vertices 0, $1 + 2i$, and $-1 + 2i$

13. the square with vertices 0, 1, $1 + i$, and i

14. the polygon with vertices 0, 1, $1 + i$, and $-1 + i$

In Problems 15–20, find the image of the given set under the given composition of
a linear function with the squaring function.

15. the ray $\arg(z) = \dfrac{\pi}{3}$; $f(z) = 2z^2 + 1 - i$

16. the line segment from 0 to $-1 + i$; $f(z) = \sqrt{2}z^2 + 2 - i$

17. the line $x = 2$; $f(z) = iz^2 - 3$

18. the line $y = -3$; $f(z) = -z^2 + i$

19. the circular arc $|z| = 2$, $0 \leq \arg(z) \leq \dfrac{\pi}{2}$; $f(z) = \frac{1}{4}e^{i\pi/4}z^2$

20. the triangle with vertices 0, 1, and $1 + i$; $f(z) = -\frac{1}{4}iz^2 + 1$

21. Find the image of the ray $\arg(z) = \pi/6$ under each of the following mappings.

   (a) $f(z) = z^3$ **(b)** $f(z) = z^4$ **(c)** $f(z) = z^5$

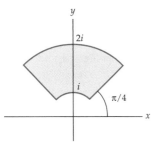

Figure 2.4.20 Figure for Problems 23 and 24

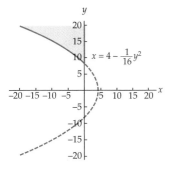

Figure 2.4.21 Figure for Problem 39

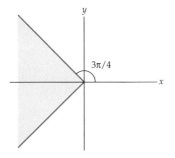

Figure 2.4.22 Figure for Problem 40

22. Find the image of the first quadrant of the complex plane under each of the following mappings.

   (a) $f(z) = z^2$ (b) $f(z) = z^3$ (c) $f(z) = z^4$

23. Find the image of the region $1 \leq |z| \leq 2$, $\pi/4 \leq \arg(z) \leq 3\pi/4$, shown in Figure 2.4.20 under each of the following mappings.

   (a) $f(z) = z^2$ (b) $f(z) = z^3$ (c) $f(z) = z^4$

24. Find the image of the region shown in Figure 2.4.20 under each of the following mappings.

   (a) $f(z) = 3z^2 + i$ (b) $f(z) = (i+1)z^3 + 1$ (c) $f(z) = \frac{1}{2}z^4 - i$

2.4.2 The Power Function $z^{1/n}$

In Problems 25–30, use (14) to find the value of the given principal nth root function at the given value of z.

25. $z^{1/2}$, $z = -i$

26. $z^{1/2}$, $z = 2 + i$

27. $z^{1/3}$, $z = -1$

28. $z^{1/3}$, $z = -3 + 3i$

29. $z^{1/4}$, $z = -1 + \sqrt{3}i$

30. $z^{1/5}$, $z = -4\sqrt{3} + 4i$

In Problems 31–38, find the image of the given set under the principal square root mapping $w = z^{1/2}$. Represent the mapping by drawing the set and its image.

31. the ray $\arg(z) = \dfrac{\pi}{4}$

32. the ray $\arg(z) = -\dfrac{2\pi}{3}$

33. the positive imaginary axis

34. the negative real axis

35. the arc $|z| = 9$, $-\dfrac{\pi}{2} \leq \arg(z) \leq \pi$

36. the arc $|z| = \dfrac{4}{7}$, $-\dfrac{\pi}{2} \leq \arg(z) \leq \dfrac{\pi}{4}$

37. the parabola $x = \dfrac{9}{4} - \dfrac{y^2}{9}$

38. the parabola $x = \dfrac{y^2}{10} - \dfrac{5}{2}$

39. Find the image of the region shown in Figure 2.4.21 under the principal square root function $w = z^{1/2}$.

40. Find the image of the region shown in Figure 2.4.22 under the principal square root function $w = z^{1/2}$. (Be careful near the negative real axis!)

Focus on Concepts

41. Use a procedure similar to that used in Example 2 to find the image of the hyperbola $xy = k$, $k \neq 0$, under $w = z^2$.

42. Use a procedure similar to that used in Example 2 to find the image of the hyperbola $x^2 - y^2 = k$, $k \neq 0$, under the mapping $w = z^2$.

43. Find two sets in the complex plane that are mapped onto the ray $\arg(w) = \pi/2$ by the function $w = z^2$.

44. Find two sets in the complex plane that are mapped onto the set bounded by the curves $u = -v$, $u = 1 - \frac{1}{4}v^2$, and the real axis $v = 0$ by the function $w = z^2$.

45. In Example 2 it was shown that the image of a vertical line $x = k$, $k \neq 0$, under $w = z^2$ is the parabola $u = k^2 - \dfrac{v^2}{4k^2}$. Use this result, your knowledge of linear mappings, and the fact that $w = -(iz)^2$ to prove that the image of a horizontal line $y = k$, $k \neq 0$, is the parabola $u = -\left(k^2 - \dfrac{v^2}{4k^2}\right)$.

46. Find three sets in the complex plane that map onto the set $\arg(w) = \pi$ under the mapping $w = z^3$.

47. Find four sets in the complex plane that map onto the circle $|w| = 4$ under the mapping $w = z^4$.

48. Find the image of the line $y = mx$ under the mapping $w = z^n$ for $n \geq 2$.

49. (a) Proceed as in Example 6 to show that the complex linear function $f(z) = az + b$, $a \neq 0$, is one-to-one on the entire complex plane.

    (b) Find a formula for the inverse function of the function in (a).

50. (a) Proceed as in Example 6 to show that the complex function $f(z) = \dfrac{a}{z} + b$, $a \neq 0$, is one-to-one on the set $|z| > 0$.

    (b) Find a formula for the inverse function of the function in (a).

51. Find the image of the half-plane $\text{Im}(z) \geq 0$ under each of the following principal nth root functions.

    (a) $f(z) = z^{1/2}$　　　(b) $f(z) = z^{1/3}$　　　(c) $f(z) = z^{1/4}$

52. Find the image of the region $|z| \leq 8$, $\pi/2 \leq \arg(z) \leq 3\pi/4$, under each of the following principal nth root functions.

    (a) $f(z) = z^{1/2}$　　　(b) $f(z) = z^{1/3}$　　　(c) $f(z) = z^{1/4}$

53. Find a function that maps the entire complex plane excluding 0 onto the set $2\pi/3 < \arg(w) \leq 4\pi/3$.

54. Read part (ii) of the Remarks, and then describe how to construct a Riemann surface for the function $f(z) = z^3$.

In Problems 55 and 56, (a) use mappings to determine upper and lower bounds on the modulus of the given function $f(z)$ defined on the given set S. That is, find real values L and M such that $L \leq |f(z)| \leq M$ for all z in S and (b) find complex values z_0 and z_1 in S such that $f(z_0) = L$ and $f(z_1) = M$.

55. $f(z) = 2iz^2 - i$; S is the quarter disk $|z| \leq 2$, $0 \leq \arg(z) \leq \pi/2$

56. $f(z) = \frac{1}{3}z^2 + 1 - i$; S is the set defined by $2 \leq |z| \leq 3$, $0 \leq \arg(z) \leq \pi$

2.5   Reciprocal Function

To The Instructor: In this section, we study the complex mapping $w = 1/z$. This section can be skipped without affecting the development of topics in Chapters 3–6. However, topics presented in this section will be used in the study of conformal mappings in Chapter 7.

In Sections 2.3 and 2.4 we examined some special types of complex polynomial functions as mappings of the complex plane. Analogous to real functions, we define a **complex rational function** to be a function of the form $f(z) = p(z)/q(z)$ where both $p(z)$ and $q(z)$ are complex polynomial functions. In this section, we study the most basic complex rational function, the reciprocal function $1/z$, as a mapping of the complex plane. An important property of the reciprocal mapping is that it maps certain lines onto circles.

Reciprocal Function   The function $1/z$, whose domain is the set of all nonzero complex numbers, is called the **reciprocal function**. To study the reciprocal function as a complex mapping $w = 1/z$, we begin by expressing this function in exponential notation. Given $z \neq 0$, if we set $z = re^{i\theta}$, then we obtain:

$$w = \frac{1}{z} = \frac{1}{re^{i\theta}} = \frac{1}{r}e^{-i\theta}. \tag{1}$$

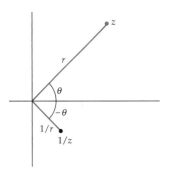

Figure 2.5.1 The reciprocal mapping

From (1), we see that the modulus of w is the reciprocal of the modulus of z and that the argument of w is the negative of the argument of z. Therefore, the reciprocal function maps a point in the z-plane with polar coordinates (r, θ) onto a point in the w-plane with polar coordinates $(1/r, -\theta)$. In Figure 2.5.1, we illustrate the relationship between z and $w = 1/z$ in a single copy of the complex plane. As we shall see, a simple way to visualize the reciprocal function as a complex mapping is as a composition of *inversion in the unit circle* followed by *reflection across the real axis*. We now proceed to define and analyze each of these mappings.

Inversion in the Unit Circle The function

$$g(z) = \frac{1}{r}e^{i\theta}, \tag{2}$$

whose domain is the set of all nonzero complex numbers, is called **inversion in the unit circle**. We will describe this mapping by considering separately the images of points *on* the unit circle, points *outside* the unit circle, and points *inside* the unit circle. Consider first a point z on the unit circle. Since $z = 1 \cdot e^{i\theta}$, it follows from (2) that $g(z) = \frac{1}{1}e^{i\theta} = z$. Therefore, each point on the unit circle is mapped onto itself by g. If, on the other hand, z is a nonzero complex number that does not lie on the unit circle, then we can write z as $z = re^{i\theta}$ with $r \neq 1$. When $r > 1$ (that is, when z is outside of the unit circle), we have that $|g(z)| = \left|\frac{1}{r}e^{i\theta}\right| = \frac{1}{r} < 1$. So, the image under g of a point z outside the unit circle is a point inside the unit circle. Conversely, if $r < 1$ (that is, if z is inside the unit circle), then $|g(z)| = \frac{1}{r} > 1$, and we conclude that if z is inside the unit circle, then its image under g is outside the unit circle. The mapping $w = e^{i\theta}/r$ is represented in Figure 2.5.2. The circle $|z| = 1$ shown in color in Figure 2.5.2(a) is mapped onto the circle $|w| = 1$ shown in black in Figure 2.5.2(b). In addition, $w = e^{i\theta}/r$ maps the region shown in light color in Figure 2.5.2(a) into the region shown in light gray in Figure 2.5.2(b), and it maps the region shown in dark color in Figure 2.5.2(a) into the region shown in dark gray in Figure 2.5.2(b).

We end our discussion of inversion in the unit circle by observing from (2) that the arguments of z and $g(z)$ are equal. It follows that if $z_1 \neq 0$ is a point with modulus r in the z-plane, then $g(z_1)$ is the unique point in the w-plane with modulus $1/r$ lying on a ray emanating from the origin making

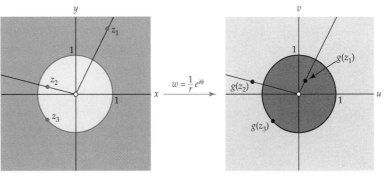

(a) Points z_1, z_2, and z_3 in the z-plane   **(b)** The images of z_1, z_2, and z_3 in the w-plane

Figure 2.5.2 Inversion in the unit circle

an angle of $\arg(z_0)$ with the positive u-axis. See Figure 2.5.2. In addition, since the moduli of z and $g(z)$ are inversely proportional, the farther a point z is from 0 in the z-plane, the closer its image $g(z)$ is to 0 in the w-plane, and, conversely, the closer z is to 0, the farther $g(z)$ is from 0.

Complex Conjugation   The second complex mapping that is helpful for describing the reciprocal mapping is a reflection across the real axis. Under this mapping the image of the point $(x,\ y)$ is $(x,\ -y)$. It is easy to verify that this complex mapping is given by the function $c(z) = \bar{z}$, which we call the **complex conjugation function**. In Figure 2.5.3, we illustrate the relationship between z and its image $c(z)$ in a single copy of the complex plane. By replacing the symbol z with $re^{i\theta}$ we can also express the complex conjugation function as $c(z) = \overline{re^{i\theta}} = \bar{r}\overline{e^{i\theta}}$. Because r is real, we have $\bar{r} = r$. Furthermore, from Problem 34 in Exercises 2.1, we have $\overline{e^{i\theta}} = e^{-i\theta}$. Therefore, the complex conjugation function can be written as $c(z) = \bar{z} = re^{-i\theta}$.

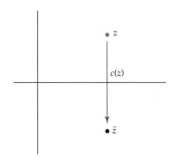

Figure 2.5.3 Complex conjugation

Reciprocal Mapping   The reciprocal function $f(z) = 1/z$ can be written as the composition of inversion in the unit circle and complex conjugation. Using the exponential forms $c(z) = re^{-i\theta}$ and $g(z) = e^{i\theta}/r$ of these functions we find that the composition $c \circ g$ is given by:

$$c(g(z)) = c\left(\frac{1}{r}e^{i\theta}\right) = \frac{1}{r}e^{-i\theta}.$$

By comparing this expression with (1), we see that $c(g(z)) = f(z) = 1/z$. This implies that, as a mapping, the reciprocal function first inverts in the unit circle, then reflects across the real axis.

> *Image of a Point under the Reciprocal Mapping*
>
> *Let z_0 be a nonzero point in the complex plane. If the point $w_0 = 1/z_0$ is plotted in the same copy of the complex plane as z_0, then w_0 is the point obtained by:*
>
> *(i) inverting z_0 in the unit circle, then*
>
> *(ii) reflecting the result across the real axis.*

EXAMPLE 1   Image of a Semicircle under $w = 1/z$

Find the image of the semicircle $|z| = 2$, $0 \leq \arg(z) \leq \pi$, under the reciprocal mapping $w = 1/z$.

Solution Let C denote the semicircle and let C' denote its image under $w = 1/z$. In order to find C', we first invert C in the unit circle, then we reflect the result across the real axis. Under inversion in the unit circle, points with modulus 2 have images with modulus $\frac{1}{2}$. Moreover, inversion in the unit circle does not change arguments. So, the image of the C under inversion in the unit circle is the semicircle $|w| = \frac{1}{2}$, $0 \leq \arg(w) \leq \pi$. Reflecting this set across the real axis negates the argument of a point but does not change its modulus. Hence, the image after reflection across the real axis is the semicircle given by $|w| = \frac{1}{2}$, $-\pi \leq \arg(w) \leq 0$. We represent this mapping in Figure

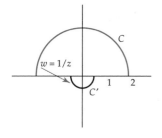

Figure 2.5.4 The reciprocal mapping

2.5.4 using a single copy of the complex plane. The semicircle C shown in color is mapped onto the semicircle C' shown in black in Figure 2.5.4 by $w = 1/z$. ❏

Using reasoning similar to that in Example 1 we can show that the reciprocal function maps the circle $|z| = k$, $k \neq 0$, onto the circle $|w| = 1/k$. As the next example illustrates, the reciprocal function also maps certain lines onto circles.

EXAMPLE 2   Image of a Line under $w = 1/z$

Find the image of the vertical line $x = 1$ under the reciprocal mapping $w = 1/z$.

Solution The vertical line $x = 1$ consists of the set of points $z = 1 + iy$, $-\infty < y < \infty$. After replacing the symbol z with $1 + iy$ in $w = 1/z$ and simplifying, we obtain:

$$w = \frac{1}{1 + iy} = \frac{1}{1 + y^2} - \frac{y}{1 + y^2}i.$$

It follows that the image of the vertical line $x = 1$ under $w = 1/z$ consists of all points $u + iv$ satisfying:

$$u = \frac{1}{1 + y^2}, \quad v = \frac{-y}{1 + y^2}, \quad \text{and} \quad -\infty < y < \infty. \tag{3}$$

We can describe this image with a single Cartesian equation by eliminating the variable y. Observe from (3) that $v = -yu$. The first equation in (3) implies that $u \neq 0$, and so we can rewrite this equation as $y = -v/u$. Now we substitute $y = -v/u$ into the first equation of (3) and simplify to obtain the quadratic equation $u^2 - u + v^2 = 0$. After completing the square in the variable u, we see that the image given in (3) is given by:

$$\left(u - \frac{1}{2}\right)^2 + v^2 = \frac{1}{4}, \ u \neq 0. \tag{4}$$

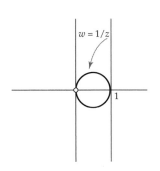

Figure 2.5.5 The reciprocal mapping

The equation in (4) defines a circle centered at $\left(\frac{1}{2}, 0\right)$ with radius $\frac{1}{2}$. However, because $u \neq 0$, the point $(0, 0)$ is *not* in the image. Using the complex variable $w = u + iv$, we can describe this image by $\left|w - \frac{1}{2}\right| = \frac{1}{2}$, $w \neq 0$. We represent this mapping using a single copy of the complex plane. In Figure 2.5.5, the line $x = 1$ shown in color is mapped onto the circle $\left|w - \frac{1}{2}\right| = \frac{1}{2}$ excluding the point $w = 0$ shown in black by $w = 1/z$. ❏

The solution in Example 2 is somewhat unsatisfactory since the image is not the entire circle $\left|w - \frac{1}{2}\right| = \frac{1}{2}$. This occurred because points on the line $x = 1$ with extremely large modulus map onto points on the circle $\left|w - \frac{1}{2}\right| = \frac{1}{2}$ that are extremely close to 0, but there is no point on the line $x = 1$ that actually maps onto 0. In order to obtain the entire circle as the image of the line we must consider the reciprocal function defined on the extended complex-number system.

In the Remarks in Section 1.5 we saw that the extended complex-number system consists of all the points in the complex plane adjoined with the ideal point ∞. In the context of mappings this set of points is commonly referred to as the **extended complex plane**. The important property of the extended complex plane for our discussion here is the correspondence, described in Section 1.5, between points on the extended complex plane and the points on the complex plane. In particular, points in the extended complex plane that are near the ideal point ∞ correspond to points with extremely large modulus in the complex plane.

We use this correspondence to extend the reciprocal function to a function whose domain and range are the extended complex plane. Since (1) already defines the reciprocal function for all points $z \neq 0$ or ∞ in the extended complex plane, we extend this function by specifying the images of 0 and ∞. A natural way to determine the image of these points is to consider the images of points nearby. Observe that if $z = re^{i\theta}$ is a point that is close to 0, then r is a small positive real number. It follows that $w = \dfrac{1}{z} = \dfrac{1}{r}e^{-i\theta}$ is a point whose modulus $1/r$ is large. That is, in the extended complex plane, if z is a point that is near 0, then $w = 1/z$ is a point that is near the ideal point ∞. It is therefore reasonable to define the reciprocal function $f(z) = 1/z$ on the extended complex plane so that $f(0) = \infty$. In a similar manner, we note that if z is a point that is near ∞ in the extended complex plane, then $f(z)$ is a point that is near 0. Thus, it is also reasonable to define the reciprocal function on the extended complex plane so that $f(\infty) = 0$.

Definition 2.5.1   The Reciprocal Function on the Extended Complex Plane

The **reciprocal function on the extended complex plane** is the function defined by:

$$f(z) = \begin{cases} 1/z, & \text{if } z \neq 0 \text{ or } \infty \\ \infty, & \text{if } z = 0 \\ 0, & \text{if } z = \infty. \end{cases}$$

Rather than introduce new notation, we shall let the notation $1/z$ represent both the reciprocal function and the reciprocal function on the extended complex plane. Whenever the ideal point ∞ is mentioned, you should assume that $1/z$ represents the reciprocal function defined on the extended complex plane.

EXAMPLE 3   Image of a Line under $w = 1/z$

Find the image of the vertical line $x = 1$ under the reciprocal function on the extended complex plane.

Solution We begin by noting that since the line $x = 1$ is an unbounded set in the complex plane, it follows that the ideal point ∞ is on the line in the extended complex plane. In Example 2 we found that the image of the points $z \neq \infty$ on the line $x = 1$ is the circle $\left| w - \frac{1}{2} \right| = \frac{1}{2}$ excluding the point $w = 0$. Thus, we need only find the image of the ideal point to determine the image of the line under the reciprocal function on the extended complex plane. From Definition 2.5.1 we have that $f(\infty) = 0$, and so $w = 0$ is the image of the ideal point. This "fills in" the missing point in the circle $\left| w - \frac{1}{2} \right| = \frac{1}{2}$. Therefore, the vertical line $x = 1$ is mapped onto the entire circle $\left|w - \frac{1}{2}\right| = \frac{1}{2}$ by the reciprocal mapping on the extended complex plane. This mapping can be represented by Figure 2.5.5 with the "hole" at $w = 0$ filled in. ❏

Because the ideal point ∞ is on every vertical line in the extended complex plane, we have that the image of any vertical line $x = k$ with $k \neq 0$ is the entire circle $\left|w - \dfrac{1}{2k}\right| = \left|\dfrac{1}{2k}\right|$ under the reciprocal function on the extended complex plane. See Problem 23 in Exercises 2.5. In a similar manner, we can also show that horizontal lines are mapped to circles by $w = 1/z$. We now summarize these mapping properties of $w = 1/z$.

Mapping Lines to Circles with $w = 1/z$

The reciprocal function on the extended complex plane maps:

(i) the vertical line $x = k$ with $k \neq 0$ onto the circle

$$\left|w - \frac{1}{2k}\right| = \left|\frac{1}{2k}\right|, \; and \tag{5}$$

(ii) the horizontal line $y = k$ with $k \neq 0$ onto the circle

$$\left|w + \frac{1}{2k}i\right| = \left|\frac{1}{2k}\right|. \tag{6}$$

These two mapping properties of the reciprocal function are illustrated in Figure 2.5.6. The vertical lines $x = k$, $k \neq 0$, shown in color in Figure 2.5.6(a) are mapped by $w = 1/z$ onto the circles centered on the real axis shown in black in Figure 2.5.6(b). The image of the line $x = k$, $k \neq 0$, contains the point $(1/k, 0)$. Thus, we see that the vertical line $x = 2$ shown in Figure 2.5.6(a) maps onto the circle centered on the real axis containing $\left(\frac{1}{2}, 0\right)$ shown in Figure 2.5.6(b), and so on. Similarly, the horizontal lines $y = k$, $k \neq 0$, shown in color in Figure 2.5.6(a) are mapped by $w = 1/z$ onto the circles centered on the imaginary axis shown in black in Figure 2.5.6(b). Since the image of the line $y = k$, $k \neq 0$, contains the point $(0, -1/k)$, we have that the line $y = 2$ shown in Figure 2.5.6(a) is the circle centered on the

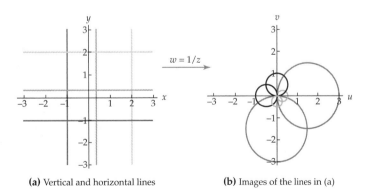

(a) Vertical and horizontal lines   (b) Images of the lines in (a)

Figure 2.5.6 Images of vertical and horizontal lines under the reciprocal mapping

imaginary axis containing the point $\left(0, -\frac{1}{2}\right)$ shown in Figure 2.5.6(b), and so on.

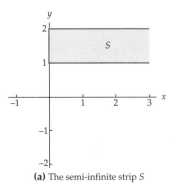

(a) The semi-infinite strip S

$w = 1/z$

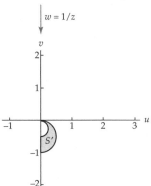

(b) The image of S

Figure 2.5.7 The reciprocal mapping

EXAMPLE 4   Mapping of a Semi-infinite Strip

Find the image of the semi-infinite horizontal strip defined by $1 \leq y \leq 2$, $x \geq 0$, under $w = 1/z$.

Solution Let S denote the semi-infinite horizontal strip defined by $1 \leq y \leq 2$, $x \geq 0$. The boundary of S consists of the line segment $x = 0$, $1 \leq y \leq 2$, and the two half-lines $y = 1$ and $y = 2$, $0 \leq x < \infty$. We first determine the images of these boundary curves. The line segment $x = 0$, $1 \leq y \leq 2$, can also be described as the set $1 \leq |z| \leq 2$, $\arg(z) = \pi/2$. Since $w = 1/z$, it follows that $\frac{1}{2} \leq |w| \leq 1$. In addition, from (1) we have that $\arg(w) = \arg(1/z) = -\arg(z)$, and so, $\arg(w) = -\pi/2$. Thus, the image of the line segment $x = 0$, $1 \leq y \leq 2$, is the line segment on the v-axis from $-\frac{1}{2}i$ to $-i$. We now consider the horizontal half-line $y = 1$, $0 \leq x < \infty$. By identifying $k = 1$ in (6), we see that the image of this half-line is an arc in the circle $\left|w + \frac{1}{2}i\right| = \frac{1}{2}$. Because the arguments of the points on the half-line satisfy $0 < \arg(z) \leq \pi/2$, it follows that the arguments of the points in its image satisfy $-\pi/2 \leq \arg(w) < 0$. Moreover, the ideal point ∞ is on the half-line, and so the point $w = 0$ is in its image. Thus, we see that the image of the half-line $y = 1$, $0 \leq x < \infty$, is the circular arc defined by $\left|w + \frac{1}{2}i\right| = \frac{1}{2}$, $-\pi/2 \leq \arg(w) \leq 0$. In a similar manner, we find the image of the horizontal half-line $y = 2$, $0 \leq x < \infty$, is the circular arc $\left|w + \frac{1}{4}i\right| = \frac{1}{4}$, $-\pi/2 \leq \arg(w) \leq 0$. We conclude by observing that, from (6), every half-line $y = k$, $1 \leq k \leq 2$, lying between the boundary half-lines $y = 1$ and $y = 2$ in the strip S maps onto a circular arc $\left|w + \frac{1}{2k}i\right| = \frac{1}{2k}$, $-\pi/2 \leq \arg(w) \leq 0$, lying between the circular arcs $\left|w + \frac{1}{2}i\right| = \frac{1}{2}$ and $\left|w + \frac{1}{4}i\right| = \frac{1}{4}$, $-\pi/2 \leq \arg(w) \leq 0$. Therefore, the semi-infinite strip S shown in color in Figure 2.5.7(a) is mapped on to the set S' shown in gray in Figure 2.5.7(b) by the complex mapping $w = 1/z$. ❑

Remarks

It is easy to verify that the reciprocal function $f(z) = 1/z$ is one-to-one. Therefore, f has a well-defined inverse function f^{-1}. We find a formula for the inverse function $f^{-1}(z)$ by solving the equation $z = f(w)$ for w. Clearly, this gives $f^{-1}(z) = 1/z$. This observation extends our understanding of the complex mapping $w = 1/z$. For example, we have seen that the image of the line $x = 1$ under the reciprocal mapping is the circle $\left|w - \frac{1}{2}\right| = \frac{1}{2}$. Since $f^{-1}(z) = 1/z = f(z)$, it then follows that the image of the circle $\left|z - \frac{1}{2}\right| = \frac{1}{2}$ under the reciprocal mapping is the line $u = 1$. In a similar manner, we see that the circles $\left|z - \frac{1}{2k}\right| = \left|\frac{1}{2k}\right|$ and $\left|z + \frac{1}{2k}i\right| = \left|\frac{1}{2k}\right|$ are mapped onto the lines $x = k$ and $y = k$, respectively.

EXERCISES 2.5   *Answers to selected odd-numbered problems begin on page ANS-8.*

In Problems 1–10, find the image of the given set under the reciprocal mapping $w = 1/z$ on the extended complex plane.

1. the circle $|z| = 5$

2. the semicircle $|z| = \frac{1}{2}$, $\pi/2 \leq \arg(z) \leq 3\pi/2$

3. the semicircle $|z| = 3$, $-\pi/4 \leq \arg(z) \leq 3\pi/4$

4. the quarter circle $|z| = \frac{1}{4}$, $\pi/2 \leq \arg(z) \leq \pi$

5. the annulus $\frac{1}{3} \leq |z| \leq 2$

6. the region $1 \leq |z| \leq 4$, $0 \leq \arg(z) \leq 2\pi/3$

7. the ray $\arg(z) = \pi/4$

8. the line segment from -1 to 1 on the real axis excluding the point $z = 0$

9. the line $y = 4$

10. the line $x = \frac{1}{6}$

In Problems 11–14, use the Remarks at the end of Section 2.5 to find the image of the given set under the reciprocal mapping $w = 1/z$ on the extended complex plane.

11. the circle $|z + i| = 1$ **12.** the circle $|z + \frac{1}{3}i| = \frac{1}{3}$

13. the circle $|z - 2| = 2$ **14.** the circle $|z + \frac{1}{4}| = \frac{1}{4}$

In Problems 15–18, find the image of the given set S under the mapping $w = 1/z$ on the extended complex plane.

15.

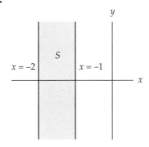

16.

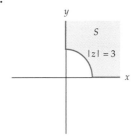

Figure 2.5.8 Figure for Problem 15   **Figure 2.5.9** Figure for Problem 16

17.

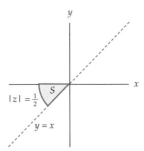

18.

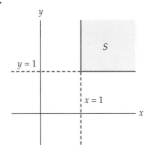

Figure 2.5.10 Figure for Problem 17   **Figure 2.5.11** Figure for Problem 18

19. Consider the function $h(z) = \dfrac{2i}{z} + 1$ defined on the extended complex plane.

    (a) Using the fact that h is a composition of the reciprocal function $f(z) = 1/z$ and the linear function $g(z) = 2iz + 1$, that is, $h(z) = g(f(z))$, describe in words the action of the mapping $w = h(z)$.

    (b) Determine the image of the line $x = 4$ under $w = h(z)$.

    (c) Determine the image of the circle $|z + 2| = 2$ under $w = h(z)$.

20. Consider the function $h(z) = \dfrac{1}{2iz - 1}$ defined on the extended complex plane.

  (a) Using the fact that h is a composition of the linear function $g(z) = 2iz - 1$ and the reciprocal function $f(z) = 1/z$, that is, $h(z) = f(g(z))$, describe in words the action of the mapping $w = h(z)$.

  (b) Determine the image of the line $y = 1$ under $w = h(z)$.

  (c) Determine the image of the circle $|z + i| = \frac{1}{2}$ under $w = h(z)$.

21. Consider the function $h(z) = 1/z^2$ defined on the extended complex plane.

  (a) Write h as a composition of the reciprocal function and the squaring function.

  (b) Determine the image of the circle $\left|z + \frac{1}{2}i\right| = \frac{1}{2}$ under the mapping $w = h(z)$.

  (c) Determine the image of the circle $|z - 1| = 1$ under the mapping $w = h(z)$.

22. Consider the mapping $h(z) = \dfrac{3i}{z^2} + 1 + i$ defined on the extended complex plane.

  (a) Write h as a composition of a linear, the reciprocal, and the squaring function.

  (b) Determine the image of the circle $\left|z + \frac{1}{2}i\right| = \frac{1}{2}$ under the mapping $w = h(z)$.

  (c) Determine the image of the circle $|z - 1| = 1$ under the mapping $w = h(z)$.

Focus on Concepts

23. Show that the image of the line $x = k$, $x \neq 0$, under the reciprocal map defined on the extended complex plane is the circle $\left|w - \dfrac{1}{2k}\right| = \left|\dfrac{1}{2k}\right|$.

24. According to the Remarks in Section 2.5, since $f(z) = 1/z$ is its own inverse function, the mapping $w = 1/z$ on the extended complex plane maps the circle $\left|z - \frac{1}{2}\right| = \frac{1}{2}$ to the line $\operatorname{Re}(w) = 1$. Verify this fact directly using the real and imaginary parts of f as in Example 1.

25. If A, B, C, and D are real numbers, then the set of points in the plane satisfying the equation:

$$A\left(x^2 + y^2\right) + Bx + Cy + D = 0 \tag{7}$$

is called a **generalized circle**.

  (a) Show that if $A = 0$, then the generalized circle is a line.

  (b) Suppose that $A \neq 0$ and let $\Delta = B^2 + C^2 - 4AD$. Complete the square in x and y to show that a generalized circle is a circle centered at $\left(\dfrac{-B}{2A}, \dfrac{-C}{2A}\right)$ with radius $\dfrac{\sqrt{\Delta}}{2A}$ provided $\Delta > 0$. (If $\Delta < 0$, the generalized circle is often called an *imaginary circle*.)

26. In this problem we will show that the image of a generalized circle (7) under the reciprocal mapping $w = 1/z$ is a generalized circle.

  (a) Rewrite (7) in polar coordinates using the equations $x = r\cos\theta$ and $y = r\sin\theta$.

  (b) Show that, in polar form, the reciprocal function $w = 1/z$ is given by:

$$w = \frac{1}{r}\left(\cos\theta - i\sin\theta\right).$$

(c) Let $w = u + iv$. Note that, from part (b), $u = \dfrac{1}{r}\cos\theta$ and $v = -\dfrac{1}{r}\sin\theta$. Now rewrite the equation from part (a) in terms of u and v using these equations.

(d) Conclude from parts (a)–(c) that the image of the generalized circle (7) under $w = 1/z$ is the generalized circle given by:

$$D\left(u^2 + v^2\right) + Bu - Cv + A = 0. \qquad (8)$$

27. Consider the line L given by the equation $Bx + Cy + D = 0$.

(a) Use Problems 25 and 26 to determine when the image of the line L under the reciprocal mapping $w = 1/z$ is a line.

(b) If the image of L is a line L', then what is the slope of L'? How does this slope compare with the slope of L?

(c) Use Problems 25 and 26 to determine when the image of the line L is a circle.

(d) If the image of L is a circle S', then what are the center and radius of S'?

28. Consider the circle S given by the equation $A\left(x^2 + y^2\right) + Bx + Cy + D = 0$ with $B^2 + C^2 - 4AD > 0$.

(a) Use Problems 25 and 26 to determine when the image of the circle S is a line.

(b) Use Problems 25 and 26 to determine when the image of the circle S is a circle.

(c) If the image of S is a circle S', then what are center and radius of S'? How do these values compare with the center and radius of S?

In Problems 29 and 30, (a) use mappings to determine upper and lower bounds on the modulus of the given function $f(z)$ defined on the given set S. That is, find real values L and M such that $L \le |f(z)| \le M$ for all z in S and (b) find complex values z_0 and z_1 in S such that $f(z_0) = L$ and $f(z_1) = M$.

29. $f(z) = \dfrac{1+i}{z} + 2$; S is the annulus $1 \le |z| \le 2$

30. $f(z) = \dfrac{1}{z} + i$; S is the half-plane $x \ge 2$

2.6   Limits and Continuity

The most important concept in elementary calculus is that of the limit. Recall that $\lim_{x \to x_0} f(x) = L$ intuitively means that values $f(x)$ of the function f can be made arbitrarily close to the real number L if values of x are chosen sufficiently close to, but not equal to, the real number x_0. In real analysis, the concepts of continuity, the derivative, and the definite integral were *all* defined using the concept of a limit. Complex limits play an equally important role in study of complex analysis. The concept of a complex limit is similar to that of a real limit in the sense that $\lim_{z \to z_0} f(z) = L$ will mean that the values $f(z)$ of the complex function f can be made arbitrarily close the complex number L if values of z are chosen sufficiently close to, but not equal to, the complex number z_0. Although outwardly similar, there is an important difference between these two concepts of limit. In a real limit, there are two directions from which x can approach x_0 on the real line, namely, from the left or from the right. In a complex limit, however, there are infinitely many directions from which z can approach z_0 in the complex plane. In order for a complex limit to exist, each way in which z can approach z_0 must yield the same limiting value.

In this section we will define the limit of a complex function, examine some of its properties, and introduce the concept of continuity for functions of a complex variable.

2.6.1 Limits

Real Limits The description of a real limit given in the section introduction is only an intuitive definition of this concept. In order to give the rigorous definition of a real limit, we must precisely state what is meant by the phrases "arbitrarily close to" and "sufficiently close to." The first thing to recognize is that a precise statement of these terms uses absolute values since $|a - b|$ measures the distance between two points on the real number line. On the real line, the points x and x_0 are "close" if $|x - x_0|$ is a small positive number. Similarly, the points $f(x)$ and L are "close" if $|f(x) - L|$ is a small positive number. In mathematics, it is customary to let the Greek letters ε and δ represent small positive real numbers. Hence, the expression "$f(x)$ can be made arbitrarily close to L" can be made precise by stating that for any real number $\varepsilon > 0$, x can be chosen so that $|f(x) - L| < \varepsilon$. In our intuitive definition we require that $|f(x) - L| < \varepsilon$ whenever values of x are "sufficiently close to, but not equal to, x_0." This means that there is some distance $\delta > 0$ with the property that if x is within distance δ of x_0 and $x \neq x_0$, then $|f(x) - L| < \varepsilon$. In other words, if $0 < |x - x_0| < \delta$, then $|f(x) - L| < \varepsilon$. The real number δ is not unique and, in general, depends on the choice of ε, the function f, and the point x_0. In summary, we have the following precise definition of the real limit:

Limit of a Real Function $f(x)$

The limit of f as x tends x_0 exists and is equal to L

if for every $\varepsilon > 0$ there exists a $\delta > 0$ such that $|f(x) - L| < \varepsilon$ (1)

whenever $0 < |x - x_0| < \delta$.

The geometric interpretation of (1) is shown in Figure 2.6.1. In this figure we see that the graph of the function $y = f(x)$ over the interval $(x_0 - \delta, x_0 + \delta)$, excluding the point x_0, lies between the lines $y = L - \varepsilon$ and $y = L + \varepsilon$ shown dashed in Figure 2.6.1. In the terminology of mappings, the interval $(x_0 - \delta, x_0 + \delta)$, excluding the point $x = x_0$, shown in color on the x-axis is mapped onto the set shown in black in the interval $(L - \varepsilon, L + \varepsilon)$ on the y-axis. For the limit to exist, the relationship exhibited in Figure 2.6.1 must exist for any choice of $\varepsilon > 0$. We also see in Figure 2.6.1 that if a smaller ε is chosen, then a smaller δ may be needed.

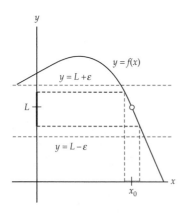

Figure 2.6.1 Geometric meaning of a real limit

Complex Limits A complex limit is, in essence, the same as a real limit except that it is based on a notion of "close" in the complex plane. Because the distance in the complex plane between two points z_1 and z_2 is given by the modulus of the difference of z_1 and z_2, the precise definition of a complex limit will involve $|z_2 - z_1|$. For example, the phrase "$f(z)$ can be made arbitrarily close to the complex number L," can be stated precisely as: for every $\varepsilon > 0$, z can be chosen so that $|f(z) - L| < \varepsilon$. Since the modulus of a complex number is a *real* number, both ε and δ still represent small positive *real* numbers in the following definition of a complex limit. The complex analogue of (1) is:

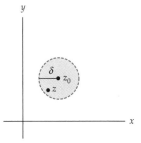

(a) Deleted δ-neighborhood of z_0

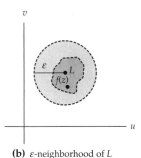

(b) ε-neighborhood of L

Figure 2.6.2 The geometric meaning of a complex limit

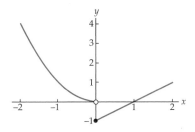

Figure 2.6.3 The limit of f does not exist as x approaches 0

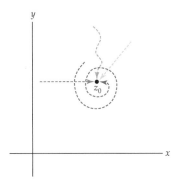

Figure 2.6.4 Different ways to approach z_0 in a limit

> **Definition 2.6.1   Limit of a Complex Function**
>
> Suppose that a complex function f is defined in a deleted neighborhood of z_0 and suppose that L is a complex number. The **limit of f as z tends to z_0 exists and is equal to L**, written as $\lim_{z \to z_0} f(z) = L$, if for every $\varepsilon > 0$ there exists a $\delta > 0$ such that $|f(z) - L| < \varepsilon$ whenever $0 < |z - z_0| < \delta$.

For complex functions, we rely on the concept of complex mappings to gain a geometric understanding of Definition 2.6.1. Recall from Section 1.5 that the set of points w in the complex plane satisfying $|w - L| < \varepsilon$ is called a neighborhood of L, and that this set consists of all points in the complex plane lying within, but not on, a circle of radius ε centered at the point L. Also recall from Section 1.5 that the set of points satisfying the inequalities $0 < |z - z_0| < \delta$ is called a deleted neighborhood of z_0 and consists of all points in the neighborhood $|z - z_0| < \delta$ excluding the point z_0. By Definition 2.6.1, if $\lim_{z \to z_0} f(z) = L$ and if ε is any positive number, then there is a deleted neighborhood of z_0 of radius δ with the property that for every point z in this deleted neighborhood, $f(z)$ is in the ε neighborhood of L. That is, f maps the deleted neighborhood $0 < |z - z_0| < \delta$ in the z-plane into the neighborhood $|w - L| < \varepsilon$ in the w-plane. In Figure 2.6.2(a), the deleted neighborhood of z_0 shown in color is mapped onto the set shown in dark gray in Figure 2.6.2(b). As required by Definition 2.6.1, the image lies within the ε-neighborhood of L shown in light gray in Figure 2.6.2(b).

Complex and real limits have many common properties, but there is at least one very important difference. For real functions, $\lim_{x \to x_0} f(x) = L$ if and only if $\lim_{x \to x_0^+} f(x) = L$ and $\lim_{x \to x_0^-} f(x) = L$. That is, there are two directions from which x can approach x_0 on the real line, from the right (denoted by $x \to x_0^+$) or from the left (denoted by $x \to x_0^-$). The real limit exists if and only if these two one-sided limits have the same value. For example, consider the real function defined by:

$$f(x) = \begin{cases} x^2, & x < 0 \\ x - 1, & x \geq 0 \end{cases}.$$

The limit of f as x approaches to 0 does not exist since $\lim_{x \to 0^-} f(x) = \lim_{x \to 0^-} x^2 = 0$, but $\lim_{x \to 0^+} f(x) = \lim_{x \to 0^+} (x - 1) = -1$. See Figure 2.6.3.

For limits of complex functions, z is allowed to approach z_0 from *any* direction in the complex plane, that is, along any curve or path through z_0. See Figure 2.6.4. In order that $\lim_{z \to z_0} f(z)$ exists and equals L, we require that $f(z)$ approach the same complex number L along every possible curve through z_0. Put in a negative way:

> ### *Criterion for the Nonexistence of a Limit*
>
> *If f approaches two complex numbers $L_1 \neq L_2$ along two different curves or paths through z_0, then $\lim_{z \to z_0} f(z)$ does not exist.*

EXAMPLE 1   A Limit That Does Not Exist

Show that $\lim\limits_{z \to 0} \dfrac{z}{\bar{z}}$ does not exist.

Solution We show that this limit does not exist by finding two different ways of letting z approach 0 that yield different values for $\lim\limits_{z \to 0} \dfrac{z}{\bar{z}}$. First, we let z approach 0 along the real axis. That is, we consider complex numbers of the form $z = x + 0i$ where the real number x is approaching 0. For these points we have:

$$\lim_{z \to 0} \frac{z}{\bar{z}} = \lim_{x \to 0} \frac{x + 0i}{x - 0i} = \lim_{x \to 0} 1 = 1. \tag{2}$$

On the other hand, if we let z approach 0 along the imaginary axis, then $z = 0 + iy$ where the real number y is approaching 0. For this approach we have:

$$\lim_{z \to 0} \frac{z}{\bar{z}} = \lim_{y \to 0} \frac{0 + iy}{0 - iy} = \lim_{y \to 0} (-1) = -1. \tag{3}$$

Since the values in (2) and (3) are not the same, we conclude that $\lim\limits_{z \to 0} \dfrac{z}{\bar{z}}$ does not exist. ❏

The limit $\lim\limits_{z \to 0} \dfrac{z}{\bar{z}}$ from Example 1 did not exist because the values of $\lim\limits_{z \to 0} \dfrac{z}{\bar{z}}$ as z approached 0 along the real and imaginary axes did not agree. However, even if these two values did agree, the complex limit may still fail to exist. See Problems 19 and 20 in Exercises 2.6. In general, computing values of $\lim\limits_{z \to z_0} f(z)$ as z approaches z_0 from different directions, as in Example 1, can prove that a limit *does not exist*, but this technique cannot be used to prove that a limit *does exist*. In order to prove that a limit does exist we must use Definition 2.6.1 directly. This requires demonstrating that for every positive real number ε there is an appropriate choice of δ that meets the requirements of Definition 2.6.1. Such proofs are commonly called "epsilon-delta proofs." Even for relatively simple functions, epsilon-delta proofs can be quite involved. Since this is an introductory text, we restrict our attention to what, in our opinion, are straightforward examples of epsilon-delta proofs.

EXAMPLE 2   An Epsilon-Delta Proof of a Limit

Prove that $\lim\limits_{z \to 1+i} (2 + i)z = 1 + 3i$.

Solution According to Definition 2.6.1, $\lim\limits_{z \to 1+i} (2 + i)z = 1 + 3i$, if, for every $\varepsilon > 0$, there is a $\delta > 0$ such that $|(2+i)z - (1+3i)| < \varepsilon$ whenever $0 < |z - (1+i)| < \delta$. Proving that the limit exists requires that we find an appropriate value of δ for a given value of ε. In other words, for a given value of ε we must find a positive number δ with the property that if $0 < |z - (1+i)| < \delta$, then $|(2+i)z - (1+3i)| < \varepsilon$. One way of finding δ is to "work backward." The idea is to start with the inequality:

$$|(2+i)z - (1+3i)| < \varepsilon \tag{4}$$

and then use properties of complex numbers and the modulus to manipulate this inequality until it involves the expression $|z - (1 + i)|$. Thus, a natural first step is to factor $(2 + i)$ out of the left-hand side of (4):

$$|2 + i| \cdot \left| z - \frac{1 + 3i}{2 + i} \right| < \varepsilon. \tag{5}$$

Because $|2 + i| = \sqrt{5}$ and $\dfrac{1 + 3i}{2 + i} = 1 + i$, (5) is equivalent to:

$$\sqrt{5} \cdot |z - (1 + i)| < \varepsilon \quad \text{or} \quad |z - (1 + i)| < \frac{\varepsilon}{\sqrt{5}}. \tag{6}$$

Thus, (6) indicates that we should take $\delta = \varepsilon/\sqrt{5}$. Keep in mind that the choice of δ is not unique. Our choice of $\delta = \varepsilon/\sqrt{5}$ is a result of the particular algebraic manipulations that we employed to obtain (6). Having found δ we now present the formal proof that $\lim\limits_{z \to 1+i} (2+i)z = 1+3i$ that does not indicate how the choice of δ was made:

Given $\varepsilon > 0$, let $\delta = \varepsilon/\sqrt{5}$. If $0 < |z - (1 + i)| < \delta$, then we have $|z - (1 + i)| < \varepsilon/\sqrt{5}$. Multiplying both sides of the last inequality by $|2 + i| = \sqrt{5}$ we obtain:

$$|2 + i| \cdot |z - (1 + i)| < \sqrt{5} \cdot \frac{\varepsilon}{\sqrt{5}} \quad \text{or} \quad |(2 + i)z - (1 + 3i)| < \varepsilon.$$

Therefore, $|(2 + i)z - (1 + 3i)| < \varepsilon$ whenever $0 < |z - (1 + i)| < \delta$. So, according to Definition 2.6.1, we have proven that $\lim\limits_{z \to 1+i} (2 + i)z = 1 + 3i.$ ❑

Real Multivariable Limits The epsilon-delta proof from Example 2 illustrates the important fact that although the theory of complex limits is based on Definition 2.6.1, this definition does not provide a convenient method for *computing* limits. We now present a practical method for *computing* complex limits in Theorem 2.6.1. In addition to being a useful computational tool, this theorem also establishes an important connection between the *complex* limit of $f(z) = u(x, y) + iv(x, y)$ and the *real* limits of the real-valued functions of two real variables $u(x, y)$ and $v(x, y)$. Since every complex function is completely determined by the real functions u and v, it should not be surprising that the limit of a complex function can be expressed in terms of the real limits of u and v.

Before stating Theorem 2.6.1, we recall some of the important concepts regarding limits of real-valued functions of two real variables $F(x, y)$. The following definition of $\lim\limits_{(x,y) \to (x_0, y_0)} F(x, y) = L$ is analogous to both (1) and Definition 2.6.1.

Limit of the Real Function $F(x, y)$

The limit of F as (x, y) tends to (x_0, y_0) exists and is equal to the real number L if for every $\varepsilon > 0$ there exists a $\delta > 0$ such that $\quad$ (7) *$|F(x, y) - L| < \varepsilon$ whenever $0 < \sqrt{(x - x_0)^2 + (y - y_0)^2} < \delta$.*

The expression $\sqrt{(x - x_0)^2 + (y - y_0)^2}$ in (7) represents the distance between the points (x, y) and (x_0, y_0) in the Cartesian plane. Using (7), it is relatively easy to prove that:

$$\lim_{(x,y)\to(x_0,y_0)} 1 = 1, \quad \lim_{(x,y)\to(x_0,y_0)} x = x_0, \text{ and } \lim_{(x,y)\to(x_0,y_0)} y = y_0. \quad (8)$$

If $\lim_{(x,y)\to(x_0,y_0)} F(x,y) = L$ and $\lim_{(x,y)\to(x_0,y_0)} G(x,y) = M$, then (7) can also be used to show:

$$\lim_{(x,y)\to(x_0,y_0)} cF(x, y) = cL, \quad c \text{ a real constant}, \quad (9)$$

$$\lim_{(x,y)\to(x_0,y_0)} (F(x, y) \pm G(x, y)) = L \pm M, \quad (10)$$

$$\lim_{(x,y)\to(x_0,y_0)} F(x, y) \cdot G(x, y) = L \cdot M, \quad (11)$$

and
$$\lim_{(x,y)\to(x_0,y_0)} \frac{F(x, y)}{G(x, y)} = \frac{L}{M}, \; M \neq 0. \quad (12)$$

Limits involving polynomial expressions in x and y can be easily computed using the limits in (8) combined with properties (9)–(12). For example,

$$\lim_{(x,y)\to(1,2)} \left(3xy^2 - y\right) = 3 \left(\lim_{(x,y)\to(1,2)} x\right)\left(\lim_{(x,y)\to(1,2)} y\right)\left(\lim_{(x,y)\to(1,2)} y\right) - \lim_{(x,y)\to(1,2)} y$$
$$= 3 \cdot 1 \cdot 2 \cdot 2 - 2 = 10.$$

In general, if $p(x, y)$ is a two-variable polynomial function, then (8)–(12) can be used to show that

$$\lim_{(x,y)\to(x_0,y_0)} p(x, y) = p(x_0, y_0). \quad (13)$$

If $p(x, y)$ and $q(x, y)$ are two-variable polynomial functions and $q(x_0, y_0) \neq 0$, then (13) and (12) give:

$$\lim_{(x,y)\to(x_0,y_0)} \frac{p(x,y)}{q(x,y)} = \frac{p(x_0, y_0)}{q(x_0, y_0)}. \quad (14)$$

We now present Theorem 2.6.1, which relates real limits of $u(x, y)$ and $v(x, y)$ with the complex limit of $f(z) = u(x, y) + iv(x, y)$. An epsilon-delta proof of Theorem 2.6.1 can be found in Appendix I.

Theorem 2.6.1   Real and Imaginary Parts of a Limit

Suppose that $f(z) = u(x, y) + iv(x, y)$, $z_0 = x_0 + iy_0$, and $L = u_0 + iv_0$. Then $\lim_{z\to z_0} f(z) = L$ if and only if

$$\lim_{(x,y)\to(x_0,y_0)} u(x, y) = u_0 \quad \text{and} \quad \lim_{(x,y)\to(x_0,y_0)} v(x, y) = v_0.$$

Theorem 2.6.1 has many uses. First and foremost, it allows us to compute many complex limits by simply computing a pair of real limits.

EXAMPLE 3   Using Theorem 2.6.1 to Compute a Limit

Use Theorem 2.6.1 to compute $\lim\limits_{z \to 1+i} \left(z^2 + i\right)$.

Solution Since $f(z) = z^2 + i = x^2 - y^2 + (2xy + 1)\,i$, we can apply Theorem 2.6.1 with $u(x,\ y) = x^2 - y^2$, $v(x,\ y) = 2xy + 1$, and $z_0 = 1 + i$. Identifying $x_0 = 1$ and $y_0 = 1$, we find u_0 and v_0 by computing the two real limits:

$$u_0 = \lim_{(x,y) \to (1,1)} \left(x^2 - y^2\right) \quad \text{and} \quad v_0 = \lim_{(x,y) \to (1,1)} (2xy + 1).$$

Since both of these limits involve only multivariable polynomial functions, we can use (13) to obtain:

$$u_0 = \lim_{(x,y) \to (1,1)} \left(x^2 - y^2\right) = 1^2 - 1^2 = 0$$

and
$$v_0 = \lim_{(x,y) \to (1,1)} (2xy + 1) = 2 \cdot 1 \cdot 1 + 1 = 3,$$

and so $L = u_0 + iv_0 = 0 + i(3) = 3i$. Therefore, $\lim\limits_{z \to 1+i} \left(z^2 + i\right) = 3i$. ❑

In addition to computing specific limits, Theorem 2.6.1 is also an important theoretical tool that allows us to derive many properties of complex limits from properties of real limits. The following theorem gives an example of this procedure.

Theorem 2.6.2   Properties of Complex Limits

Suppose that f and g are complex functions. If $\lim\limits_{z \to z_0} f(z) = L$ and $\lim\limits_{z \to z_0} g(z) = M$, then

(i) $\lim\limits_{z \to z_0} cf(z) = cL$, c a complex constant,

(ii) $\lim\limits_{z \to z_0} (f(z) \pm g(z)) = L \pm M$,

(iii) $\lim\limits_{z \to z_0} f(z) \cdot g(z) = L \cdot M$, and

(iv) $\lim\limits_{z \to z_0} \dfrac{f(z)}{g(z)} = \dfrac{L}{M}$, provided $M \neq 0$.

Proof of (i) Each part of Theorem 2.6.2 follows from Theorem 2.6.1 and the analogous property (9)–(12). We will prove part (i) and leave the remaining parts as exercises.

Let $f(z) = u(x,\ y) + iv(x,\ y)$, $z_0 = x_0 + iy_0$, $L = u_0 + iv_0$, and $c = a + ib$. Since $\lim\limits_{z \to z_0} f(z) = L$, it follows from Theorem 2.6.1 that

$$\lim_{(x,y) \to (x_0,y_0)} u(x,\ y) = u_0 \quad \text{and} \quad \lim_{(x,y) \to (x_0,y_0)} v(x,\ y) = v_0.$$

By (9) and (10), we have

$$\lim_{(x,y) \to (x_0,y_0)} (au(x,\ y) - bv(x,\ y)) = au_0 - bv_0$$

and
$$\lim_{(x,y)\to(x_0,y_0)} (bu(x,\ y) + av(x,\ y)) = bu_0 + av_0.$$

However, $\operatorname{Re}(cf(z)) = au(x,\ y) - bv(x,\ y)$ and $\operatorname{Im}(cf(z)) = bu(x,\ y) + av(x,\ y)$. Therefore, by Theorem 2.6.1,

$$\lim_{z\to z_0} cf(z) = au_0 - bv_0 + i\,(bu_0 + av_0) = cL. \qquad\qquad \square$$

Of course the results in Theorems 2.6.2(ii) and 2.6.2(iii) hold for any finite sum of functions or finite product of functions, respectively. After establishing a couple of basic complex limits, we can use Theorem 2.6.2 to compute a large number of limits in a very direct manner. The two basic limits that we need are those of the **complex constant function** $f(z) = c$, where c is a complex constant, and the **complex identity function** $f(z) = z$. In Problem 45 in Exercises 2.6 you will be asked to show that:

$$\lim_{z\to z_0} c = c,\ \ c \text{ a complex constant}, \tag{15}$$

and
$$\lim_{z\to z_0} z = z_0. \tag{16}$$

The following example illustrates how these basic limits can be combined with the Theorem 2.6.2 to compute limits of complex rational functions.

EXAMPLE 4   **Computing Limits with Theorem 2.6.2**

Use Theorem 2.6.2 and the basic limits (15) and (16) to compute the limits

(a) $\displaystyle \lim_{z\to i} \frac{(3+i)z^4 - z^2 + 2z}{z+1}$

(b) $\displaystyle \lim_{z\to 1+\sqrt{3}i} \frac{z^2 - 2z + 4}{z - 1 - \sqrt{3}i}$

Solution

(a) By Theorem 2.6.2(iii) and (16), we have:

$$\lim_{z\to i} z^2 = \lim_{z\to i} z \cdot z = \left(\lim_{z\to i} z\right) \cdot \left(\lim_{z\to i} z\right) = i \cdot i = -1.$$

Similarly, $\lim\limits_{z\to i} z^4 = i^4 = 1$. Using these limits, Theorems 2.6.2(i), 2.6.2(ii), and the limit in (16), we obtain:

$$\lim_{z\to i} \left((3+i)z^4 - z^2 + 2z\right) = (3+i)\lim_{z\to i} z^4 - \lim_{z\to i} z^2 + 2\lim_{z\to i} z$$
$$= (3+i)(1) - (-1) + 2(i)$$
$$= 4 + 3i,$$

and $\lim\limits_{z\to i}(z+1) = 1+i$. Therefore, by Theorem 2.6.2(iv), we have:

$$\lim_{z\to i} \frac{(3+i)z^4 - z^2 + 2z}{z+1} = \frac{\displaystyle\lim_{z\to i}\left((3+i)z^4 - z^2 + 2z\right)}{\displaystyle\lim_{z\to i}(z+1)} = \frac{4+3i}{1+i}.$$

After carrying out the division, we obtain $\lim\limits_{z \to i} \dfrac{(3+i)z^4 - z^2 + 2z}{z+1} = \dfrac{7}{2} - \dfrac{1}{2}i.$

(b) In order to find $\lim\limits_{z \to 1 + \sqrt{3}i} \dfrac{z^2 - 2z + 4}{z - 1 - \sqrt{3}i}$, we proceed as in (a):

$$\lim_{z \to 1 + \sqrt{3}i} \left(z^2 - 2z + 4\right) = \left(1 + \sqrt{3}i\right)^2 - 2\left(1 + \sqrt{3}i\right) + 4$$

$$= -2 + 2\sqrt{3}i - 2 - 2\sqrt{3}i + 4 = 0,$$

and $\lim\limits_{z \to 1 + \sqrt{3}i} \left(z - 1 - \sqrt{3}i\right) = 1 + \sqrt{3}i - 1 - \sqrt{3}i = 0$. It appears that we cannot apply Theorem 2.6.2(iv) since the limit of the denominator is 0. However, in the previous calculation we found that $1 + \sqrt{3}i$ is a root of the quadratic polynomial $z^2 - 2z + 4$. From Section 1.6, recall that if z_1 is a root of a quadratic polynomial, then $z - z_1$ is a factor of the polynomial. Using long division, we find that

$$z^2 - 2z + 4 = \left(z - 1 + \sqrt{3}i\right)\left(z - 1 - \sqrt{3}i\right).$$

See (5) in Section 1.6. Because z is not allowed to take on the value $1 + \sqrt{3}i$ in the limit, we can cancel the common factor in the numerator and denominator of the rational function. That is,

$$\lim_{z \to 1 + \sqrt{3}i} \frac{z^2 - 2z + 4}{z - 1 - \sqrt{3}i} = \lim_{z \to 1 + \sqrt{3}i} \frac{\left(z - 1 + \sqrt{3}i\right)\left(z - 1 - \sqrt{3}i\right)}{z - 1 - \sqrt{3}i}$$

$$= \lim_{z \to 1 + \sqrt{3}i} \left(z - 1 + \sqrt{3}i\right).$$

By Theorem 2.6.2(ii) and the limits in (15) and (16), we then have

$$\lim_{z \to 1 + \sqrt{3}i} \left(z - 1 + \sqrt{3}i\right) = 1 + \sqrt{3}i - 1 + \sqrt{3}i = 2\sqrt{3}i.$$

Therefore, $\lim\limits_{z \to 1 + \sqrt{3}i} \dfrac{z^2 - 2z + 4}{z - 1 - \sqrt{3}i} = 2\sqrt{3}i.$ ❏

In Section 3.1 we will calculate the limit in part (b) of Example 4 in a different manner.

2.6.2 Continuity

Continuity of Real Functions Recall that if the limit of a real function f as x approaches the point x_0 exists and agrees with the value of the function f at x_0, then we say that f is *continuous* at the point x_0. In symbols, this definition is given by:

> ### Continuity of a Real Function $f(x)$
> A function f is continuous at a point x_0 if $\lim\limits_{x \to x_0} f(x) = f(x_0)$. (17)

Observe that in order for the equation $\lim\limits_{x \to x_0} f(x) = f(x_0)$ in (17) to be satisfied, three things must be true. The limit $\lim\limits_{x \to x_0} f(x)$ must exist, f must be defined at x_0, and these two values must be equal. If any one of these three conditions fail, then f cannot be continuous at x_0. For example, the function

$$f(x) = \begin{cases} x^2, & x < 0 \\ x - 1, & x \geq 0 \end{cases},$$

illustrated in Figure 2.6.3 is not continuous at the point $x = 0$ since $\lim\limits_{x \to 0} f(x)$ does not exist. On a similar note, even though $\lim\limits_{x \to 1} \dfrac{x^2 - 1}{x - 1} = 2$, the function $f(x) = \dfrac{x^2 - 1}{x - 1}$ is not continuous at $x = 1$ because $f(1)$ is not defined.

In real analysis, we visualize the concept of continuity using the graph of the function f. Informally, a function f is continuous if there are no breaks or holes in the graph of f. Since we cannot graph a complex function, our discussion of the continuity of complex functions will be primarily algebraic in nature.

Continuity of Complex Functions

The definition of continuity for a complex function is, in essence, the same as that for a real function. That is, a complex function f is continuous at a point z_0 if the limit of f as z approaches z_0 exists and is the same as the value of f at z_0. This gives the following definition for complex functions, which is analogous to (17).

Definition 2.6.2   Continuity of a Complex Function

A complex function f is **continuous at a point z_0** if

$$\lim_{z \to z_0} f(z) = f(z_0).$$

Analogous to real functions, if a complex f is continuous at a point, then the following three conditions must be met.

Criteria for Continuity at a Point

A complex function f is continuous at a point z_0 if each of the following three conditions hold:

(i) $\lim\limits_{z \to z_0} f(z)$ *exists,*

(ii) f *is defined at z_0, and*

(iii) $\lim\limits_{z \to z_0} f(z) = f(z_0)$.

If a complex function f is not continuous at a point z_0 then we say that f is **discontinuous** at z_0. For example, the function $f(z) = \dfrac{1}{1 + z^2}$ is discontinuous at $z = i$ and $z = -i$.

EXAMPLE 5   Checking Continuity at a Point

Consider the function $f(z) = z^2 - iz + 2$. In order to determine if f is continuous at, say, the point $z_0 = 1 - i$, we must find $\lim\limits_{z \to z_0} f(z)$ and $f(z_0)$, then check to see whether these two complex values are equal. From Theorem 2.6.2 and the limits in (15) and (16) we obtain:

$$\lim_{z \to z_0} f(z) = \lim_{z \to 1-i} \left(z^2 - iz + 2\right) = (1-i)^2 - i(1-i) + 2 = 1 - 3i.$$

Furthermore, for $z_0 = 1 - i$ we have:

$$f(z_0) = f(1-i) = (1-i)^2 - i(1-i) + 2 = 1 - 3i.$$

Since $\lim\limits_{z \to z_0} f(z) = f(z_0)$, we conclude that $f(z) = z^2 - iz + 2$ is continuous at the point $z_0 = 1 - i$.   ❏

As Example 5 indicates, the continuity of a complex polynomial or a rational function is easily determined using Theorem 2.6.2 and the limits in (15) and (16). More complicated functions, however, often require other techniques.

EXAMPLE 6   Discontinuity of Principal Square Root Function

Show that the principal square root function $f(z) = z^{1/2}$ defined by (7) of Section 2.4 is discontinuous at the point $z_0 = -1$.

Solution We show that $f(z) = z^{1/2}$ is discontinuous at $z_0 = -1$ by demonstrating that the limit $\lim\limits_{z \to -1} z^{1/2}$ does not exist. By the criterion on page 101, it suffices to find two different paths through $z_0 = -1$ for which $z^{1/2}$ approaches different values. Before we begin, recall from (7) of Section 2.4 that the principal square root function is defined by $z^{1/2} = \sqrt{|z|}e^{i\,\mathrm{Arg}(z)/2}$. Now consider z approaching -1 along the quarter of the unit circle lying in the second quadrant. See Figure 2.6.5. That is, consider the points $|z| = 1$, $\pi/2 < \arg(z) < \pi$. In exponential form, this path can be described as $z = e^{i\theta}$, $\pi/2 < \theta < \pi$, with θ approaching π. Thus, by setting $|z| = 1$ and letting $\mathrm{Arg}(z) = \theta$ approach π, we obtain:

$$\lim_{z \to -1} z^{1/2} = \lim_{z \to -1} \sqrt{|z|}e^{i\,\mathrm{Arg}(z)/2} = \lim_{\theta \to \pi} \sqrt{1}e^{i\theta/2}.$$

Since $e^{i\theta/2} = \cos(\theta/2) + i\sin(\theta/2)$, this simplifies to:

$$\lim_{z \to -1} z^{1/2} = \lim_{\theta \to \pi} \left(\cos\frac{\theta}{2} + i\sin\frac{\theta}{2}\right) = \cos\frac{\pi}{2} + i\sin\frac{\pi}{2} = 0 + i(1) = i. \quad (18)$$

Next, we let z approach -1 along the quarter of the unit circle lying in the third quadrant. Again refer to Figure 2.6.5. Along this curve we have the points $z = e^{i\theta}$, $-\pi < \theta < -\pi/2$, with θ approaching $-\pi$. By setting $|z| = 1$ and letting $\mathrm{Arg}(z) = \theta$ approach $-\pi$ we find:

$$\lim_{z \to -1} z^{1/2} = \lim_{\theta \to -\pi} e^{i\theta/2} = \lim_{\theta \to -\pi} \left(\cos\frac{\theta}{2} + i\sin\frac{\theta}{2}\right) = -i. \quad (19)$$

Because the complex values in (18) and (19) do not agree, we conclude that $\lim\limits_{z \to -1} z^{1/2}$ does not exist. Therefore, the principal square root function $f(z) = z^{1/2}$ is discontinuous at the point $z_0 = -1$.   ❏

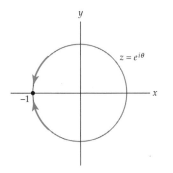

Figure 2.6.5 Figure for Example 6

In Definition 2.6.2 we defined continuity of a complex function f at a single point z_0 in the complex plane. We are often also interested in the continuity of a function on a set of points in the complex plane. A complex function f is **continuous on a set S** if f is continuous at z_0 for each z_0 in S. For example, using Theorem 2.6.2 and the limits in (15) and (16) we can show that $f(z) = z^2 - iz + 2$ is continuous at any point z_0 in the complex plane. Therefore, $f(z) = z^2 - iz + 2$ is continuous on **C**. The function $f(z) = \dfrac{1}{z^2 + 1}$, on the other hand, is continuous on the set consisting of all complex z such that $z \neq \pm i$.

Properties of Continuous Functions   Because the concept of continuity is defined using the complex limit, various properties of complex limits can be translated into statements about continuity. Consider Theorem 2.6.1, which describes the connection between the complex limit of $f(z) = u(x,\, y) + iv(x,\, y)$ and the real limits of u and v. Using the following definition of continuity for real functions $F(x,\, y)$, we can restate this theorem about limits as a theorem about continuity.

Continuity of a Real Function $F(x, y)$

A function F is continuous at a point $(x_0,\, y_0)$ if

$$\lim_{(x,y)\to(x_0,y_0)} F(x,\, y) = F(x_0,\, y_0). \tag{20}$$

Again, this definition of continuity is analogous to (17). From (20) and Theorem 2.6.1, we obtain the following result.

Theorem 2.6.3   Real and Imaginary Parts of a Continuous Function

If $f(z) = u(x,\, y) + iv(x,\, y)$ and $z_0 = x_0 + iy_0$, then the complex function f is continuous at the point z_0 if and only if both real functions u and v are continuous at the point $(x_0,\, y_0)$.

Proof   If f is continuous at z_0, then from Definition 2.6.2 we have:

$$\lim_{z\to z_0} f(z) = f(z_0) = u(x_0, y_0) + iv(x_0, y_0). \tag{21}$$

After making the identifications $u_0 = u(x_0, y_0)$ and $v_0 = v(x_0, y_0)$, Theorem 2.6.1 gives:

$$\lim_{(x,y)\to(x_0,y_0)} u(x,\, y) = u(x_0,\, y_0) \quad \text{and} \quad \lim_{(x,y)\to(x_0,y_0)} v(x,\, y) = v(x_0,\, y_0). \tag{22}$$

Therefore, from (20), both u and v are continuous at $(x_0,\, y_0)$. Conversely, if u and v are continuous at $(x_0,\, y_0)$, then

$$\lim_{(x,y)\to(x_0,y_0)} u(x,\, y) = u(x_0, y_0) \quad \text{and} \quad \lim_{(x,y)\to(x_0,y_0)} v(x,\, y) = v(x_0,\, y_0).$$

It then follows from Theorem 2.6.1 that $\lim\limits_{z\to z_0} f(z) = u(x_0,\, y_0) + iv(x_0,\, y_0) = f(z_0)$. Therefore, f is continuous by Definition 2.6.2.   ❑

EXAMPLE 7 Checking Continuity Using Theorem 2.6.3

Show that the function $f(z) = \bar{z}$ is continuous on **C**.

Solution According to Theorem 2.6.3, $f(z) = \bar{z} = \overline{x + iy} = x - iy$ is continuous at $z_0 = x_0 + iy_0$ if both $u(x, y) = x$ and $v(x, y) = -y$ are continuous at (x_0, y_0). Because u and v are two-variable polynomial functions, it follows from (13) that:

$$\lim_{(x,y)\to(x_0,y_0)} u(x, y) = x_0 \quad \text{and} \quad \lim_{(x,y)\to(x_0,y_0)} v(x, y) = -y_0.$$

This implies that u and v are continuous at (x_0, y_0), and, therefore, that f is continuous at $z_0 = x_0 + iy_0$ by Theorem 2.6.3. Since $z_0 = x_0 + iy_0$ was an arbitrary point, we conclude that the function $f(z) = \bar{z}$ is continuous on **C**. ❏

The algebraic properties of complex limits from Theorem 2.6.2 can also be restated in terms of continuity of complex functions.

Theorem 2.6.4 Properties of Continuous Functions

If f and g are continuous at the point z_0, then the following functions are continuous at the point z_0:

(i) cf, c a complex constant,

(ii) $f \pm g$,

(iii) $f \cdot g$, and

(iv) $\dfrac{f}{g}$, provided $g(z_0) \neq 0$.

Proof of (ii) We prove only (ii); proofs of the remaining parts are similar. Since f and g are continuous at z_0 we have that $\lim_{z \to z_0} f(z) = f(z_0)$ and $\lim_{z \to z_0} g(z) = g(z_0)$. From Theorem 2.6.2(ii), it follows that $\lim_{z \to z_0} (f(z) + g(z)) = f(z_0) + g(z_0)$. Therefore, $f + g$ is continuous at z_0 by Definition 2.6.2. ❏

Of course, the results of Theorems 2.6.4(ii) and 2.6.4(iii) extend to any finite sum or finite product of continuous functions, respectively. We can use these facts to show that complex polynomials are continuous functions.

Theorem 2.6.5 Continuity of Polynomial Functions

Complex polynomial functions are continuous on the entire complex plane.

Proof Let $p(z) = a_n z^n + a_{n-1} z^{n-1} + \cdots + a_1 z + a_0$ be a complex polynomial function and let z_0 be any point in the complex plane **C**. From (16) we have that the identity function $f(z) = z$ is continuous at z_0, and by repeated application of Theorem 2.6.4(iii), this implies that the power function $f(z) = z^n$, where n is an integer and $n \geq 1$, is continuous at this point as well. Moreover, (15) implies that every complex constant function $f(z) = c$ is continuous at

z_0, and so it follows from Theorem 2.6.4(i) that each of the functions $a_n z^n$, $a_{n-1} z^{n-1}, \ldots, a_1 z$, and a_0 are continuous at z_0. Now from repeated application of Theorem 2.6.4(ii) we see that $p(z) = a_n z^n + a_{n-1} z^{n-1} + \cdots + a_1 z + a_0$ is continuous at z_0. Since z_0 was allowed to be any point in the complex plane, we have shown that the polynomial function p is continuous on the entire complex plane **C**.   ❏

Since a rational function $f(z) = p(z)/q(z)$ is quotient of the polynomial functions p and q, it follows from Theorem 2.6.5 and Theorem 2.6.4(iv) that f is continuous at every point z_0 for which $q(z_0) \neq 0$. In other words,

> ## *Continuity of Rational Functions*
>
> *Rational functions are continuous on their domains.*

Bounded Functions   Continuous complex functions have many important properties that are analogous to properties of continuous real functions. For instance, if a real function f is continuous on a closed interval I on the real line, then f is bounded on I. This means that there is a real number $M > 0$ such that $|f(x)| \leq M$ for all x in I. A similar result for real multivariable functions states that if $F(x, y)$ is continuous on a closed and bounded region R of the Cartesian plane, then there is a real number $M > 0$ such that $|F(x,y)| \leq M$ for all (x, y) in R, and we say F is bounded on R.

Now suppose that the complex function $f(z) = u(x, y) + iv(x, y)$ is defined on a closed and bounded region R in the complex plane. As with real functions, we say that f is **bounded** on R if there exists a real constant $M > 0$ such that $|f(z)| < M$ for all z in R. If f is continuous on R, then Theorem 2.6.3 tells us that $\underline{u \text{ and } v \text{ are continuous}}$ real functions on R. It follows that $F(x,y) = \sqrt{[u(x,y)]^2 + [v(x,y)]^2}$ is also continuous on R. Because F is continuous on the closed and bounded region R, we conclude that F is bounded on R. That is, there is a real constant $M > 0$ such that $|F(x,y)| \leq M$ for all (x, y) in R. However, since $|f(z)| = F(x, y)$, we have that $|f(z)| \leq M$ for all z in R. In other words, the complex function f is bounded on R. This establishes the following important property of continuous complex functions.

> ## *A Bounding Property*
>
> *If a complex function f is continuous on a closed and bounded region R, then f is bounded on R. That is, there is a real constant $M > 0$ such that $|f(z)| \leq M$ for all z in R.*

While this result assures us that a bound M exists for f on R, it offers no practical approach to find it. One approach to find a bound is to use the triangle inequality. See Example 3 in Section 1.2. Another approach to determine a bound is to use complex mappings. See Problems 38 and 39 in Exercises 2.3, Problems 55 and 56 in Exercises 2.4, and Problems 29 and 30 in Exercises 2.5. In Chapter 5, we will see that for a special class of important complex functions, the bound can only be attained by a point in the boundary of R.

Branches   In Section 2.4 we discussed, briefly, the concept of a multiple-valued function $F(z)$ that assigns a set of complex numbers to the input z. (Recall that our convention is to always use uppercase letters such as F, G, and H to represent multiple-valued functions.) Examples of multiple-valued functions include $F(z) = z^{1/n}$, which assigns to the input z the set of n nth roots of z, and $G(z) = \arg(z)$, which assigns to the input z the infinite set of arguments of z. In practice, it is often the case that we need a consistent way of choosing just one of the roots of a complex number or, maybe, just one of the arguments of a complex number. That is, we are usually interested in computing just one of the values of a multiple-valued function. If we make this choice of value with the concept of continuity in mind, then we obtain a *function* that is called a *branch* of a multiple-valued function. In more rigorous terms, a **branch** of a multiple-valued function F is a function f_1 that is continuous on some domain and that assigns exactly one of the multiple-values of F to each point z in that domain.

Notation: Branches

When representing branches of a multiple-valued function F with functional notation, we will use lowercase letters with a numerical subscript such as f_1, f_2, and so on.

The requirement that a branch be continuous means that the domain of a branch is different from the domain of the multiple-valued function. For example, the multiple-valued function $F(z) = z^{1/2}$ that assigns to each input z the set of two square roots of z is defined for all nonzero complex numbers z. Even though the principal square root function $f(z) = z^{1/2}$ does assign exactly one value of F to each input z (namely, it assigns to z the principal square root of z), f *is not* a branch of F. The reason for this is that the principal square root function is not continuous on its domain. In particular, in Example 6 we showed that $f(z) = z^{1/2}$ is not continuous at $z_0 = -1$. The argument used in Example 6 can be easily modified to show that $f(z) = z^{1/2}$ is discontinuous at every point on the negative real axis. Therefore, in order to obtain a branch of $F(z) = z^{1/2}$ that agrees with the principal square root function, we must restrict the domain to exclude points on the negative real axis. This gives the function

$$f_1(z) = \sqrt{r}e^{i\theta/2}, \quad -\pi < \theta < \pi. \tag{23}$$

We call the function f_1 defined by (23) the **principal branch** of $F(z) = z^{1/2}$ because the value of θ represents the principal argument of z for all z in $\mathrm{Dom}\,(f_1)$. In the following example we show that f_1 is, in fact, a branch of F.

EXAMPLE 8   A Branch of $F(z) = z^{1/2}$

Show that the function f_1 defined by (23) is a branch of the multiple-valued function $F(z) = z^{1/2}$.

Solution   The domain of the function f_1 is the set defined by $|z| > 0$, $-\pi < \arg(z) < \pi$, shown in gray in Figure 2.6.6. From (8) of Section 2.4, we see that the function f_1 agrees with the principal square root function f on this set. Thus, f_1 does assign to the input z exactly one of the values of

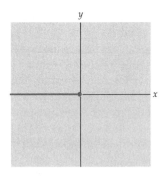

Figure 2.6.6 The domain D of the branch f_1

$F(z) = z^{1/2}$. It remains to show that f_1 is a continuous function on its domain. In order to see that this is so, let z be a point with $|z| > 0$, $-\pi < \arg(z) < \pi$. If $z = x + iy$ and $x > 0$, then $z = re^{i\theta}$ where $r = \sqrt{x^2 + y^2}$ and $\theta = \tan^{-1}(y/x)$. Since $-\pi/2 < \tan^{-1}(y/x) < \pi/2$, the inequality $-\pi < \theta < \pi$ is satisfied. Thus, substituting the expressions for r and θ in (23) we obtain:

$$f_1(z) = \sqrt[4]{x^2 + y^2}\, e^{i\tan^{-1}(y/x)/2}$$
$$= \sqrt[4]{x^2 + y^2} \cos\left(\frac{\tan^{-1}(y/x)}{2}\right) + i\sqrt[4]{x^2 + y^2}\sin\left(\frac{\tan^{-1}(y/x)}{2}\right).$$

Because the real and imaginary parts of f_1 are continuous real functions for $x > 0$, we conclude from Theorem 2.6.3 that f_1 is continuous for $x > 0$. A similar argument can be made for points with $y > 0$ using $\theta = \cot^{-1}(x/y)$ and for points with $y < 0$ using $\theta = -\cot^{-1}(x/y)$. In each case, we conclude from Theorem 2.6.3 that f_1 is continuous. Therefore, the function f_1 defined in (23) is a branch of the multiple-valued function $F(z) = z^{1/2}$. ❑

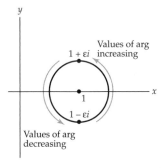

(a) $z = 1$ is not a branch point

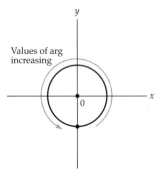

(b) $z = 0$ is a branch point

Figure 2.6.7 $G(z) = \arg(z)$

Branch Cuts and Points Although the multiple-valued function $F(z) = z^{1/2}$ is defined for all nonzero complex numbers **C**, the principal branch f_1 is defined only on the domain $|z| > 0$, $-\pi < \arg(z) < \pi$. In general, a **branch cut** for a branch f_1 of a multiple-valued function F is a portion of a curve that is excluded from the domain of F so that f_1 is continuous on the remaining points. Therefore, the nonpositive real axis, shown in color in Figure 2.6.6, is a branch cut for the principal branch f_1 given by (23) of the multiple-valued function $F(z) = z^{1/2}$. A different branch of F with the same branch cut is given by $f_2(z) = \sqrt{r}e^{i\theta/2}$, $\pi < \theta < 3\pi$. These branches are distinct because for, say, $z = i$ we have $f_1(i) = \frac{1}{2}\sqrt{2} + \frac{1}{2}\sqrt{2}i$, but $f_2(i) = -\frac{1}{2}\sqrt{2} - \frac{1}{2}\sqrt{2}i$. Notice that if we set $\phi = \theta - 2\pi$, then the branch f_2 can be expressed as $f_2(z) = \sqrt{r}e^{i(\phi+2\pi)/2} = \sqrt{r}e^{i\phi/2}e^{i\pi}$, $-\pi < \phi < \pi$. Since $e^{i\pi} = -1$, this simplifies to $f_2(z) = -\sqrt{r}e^{i\phi/2}$, $-\pi < \phi < \pi$. Thus, we have shown that $f_2 = -f_1$. You can think of these two branches of $F(z) = z^{1/2}$ as being analogous to the positive and negative square roots of a positive real number.

Other branches of $F(z) = z^{1/2}$ can be defined in a manner similar to (23) using any ray emanating from the origin as a branch cut. For example, $f_3(z) = \sqrt{r}e^{i\theta/2}$, $-3\pi/4 < \theta < 5\pi/4$, defines a branch of $F(z) = z^{1/2}$. The branch cut for f_3 is the ray $\arg(z) = -3\pi/4$ together with the point $z = 0$.

It is not a coincidence that the point $z = 0$ is on the branch cut for f_1, f_2, and f_3. The point $z = 0$ must be on the branch cut of *every* branch of the multiple-valued function $F(z) = z^{1/2}$. In general, a point with the property that it is on the branch cut of every branch is called a **branch point** of F. Alternatively, a branch point is a point z_0 with the following property: If we traverse any circle centered at z_0 with sufficiently small radius starting at a point z_1, then the values of any branch do not return to the value at z_1. For example, consider any branch of the multiple-valued function $G(z) = \arg(z)$. At the point, say, $z_0 = 1$, if we traverse the small circle $|z - 1| = \varepsilon$ counterclockwise from the point $z_1 = 1 - \varepsilon i$, then the values of the branch increase until we reach the point $1 + \varepsilon i$; then the values of the branch decrease back down to the value of the branch at z_1. See Figure 2.6.7(a). This means that the point $z_0 = 1$ is not a branch point. On the other hand, suppose we repeat this process for the point $z_0 = 0$. For the small circle $|z| = \varepsilon$, the values of the branch increase along the entire circle. See Figure 2.6.7(b). By the time we have returned to our starting point, the value of the branch is no longer the same; it has increased by 2π. Therefore, $z_0 = 0$ is a branch point of $G(z) = \arg(z)$.

> **Remarks** *Comparison with Real Analysis*
>
> (*i*) As with real analysis, we can define the concepts of infinite limits and limits at infinity for complex functions. Intuitively, the limit $\lim_{z \to \infty} f(z) = L$ means that values of the function f can be made arbitrarily close to L if values of z are chosen so that $|z|$ is sufficiently large. A precise statement of a **limit at infinity** is:
>
> *The limit of f as z tends to ∞ exists and is equal to L if for every $\varepsilon > 0$ there exists a $\delta > 0$ such that $|f(z) - L| < \varepsilon$ whenever $|z| > 1/\delta$.*
>
> Using this definition it is not hard to show that:
>
> $$\lim_{z \to \infty} f(z) = L \text{ if and only if } \lim_{z \to 0} f\left(\frac{1}{z}\right) = L. \tag{24}$$
>
> Similarly, the **infinite limit** $\lim_{z \to z_0} f(z) = \infty$ is defined by:
>
> *The limit of f as z tends to z_0 is ∞ if for every $\varepsilon > 0$ there is a $\delta > 0$ such that $|f(z)| > 1/\varepsilon$ whenever $0 < |z - z_0| < \delta$.*
>
> From this definition we obtain the following result:
>
> $$\lim_{z \to z_0} f(z) = \infty \text{ if and only if } \lim_{z \to z_0} \frac{1}{f(z)} = 0. \tag{25}$$
>
> See Problems 21–26 in Exercises 2.6.
>
> (*ii*) In real analysis we visualize a continuous function as a function whose graph has no breaks or holes in it. It is natural to ask if there is an analogous property for continuous complex functions. The answer is yes, but this property must be stated in terms of complex mappings. We begin by recalling that a parametric curve defined by parametric equations $x = x(t)$ and $y = y(t)$ is called continuous if the real functions x and y are continuous. In a similar manner, we say that a complex parametric curve defined by $z(t) = x(t) + iy(t)$ is **continuous** if both $x(t)$ and $y(t)$ are continuous real functions. As with parametric curves in the Cartesian plane, a continuous parametric curve in the complex plane has no breaks or holes in it. Such curves provide a means to visualize continuous complex functions.
>
> *If a complex function f is continuous on a set S, then the image of every continuous parametric curve in S must be a continuous curve.*
>
> To see why this is so, consider a continuous complex function $f(z) = u(x, y) + iv(x, y)$ and a continuous parametric curve defined by $z(t) = x(t) + iy(t)$. From Theorem 2.6.3, $u(x, y)$ and $v(x, y)$ are continuous real functions. Moreover, since $x(t)$ and $y(t)$ are continuous functions, it follows from multivariable calculus that the compositions $u(x(t), y(t))$ and $v(x(t), y(t))$ are continuous functions. Therefore, the image of the parametric curve given by $w(t) = f(z(t)) = u(x(t), y(t)) + iv(x(t), y(t))$ is continuous. See Problems 59–62 in Exercises 2.6.

EXERCISES 2.6   *Answers to selected odd-numbered problems begin on page ANS-9.*

2.6.1 Limits

In Problems 1–8, use Theorem 2.6.1 and the properties of real limits on page 104 to compute the given complex limit.

1. $\lim\limits_{z\to 2i} \left(z^2 - \bar{z}\right)$

2. $\lim\limits_{z\to 1+i} \dfrac{z - \bar{z}}{z + \bar{z}}$

3. $\lim\limits_{z\to 1-i} \left(|z|^2 - i\bar{z}\right)$

4. $\lim\limits_{z\to 3i} \dfrac{\operatorname{Im}\left(z^2\right)}{z + \operatorname{Re}\left(z\right)}$

5. $\lim\limits_{z\to \pi i} e^z$

6. $\lim\limits_{z\to 0} \dfrac{z^2 + \bar{z}^2}{\operatorname{Re}(z) + \operatorname{Im}(z)}$

7. $\lim\limits_{z\to 0} \dfrac{e^z - e^{\bar{z}}}{\operatorname{Im}(z)}$

8. $\lim\limits_{z\to 1+i} \left(\log_e \left|x^2 + y^2\right| + i\arctan\dfrac{y}{x}\right)$

In Problems 9–16, use Theorem 2.6.2 and the basic limits (15) and (16) to compute the given complex limit.

9. $\lim\limits_{z\to 2-i} \left(z^2 - z\right)$

10. $\lim\limits_{z\to i} \left(z^5 - z^2 + z\right)$

11. $\lim\limits_{z\to e^{i\pi/4}} \left(z + \dfrac{1}{z}\right)$

12. $\lim\limits_{z\to 1+i} \dfrac{z^2 + 1}{z^2 - 1}$

13. $\lim\limits_{z\to -i} \dfrac{z^4 - 1}{z + i}$

14. $\lim\limits_{z\to 2+i} \dfrac{z^2 - (2+i)^2}{z - (2+i)}$

15. $\lim\limits_{z\to z_0} \dfrac{(az + b) - (az_0 + b)}{z - z_0}$

16. $\lim\limits_{z\to -3+i\sqrt{2}} \dfrac{z + 3 - i\sqrt{2}}{z^2 + 6z + 11}$

17. Consider the limit $\lim\limits_{z\to 0} \dfrac{\operatorname{Re}(z)}{\operatorname{Im}(z)}$.

  (a) What value does the limit approach as z approaches 0 along the line $y = x$?

  (b) What value does the limit approach as z approaches 0 along the imaginary axis?

  (c) Based on your answers for (a) and (b), what can you say about $\lim\limits_{z\to 0} \dfrac{\operatorname{Re}(z)}{\operatorname{Im}(z)}$?

18. Consider the limit $\lim\limits_{z\to i} \left(|z| + i\operatorname{Arg}\left(iz\right)\right)$.

  (a) What value does the limit approach as z approaches i along the unit circle $|z| = 1$ in the first quadrant?

  (b) What value does the limit approach as z approaches i along the unit circle $|z| = 1$ in the second quadrant?

  (c) Based on your answers for (a) and (b), what can you say about $\lim\limits_{z\to i} \left(|z| + i\operatorname{Arg}\left(iz\right)\right)$?

19. Consider the limit $\lim\limits_{z\to 0} \left(\dfrac{z}{\bar{z}}\right)^2$.

  (a) What value does the limit approach as z approaches 0 along the real axis?

  (b) What value does the limit approach as z approaches 0 along the imaginary axis?

  (c) Do the answers from (a) and (b) imply that $\lim\limits_{z\to 0} \left(\dfrac{z}{\bar{z}}\right)^2$ exists? Explain.

  (d) What value does the limit approach as z approaches 0 along the line $y = x$?

  (e) What can you say about $\lim\limits_{z\to 0} \left(\dfrac{z}{\bar{z}}\right)^2$?

20. Consider the limit $\lim\limits_{z \to 0} \left(\dfrac{2y^2}{x^2} - \dfrac{x^2 - y^2}{y^2} i \right)$.

(a) What value does the limit approach as z approaches 0 along the line $y = x$?

(b) What value does the limit approach as z approaches 0 along the line $y = -x$?

(c) Do the answers from (a) and (b) imply that $\lim\limits_{z \to 0} \left(\dfrac{2y^2}{x^2} - \dfrac{x^2 - y^2}{y^2} i \right)$ exists? Explain.

(d) What value does the limit approach as z approaches 0 along the line $y = 2x$?

(e) What can you say about $\lim\limits_{z \to 0} \left(\dfrac{2y^2}{x^2} - \dfrac{x^2 - y^2}{y^2} i \right)$?

Problems 21–26 involve concepts of infinite limits and limits at infinity discussed in (*i*) of the Remarks. In Problems 21–26, use (24) or (25), Theorem 2.6.2, and the basic limits (15) and (16) to compute the given complex limit.

21. $\lim\limits_{z \to \infty} \dfrac{z^2 + iz - 2}{(1 + 2i)z^2}$

22. $\lim\limits_{z \to \infty} \dfrac{iz + 1}{2z - i}$

23. $\lim\limits_{z \to i} \dfrac{z^2 - 1}{z^2 + 1}$

24. $\lim\limits_{z \to -i/2} \dfrac{(1 - i)z + i}{2z + i}$

25. $\lim\limits_{z \to \infty} \dfrac{z^2 - (2 + 3i)z + 1}{iz - 3}$

26. $\lim\limits_{z \to i} \dfrac{z^2 + 1}{z^2 + z + 1 - i}$

2.6.2 Continuity

In Problems 27–34, show that the function f is continuous at the given point.

27. $f(z) = z^2 - iz + 3 - 2i;\ z_0 = 2 - i$

28. $f(z) = z^3 - \dfrac{1}{z};\ z_0 = 3i$

29. $f(z) = \dfrac{z^3}{z^3 + 3z^2 + z};\ z_0 = i$

30. $f(z) = \dfrac{z - 3i}{z^2 + 2z - 1};\ z_0 = 1 + i$

31. $f(z) = \begin{cases} \dfrac{z^3 - 1}{z - 1}, & |z| \neq 1 \\ 3, & |z| = 1 \end{cases};\ z_0 = 1$

32. $f(z) = \begin{cases} \dfrac{z^3 - 1}{z^2 + z + 1}, & |z| \neq 1 \\ \dfrac{-1 + i\sqrt{3}}{2}, & |z| = 1 \end{cases};\ z_0 = \dfrac{1 + i\sqrt{3}}{2}$

33. $f(z) = \bar{z} - 3\,\mathrm{Re}(z) + i;\ z_0 = 3 - 2i$

34. $f(z) = \dfrac{\mathrm{Re}(z)}{z + iz} - 2z^2;\ z_0 = e^{i\pi/4}$

In Problems 35–40, show that the function f is discontinuous at the given point.

35. $f(z) = \dfrac{z^2 + 1}{z + i};\ z_0 = -i$

36. $f(z) = \dfrac{1}{|z| - 1};\ z_0 = i$

37. $f(z) = \operatorname{Arg}(z); \ z = -1$ **38.** $f(z) = \operatorname{Arg}(iz); \ z_0 = i$

39. $f(z) = \begin{cases} \dfrac{z^3 - 1}{z - 1}, & |z| \neq 1 \\ 3, & |z| = 1 \end{cases}; \ z_0 = i$ **40.** $f(z) = \begin{cases} \dfrac{z}{|z|}, & z \neq 0 \\ 1, & z = 0 \end{cases}; \ z_0 = 0$

In Problems 41–44, use Theorem 2.6.3 to determine the largest region in the complex plane on which the function f is continuous.

41. $f(z) = \operatorname{Re}(z)\operatorname{Im}(z)$ **42.** $f(z) = \bar{z}$

43. $f(z) = \dfrac{z - 1}{z\bar{z} - 4}$ **44.** $f(z) = \dfrac{z^2}{(|z| - 1)\operatorname{Im}(z)}$

Focus on Concepts

45. Use Theorem 2.6.1 to prove:

  (a) $\lim\limits_{z \to z_0} c = c$, where c is a constant.   (b) $\lim\limits_{z \to z_0} z = z_0$.

46. Use Theorem 2.6.1 to show that $\lim\limits_{z \to z_0} \bar{z} = \bar{z}_0$.

47. Use Theorem 2.6.2 and Problem 46 to show that

  (a) $\lim\limits_{z \to z_0} \operatorname{Re}(z) = \operatorname{Re}(z_0)$.

  (b) $\lim\limits_{z \to z_0} \operatorname{Im}(z) = \operatorname{Im}(z_0)$.

  (c) $\lim\limits_{z \to z_0} |z| = |z_0|$.

48. Use Theorem 2.6.1 to prove part (*ii*) of Theorem 2.6.2.

49. The following is an epsilon-delta proof that $\lim\limits_{z \to z_0} z = z_0$. Fill in the missing parts.

  Proof By Definition 2.6.1, $\lim\limits_{z \to z_0} z = z_0$ if for every $\varepsilon > 0$ there is a $\delta > 0$ such that $|\underline{\quad\quad}| < \varepsilon$ whenever $0 < |\underline{\quad\quad}| < \delta$. Setting $\delta = \underline{\quad\quad}$ will ensure that the previous statement is true.

50. The following is an epsilon-delta proof that $\lim\limits_{z \to z_0} \bar{z} = \bar{z}_0$. Provide the missing justifications in the proof.

  Proof By Definition 2.6.1, $\lim\limits_{z \to z_0} \bar{z} = \bar{z}_0$ if for every $\varepsilon > 0$ there is a $\delta > 0$ such that $|\underline{\quad\quad}| < \varepsilon$ whenever $0 < |\underline{\quad\quad}| < \delta$. By properties of complex modulus and conjugation, $|z - z_0| = |\overline{z - z_0}| = |\underline{\quad\quad}|$. Therefore, if $0 < |z - z_0| < \delta$ and $\delta = \underline{\quad\quad}$, then $|\bar{z} - \bar{z}_0| < \varepsilon$.

51. In this problem we will develop an epsilon-delta proof that
  $\lim\limits_{z \to 1+i} ((1 - i)z + 2i) = 2 + 2i$.

  (a) Write down the epsilon-delta definition (Definition 2.6.1) of
  $\lim\limits_{z \to 1+i} [(1 - i)z + 2i] = 2 + 2i$.

  (b) Factor out $(1 - i)$ from the inequality involving ε (from part (a)) and simplify. Now rewrite this inequality in the form $|z - (1 + i)| < \underline{\quad\quad}$.

  (c) Based on your work from part (b), what should δ be set equal to?

  (d) Write an epsilon-delta proof that $\lim\limits_{z \to 1+i} [(1 - i)z + 2i] = 2 + 2i$.

52. In this problem we will develop an epsilon-delta proof that
$$\lim_{z \to 2i} \frac{2z^2 - 3iz + 2}{z - 2i} = 5i.$$

(a) Write down the epsilon-delta definition (Definition 2.6.1) of
$$\lim_{z \to 2i} \frac{2z^2 - 3iz + 2}{z - 2i} = 5i.$$

(b) Simplify the inequality involving ε (from part (a)), then rewrite this inequality in the form $|z - 2i| < \underline{\hspace{1cm}}$.

(c) Based on your work from part (b), what should δ be set equal to?

(d) Write an epsilon-delta proof that $\lim_{z \to 2i} \dfrac{2z^2 - 3iz + 2}{z - 2i} = 5i.$

53. (a) Is it true that $\lim_{z \to z_0} \overline{f(z)} = \lim_{z \to z_0} f(\bar{z})$ for any complex function f? If so, then give a brief justification; if not, then find a counterexample.

(b) If $f(z)$ is a continuous function at z_0, then is it true that $\overline{f(z)}$ is continuous at z_0?

54. If f is a function for which $\lim_{x \to 0} f(x + i0) = 0$ and $\lim_{y \to 0} f(0 + iy) = 0$, then can you conclude that $\lim_{z \to 0} f(z) = 0$? Explain.

55. (a) Prove that the function $f(z) = \text{Arg}(z)$ is discontinuous at every point on the negative real axis.

(b) Prove that the function f_1 defined by

$$f_1(z) = \theta, \quad -\pi < \theta < \pi$$

is a branch of the multiple-valued function $F(z) = \arg(z)$. [*Hint:* See Example 8.]

56. Consider the multiple-valued function $F(z) = z^{1/3}$ that assigns to z the set of three cube roots of z. Explicitly define three distinct branches f_1, f_2, and f_3 of F, all of which have the nonnegative real axis as a branch cut.

57. Consider the multiple-valued function $F(z) = (z - 1 + i)^{1/2}$.

(a) What is the branch point of F? Explain.

(b) Explicitly define two distinct branches of f_1 and f_2 of F. In each case, state the branch cut.

58. Consider the multiple-valued function $F(z) = (z^2 + 1)^{1/2}$. What are the branch points (there are two of them) of F? Explain.

Computer Lab Assignments

Reread part (ii) of the Remarks at the end of Section 2.6. In Problems 57–60, use a CAS to show that the given function is not continuous inside the unit circle by plotting the image of the given continuous parametric curve. (Be careful, *Mathematica* and *Maple* plots can sometimes be misleading.)

59. $f(z) = z + \text{Arg}(z)$, $z(t) = -\frac{1}{2} + \frac{1}{2}\sqrt{3}it$, $-1 \leq t \leq 1$

60. $f(z) = \sqrt[4]{r}e^{i\theta/4}$, $\theta = \text{Arg}(z)$, $z(t) = -\frac{1}{2} + \frac{1}{2}\sqrt{3}it$, $-1 \leq t \leq 1$

61. $f(z) = \sqrt{r}e^{i\theta/2}$, $\theta = \text{Arg}(z)$, $z(t) = -\frac{1}{2} + \frac{1}{4}e^{it}$, $0 \leq t \leq 2\pi$

62. $f(z) = |z - 1|\text{Arg}(-z) + i\text{Arg}(iz)$, $z(t) = \frac{1}{2}e^{it}$, $0 \leq t \leq 2\pi$

2.7 Applications

To the Instructor: Applications to streamlines presented in this section assume a familiarity with solving differential equations.

In this chapter we saw that one of the main differences between real and complex functions is the inability to draw the graph of a complex function. This motivated the introduction of mappings as an alternative method for graphically representing complex functions. There are, however, other ways to visualize complex functions. In this section we will show that complex functions give complex representations of two-dimensional vector fields. In later chapters, we will use the complex representation of a vector field to solve applied problems in the areas of fluid flow, heat flow, gravitation, and electrostatics.

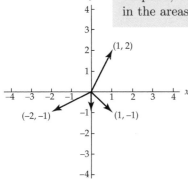

(a) Values of **F** plotted as position vectors

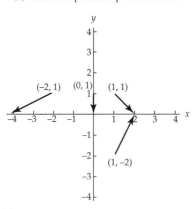

(b) Values of **F** plotted with initial point at (x, y)

Figure 2.7.1 Some vector values of the function $\mathbf{F}(x, y) = x\mathbf{i} - y\mathbf{j}$

Vector Fields   In multivariable calculus, a vector-valued function of two real variables

$$\mathbf{F}(x, y) = (P(x, y), Q(x, y)) \tag{1}$$

is also called a **two-dimensional vector field**. Using the standard orthogonal unit basis vectors **i** and **j**, we can also express the vector field in (1) as:

$$\mathbf{F}(x, y) = P(x, y)\mathbf{i} + Q(x, y)\mathbf{j}. \tag{2}$$

For example, the function $\mathbf{F}(x, y) = (x + y)\mathbf{i} + (2xy)\mathbf{j}$ is a two-dimensional vector field, for which, say, $\mathbf{F}(1, 3) = (1 + 3)\mathbf{i} + (2 \cdot 1 \cdot 3)\mathbf{j} = 4\mathbf{i} + 6\mathbf{j}$. Values of a function **F** given by (2) are vectors that can be plotted as position vectors with initial point at the origin. However, in order to obtain a graphical representation of the vector field (2) that displays the relation between the input (x, y) and the output $\mathbf{F}(x, y)$, we plot the vector $\mathbf{F}(x, y)$ with initial point (x, y) and terminal point $(x + P(x, y), y + Q(x, y))$. For example, in Figure 2.7.1(a) the four functional values $\mathbf{F}(1, 1) = \mathbf{i} - \mathbf{j}$, $\mathbf{F}(0, 1) = -\mathbf{j}$, $\mathbf{F}(1, -2) = \mathbf{i} + 2\mathbf{j}$, and $\mathbf{F}(-2, 1) = -2\mathbf{i} - \mathbf{j}$ of the vector field $\mathbf{F}(x, y) = x\mathbf{i} - y\mathbf{j}$ are plotted as position vectors, whereas in Figure 2.7.1(b) we have a portion of the graphical representation of the vector field obtained by plotting these four vectors with initial point at (x, y). Specifically, Figure 2.7.1(b) consists of the four vectors $\mathbf{i} - \mathbf{j}$, $-\mathbf{j}$, $\mathbf{i} + 2\mathbf{j}$, and $-2\mathbf{i} - \mathbf{j}$ plotted with initial points at $(1, 1)$, $(0, 1)$, $(1, -2)$, and $(-2, 1)$, and terminal points $(2, 0)$, $(0, 0)$, $(2, 0)$, and $(-4, 0)$, respectively.

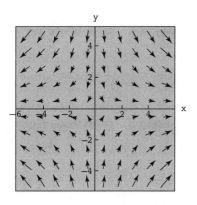

Figure 2.7.2 The vector field $f(z) = \bar{z}$

Complex Functions as Vector Fields   There is a natural way to represent a vector field $\mathbf{F}(x, y) = P(x, y)\mathbf{i} + Q(x, y)\mathbf{j}$ with a complex function f. Namely, we use the functions P and Q as the real and imaginary parts of f, in which case, we say that the complex function $f(z) = P(x, y) + iQ(x, y)$ is the **complex representation** of the vector field $\mathbf{F}(x, y) = P(x, y)\mathbf{i} + Q(x, y)\mathbf{j}$. Conversely, any complex function $f(z) = u(x, y) + iv(x, y)$ has an associated vector field $\mathbf{F}(x, y) = u(x, y)\mathbf{i} + v(x, y)\mathbf{j}$. From this point on we shall refer to both $\mathbf{F}(x, y) = P(x, y)\mathbf{i} + Q(x, y)\mathbf{j}$ and $f(x, y) = u(x, y) + iv(x, y)$ as **vector fields**. As an example of this discussion, consider the vector field $f(z) = \bar{z}$. Since $f(z) = x - iy$, the function f is the complex representation of the vector field $\mathbf{F}(x, y) = x\mathbf{i} - y\mathbf{j}$. Part of this vector field was shown in Figure 2.7.1(b). A more complete representation of the vector field $f(z) = \bar{z}$ is shown in Figure 2.7.2 (this plot was created in *Mathematica*). Observe that the vector field plotted in Figure 2.7.2 gives a graphical representation of the complex function $f(z) = \bar{z}$ that is different from a mapping. Compare with Figure 2.5.3 in Section 2.5.

When plotting a vector field **F** associated with a complex function f it is helpful to note that plotting the vector $\mathbf{F}(x, y)$ with initial point (x, y) is equivalent to plotting the vector representation of the complex number $f(z)$ with initial point z. We illustrate this remark in the following example.

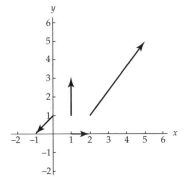

Figure 2.7.3 Vectors in the vector field $f(z) = z^2$

EXAMPLE 1   Plotting Vectors in a Vector Field

Plot the vectors in the vector field $f(z) = z^2$ corresponding to the points $z = 1,\ 2+i,\ 1+i$, and i.

Solution   By a straightforward computation we find that:

$$f(1) = 1^2 = 1, \qquad\qquad f(2+i) = (2+i)^2 = 3 + 4i,$$
$$f(i) = i^2 = -1, \qquad \text{and} \qquad f(1+i) = (1+i)^2 = 2i.$$

This implies that in the vector field $f(z) = z^2$ we have the vector representations of the complex numbers 1, $3+4i$, -1, and $2i$ plotted with initial points at 1, $2+i$, i, and $1+i$, respectively. These vectors are shown in Figure 2.7.3. ❑

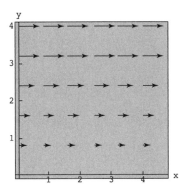

Figure 2.7.4 *Mathematica* plot of the vector field $f(z) = iy$

Use of Computers   Plotting vector fields by hand is a simple but tedious procedure. Fortunately, computer algebra systems such as *Mathematica* and *Maple* have built-in commands to plot two-dimensional vector fields. In Figure 2.7.4, the vector field $f(z) = iy$ has been plotted using the **PlotVectorField** command in *Mathematica*. Observe that the lengths of the vectors in the *Mathematica* plot are much smaller than they should be. For example at, say, $z = 1 + i$ we have $f(1 + i) = i$, but the vector plotted at $z = 1 + i$ does not have length 1. The reason for this is that *Mathematica* scales the vectors in a vector field in order to create a nicer image (in particular, *Mathematica* scales to ensure that no vectors overlap). Therefore, the lengths of the vectors in Figure 2.7.4 do not accurately represents the *absolute* lengths of vectors in this vector field. The vectors in Figure 2.7.4 do, however, accurately represent the *relative* lengths of the vectors in the vector field.*

In many applications the primary interest is in the directions and not the magnitudes of the vectors in a vector field. For example, in the forthcoming discussion we will be concerned with determining the paths along which particles move in a fluid flow. For this type of application, we can use a normalized vector field. In a **normalized vector field** all vectors are scaled to have the same length. Figure 2.7.5 displays a normalized vector field for $f(z) = iy$ created using the **ScaleFunction** option with the **PlotVectorField** command in *Mathematica*. Compare with Figure 2.7.4.

Keep in mind that in many applications the magnitudes of the vectors are important and, in such cases, a normalized vector field is inappropriate. We will see examples of this in Chapter 5 when we discuss the circulation and net flux of a fluid flow. In this text, whenever we use a normalized vector field, we will explicitly state so. Therefore, in a graphical sense, the term vector field will refer to a plot of a set of vectors that has *not* been normalized.

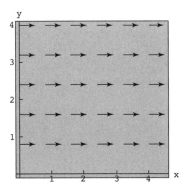

Figure 2.7.5 *Mathematica* plot of the normalized vector field $f(z) = iy$

Fluid Flow   One of the many uses of vector fields in applied mathematics is to model fluid flow. Because we are confined to two dimensions in complex analysis, let us consider only **planar flows** of a fluid. This means that the

---

*For more information on plotting vector fields in *Mathematica*, refer to the technical report *Guide to Standard Mathematical Packages* published by Wolfram Research.

movement of the fluid takes place in planes that are parallel to the xy-plane and that the motion and the physical traits of the fluid are identical in all planes. These assumptions allow us to analyze the flow of a single sheet of the fluid. Suppose that $f(z) = P(x, y) + iQ(x, y)$ represents a **velocity field** of a planar flow in the complex plane. Then $f(z)$ specifies the velocity of a particle of the fluid located at the point z in the plane. The modulus $|f(z)|$ is the **speed** of the particle and the vector $f(z)$ gives the direction of the flow at that point.

For a velocity field $f(z) = P(x, y) + iQ(x, y)$ of a planar flow, the functions P and Q represent the components of the velocity in the x- and y-directions, respectively. If $z(t) = x(t) + iy(t)$ is a parametrization of the path that a particle follows in the fluid flow, then the tangent vector $z'(t) = x'(t) + iy'(t)$ to the path must coincide with $f(z(t))$. Therefore, the real and imaginary parts of the tangent vector to the path of a particle in the fluid must satisfy the system of differential equations

$$\frac{dx}{dt} = P(x, y)$$

$$\frac{dy}{dt} = Q(x, y). \tag{3}$$

The family of solutions to the system of first-order differential equations (3) is called the **streamlines** of the planar flow associated with $f(z)$.

EXAMPLE 2   Streamlines

Find the streamlines of the planar flow associated with $f(z) = \bar{z}$.

Solution Since $f(z) = \bar{z} = x - iy$, we identify $P(x, y) = x$ and $Q(x, y) = -y$. From (3) the streamlines of f are the family of solutions to the system of differential equations

$$\frac{dx}{dt} = x$$

$$\frac{dy}{dt} = -y.$$

These differential equations are independent of each other and so each can be solved by separation of variables. This gives the general solutions $x(t) = c_1 e^t$ and $y(t) = c_2 e^{-t}$ where c_1 and c_2 are real constants. In order to plot the curve $z(t) = x(t) + iy(t)$, we eliminate the parameter t to obtain a Cartesian equation in x and y. This is easily done by multiplying the two solutions to obtain $xy = c_1 c_2$. Because c_1 and c_2 can be any real constants, this family of curves can be given by $xy = c$ where c is a real constant. In conclusion, we have shown that particles in the planar flow associated with $f(z) = \bar{z}$ move along curves in the family of hyperbolas $xy = c$. In Figure 2.7.6, we have used *Mathematica* to plot the streamlines corresponding to $c = \pm 1, \pm 4$, and ± 9 for this flow. These streamlines are shown in black superimposed over the plot of the normalized vector field of $f(z) = \bar{z}$.   ❑

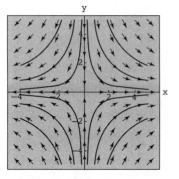

Figure 2.7.6 Streamlines in the planar flow associated with $f(z) = \bar{z}$

EXAMPLE 3   Streamlines

Find the streamlines of the planar flow associated with $f(z) = \bar{z}^2$.

Solution We proceed as in Example 2. Since the function f can be expressed as $f(z) = \bar{z}^2 = x^2 - y^2 - 2xyi$, we identify $P(x, y) = x^2 - y^2$ and $Q(x, y) = -2xy$. Thus, the streamlines of this flow satisfy the system of differential equations:

$$\frac{dx}{dt} = x^2 - y^2$$

$$\frac{dy}{dt} = -2xy. \tag{4}$$

The chain rule of elementary calculus states that $(dy/dx) \cdot (dx/dt) = dy/dt$, and so after solving for dy/dx we have that

$$\left(\frac{dy}{dt}\right) \bigg/ \left(\frac{dx}{dt}\right) = \frac{dy}{dx}.$$

Therefore, by dividing $dy/dt = -2xy$ by $dx/dt = x^2 - y^2$, we find that the system in (4) is equivalent to the first-order differential equation:

$$\frac{dy}{dx} = \frac{-2xy}{x^2 - y^2} \quad \text{or} \quad 2xy\, dx + \left(x^2 - y^2\right) dy = 0. \tag{5}$$

Recall that a differential equation of the form $M(x, y)dx + N(x, y)dy = 0$ is called exact if $\partial M/\partial y = \partial N/\partial x$. Given an exact differential equation, if we can find a function $F(x, y)$ for which $\partial F/\partial x = M$ and $\partial F/\partial y = N$, then $F(x, y) = c$ is an implicit solution to the differential equation. Identifying $M(x, y) = 2xy$ and $N(x, y) = x^2 - y^2$, we see that our differential equation in (5) is exact since

$$\frac{\partial M}{\partial y} = \frac{\partial}{\partial y}\left(2xy\right) = 2x = \frac{\partial}{\partial x}\left(x^2 - y^2\right) = \frac{\partial N}{\partial x}.$$

To find a function F for which $\partial F/\partial x = M$ and $\partial F/\partial y = N$, we first partially integrate the function $M(x, y) = 2xy$ with respect to the variable x:

$$F(x, y) = \int 2xy\, dx = x^2 y + g(y).$$

The function $g(y)$ is then determined by taking the partial derivative of F with respect to the variable y and setting this expression equal to $N(x, y) = x^2 - y^2$:

$$\frac{\partial F}{\partial y} = x^2 + g'(y) = x^2 - y^2.$$

This implies that $g'(y) = -y^2$, and so we can take $g(y) = -\frac{1}{3}y^3$. In conclusion, $F(x, y) = x^2 y - \frac{1}{3}y^3 = c$ is an implicit solution of the differential equation in (5), and so the streamlines of the planar flow associated with $f(z) = \bar{z}^2$ are given by:

$$x^2 y - \frac{1}{3}y^3 = c$$

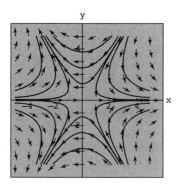

Figure 2.7.7 Streamlines in the planar flow associated with $f(z) = \bar{z}^2$

where c is a real constant. In Figure 2.7.7, *Mathematica* has been used to plot the streamlines corresponding to $c = \pm\frac{2}{3}$, $\pm\frac{16}{3}$, ± 18. These streamlines are shown in black superimposed over the plot of the normalized vector field for the flow. ❑

EXERCISES 2.7   *Answers to selected odd-numbered problems begin on page ANS-9.*

In Problems 1–8, (**a**) plot the images of the complex numbers $z = 1$, $1+i$, $1-i$, and i under the given function f as position vectors, and (**b**) plot the images as vectors in the vector field associated with f.

1. $f(z) = 2z - i$

2. $f(z) = z^3$

3. $f(z) = \overline{1 - z^2}$

4. $f(z) = \dfrac{1}{z}$

5. $f(z) = z - \dfrac{1}{z}$

6. $f(z) = z^{1/2}$, the principal square root function given by (7) of Section 2.4

7. $f(z) = \dfrac{1}{\bar{z}}$

8. $f(z) = \log_e |z| + i\mathrm{Arg}(z)$

In Problems 9–12, (**a**) find the streamlines of the planar flow associated with the given complex function f and (**b**) sketch the streamlines.

9. $f(z) = 1 - 2i$

10. $f(z) = \dfrac{1}{\bar{z}}$

11. $f(z) = iz$

12. $f(z) = (1+i)\bar{z}$

Focus on Concepts

13. Let f be a complex function. Explain the relationship between the vector field associated with $f(z)$ and the vector field associated with $g(z) = f(z - 1)$. Illustrate with sketches using a simple function for f.

14. Let f be a complex function. Explain the relationship between the vector field associated with $f(z)$ and the vector field associated with $g(z) = if(z)$. Illustrate with sketches using a simple function for f.

15. Consider the planar flow associated with $f(z) = c$ where c is a complex constant.

(**a**) Find the streamlines of this flow.

(**b**) Explain why this flow is called a **uniform flow**.

16. Consider the planar flow associated with $f(z) = 1 - 1/\bar{z}^2$.

(**a**) Use a CAS to plot the vector field associated with f in the region $|z| > 1$.

(**b**) Verify analytically that the unit circle $x^2 + y^2 = 1$ is a streamline in this flow.

(**c**) Explain why $f(z) = 1 - 1/\bar{z}^2$ is called a **flow around the unit circle**.

Computer Lab Assignments

In Problems 17–22, use a CAS to plot the vector field associated with the given complex function f.

17. $f(z) = 2z - i$

18. $f(z) = z^3$

19. $f(z) = 1 - z^2$

20. $f(z) = \dfrac{1}{z}$

21. $f(z) = 2 + i$

22. $f(z) = 1 - \dfrac{1}{\bar{z}^2}$

CHAPTER 2 REVIEW QUIZ *Answers to selected odd-numbered problems begin on page ANS-10.*

In Problems 1–20, answer true or false. If the statement is false, justify your answer by either explaining why it is false or giving a counterexample; if the statement is true, justify your answer by either proving the statement or citing an appropriate result in this chapter.

1. If $f(z)$ is a complex function, then $f(x + 0i)$ must be a real number.

2. $\arg(z)$ is a complex function.

3. The domain of the function $f(z) = \dfrac{1}{z^2 + i}$ is all complex numbers.

4. The domain of the function $f(z) = e^{z^2 - (1+i)z + 2}$ is all complex numbers.

5. If $f(z)$ is a complex function with $u(x,\,y) = 0$, then the range of f lies in the imaginary axis.

6. The entire complex plane is mapped onto the real axis $v = 0$ by $w = z + \bar{z}$.

7. The entire complex plane is mapped onto the unit circle $|w| = 1$ by $w = \dfrac{z}{|z|}$.

8. The range of the function $f(z) = \operatorname{Arg}(z)$ is all real numbers.

9. The image of the circle $|z - z_0| = \rho$ under a linear mapping is a circle with a (possibly) different center, but the same radius.

10. The linear mapping $w = \left(1 - \sqrt{3}i\right)z + 2$ acts by rotating through an angle of $\pi/3$ radians clockwise about the origin, magnifying by a factor of 2, then translating by 2.

11. There is more than one linear mapping that takes the circle $|z - 1| = 1$ to the circle $|z + i| = 1$.

12. The lines $x = 3$ and $x = -3$ are mapped onto to the same parabola by $w = z^2$.

13. There are no solutions to the equation $\operatorname{Arg}(z) = \operatorname{Arg}\left(z^3\right)$.

14. If $f(z) = z^{1/4}$ is the principal fourth root function, then $f(-1) = -\frac{1}{2}\sqrt{2} + \frac{1}{2}\sqrt{2}i$.

15. The complex number i is not in the range of the principal cube root function.

16. Under the mapping $w = 1/z$ on the extended complex plane, the domain $|z| > 3$ is mapped onto the domain $|w| < \frac{1}{3}$.

17. If f is a complex function for which $\lim\limits_{z \to 2+i} \operatorname{Re}(f(z)) = 4$ and $\lim\limits_{z \to 2+i} \operatorname{Im}(f(z)) = -1$, then $\lim\limits_{z \to 2+i} f(z) = 4 - i$.

18. If f is a complex function for which $\lim\limits_{x \to 0} f(x + 0i) = 0$ and $\lim\limits_{y \to 0} f(0 + iy) = 0$, then $\lim\limits_{z \to 0} f(z) = 0$.

19. If f is a complex function that is continuous at the point $z = 1 + i$, then the function $g(z) = 3\left[f(z)\right]^2 - (2 + i)f(z) + i$ is continuous at $z = 1 + i$.

20. If f is a complex function that is continuous on the entire complex plane, then the function $g(z) = \overline{f(z)}$ is continuous on the entire complex plane.

In Problems 21–40, try to fill in the blanks without referring back to the text.

21. If $f(z) = z^2 + i\bar{z}$ then the real and imaginary parts of f are $u(x,\,y) =$ _____ and $v(x,\,y) =$ _____ .

22. If $f(z) = \dfrac{|z - 1|}{z^2 + 2iz + 2}$, then the natural domain of f is _____ .

23. If $f(z) = z - \bar{z}$, then the range of f is contained in the _____ axis.

24. The exponential function e^z has real and imaginary parts $u(x,y) = $ _____ and $v(x,y) = $ _____ .

25. A parametrization of the line segment from $1 + i$ to $2i$ is $z(t) = $ _____ .

26. A parametrization of the circle centered at $1 - i$ with radius 3 is $z(t) = $ _____ .

27. Every complex linear mapping is a composition of at most one _____ , one _____ , and one _____ .

28. The complex mapping $w = iz + 2$ rotates and _____ , but does not _____ .

29. The function z^2 squares the modulus of z and _____ its argument.

30. The image of the sector $0 \leq \arg(z) \leq \pi/2$ under the mapping $w = z^3$ is _____ .

31. The image of horizontal and vertical lines under the mapping $w = z^2$ is _____ .

32. The principal nth root function $z^{1/n}$ maps the complex plane onto the region _____ .

33. If $f(z) = z^{1/6}$ is the principal 6th root function, then $f(-1) = $ _____ .

34. The complex reciprocal function $1/z$ is a composition of _____ in the _____ circle followed by reflection across the _____ -axis.

35. According to the formal definition of a complex limit, $\lim\limits_{z \to 2i} \left(z^2 - i\right) = -4 - i$ if for every $\varepsilon > 0$ there is a $\delta > 0$ such that $|$ _____ $| < \varepsilon$ whenever $0 < |z- $ _____ $| < \delta$.

36. If $f(z) = \dfrac{z + \bar{z}}{z}$, then $\lim\limits_{x \to 0} f(x + 0i) = $ _____ and $\lim\limits_{y \to 0} f(0 + iy) = $ _____ . Therefore, $\lim\limits_{z \to 0} f(z)$ _____ .

37. A complex function f is continuous at $z = z_0$ if $\lim\limits_{z \to z_0} f(z) = $ _____ .

38. The function $f(z) = $ _____ is an example of a function that is continuous on the domain $|z| > 0$, $-\pi < \arg(z) < \pi$.

39. The complex function $f(z) = \dfrac{x}{y} + i \, \log_e x$ is continuous on the region _____ .

40. Both _____ and _____ are examples of multiple-valued functions.

Analytic Functions

Introduction In the preceding chapter we introduced the notion of a complex function. Analogous to the calculus of real functions we can develop the notions of derivatives and integrals of complex functions based on the fundamental concept of a limit. In this chapter our principal focus will be on the definition and the properties of the derivative of a complex function.

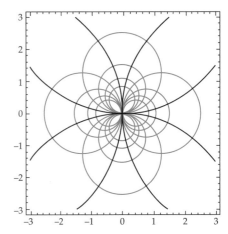

Level curves for $f(z) = 1/z$.
See page 153.

3.1   Differentiability and Analyticity

The calculus of complex functions deals with the usual concepts of derivatives and integrals of these functions. In this section we shall give the limit definition of the derivative of a complex function $f(z)$. Although many of the concepts in this section will seem familiar to you, such as the product, quotient, and chain rules of differentiation, there are important differences between this material and the calculus of real functions $f(x)$. As the subsequent chapters of this text unfold, you will see that except for familiarity of names and definitions, there is little similarity between the *interpretations* of quantities such as $f'(x)$ and $f'(z)$.

The Derivative   Suppose $z = x + iy$ and $z_0 = x_0 + iy_0$; then the change in z_0 is the difference $\Delta z = z - z_0$ or $\Delta z = x - x_0 + i(y - y_0) = \Delta x + i\Delta y$. If a complex function $w = f(z)$ is defined at z and z_0, then the corresponding change in the function is the difference $\Delta w = f(z_0 + \Delta z) - f(z_0)$. The **derivative** of the function f is defined in terms of a limit of the difference quotient $\Delta w / \Delta z$ as $\Delta z \to 0$.

Definition 3.1.1   Derivative of a Complex Function

Suppose the complex function f is defined in a neighborhood of a point z_0. The **derivative** of f at z_0, denoted by $f'(z_0)$, is

$$f'(z_0) = \lim_{\Delta z \to 0} \frac{f(z_0 + \Delta z) - f(z_0)}{\Delta z} \tag{1}$$

provided this limit exists.

If the limit in (1) exists, then the function f is said to be **differentiable** at z_0. Two other symbols denoting the derivative of $w = f(z)$ are w' and dw/dz. If the latter notation is used, then the value of a derivative at a specified point z_0 is written $\left. \dfrac{dw}{dz} \right|_{z=z_0}$.

EXAMPLE 1   Using Definition 3.1.1

Use Definition 3.1.1 to find the derivative of $f(z) = z^2 - 5z$.

Solution Because we are going to compute the derivative of f at any point, we replace z_0 in (1) by the symbol z. First,

$$f(z + \Delta z) = (z + \Delta z)^2 - 5(z + \Delta z) = z^2 + 2z\Delta z + (\Delta z)^2 - 5z - 5\Delta z.$$

Second,

$$f(z + \Delta z) - f(z) = z^2 + 2z\Delta z + (\Delta z)^2 - 5z - 5\Delta z - (z^2 - 5z)$$
$$= 2z\Delta z + (\Delta z)^2 - 5\Delta z.$$

Then, finally, (1) gives

$$f'(z) = \lim_{\Delta z \to 0} \frac{2z\Delta z + (\Delta z)^2 - 5\Delta z}{\Delta z}$$

$$= \lim_{\Delta z \to 0} \frac{\Delta z(2z + \Delta z - 5)}{\Delta z}$$

$$= \lim_{\Delta z \to 0} (2z + \Delta z - 5).$$

The limit is $f'(z) = 2z - 5$. ❏

Rules of Differentiation The familiar rules of differentiation in the calculus of real variables carry over to the calculus of complex variables. If f and g are differentiable at a point z, and c is a complex constant, then (1) can be used to show:

Differentiation Rules

Constant Rules: $\dfrac{d}{dz}c = 0$ and $\dfrac{d}{dz}cf(z) = cf'(z)$ (2)

Sum Rule: $\dfrac{d}{dz}[f(z) \pm g(z)] = f'(z) \pm g'(z)$ (3)

Product Rule: $\dfrac{d}{dz}[f(z)g(z)] = f(z)g'(z) + f'(z)g(z)$ (4)

Quotient Rule: $\dfrac{d}{dz}\left[\dfrac{f(z)}{g(z)}\right] = \dfrac{g(z)f'(z) - f(z)g'(z)}{[g(z)]^2}$ (5)

Chain Rule: $\dfrac{d}{dz}f(g(z)) = f'(g(z))\,g'(z).$ (6)

The **power rule** for differentiation of powers of z is also valid:

$$\frac{d}{dz}z^n = nz^{n-1}, \quad n \text{ an integer.} \tag{7}$$

Combining (7) with (6) gives the **power rule for functions**:

$$\frac{d}{dz}[g(z)]^n = n[g(z)]^{n-1}g'(z), \quad n \text{ an integer.} \tag{8}$$

EXAMPLE 2   **Using the Rules of Differentiation**

Differentiate:

(a) $f(z) = 3z^4 - 5z^3 + 2z$ **(b)** $f(z) = \dfrac{z^2}{4z+1}$ **(c)** $f(z) = \left(iz^2 + 3z\right)^5$

Solution

(a) Using the power rule (7), the sum rule (3), along with (2), we obtain

$$f'(z) = 3 \cdot 4z^3 - 5 \cdot 3z^2 + 2 \cdot 1 = 12z^3 - 15z^2 + 2.$$

(b) From the quotient rule (5),

$$f'(z) = \frac{(4z+1) \cdot 2z - z^2 \cdot 4}{(4z+1)^2} = \frac{4z^2 + 2z}{(4z+1)^2}.$$

(c) In the power rule for functions (8) we identify $n = 5$, $g(z) = iz^2 + 3z$, and $g'(z) = 2iz + 3$, so that

$$f'(z) = 5(iz^2 + 3z)^4(2iz + 3).$$

❑

For a complex function f to be differentiable at a point z_0, we know from the preceding chapter that the limit $\lim\limits_{\Delta z \to 0} \dfrac{f(z_0 + \Delta z) - f(z_0)}{\Delta z}$ must exist and equal the same complex number from any direction; that is, the limit must exist regardless of how Δz approaches 0. This means that in complex analysis, the requirement of differentiability of a function $f(z)$ at a point z_0 is a far greater demand than in real calculus of functions $f(x)$ where we can approach a real number x_0 on the number line from only two directions. If a complex function is made up by specifying its real and imaginary parts u and v, such as $f(z) = x + 4iy$, there is a good chance that it is not differentiable.

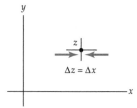

(a) $\Delta z \to 0$ along a line parallel to x-axis

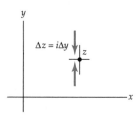

(b) $\Delta z \to 0$ along a line parallel to y-axis

Figure 3.1.1 Approaching z along a horizontal line and then along a vertical line

---

EXAMPLE 3 A Function That Is Nowhere Differentiable

Show that the function $f(z) = x + 4iy$ is not differentiable at any point z.

Solution Let z be any point in the complex plane. With $\Delta z = \Delta x + i\Delta y$,

$$f(z + \Delta z) - f(z) = (x + \Delta x) + 4i(y + \Delta y) - x - 4iy = \Delta x + 4i\Delta y$$

and so
$$\lim_{\Delta z \to 0} \frac{f(z + \Delta z) - f(z)}{\Delta z} = \lim_{\Delta z \to 0} \frac{\Delta x + 4i\Delta y}{\Delta x + i\Delta y}. \tag{9}$$

Now, as shown in Figure 3.1.1(a), if we let $\Delta z \to 0$ along a line parallel to the x-axis, then $\Delta y = 0$ and $\Delta z = \Delta x$. Along this path we have:

$$\lim_{\Delta z \to 0} \frac{f(z + \Delta z) - f(z)}{\Delta z} = \lim_{\Delta z \to 0} \frac{\Delta x}{\Delta x} = 1. \tag{10}$$

On the other hand, if we let $\Delta z \to 0$ along a line parallel to the y-axis as shown in Figure 3.1.1(b), then $\Delta x = 0$ and $\Delta z = i\Delta y$ so that

$$\lim_{\Delta z \to 0} \frac{f(z + \Delta z) - f(z)}{\Delta z} = \lim_{\Delta z \to 0} \frac{4i\Delta y}{i\Delta y} = 4. \tag{11}$$

In view of the obvious fact that the values in (10) and (11) are different, we conclude that the limit in (9) does not exist. Therefore, $f(z) = x + 4iy$ is nowhere differentiable; that is, f is not differentiable at any point z. ❑

The basic power rule (7) does not apply to powers of the conjugate of z because, like the function in Example 3, the function $f(z) = \bar{z}$ is nowhere differentiable. See Problem 21 in Exercises 3.1.

Analytic Functions Even though the requirement of differentiability is a stringent demand, there is a class of functions that is of great importance whose members satisfy even more severe requirements. These functions are called **analytic functions**.

> **Definition 3.1.2 Analyticity at a Point**
>
> A complex function $w = f(z)$ is said to be **analytic at a point** z_0 if f is differentiable at z_0 and at every point in some neighborhood of z_0.

A function **f is analytic in a domain D** if it is analytic at every point in D. The phrase "analytic *on* a domain D" is also used. Although we shall not use these terms in this text, a function f that is analytic throughout a domain D is called **holomorphic** or **regular**.

Very Important ➡ You should reread Definition 3.1.2 carefully. Analyticity at a point is *not* the same as differentiability at a point. Analyticity *at a point is a neighborhood property*; in other words, analyticity is a property that is defined over an open set. It is left as an exercise to show that the function $f(z) = |z|^2$ is differentiable at $z = 0$ but is not differentiable anywhere else. Even though $f(z) = |z|^2$ is differentiable at $z = 0$, it is not analytic at that point because there exists no neighborhood of $z = 0$ *throughout which f is differentiable*; hence the function $f(z) = |z|^2$ is nowhere analytic. See Problem 19 in Exercises 3.1.

In contrast, the simple polynomial $f(z) = z^2$ is differentiable at every point z in the complex plane. Hence, $f(z) = z^2$ is analytic everywhere.

Entire Functions A function that is analytic at every point z in the complex plane is said to be an **entire function**. In view of differentiation rules (2), (3), (7), and (5), we can conclude that polynomial functions are differentiable at every point z in the complex plane and rational functions are analytic throughout any domain D that contains no points at which the denominator is zero. The following theorem summarizes these results.

> **Theorem 3.1.1 Polynomial and Rational Functions**
>
> (*i*) A complex polynomial function $p(z) = a_n z^n + a_{n-1} z^{n-1} + \cdots + a_1 z + a_0$, where n is a nonnegative integer, is an entire function.
>
> (*ii*) A complex rational function $f(z) = \dfrac{p(z)}{q(z)}$, where p and q are polynomial functions, is analytic in any domain D that contains no point z_0 for which $q(z_0) = 0$.

Singular Points Since the rational function $f(z) = 4z/\left(z^2 - 2z + 2\right)$ is undefined at $1 + i$ and $1 - i$, f fails to be analytic at these points. Thus by (*ii*) of Theorem 3.1.1, f is not analytic in any domain containing one or both of these points. In general, a point z at which a complex function $w = f(z)$

$1 \pm i$ are zeros of the denominator of f. ➡

fails to be analytic is called a **singular point** of f. We will discuss singular points in greater depth in Chapter 6.

Another consequence of the differentiation rules on page 129 is:

> ### Analyticity of Sum, Product, and Quotient
>
> *If the functions f and g are analytic in a domain D, then the sum $f(z) + g(z)$, difference $f(z) - g(z)$, and product $f(z)g(z)$ are analytic in D. The quotient $f(z)/g(z)$ is analytic provided $g(z) \neq 0$ in D.*

An Alternative Definition of $f'(z)$ Sometimes it is convenient to define the derivative of a function f using an alternative form of the difference quotient $\Delta w/\Delta z$. Since $\Delta z = z - z_0$, then $z = z_0 + \Delta z$, and so (1) can be written as

$$f'(z_0) = \lim_{z \to z_0} \frac{f(z) - f(z_0)}{z - z_0}. \tag{12}$$

In contrast to what we did in Example 1, if we wish to compute f' at a general point z using (12), then we replace z_0 by the symbol z *after* the limit is computed. See Problems 7–10 in Exercises 3.1.

As in real analysis, if a complex function f is differentiable at a point, then the function is necessarily continuous at the point. We use the form of the derivative given in (12) to prove the last statement.

> **Theorem 3.1.2   Differentiability Implies Continuity**
>
> If f is differentiable at a point z_0 in a domain D, then f is continuous at z_0.

Proof   The limits $\displaystyle\lim_{z \to z_0} \frac{f(z) - f(z_0)}{z - z_0}$ and $\displaystyle\lim_{z \to z_0} (z - z_0)$ exist and equal $f'(z_0)$ and 0, respectively. Hence by Theorem 2.6.2(*iii*) of Section 2.6, we can write the following limit of a product as the product of the limits:

$$\lim_{z \to z_0} (f(z) - f(z_0)) = \lim_{z \to z_0} \frac{f(z) - f(z_0)}{z - z_0} \cdot (z - z_0)$$

$$= \lim_{z \to z_0} \frac{f(z) - f(z_0)}{z - z_0} \cdot \lim_{z \to z_0} (z - z_0) = f'(z_0) \cdot 0 = 0.$$

From $\displaystyle\lim_{z \to z_0} (f(z) - f(z_0)) = 0$ we conclude that $\displaystyle\lim_{z \to z_0} f(z) = f(z_0)$. In view of Definition 2.6.2, f is continuous at z_0.   ❏

Of course the converse of Theorem 3.1.2 is not true; continuity of a function f at a point does not guarantee that f is differentiable at the point. It follows from Theorem 2.6.3 that the simple function $f(z) = x + 4iy$ is continuous everywhere because the real and imaginary parts of f, $u(x, y) = x$ and $v(x, y) = 4y$ are continuous at any point (x, y). Yet we saw in Example 3 that $f(z) = x + 4iy$ is not differentiable at any point z.

As another consequence of differentiability, L'Hôpital's rule for computing limits of the indeterminate form $0/0$, carries over to complex analysis.

> **Theorem 3.1.3 L'Hôpital's Rule**
>
> Suppose f and g are functions that are analytic at a point z_0 and $f(z_0) = 0$, $g(z_0) = 0$, but $g'(z_0) \neq 0$. Then
>
> $$\lim_{z \to z_0} \frac{f(z)}{g(z)} = \frac{f'(z_0)}{g'(z_0)}. \tag{13}$$

The task of establishing (13) is neither long nor difficult. You are guided through the steps of a proof in Problem 33 in Exercises 3.1.

EXAMPLE 4   Using L'Hôpital's Rule

Compute $\displaystyle\lim_{z \to 2+i} \frac{z^2 - 4z + 5}{z^3 - z - 10i}$.

Solution If we identify $f(z) = z^2 - 4z + 5$ and $g(z) = z^3 - z - 10i$, you should verify that $f(2+i) = 0$ and $g(2+i) = 0$. The given limit has the indeterminate form $0/0$. Now since f and g are polynomial functions, both functions are necessarily analytic at $z_0 = 2 + i$. Using

$$f'(z) = 2z - 4, \quad g'(z) = 3z^2 - 1, \quad f'(2+i) = 2i, \quad g'(2+i) = 8 + 12i,$$

we see that (13) gives

$$\lim_{z \to 2+i} \frac{z^2 - 4z + 5}{z^3 - z - 10i} = \frac{f'(2+i)}{g'(2+i)} = \frac{2i}{8 + 12i} = \frac{3}{26} + \frac{1}{13}i.$$

❑

In part (b) of Example 4 in Section 2.6 we resorted to the lengthy procedure of factoring and cancellation to compute the limit

$$\lim_{z \to 1+\sqrt{3}i} \frac{z^2 - 2z + 4}{z - 1 - \sqrt{3}i}. \tag{14}$$

A rereading of that example shows that the limit (14) has the indeterminate form $0/0$. With $f(z) = z^2 - 2z + 4$, $g(z) = z - 1 - \sqrt{3}i$, $f'(z) = 2z - 2$, and $g'(z) = 1$, L'Hôpital's rule (13) gives immediately

$$\lim_{z \to 1+\sqrt{3}i} \frac{z^2 - 2z + 4}{z - 1 - \sqrt{3}i} = \frac{f'(1+\sqrt{3}i)}{1} = 2\left(1 + \sqrt{3}i - 1\right) = 2\sqrt{3}i.$$

Remarks *Comparison with Real Analysis*

(*i*) In real calculus the derivative of a function $y = f(x)$ at a point x has many interpretations. For example, $f'(x)$ is the slope of the tangent line to the graph of f at the point $(x, f(x))$. When the slope is positive, negative, or zero, the function, in turn, is increasing, decreasing, and possibly has a maximum or minimum. Also, $f'(x)$ is the instantaneous rate of change of f at x. In a physical setting, this rate of change can be interpreted as velocity of a moving object.

None of these interpretations carry over to complex calculus. Thus it is fair to ask: What does the derivative of a complex function $w = f(z)$ represent? Here is the answer: In complex analysis the primary concern is *not* what a derivative of function *is* or *represents*, but rather, it is whether a function f actually *has* a derivative. The fact that a complex function f possesses a derivative tells us a lot about the function. As we have just seen, when f is differentiable at z and at every point in some neighborhood of z, then f is analytic at the point z. You will see the importance of analytic functions in the remaining chapters of this book. For example, the derivative plays an important role in the theory of mappings by complex functions. Roughly, under a mapping defined by an analytic function f, the magnitude and sense of an angle between two curves that intersect a point z_0 in the z-plane is preserved in the w-plane at all points at which $f'(z) \neq 0$. See Chapter 7.

(ii) We pointed out in the foregoing discussion that $f(z) = |z|^2$ was differentiable only at the single point $z = 0$. In contrast, the real function $f(x) = |x|^2$ is differentiable everywhere. The real function $f(x) = x$ is differentiable everywhere, but the complex function $f(z) = x = \operatorname{Re}(z)$ is nowhere differentiable.

(iii) The differentiation formulas (2)–(8) are important, but not nearly as important as in real analysis. In complex analysis we deal with functions such as $f(z) = 4x^2 - iy$ and $g(z) = xy + i(x + y)$, which, even if they possess derivatives, cannot be differentiated by formulas (2)–(8).

(iv) In this section we have not mentioned the concept of higher-order derivatives of complex functions. We will pursue this topic in depth in Section 5.5. There is nothing surprising about the definitions of higher derivatives; they are defined in exactly the same manner as in real analysis. For example, the second derivative is the derivative of the first derivative. In the case $f(z) = 4z^3$ we see that $f'(z) = 12z^2$ and so the second derivative is $f''(z) = 24z$. But there is a major difference between real and complex variables concerning the existence of higher-order derivatives. In real analysis, if a function f possesses, say, a first derivative, there is no guarantee that f possesses any other higher derivatives. For example, on the interval $(-1, 1)$, $f(x) = x^{3/2}$ is differentiable at $x = 0$, but $f'(x) = \frac{3}{2}x^{1/2}$ is not differentiable at $x = 0$. In complex analysis, if a function f is *analytic* in a domain D then, by assumption, f possesses a derivative at each point in D. We will see in Section 5.5 that this fact alone guarantees that f possesses higher-order derivatives at all points in D. Indeed, an analytic function f is infinitely differentiable in D.

(v) The definition of "analytic at a point a" in real analysis differs from the usual definition of that concept in complex analysis (Definition 3.1.2). In real analysis, analyticity of a function is defined in terms of power series: A function $y = f(x)$ is analytic at a point a if f has a Taylor series at a that represents f in some neighborhood of a. In view of Remark (iv), why are these two definitions not really that different?

(vi) As in real calculus, it may be necessary to apply L'Hôpital's rule several times in succession to calculate a limit. In other words, if $f(z_0)$, $g(z_0)$, $f'(z_0)$, and $g'(z_0)$ are all zero, the limit $\lim\limits_{z \to z_0} f(z)/g(z)$

may still exist. In general, if f, g, and their first $n-1$ derivatives are zero at z_0 and $g^{(n)}(z_0) \neq 0$, then

$$\lim_{z \to z_0} \frac{f(z)}{g(z)} = \frac{f^{(n)}(z_0)}{g^{(n)}(z_0)}.$$

EXERCISES 3.1 *Answers to selected odd-numbered problems begin on page ANS-11.*

In Problems 1–6, use (1) of Definition 3.1.1 to find $f'(z)$ for the given function.

1. $f(z) = 9iz + 2 - 3i$
2. $f(z) = 15z^2 - 4z + 1 - 3i$

3. $f(z) = iz^3 - 7z^2$
4. $f(z) = \dfrac{1}{z}$

5. $f(z) = z - \dfrac{1}{z}$
6. $f(z) = -z^{-2}$

In Problems 7–10, use the alternative definition (12) to find $f'(z)$ for the given function.

7. $f(z) = 5z^2 - 10z + 8$
8. $f(z) = z^3$

9. $f(z) = z^4 - z^2$
10. $f(z) = \dfrac{1}{2iz}$

In Problems 11–18, use the rules of differentiation to find $f'(z)$ for the given function.

11. $f(z) = (2 - i)z^5 + iz^4 - 3z^2 + i^6$
12. $f(z) = 5(iz)^3 - 10z^2 + 3 - 4i$

13. $f(z) = (z^6 - 1)(z^2 - z + 1 - 5i)$
14. $f(z) = (z^2 + 2z - 7i)^2(z^4 - 4iz)^3$

15. $f(z) = \dfrac{iz^2 - 2z}{3z + 1 - i}$
16. $f(z) = -5iz^2 + \dfrac{2 + i}{z^2}$

17. $f(z) = \left(z^4 - 2iz^2 + z\right)^{10}$
18. $f(z) = \left(\dfrac{(4 + 2i)z}{(2 - i)z^2 + 9i}\right)^3$

19. The function $f(z) = |z|^2$ is continuous at the origin.

 (a) Show that f *is* differentiable at the origin.

 (b) Show that f *is not* differentiable at any point $z \neq 0$.

20. Show that the function

$$f(z) = \begin{cases} 0, & z = 0 \\ \dfrac{x^3 - y^3}{x^2 + y^2} + i\dfrac{x^3 + y^3}{x^2 + y^2}, & z \neq 0 \end{cases}$$

is not differentiable at $z = 0$ by letting $\Delta z \to 0$ first along the x-axis and then along the line $y = x$.

In Problems 21 and 22, proceed as in Example 3 to show that the given function is nowhere differentiable.

21. $f(z) = \bar{z}$
22. $f(z) = |z|$

In Problems 23–26, use L'Hôpital's rule to compute the given limit.

23. $\displaystyle\lim_{z \to i} \dfrac{z^7 + i}{z^{14} + 1}$
24. $\displaystyle\lim_{z \to \sqrt{2} + \sqrt{2}i} \dfrac{z^4 + 16}{z^2 - 2\sqrt{2}z + 4}$

25. $\displaystyle\lim_{z \to 1 + i} \dfrac{z^5 + 4z}{z^2 - 2z + 2}$
26. $\displaystyle\lim_{z \to \sqrt{2}i} \dfrac{z^3 + 5z^2 + 2z + 10}{z^5 + 2z^3}$

In Problems 27–30, determine the points at which the given function is not analytic.

27. $f(z) = \dfrac{iz^2 - 2z}{3z + 1 - i}$
28. $f(z) = -5iz^2 + \dfrac{2 + i}{z^2}$

29. $f(z) = \left(z^4 - 2iz^2 + z\right)^{10}$
30. $f(z) = \left(\dfrac{(4 + 2i)z}{(2 - i)z^2 + 9i}\right)^3$

Focus on Concepts

31. Use Definition 2.6.2 and Theorem 2.6.2 to prove the constant rule $\dfrac{d}{dz} cf(z) = cf'(z)$.

32. Use Definition 2.6.2 and Theorem 2.6.2 to prove the sum rule
$$\frac{d}{dz}\left[f(z) + g(z)\right] = f'(z) + g'(z).$$

33. In this problem you are guided through the start of the proof of the proposition:

If functions f and g are analytic at a point z_0 and $f(z_0) = 0$, $g(z_0) = 0$,
but $g'(z_0) \neq 0$, then $\displaystyle\lim_{z \to z_0} \frac{f(z)}{g(z)} = \frac{f'(z_0)}{g'(z_0)}$.

Proof We begin with the hypothesis that f and g are analytic at a point z_0. Analyticity at z_0 implies f and g are differentiable at z_0. Hence from (12) both limits,

$$f'(z_0) = \lim_{z \to z_0} \frac{f(z) - f(z_0)}{z - z_0} \quad \text{and} \quad g'(z_0) = \lim_{z \to z_0} \frac{g(z) - g(z_0)}{z - z_0}$$

exist. But since $f(z_0) = 0$, $g(z_0) = 0$, the foregoing limits are the same as

$$f'(z_0) = \lim_{z \to z_0} \frac{f(z)}{z - z_0} \quad \text{and} \quad g'(z_0) = \lim_{z \to z_0} \frac{g(z)}{z - z_0}.$$

Now examine $\displaystyle\lim_{z \to z_0} \frac{f(z)}{g(z)}$ and finish the proof.

34. In this problem you are guided through the start of the proof of the product rule.

Proof We begin with the hypothesis that f and g are differentiable at a point z; that is, each of the following limits exist:

$$f'(z) = \lim_{\Delta z \to 0} \frac{f(z + \Delta z) - f(z)}{\Delta z} \quad \text{and} \quad g'(z) = \lim_{\Delta z \to 0} \frac{g(z + \Delta z) - g(z)}{\Delta z}$$

(a) Justify the equality

$$\frac{d}{dz}\left[f(z)g(z)\right] = \lim_{\Delta z \to 0} \frac{f(z + \Delta z)g(z + \Delta z) - f(z)g(z)}{\Delta z}$$

$$= \lim_{\Delta z \to 0} \left[\frac{f(z + \Delta z) - f(z)}{\Delta z} g(z + \Delta z) + f(z)\frac{g(z + \Delta z) - g(z)}{\Delta z}\right].$$

(b) Use Definition 2.6.2 to justify $\displaystyle\lim_{\Delta z \to 0} g(z + \Delta z) = g(z)$.

(c) Use Theorems 2.6.2(ii) and 2.6.2(iii) to finish the proof.

35. In this problem you will prove that $f(z) = \bar{z}$ is nowhere differentiable using the polar form $\Delta z = r(\cos\theta + i\sin\theta)$.

(a) What must r approach if $\Delta z \to 0$? Must θ approach a specific value?

(b) If $f(z) = \bar{z}$, show that

$$\lim_{\Delta z \to 0} \frac{f(z + \Delta z) - f(z)}{\Delta z} = \frac{\cos\theta - i\sin\theta}{\cos\theta + i\sin\theta}.$$

(c) Explain succinctly why the result in part (b) shows that f is nowhere differentiable.

36. In this problem you will prove that $f(z) = |z|$ is nowhere differentiable using the alternative definition (12) and polar forms of z and z_0.

(a) Let $z_0 = r(\cos\theta + i\sin\theta)$ and $z = (r + \Delta r)(\cos\theta + i\sin\theta)$. Along this path determine $\displaystyle\lim_{z \to z_0} \frac{f(z) - f(z_0)}{z - z_0}$.

(b) Let $z_0 = r(\cos\theta + i\sin\theta)$ and $z = r\left[\cos(\theta + \Delta\theta) + i\sin(\theta + \Delta\theta)\right]$. Along this path determine $\lim\limits_{z \to z_0} \dfrac{f(z) - f(z_0)}{z - z_0}$.

(c) Explain succinctly why the results in parts (a) and (b) show that f is nowhere differentiable.

37. Suppose $f'(z)$ exists at a point z. Is $f'(z)$ continuous at z?

38. **(a)** Let $f(z) = z^2$. Write down the real and imaginary parts of f and f'. What do you observe?

(b) Repeat part (a) for $f(z) = 3iz + 2$.

(c) Make a conjecture about the relationship between real and imaginary parts of f versus f'.

3.2   Cauchy-Riemann Equations

In the preceding section we saw that a function f of a complex variable z is *analytic at a point z* when f is differentiable at z and differentiable at every point in some neighborhood of z. This requirement is more stringent than simply *differentiability at a point* because a complex function can be differentiable at a point z but yet be differentiable nowhere else. A function f is *analytic in a domain D* if f is differentiable at all points in D. We shall now develop a test for analyticity of a complex function $f(z) = u(x, y) + iv(x, y)$ that is based on partial derivatives of its real and imaginary parts u and v.

A Necessary Condition for Analyticity   In the next theorem we see that if a function $f(z) = u(x, y) + iv(x, y)$ is differentiable at a point z, then the functions u and v must satisfy a pair of equations that relate their first-order partial derivatives.

Theorem 3.2.1   Cauchy-Riemann Equations

Suppose $f(z) = u(x, y) + iv(x, y)$ is differentiable at a point $z = x + iy$. Then at z the first-order partial derivatives of u and v exist and satisfy the **Cauchy-Riemann equations**

$$\frac{\partial u}{\partial x} = \frac{\partial v}{\partial y} \qquad \text{and} \qquad \frac{\partial u}{\partial y} = -\frac{\partial v}{\partial x}. \tag{1}$$

Proof   The derivative of f at z is given by

$$f'(z) = \lim_{\Delta z \to 0} \frac{f(z + \Delta z) - f(z)}{\Delta z}. \tag{2}$$

By writing $f(z) = u(x, y) + iv(x, y)$ and $\Delta z = \Delta x + i\Delta y$, (2) becomes

$$f'(z) = \lim_{\Delta z \to 0} \frac{u(x + \Delta x, y + \Delta y) + iv(x + \Delta x, y + \Delta y) - u(x, y) - iv(x, y)}{\Delta x + i\Delta y}. \tag{3}$$

Since the limit (2) is assumed to exist, Δz can approach zero from any convenient direction. In particular, if we choose to let $\Delta z \to 0$ along a horizontal line, then $\Delta y = 0$ and $\Delta z = \Delta x$. We can then write (3) as

$$\begin{aligned} f'(z) &= \lim_{\Delta x \to 0} \frac{u(x + \Delta x, y) - u(x, y) + i\left[v(x + \Delta x, y) - v(x, y)\right]}{\Delta x} \\ &= \lim_{\Delta x \to 0} \frac{u(x + \Delta x, y) - u(x, y)}{\Delta x} + i \lim_{\Delta x \to 0} \frac{v(x + \Delta x, y) - v(x, y)}{\Delta x}. \end{aligned} \tag{4}$$

The existence of $f'(z)$ implies that each limit in (4) exists. These limits are the definitions of the first-order partial derivatives with respect to x of u and v, respectively. Hence, we have shown two things: both $\partial u/\partial x$ and $\partial v/\partial x$ exist at the point z, and that the derivative of f is

$$f'(z) = \frac{\partial u}{\partial x} + i\frac{\partial v}{\partial x}. \tag{5}$$

We now let $\Delta z \to 0$ along a vertical line. With $\Delta x = 0$ and $\Delta z = i\Delta y$, (3) becomes

$$f'(z) = \lim_{\Delta y \to 0} \frac{u(x, y + \Delta y) - u(x, y)}{i\Delta y} + i \lim_{\Delta y \to 0} \frac{v(x, y + \Delta y) - v(x, y)}{i\Delta y}. \tag{6}$$

In this case (6) shows us that $\partial u/\partial y$ and $\partial v/\partial y$ exist at z and that

$$f'(z) = -i\frac{\partial u}{\partial y} + \frac{\partial v}{\partial y}. \tag{7}$$

By equating the real and imaginary parts of (5) and (7) we obtain the pair of equations in (1). ❏

Because Theorem 3.2.1 states that the Cauchy-Riemann equations (1) hold at z as a *necessary* consequence of f being differentiable at z, we cannot use the theorem to help us determine where f is differentiable. *But* it is important to realize that Theorem 3.2.1 can tell us where a function f does not possess a derivative. If the equations in (1) are *not* satisfied at a point z, then f cannot be differentiable at z. We have already seen in Example 3 of Section 3.1 that $f(z) = x + 4iy$ is not differentiable at any point z. If we identify $u = x$ and $v = 4y$, then $\partial u/\partial x = 1$, $\partial v/\partial y = 4$, $\partial u/\partial y = 0$, and $\partial v/\partial x = 0$. In view of

$$\frac{\partial u}{\partial x} = 1 \neq \frac{\partial v}{\partial y} = 4$$

the two equations in (1) cannot be simultaneously satisfied at any point z. Therefore, by Theorem 3.2.1, f is nowhere differentiable.

It also follows from Theorem 3.2.1 that if a complex function $f(z) = u(x, y) + iv(x, y)$ is analytic throughout a domain D, then the real functions u and v satisfy the Cauchy-Riemann equations (1) at every point in D.

EXAMPLE 1   Verifying Theorem 3.2.1

The polynomial function $f(z) = z^2 + z$ is analytic for all z and can be written as $f(z) = x^2 - y^2 + x + i(2xy + y)$. Thus, $u(x, y) = x^2 - y^2 + x$ and $v(x, y) = 2xy + y$. For any point (x, y) in the complex plane we see that the Cauchy-Riemann equations are satisfied:

$$\frac{\partial u}{\partial x} = 2x + 1 = \frac{\partial v}{\partial y} \quad \text{and} \quad \frac{\partial u}{\partial y} = -2y = -\frac{\partial v}{\partial x}.$$

❏

The contrapositive* form of the sentence preceding Example 1 is:

Criterion for Non-analyticity

If the Cauchy-Riemann equations are not satisfied at every point z in a domain D, then the function $f(z) = u(x,\ y) + iv(x,\ y)$ cannot be analytic in D.

EXAMPLE 2   Using the Cauchy-Riemann Equations

Show that the complex function $f(z) = 2x^2 + y + i(y^2 - x)$ is not analytic at any point.

Solution We identify $u(x,\ y) = 2x^2 + y$ and $v(x,\ y) = y^2 - x$. From

$$
\begin{aligned}
\frac{\partial u}{\partial x} &= 4x \quad \text{and} \quad \frac{\partial v}{\partial y} = 2y \\[2mm]
\frac{\partial u}{\partial y} &= 1 \quad\ \text{and} \quad \frac{\partial v}{\partial x} = -1
\end{aligned}
\tag{8}
$$

we see that $\partial u/\partial y = -\partial v/\partial x$ but that the equality $\partial u/\partial x = \partial v/\partial y$ is satisfied only on the line $y = 2x$. However, for any point z on the line, there is no neighborhood or open disk about z in which the Cauchy-Riemann equations are satisfied. We conclude that f is nowhere analytic.   ❏

A Sufficient Condition for Analyticity By themselves, the Cauchy-Riemann equations do not ensure analyticity of a function $f(z) = u(x,\ y) + iv(x,\ y)$ at a point $z = x + iy$. It is possible for the Cauchy-Riemann equations to be satisfied at z and yet $f(z)$ may not be differentiable at z, or $f(z)$ may be differentiable at z but nowhere else. In either case, f is not analytic at z. See Problem 36 in Exercises 3.2. However, when we add the condition of continuity to u and v *and* to the four partial derivatives $\partial u/\partial x$, $\partial u/\partial y$, $\partial v/\partial x$, and $\partial v/\partial y$, it can be shown that the Cauchy-Riemann equations are not only necessary but also sufficient to guarantee analyticity of $f(z) = u(x,\ y) + iv(x,\ y)$ at z. The proof is long and complicated and so we state only the result.

Theorem 3.2.2   Criterion for Analyticity

Suppose the real functions $u(x,\ y)$ and $v(x,\ y)$ are continuous and have continuous first-order partial derivatives in a domain D. If u and v satisfy the Cauchy-Riemann equations (1) at all points of D, then the complex function $f(z) = u(x,\ y) + iv(x,\ y)$ is analytic in D.

---

*A proposition "If P, then Q" is logically equivalent to its contrapositive "If not Q, then not P."

EXAMPLE 3   Using Theorem 3.2.2

For the function $f(z) = \dfrac{x}{x^2 + y^2} - i\dfrac{y}{x^2 + y^2}$, the real functions $u(x,\ y) = \dfrac{x}{x^2+y^2}$ and $v(x,\ y) = -\dfrac{y}{x^2+y^2}$ are continuous except at the point where $x^2+y^2 = 0$, that is, at $z = 0$. Moreover, the four first-order partial derivatives

$$\frac{\partial u}{\partial x} = \frac{y^2 - x^2}{(x^2+y^2)^2}, \qquad \frac{\partial u}{\partial y} = -\frac{2xy}{(x^2+y^2)^2},$$
$$\frac{\partial v}{\partial x} = \frac{2xy}{(x^2+y^2)^2}, \quad \text{and} \quad \frac{\partial v}{\partial y} = \frac{y^2 - x^2}{(x^2+y^2)^2}$$

are continuous except at $z = 0$. Finally, we see from

$$\frac{\partial u}{\partial x} = \frac{y^2-x^2}{(x^2+y^2)^2} = \frac{\partial v}{\partial y} \quad \text{and} \quad \frac{\partial u}{\partial y} = -\frac{2xy}{(x^2+y^2)^2} = -\frac{\partial v}{\partial x}$$

that the Cauchy-Riemann equations are satisfied except at $z = 0$. Thus we conclude from Theorem 3.2.2 that f is analytic in any domain D that does not contain the point $z = 0$.   ❑

The results in (5) and (7) were obtained under the basic assumption that f was differentiable at the point z. In other words, if f is differentiable, then (5) and (7) give us a formula for computing the derivative $f'(z)$:

$$f'(z) = \frac{\partial u}{\partial x} + i\frac{\partial v}{\partial x} = \frac{\partial v}{\partial y} - i\frac{\partial u}{\partial y}. \qquad (9)$$

For example, we already know from part (i) of Theorem 3.1.1 that $f(z) = z^2$ is entire and so is differentiable for all z. With $u(x,\ y) = x^2 - y^2$, $\partial u/\partial x = 2x$, $v(x,\ y) = 2xy$, and $\partial v/\partial y = 2y$, we see from (9) that

$$f'(z) = 2x + i2y = 2(x + iy) = 2z.$$

Recall that analyticity implies differentiability but not conversely. Theorem 3.2.2 has an analogue that gives the following criterion for differentiability.

Sufficient Conditions for Differentiability

If the real functions $u(x,y)$ and $v(x,y)$ are continuous and have continuous first-order partial derivatives in some neighborhood of a point z, and if u and v satisfy the Cauchy-Riemann equations (1) at z, then the complex function $f(z) = u(x,\ y) + iv(x,\ y)$ is differentiable at z and $f'(z)$ is given by (9).

EXAMPLE 4   A Function Differentiable on a Line

In Example 2 we saw that the complex function $f(z) = 2x^2 + y + i(y^2 - x)$ was nowhere analytic, but yet the Cauchy-Riemann equations were satisfied on the line $y = 2x$. Since the functions $u(x,\ y) = 2x^2 + y$, $\partial u/\partial x = 4x$, $\partial u/\partial y = 1$, $v(x,\ y) = y^2 - x$, $\partial v/\partial x = -1$, and $\partial v/\partial y = 2y$ are continuous at every point, it follows that f is differentiable on the line $y = 2x$. Moreover, from (9) we see that the derivative of f at points on this line is given by $f'(z) = 4x - i = 2y - i$. ☐

The following theorem is a direct consequence of the Cauchy-Riemann equations. Its proof is left as an exercise. See Problems 28 and 30 in Exercises 3.2.

Theorem 3.2.3   Constant Functions

Suppose the function $f(z) = u(x, y) + iv(x, y)$ is analytic in a domain D.

(i) If $|f(z)|$ is constant in D, then so is $f(z)$.

(ii) If $f'(z) = 0$ in D, then $f(z) = c$ in D, where c is a complex constant.

Polar Coordinates   In Section 2.1 we saw that a complex function can be expressed in terms of polar coordinates. Indeed, the form $f(z) = u(r, \theta) + iv(r, \theta)$ is often more convenient to use. In polar coordinates the Cauchy-Riemann equations become

$$\frac{\partial u}{\partial r} = \frac{1}{r}\frac{\partial v}{\partial \theta}, \quad \frac{\partial v}{\partial r} = -\frac{1}{r}\frac{\partial u}{\partial \theta}. \tag{10}$$

The polar version of (9) at a point z whose polar coordinates are (r, θ) is then

$$f'(z) = e^{-i\theta}\left(\frac{\partial u}{\partial r} + i\frac{\partial v}{\partial r}\right) = \frac{1}{r}e^{-i\theta}\left(\frac{\partial v}{\partial \theta} - i\frac{\partial u}{\partial \theta}\right). \tag{11}$$

See Problems 34 and 35 in Exercises 3.2.

Remarks   *Comparison with Real Analysis*

In real calculus, one of the noteworthy properties of the exponential function $f(x) = e^x$ is that $f'(x) = e^x$. In (3) of Section 2.1 we gave the definition of the complex exponential $f(z) = e^z$. We are now in a position to show that $f(z) = e^z$ is differentiable everywhere and that this complex function shares the same derivative property as its real counterpart, that is, $f'(z) = f(z)$. See Problem 23 in Exercises 3.2.

EXERCISES 3.2   *Answers to selected odd-numbered problems begin on page ANS-11.*

In Problems 1 and 2, the given function is analytic for all z. Show that the Cauchy-Riemann equations are satisfied at every point.

1. $f(z) = z^3$ **2.** $f(z) = 3z^2 + 5z - 6i$

In Problems 3–8, show that the given function is not analytic at any point.

3. $f(z) = \text{Re}(z)$ **4.** $f(z) = y + ix$

5. $f(z) = 4z - 6\bar{z} + 3$ **6.** $f(z) = \bar{z}^2$

7. $f(z) = x^2 + y^2$ **8.** $f(z) = \dfrac{x}{x^2 + y^2} + i\dfrac{y}{x^2 + y^2}$

In Problems 9–16, (a) use Theorem 3.2.2 to show that the given function is analytic in an appropriate domain, and (b) use (9) or (11) to find the derivative of the function in the domain.

9. $f(z) = e^{-x} \cos y - ie^{-x} \sin y$

10. $f(z) = x + \sin x \cosh y + i(y + \cos x \sinh y)$

11. $f(z) = e^{x^2-y^2} \cos 2xy + ie^{x^2-y^2} \sin 2xy$

12. $f(z) = 4x^2 + 5x - 4y^2 + 9 + i(8xy + 5y - 1)$

13. $f(z) = \dfrac{x-1}{(x-1)^2 + y^2} - i\dfrac{y}{(x-1)^2 + y^2}$

14. $f(z) = \dfrac{x^3 + xy^2 + x}{x^2 + y^2} + i\dfrac{x^2y + y^3 - y}{x^2 + y^2}$

15. $f(z) = \dfrac{\cos \theta}{r} - i\dfrac{\sin \theta}{r}$

16. $f(z) = 5r \cos \theta + r^4 \cos 4\theta + i(5r \sin \theta + r^4 \sin 4\theta)$

In Problems 17 and 18, find real constants a, b, c, and d so that the given function is entire.

17. $f(z) = 3x - y + 5 + i(ax + by - 3)$

18. $f(z) = x^2 + axy + by^2 + i(cx^2 + dxy + y^2)$

In Problems 19–22, (a) show that the given function is not analytic at any point but is differentiable along the indicated curve(s), and (b) use (9) or (11) to find the derivative of the function on the curve(s).

19. $f(z) = x^2 + y^2 + 2ixy$; x-axis

20. $f(z) = 3x^2y^2 - 6ix^2y^2$; coordinate axes

21. $f(z) = x^3 + 3xy^2 - x + i(y^3 + 3x^2y - y)$; coordinate axes

22. $f(z) = x^2 - x + y + i(y^2 - 5y - x)$; $y = x + 2$

23. In Section 2.1 we defined the complex exponential function $f(z) = e^z$ in the following manner $e^z = e^x \cos y + ie^x \sin y$.

   (a) Show that $f(z) = e^z$ is an entire function.

   (b) Show that $f'(z) = f(z)$.

24. In Section 2.6 we defined the principal branch of the complex square root in the following manner $z^{1/2} = \sqrt{r}e^{i\theta/2}$, $-\pi < \theta < \pi$.

   (a) Use (10) to show that $f_1(z) = z^{1/2}$ is analytic in the domain $-\pi < \theta < \pi$.

   (b) Use (11) to show that $f_1'(z) = \frac{1}{2}z^{-1/2}$ in this domain.

Focus on Concepts

25. Suppose $f(z)$ is analytic. Show that $|f'(z)|^2 = u_x^2 + v_x^2 = u_y^2 + v_y^2$.

26. Suppose $u(x, y)$ and $v(x, y)$ are the real and imaginary parts of an analytic function f. Can $g(z) = v(x, y) + iu(x, y)$ be an analytic function? Discuss and defend your answer with sound mathematics.

27. Suppose $f(z)$ is analytic. Can $g(z) = \overline{f(z)}$ be analytic? Discuss and defend your answer with sound mathematics.

28. In this problem you are guided through the start of the proof of the proposition:

> If f is analytic in a domain D, and $|f(z)| = c$, where c is constant, then f is constant throughout D.

Proof We begin with the hypothesis that $|f(z)| = c$. If $f(z) = u(x,\ y) + iv(x,\ y)$ then $|f(z)|^2 = c^2$ is the same as $u^2 + v^2 = c^2$. The partial derivatives of the last expression with respect to x and y are, respectively,

$$2uu_x + 2vv_x = 0 \quad \text{and} \quad 2uu_y + 2vv_y = 0.$$

Complete the proof by using the Cauchy-Riemann equations to replace v_x and v_y in the last pair of equations. Then solve for u_x and u_y and draw a conclusion. Use the Cauchy-Riemann equations again and solve for v_x and v_y.

29. If $f(z) = \overline{z}/z$, then show that $|f(z)|$ is constant in the domain defined by $z \neq 0$. Explain why this does not contradict Theorem 3.2.3(i).

30. In this problem you are guided through the start of the proof of the proposition:

> If f is analytic in a domain D, and $f'(z) = 0$, then f is constant throughout D.

Proof We begin with the hypothesis that f is analytic in D and hence it is differentiable throughout D. Hence by (9) of this section and the assumption that $f'(z) = 0$ in D, we have $f'(z) = \dfrac{\partial u}{\partial x} + i\dfrac{\partial v}{\partial x} = 0$. Now complete the proof.

31. Use the Cauchy-Riemann equations to prove that if f is analytic in a domain D, and if $f'(z)$ is a constant, then $f(z)$ must be a complex linear function throughout D.

32. Use the proposition in Problem 30 to show that if f and g are analytic and $f'(z) = g'(z)$, then $f(z) = g(z) + c$, where c is a constant. [*Hint*: Form $h(z) = f(z) - g(z)$.]

33. If $f(z)$ and $\overline{f(z)}$ are both analytic in a domain D, then what can be said about f throughout D?

34. Suppose $x = r\cos\theta$, $y = r\sin\theta$, and $f(z) = u(x,\ y) + iv(x,\ y)$. Show that

$$\frac{\partial u}{\partial r} = \frac{\partial u}{\partial x}\cos\theta + \frac{\partial u}{\partial y}\sin\theta, \qquad \frac{\partial u}{\partial \theta} = -\frac{\partial u}{\partial x}r\sin\theta + \frac{\partial u}{\partial y}r\cos\theta \qquad (12)$$

and

$$\frac{\partial v}{\partial r} = \frac{\partial v}{\partial x}\cos\theta + \frac{\partial v}{\partial y}\sin\theta, \qquad \frac{\partial v}{\partial \theta} = -\frac{\partial v}{\partial x}r\sin\theta + \frac{\partial v}{\partial y}r\cos\theta. \qquad (13)$$

Now use (1) in the foregoing expressions for v_r and v_θ. By comparing your results with the expressions for u_r and u_θ, deduce the Cauchy-Riemann equations in polar coordinates given in (10).

35. Suppose the function $f(z) = u(r,\ \theta) + iv(r,\ \theta)$ is differentiable at a point z whose polar coordinates are $(r,\ \theta)$. Solve the two equations in (12) for u_x and then solve the two equations in (13) for v_x. Then show that the derivative of f at $(r,\ \theta)$ is

$$f'(z) = (\cos\theta - i\sin\theta)\left(\frac{\partial u}{\partial r} + i\frac{\partial v}{\partial r}\right) = e^{-i\theta}\left(\frac{\partial u}{\partial r} + i\frac{\partial v}{\partial r}\right).$$

36. Consider the function

$$f(z) = \begin{cases} 0, & z = 0 \\ \dfrac{z^5}{|z^4|}, & z \neq 0. \end{cases}$$

(**a**) Express f in the form $f(z) = u(x,\ y) + iv(x,\ y)$,

(b) Show that f is not differentiable at the origin.

(c) Show that the Cauchy-Riemann equations are satisfied at the origin. [*Hint:* Use the limit definitions of the partial derivatives $\partial u/\partial x$, $\partial u/\partial y$, $\partial v/\partial x$, and $\partial v/\partial y$ at $(0,\ 0)$.]

3.3   Harmonic Functions

In Section 5.5 we shall see that when a complex function $f(z) = u(x,\ y) + iv(x,\ y)$ is analytic at a point z, then *all* the derivatives of f: $f'(z), f''(z), f'''(z), \ldots$ are also analytic at z. As a consequence of this remarkable fact, we can conclude that *all* partial derivatives of the real functions $u(x,\ y)$ and $v(x,\ y)$ are continuous at z. From the continuity of the partial derivatives we then know that the second-order mixed partial derivatives are equal. This last fact, coupled with the Cauchy-Riemann equations, will be used in this section to demonstrate that there is a connection between the real and imaginary parts of an analytic function $f(z) = u(x,\ y) + iv(x,\ y)$ and the second-order partial differential equation

$$\frac{\partial^2 \phi}{\partial x^2} + \frac{\partial^2 \phi}{\partial y^2} = 0. \tag{1}$$

This equation, one of the most famous in applied mathematics, is known as **Laplace's equation** in two variables. The sum $\dfrac{\partial^2 \phi}{\partial x^2} + \dfrac{\partial^2 \phi}{\partial y^2}$ of the two second-order partial derivatives in (1) is denoted by $\nabla^2 \phi$ and is called the **Laplacian** of ϕ. Laplace's equation (1) is then abbreviated as $\nabla^2 \phi = 0$.

Harmonic Functions   A solution $\phi(x, y)$ of Laplace's equation (1) in a domain D of the plane is given a special name.

Definition 3.3.1   Harmonic Functions

A real-valued function ϕ of two real variables x and y that has continuous first and second-order partial derivatives in a domain D and satisfies Laplace's equation is said to be **harmonic** in D.

Harmonic functions are encountered in the study of temperatures and potentials. These applications are introduced in Section 3.4. In the next theorem we see that both the real and the imaginary parts of an analytic function are harmonic.

Theorem 3.3.1   Harmonic Functions

Suppose the complex function $f(z) = u(x,\ y) + iv(x,\ y)$ is analytic in a domain D. Then the functions $u(x,\ y)$ and $v(x,\ y)$ are harmonic in D.

Proof Assume $f(z) = u(x,\ y) + iv(x,\ y)$ is analytic in a domain D and that u and v have continuous second-order partial derivatives in D.* Since f is analytic, the Cauchy-Riemann equations are satisfied at every point z. Differentiating both sides of $\partial u/\partial x = \partial v/\partial y$ with respect to x and differentiating

---

*The continuity of the second-order partial derivatives of u and v is not part of the hypothesis of the theorem. This fact will be proved in Chapter 5.

both sides of $\partial u/\partial y = -\partial v/\partial x$ with respect to y give, respectively,

$$\frac{\partial^2 u}{\partial x^2} = \frac{\partial^2 v}{\partial x \partial y} \quad \text{and} \quad \frac{\partial^2 u}{\partial y^2} = -\frac{\partial^2 v}{\partial y \partial x}. \tag{2}$$

With the assumption of continuity, the mixed partials $\partial^2 v/\partial x \partial y$ and $\partial^2 v/\partial y \partial x$ are equal. Hence, by adding the two equations in (2) we see that

$$\frac{\partial^2 u}{\partial x^2} + \frac{\partial^2 u}{\partial y^2} = 0 \quad \text{or} \quad \nabla^2 u = 0.$$

This shows that $u(x, y)$ is harmonic.

Now differentiating both sides of $\partial u/\partial x = \partial v/\partial y$ with respect to y and differentiating both sides of $\partial u/\partial y = -\partial v/\partial x$ with respect to x, give, in turn, $\partial^2 u/\partial y \partial x = \partial^2 v/\partial y^2$ and $\partial^2 u/\partial x \partial y = -\partial^2 v/\partial^2 x$. Subtracting the last two equations yields $\nabla^2 v = 0$. ❏

EXAMPLE 1 Harmonic Functions

The function $f(z) = z^2 = x^2 - y^2 + 2xyi$ is entire. Therefore, the functions $u(x, y) = x^2 - y^2$ and $v(x, y) = 2xy$ are necessarily harmonic in any domain D of the complex plane. This is also easy to verify directly:

$$\frac{\partial^2 u}{\partial x^2} + \frac{\partial^2 u}{\partial y^2} = 2 - 2 = 0 \quad \text{and} \quad \frac{\partial^2 v}{\partial x^2} + \frac{\partial^2 v}{\partial y^2} = 0 + 0 = 0.$$

❏

Harmonic Conjugate Functions We have just shown that if a function $f(z) = u(x, y) + iv(x, y)$ is analytic in a domain D, then its real and imaginary parts u and v are necessarily harmonic in D. Now suppose $u(x, y)$ is a given real function that is known to be harmonic in D. If it is possible to find another real harmonic function $v(x, y)$ so that u and v satisfy the Cauchy-Riemann equations throughout the domain D, then the function $v(x, y)$ is called a **harmonic conjugate** of $u(x, y)$. By combining the functions as $u(x, y) + iv(x, y)$ we obtain a complex function that is analytic in D.

EXAMPLE 2 Harmonic Conjugate

(a) Verify that the function $u(x, y) = x^3 - 3xy^2 - 5y$ is harmonic in the entire complex plane.

(b) Find the harmonic conjugate function of u.

Solution

(a) From the partial derivatives

$$\frac{\partial u}{\partial x} = 3x^2 - 3y^2, \quad \frac{\partial^2 u}{\partial x^2} = 6x, \quad \frac{\partial u}{\partial y} = -6xy - 5, \quad \frac{\partial^2 u}{\partial y^2} = -6x$$

we see that u satisfies Laplace's equation

$$\frac{\partial^2 u}{\partial x^2} + \frac{\partial^2 u}{\partial y^2} = 6x - 6x = 0.$$

(b) Since the conjugate harmonic function v must satisfy the Cauchy-Riemann equations $\partial v/\partial y = \partial u/\partial x$ and $\partial v/\partial x = -\partial u/\partial y$, we must have

$$\frac{\partial v}{\partial y} = 3x^2 - 3y^2 \quad \text{and} \quad \frac{\partial v}{\partial x} = 6xy + 5. \tag{3}$$

Partial integration of the first equation in (3) with respect to the variable y gives $v(x, y) = 3x^2y - y^3 + h(x)$. The partial derivative with respect to x of this last equation is

$$\frac{\partial v}{\partial x} = 6xy + h'(x).$$

When this result is substituted into the second equation in (3) we obtain $h'(x) = 5$, and so $h(x) = 5x + C$, where C is a real constant. Therefore, the harmonic conjugate of u is $v(x, y) = 3x^2y - y^3 + 5x + C$.   ❑

In Example 2, by combining u and its harmonic conjugate v as $u(x,y) + iv(x,y)$, the resulting complex function

$$f(z) = x^3 - 3xy^2 - 5y + i(3x^2 - y^3 + 5x + C)$$

is an analytic function throughout the domain D consisting, in this case, of the entire complex plane. In Example 1, since $f(z) = z^2 = x^2 - y^2 + 2xyi$ is entire, the real function $v(x, y) = 2xy$ is the harmonic conjugate of $u(x, y) = x^2 - y^2$. See Problem 20 in Exercises 3.3.

Remarks *Comparison with Real Analysis*

In this section we have seen if $f(z) = u(x, y) + iv(x, y)$ is an analytic function in a domain D, then both functions u and v satisfy $\nabla^2 \phi = 0$ in D. There is another important connection between analytic functions and Laplace's equation. In applied mathematics it is often the case that we wish to solve Laplace's equation $\nabla^2 \phi = 0$ in a domain D in the xy-plane, and for reasons that depend in a very fundamental manner on the shape of D, it simply may not be possible to determine ϕ. *But* it may be possible to devise a special analytic mapping $f(z) = u(x, y) + iv(x, y)$ or

$$u = u(x,y), \ v = v(x,y), \tag{4}$$

from the xy-plane to the uv-plane so that D', the image of D under (4), not only has a more convenient shape but the function $\phi(x, y)$ that satisfies Laplace's equation in D also satisfies Laplace's equation in D'. We then solve Laplace's equation in D' (the solution Φ will be a function of u and v) and then return to the xy-plane and $\phi(x, y)$ by means of (4). This *invariance* of Laplace's equation under the mapping will be utilized in Chapters 4 and 7. See Figure 3.3.1.

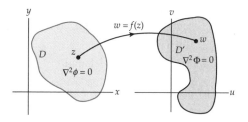

Figure 3.3.1 A solution of Laplace's equation in D is found by solving it in D'.

EXERCISES 3.3 *Answers to selected odd-numbered problems begin on page ANS-11.*

In Problems 1–10, (**a**) verify that the given function u is harmonic in an appropriate domain D, (**b**) find $v(x, y)$, the harmonic conjugate of u, and (**c**) form the corresponding analytic function $f(z) = u + iv$.

1. $u(x, y) = x$

2. $u(x, y) = 2x - 2xy$

3. $u(x, y) = x^2 - y^2$

4. $u(x, y) = x^3 - 3xy^2$

5. $u(x, y) = \log_e(x^2 + y^2)$

6. $u(x, y) = \cos x \cosh y$

7. $u(x, y) = e^x(x \cos y - y \sin y)$

8. $u(x, y) = -e^{-x} \sin y$

9. $u(x, y) = \dfrac{x}{x^2 + y^2}$

10. $u(x, y) = \arctan(-y/x)$

In Problems 11 and 12, verify that the given function u is harmonic in an appropriate domain D. Find its harmonic conjugate v and find an analytic function $f(z) = u + iv$ satisfying the indicated condition.

11. $u(x, y) = xy + x + 2y - 5;\ f(2i) = -1 + 5i$

12. $u(x, y) = 4xy^3 - 4x^3y + x;\ f(1 + i) = 5 + 4i$

13. (**a**) Show that $v(x, y) = \dfrac{x}{x^2 + y^2}$ is harmonic in a domain D not containing the origin.

    (**b**) Find a function $f(z) = u(x, y) + iv(x, y)$ that is analytic in domain D.

    (**c**) Express the function f found in part (b) in terms of the symbol z.

14. Suppose $f(z) = u(r, \theta) + iv(r, \theta)$ is analytic in a domain D not containing the origin. Use the Cauchy-Riemann equations (10) of Section 3.2 in the form $r u_r = v_\theta$ and $r v_r = -u_\theta$ to show that $u(r, \theta)$ satisfies Laplace's equation in polar coordinates:

$$r^2 \frac{\partial^2 u}{\partial r^2} + r \frac{\partial u}{\partial r} + \frac{\partial^2 u}{\partial \theta^2} = 0. \qquad (5)$$

In Problems 15 and 16, use (5) to verify that the given function u is harmonic in a domain D not containing the origin.

15. $u(r, \theta) = r^3 \cos 3\theta$

16. $u(r, \theta) = \dfrac{10r^2 - \sin 2\theta}{r^2}$

Focus on Concepts

17. (**a**) Verify that $u(x, y) = e^{x^2 - y^2} \cos 2xy$ is harmonic in an appropriate domain D.

    (b) Find its harmonic conjugate v and find analytic function $f(z) = u + iv$ satisfying $f(0) = 1$. [*Hint:* When integrating, think of reversing the product rule.]

18. Express the function f found in Problem 11 in terms of the symbol z.

19. **(a)** Show that $\phi(x, y, z) = \dfrac{1}{\sqrt{x^2 + y^2 + z^2}}$ is harmonic, that is, satisfies Laplace's equation $\dfrac{\partial^2 u}{\partial x^2} + \dfrac{\partial^2 u}{\partial y^2} + \dfrac{\partial^2 u}{\partial z^2} = 0$ in a domain D of space not containing the origin.

    (b) Is the two-dimensional analogue of the function in part (a), $\phi(x, y) = \dfrac{1}{\sqrt{x^2 + y^2}}$, harmonic in a domain D of the plane not containing the origin?

20. Construct an example accompanied by a brief explanation that illustrates the following fact:

> *If v is a harmonic conjugate of u in some domain D, then u is, in general, not a harmonic conjugate of v.*

21. If $f(z) = u(x, y) + iv(x, y)$ is an analytic function in a domain D and $f(z) \neq 0$ for all z in D, show that $\phi(x, y) = \log_e |f(z)|$ is harmonic in D.

22. In this problem you are guided through the start of the proof of the proposition:

> *If $u(x, y)$ is a harmonic function and $v(x, y)$ is its harmonic conjugate, then the function $\phi(x, y) = u(x, y)v(x, y)$ is harmonic.*

Proof Suppose $f(z) = u(x, y) + iv(x, y)$ is analytic in a domain D. We saw in Section 3.1 that the product of two analytic functions is analytic. Hence $[f(z)]^2$ is analytic. Now examine $[f(z)]^2$ and finish the proof.

3.4   Applications

In Section 3.3 we saw that if the function $f(z) = u(x, y) + iv(x, y)$ is analytic in a domain D, then the real and imaginary parts of f are harmonic; that is, both u and v have continuous second-partial derivatives and satisfy Laplace's equation in D:

$$\frac{\partial^2 u}{\partial x^2} + \frac{\partial^2 u}{\partial y^2} = 0 \quad \text{and} \quad \frac{\partial^2 v}{\partial x^2} + \frac{\partial^2 v}{\partial y^2} = 0. \tag{1}$$

Conversely, if we know that a function $u(x, y)$ is harmonic in D, we can find a unique (up to an additive constant) harmonic conjugate $v(x, y)$ and construct a function $f(z)$ that is analytic in D.

    In the physical sciences and engineering, Laplace's partial differential equation is often encountered as a mathematical model of some time-independent phenomenon, and in that context the problem we face is to *solve* the equation subject to certain physical side conditions called boundary conditions. See Problems 11–14 in Exercises 3.4. Because of the link displayed in (1), analytic functions are the source of an unlimited number of solutions of Laplace's equation, and we may be able to find one that fits the problem at hand. See Sections 4.5 and 7.5. This is just one reason why the theory of complex variables is so essential in the study of applied mathematics.

    We begin this section by showing that the level curves of the real and imaginary parts of an analytic function $f(z) = u(x, y) + iv(x, y)$ are two **orthogonal families** of curves.

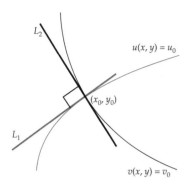

Figure 3.4.1 Tangents L_1 and L_2 at point of intersection z_0 are perpendicular.

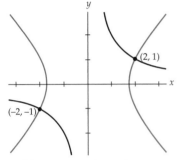

(a) Curves are orthogonal at points of intersection

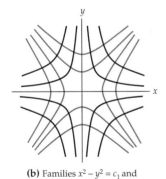

(b) Families $x^2 - y^2 = c_1$ and $2xy = c_2$

Figure 3.4.2 Orthogonal families

Orthogonal Families　Suppose the function $f(z) = u(x,y) + iv(x,y)$ is analytic in some domain D. Then the real and imaginary parts of f can be used to define two families of curves in D. The equations

$$u(x,y) = c_1 \qquad \text{and} \qquad v(x,y) = c_2, \qquad (2)$$

where c_1 and c_2 are arbitrary real constants, are called **level curves** of u and v, respectively. The level curves (2) are **orthogonal families**. Roughly, this means that each curve in one family is orthogonal to each curve in the other family. More precisely, at a point of intersection $z_0 = x_0 + iy_0$, where we shall assume that $f'(z_0) \neq 0$, the tangent line L_1 to the level curve $u(x,y) = u_0$ and the tangent line L_2 to the level curve $v(x,y) = v_0$ are perpendicular. See Figure 3.4.1. The numbers u_0 and v_0 are defined by evaluating u and v at z_0, that is, $c_1 = u(x_0, y_0) = u_0$ and $c_2 = v(x_0, y_0) = v_0$. To prove that L_1 and L_2 are perpendicular at z_0 we demonstrate that the slope of one tangent is the negative reciprocal of the slope of the other by showing that the product of the two slopes is -1. We begin by differentiating $u(x,y) = u_0$ and $v(x,y) = v_0$ with respect to x using the chain rule of partial differentiation:

$$\frac{\partial u}{\partial x} + \frac{\partial u}{\partial y}\frac{dy}{dx} = 0 \qquad \text{and} \qquad \frac{\partial v}{\partial x} + \frac{\partial v}{\partial y}\frac{dy}{dx} = 0.$$

We then solve each of the foregoing equations for dy/dx:

$$\underbrace{\frac{dy}{dx} = -\frac{\partial u/\partial x}{\partial u/\partial y},}_{\text{slope of a tangent to curve } u(x,y) = u_0} \qquad \underbrace{\frac{dy}{dx} = -\frac{\partial v/\partial x}{\partial v/\partial y},}_{\text{slope of a tangent to curve } v(x,y) = v_0} \qquad (3)$$

At (x_0, y_0) we see from (3), the Cauchy-Reimann equations $u_x = v_y$, $u_y = -v_x$, and from $f'(z_0) \neq 0$, that the product of the two slope functions is

$$\left(-\frac{\partial u/\partial x}{\partial u/\partial y}\right)\left(-\frac{\partial v/\partial x}{\partial v/\partial y}\right) = \left(\frac{\partial v/\partial y}{\partial v/\partial x}\right)\left(-\frac{\partial v/\partial x}{\partial v/\partial y}\right) = -1. \qquad (4)$$

EXAMPLE 1　Orthogonal Families

For $f(z) = z^2 = x^2 - y^2 + 2xyi$ we identify $u(x,y) = x^2 - y^2$ and $v(x,y) = 2xy$. For this function, the families of level curves $x^2 - y^2 = c_1$ and $2xy = c_2$ are two families of hyperbolas. Since f is analytic for all z, these families are orthogonal. At a specific point, say, $z_0 = 2 + i$ we find $2^2 - 1^2 = 3 = c_1$ and $2(2)(1) = 4 = c_2$ and two corresponding orthogonal curves are $x^2 - y^2 = 3$ and $xy = 2$. Inspection of Figure 3.4.2(a) shows $x^2 - y^2 = 3$ in color and $xy = 2$ in black; the curves are orthogonal at $z_0 = 2 + i$ (and at $-2 - i$, by symmetry of the curves). In Figure 3.4.2(b) both families are superimposed on the same coordinate axes, the curves in the family $x^2 - y^2 = c_1$ are drawn in color whereas the curves in family $2xy = c_2$ are in black.　❑

Gradient Vector　In vector calculus, if $f(x,y)$ is a differentiable scalar function, then the **gradient** of f, written either **grad** f or ∇f (the symbol ∇ is called a *nabla* or *del*), is defined to be the two-dimensional vector

$$\nabla f = \frac{\partial f}{\partial x}\mathbf{i} + \frac{\partial f}{\partial y}\mathbf{j}. \qquad (5)$$

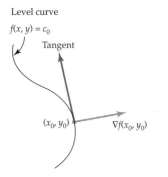

Level curve

$f(x, y) = c_0$

Tangent

(x_0, y_0) $\nabla f(x_0, y_0)$

Figure 3.4.3 Gradient is perpendicular to level curve at (x_0, y_0)

As shown in color in Figure 3.4.3, the gradient vector $\nabla f(x_0, y_0)$ at a point (x_0, y_0) is perpendicular to the level curve of $f(x, y)$ passing through that point, that is, to the level curve $f(x, y) = c_0$, where $c_0 = f(x_0, y_0)$. To see this, suppose that $x = g(t)$, $y = h(t)$, where $x_0 = g(t_0), y_0 = h(t_0)$ are parametric equations for the curve $f(x, y) = c_0$. Then the derivative of $f(x(t), y(t)) = c_0$ with respect to t is

$$\frac{\partial f}{\partial x}\frac{dx}{dt} + \frac{\partial f}{\partial y}\frac{dy}{dt} = 0. \tag{6}$$

This last result is the dot product of (5) with the tangent vector $\mathbf{r}'(t) = x'(t)\mathbf{i} + y'(t)\mathbf{j}$. Specifically, at $t = t_0$, (6) shows that if $\mathbf{r}'(t_0) \neq \mathbf{0}$, then $\nabla f(x_0, y_0) \cdot \mathbf{r}'(t_0) = 0$. This means that ∇f is perpendicular to the level curve at (x_0, y_0).

Gradient Fields   As discussed in Section 2.7, in complex analysis two-dimensional vector fields $\mathbf{F}(x, y) = P(x, y)\mathbf{i} + Q(x, y)\mathbf{j}$, defined in some domain D of the plane, are of interest to us because $\mathbf{F}$ can be represented equivalently as a complex function $f(z) = P(x, y) + iQ(x, y)$. Of particular importance in science are vector fields that can be written as the gradient of some scalar function ϕ with continuous second partial derivatives. In other words, where field $\mathbf{F}(x, y) = P(x, y)\mathbf{i} + Q(x, y)\mathbf{j}$ is the same as

$$\mathbf{F}(x, y) = \nabla\phi = \frac{\partial \phi}{\partial x}\mathbf{i} + \frac{\partial \phi}{\partial y}\mathbf{j}, \tag{7}$$

and so $P(x, y) = \partial\phi/\partial x$ and $Q(x, y) = \partial\phi/\partial y$. Such a vector field $\mathbf{F}$ is called a **gradient field** and ϕ is called a **potential function** or simply the **potential** for $\mathbf{F}$. Gradient fields occur naturally in the study of electricity and magnetism, fluid flows, gravitation, and steady-state temperatures. In a gradient force field, such as a gravitational field, the work done by the force upon a particle moving from position A to position B is the same for all paths between these points. Moreover, the work done by the force along a closed path is zero; in other words, the law of conservation of mechanical energy holds: *kinetic energy + potential energy = constant*. For this reason, gradient fields are also known as **conservative fields**.

In the study of electrostatics the electric field intensity $\mathbf{F}$ due to a collection of stationary charges in a region of the plane is given by $\mathbf{F}(x, y) = -\nabla\phi$, where the real-valued function $\phi(x, y)$ is called the **electrostatic potential**. Gauss' law asserts that the divergence of the field $\mathbf{F}$, that is, $\nabla \cdot \mathbf{F} = P_x + Q_y$ is proportional to the charge density ρ, where ρ is a scalar function. If the

Note ➡

region of the plane is free of charges, then the divergence of $\mathbf{F}$ is zero.* Since $\mathbf{F} = -\nabla\phi$, if $\nabla \cdot \mathbf{F} = 0$, then

$$\nabla \cdot \mathbf{F} = \nabla \cdot \left(-\frac{\partial \phi}{\partial x}\mathbf{i} - \frac{\partial \phi}{\partial y}\mathbf{j}\right) = -\frac{\partial^2 \phi}{\partial x^2} - \frac{\partial^2 \phi}{\partial y^2} = 0,$$

or $\nabla^2 \phi = 0$. In other words: The potential function ϕ satisfies Laplace's equation and is therefore harmonic in some domain D.

---

*The electrostatic potential is then due to charges that are either outside the charge-free region or on the boundary of the region.

Complex Potential　　In general, if a potential function $\phi(x, y)$ satisfies Laplace's equation in some domain D, it is harmonic, and we know from Section 3.3 that there exists a harmonic conjugate function $\psi(x, y)$ defined in D so that the complex function

$$\Omega(z) = \phi(x, y) + i\psi(x, y) \tag{8}$$

is analytic in D. The function $\Omega(z)$ in (8) is called the **complex potential** corresponding to the real potential ϕ. As we have seen in the initial discussion of this section, the level curves of ϕ and ψ are orthogonal families. The level curves of ϕ, $\phi(x, y) = c_1$, are called **equipotential curves**—that is, curves along which the potential is constant. In the case in which ϕ represents electrostatic potential, the electric field intensity $\mathbf{F}$ must be directed along the family of curves orthogonal to the equipotential curves because the force of the field is the gradient of the potential ϕ, $\mathbf{F}(x, y) = -\nabla\phi$, and as demonstrated in (6), the gradient vector at a point (x_0, y_0) is perpendicular to a level curve of ϕ at (x_0, y_0). For this reason the level curves $\psi(x, y) = c_2$, curves that are orthogonal to the family $\phi(x, y) = c_1$, are called **lines of force** and are the paths along which a charged particle will move in the electrostatic field. See Figure 3.4.4.

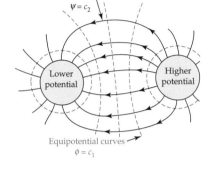

Lines of force
$\psi = c_2$

Lower potential

Higher potential

Equipotential curves
$\phi = c_1$

Figure 3.4.4 Electric field

Ideal Fluid　　In fluid mechanics a flow is said to be two-dimensional or a **planar flow** if the fluid (which could be water, or even air, moving at slow speeds) moves in planes parallel to the xy-plane and the motion and physical properties of the fluid in each plane is precisely the same manner as it is in the xy-plane. Suppose $\mathbf{F}(x, y)$ is the two-dimensional velocity field of a nonviscous fluid that is **incompressible**, that is, a fluid for which div $\mathbf{F} = 0$ or $\nabla \cdot \mathbf{F} = 0$. The flow is **irrotational** if curl $\mathbf{F} = \mathbf{0}$ or $\nabla \times \mathbf{F} = \mathbf{0}$.[*] An incompressible fluid whose planar flow is irrotational is said to be an **ideal fluid**. The velocity field $\mathbf{F}$ of an ideal fluid is a gradient field and can be represented by (7), where ϕ is a real-valued function called a **velocity potential**. The level curves $\phi(x, y) = c_1$ are called **equipotential curves** or simply **equipotentials**. Moreover, ϕ satisfies Laplace's equation because div $\mathbf{F} = 0$ is equivalent to $\nabla \cdot \mathbf{F} = \nabla \cdot (\nabla\phi) = 0$ or $\nabla^2\phi = 0$ and so ϕ is harmonic. The harmonic conjugate $\psi(x, y)$ is called the **stream function** and its level curves $\psi(x, y) = c_2$ are called **streamlines**. Streamlines represent the actual paths along which particles in the fluid will move. The function $\Omega(z) = \phi(x, y) + i\psi(x, y)$ is called the **complex velocity potential** of the flow. See Figure 3.4.5.

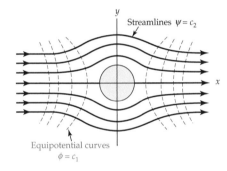

y

Streamlines $\psi = c_2$

x

Equipotential curves
$\phi = c_1$

Figure 3.4.5 Fluid flow

Heat Flow　　Finally, if $\phi(x, y)$ represents time-independent or steady-state temperature that satisfies Laplace's equation, then the level curves $\phi(x, y) = c_1$ are curves along which the temperature is constant and are called **isotherms**. The level curves $\psi(x, y) = c_2$ of the harmonic conjugate function of ϕ are the curves along which heat flows and are called **flow lines** or **flux lines**. See Figure 3.4.6.

Table 3.4.1 summarizes some of the applications of the complex potential function $\Omega(z) = \phi(x, y) + i\psi(x, y)$ and the names given to the level curves

$$\phi(x, y) = c_1 \quad \text{and} \quad \psi(x, y) = c_2.$$

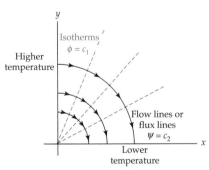

y

Isotherms
$\phi = c_1$

Higher temperature

Flow lines or flux lines
$\psi = c_2$

Lower temperature

x

Figure 3.4.6 Flow of heat

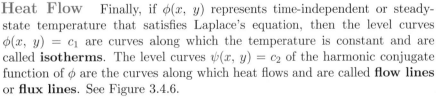

[*]We will discuss fluid flow in greater detail in Section 5.6.

Application	Level curves $\phi(x,y) = c_1$	Level curves $\psi(x,y) = c_2$
electrostatics	equipotential curves	lines of force
fluid flow	equipotential curves	streamlines of flow
gravitation	equipotential curves	lines of force
heat flow	isotherms	lines of heat flux

Table 3.4.1 Complex potential function $\mathbf{\Omega}(z) = \phi(x,y) + i\psi(x,y)$

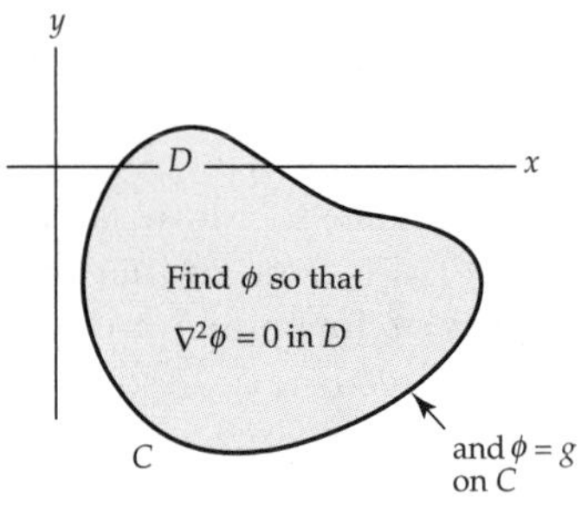

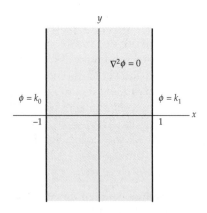

Figure 3.4.7 Dirichlet problem

Dirichlet Problems   A classical and important problem in applied mathematics involving Laplace's equation is illustrated in Figure 3.4.7 and put into words next.

> *Dirichlet Problem*
>
> *Suppose that D is a domain in the plane and that g is a function defined on the boundary C of D. The problem of finding a function $\phi(x,y)$, which satisfies Laplace's equation in D and which equals g on the boundary C of D is called a **Dirichlet problem**.*

Such problems arise frequently in the two-dimensional modeling of electrostatics, fluid flow, gravitation, and heat flow.

In the next example we solve a Dirichlet problem. Although the problem is quite simple, its solution will aid us in the solution of more complicated problems in Section 4.5.

Figure 3.4.8 Figure for Example 2

EXAMPLE 2   A Simple Dirichlet Problem

Solve the Dirichlet problem illustrated in Figure 3.4.8. The domain D is a vertical infinite strip defined by $-1 < x < 1$, $-\infty < y < \infty$; the boundaries of D are the vertical lines $x = -1$ and $x = 1$.

Solution   The Dirichlet problem in Figure 3.4.8 is:

$$Solve: \quad \frac{\partial^2 \phi}{\partial x^2} + \frac{\partial^2 \phi}{\partial y^2} = 0, \qquad -1 < x < 1, \qquad -\infty < y < \infty. \qquad (9)$$

$$Subject\ to: \quad \phi(-1, y) = k_0, \qquad \phi(1, y) = k_1, \qquad -\infty < y < \infty, \qquad (10)$$

where k_0 and k_1 are constants.

The shape of D along with the fact that the two boundary conditions are constant suggest that the function ϕ is independent of y; that is, it is reasonable to try to seek a solution of (9) of the form $\phi(x)$. With this latter assumption, Laplace's partial differential equation (9) becomes the ordinary differential equation $d^2\phi/dx^2 = 0$. Integrating twice gives the general solution $\phi(x) = ax + b$. The boundary conditions enable us to solve for the coefficients a and b. In particular, from $\phi(-1) = k_0$ and $\phi(1) = k_1$ we must have $a(-1) + b = k_0$ and $a(1) + b = k_1$, respectively. Adding the two simultaneous equations gives $2b = k_0 + k_1$, whereas subtracting the first equation from the second yields $2a = k_1 - k_0$. These two results give us a and b:

$$b = \frac{k_1 + k_0}{2} \quad \text{and} \quad a = \frac{k_1 - k_0}{2}.$$

In Section 4.5, for purposes that will be clear there, solution (11) will be denoted as $\phi(x, y)$.
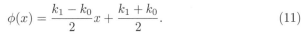

Therefore, we have the following solution of the given Dirichlet problem

$$\phi(x) = \frac{k_1 - k_0}{2}x + \frac{k_1 + k_0}{2}. \tag{11}$$

$\square$

The problem in Example 2 can be interpreted as the determination the electrostatic potential ϕ between two infinitely long parallel conducting plates that are held at constant potentials. Since it satisfies Laplace's equation in D, ϕ is a harmonic function. Hence, a harmonic conjugate ψ can be found as follows. Because ϕ and ψ must satisfy the Cauchy-Riemann equations, we have:

$$\frac{\partial \psi}{\partial y} = \frac{\partial \phi}{\partial x} = \frac{k_1 - k_0}{2} \qquad \text{and} \qquad \frac{\partial \psi}{\partial x} = -\frac{\partial \phi}{\partial y} = 0.$$

The second equation indicates that ψ is a function of y alone, and so integrating the first equation with respect to y we obtain:

$$\psi(y) = \frac{k_1 - k_0}{2}y,$$

where, for convenience, we have taken the constant of integration to be 0. From (8), a complex potential function for the Dirichlet problem in Example 2 is then $\mathbf{\Omega}(z) = \phi(x) + i\psi(y)$, or

$$\mathbf{\Omega}(z) = \frac{k_1 - k_0}{2}x + \frac{k_1 + k_0}{2} + i\frac{k_1 - k_0}{2}y = \frac{k_1 - k_0}{2}z + \frac{k_1 + k_0}{2}.$$

The level curves of ϕ or equipotential curves* are the vertical lines $x = c_1$ shown in color in Figure 3.4.9, and the level curves of ψ or the lines of force are the horizontal line segments $y = c_2$ shown in black. The figure clearly shows that the two families of level curves are orthogonal.

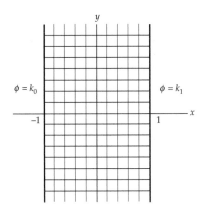

Figure 3.4.9 The equipotential curves and lines of force for Example 2

Remarks

In Section 4.5 and in Chapter 7 we shall introduce a method which enables us to solve Dirichlet problems using analytic mappings.

EXERCISES 3.4   *Answers to selected odd-numbered problems begin on page ANS-11.*

In Problems 1–4, identify the two families of level curves defined by the given analytic function f. By hand, sketch two curves from each family on the same coordinate axes.

1. $f(z) = 2iz - 3 + i$ **2.** $f(z) = (z - 1)^2$

3. $f(z) = \dfrac{1}{z}$ **4.** $f(z) = z + \dfrac{1}{z}$

In Problems 5–8, the given analytic function $f(z) = u + iv$ defines two families of level curves $u(x, y) = c_1$ and $v(x, y) = c_2$. First use implicit differentiation to compute dy/dx for each family and then verify that the families are orthogonal.

---

*The level curves of ϕ are $\phi(x) = C_1$ or $\frac{1}{2}(k_1 - k_0)x + \frac{1}{2}(k_1 + k_0) = C_1$. Solving for x gives $x = \left(C_1 - \frac{1}{2}k_1 - k_0\right)/\frac{1}{2}(k_1 - k_0)$. Set the constant on the right side of the last equation equal to c_1.

5. $f(z) = x - 2x^2 + 2y^2 + i(y - 4xy)$

6. $f(z) = x^3 - 3xy^2 + i(3x^2y - y^3)$

7. $f(z) = e^{-x}\cos y - ie^{-x}\sin y$

8. $f(z) = x + \dfrac{x}{x^2+y^2} + i\left(y - \dfrac{y}{x^2+y^2}\right)$

In Problems 9 and 10, the given real-valued function ϕ is the velocity potential for the planar flow of an incompressible and irrotational fluid. Find the velocity field **F** of the flow. Assume an appropriate domain D of the plane.

9. $\phi(x,y) = \dfrac{x}{x^2+y^2}$ 10. $\phi(x,y) = \frac{1}{2}A\log_e\left(x^2+(y+1)^2\right)$, $A>0$

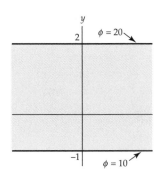

Figure 3.4.10 Figure for Problem 12

11. (a) Find the potential ϕ if the domain D in Figure 3.4.8 is replaced by $0 < x < 1$, $-\infty < y < \infty$, and the potentials on the boundaries are $\phi(0,y) = 50$, $\phi(1,y) = 0$.

    (b) Find the complex potential $\mathbf{\Omega}(z)$.

12. (a) Find the potential ϕ in the domain D between the two infinitely long conducting plates parallel to the x-axis shown in Figure 3.4.10 if the potentials on the boundaries are $\phi(x, -1) = 10$ and $\phi(x, 2) = 20$.

    (b) Find the complex potential $\mathbf{\Omega}(z)$.

    (c) Sketch the equipotential curves and the lines of force.

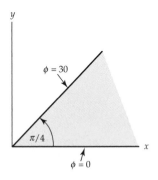

Figure 3.4.11 Figure for Problem 13

13. The potential $\phi(\theta)$ between the two infinitely long conducting plates forming an infinite wedge shown in Figure 3.4.11 satisfies Laplace's equation in polar coordinates in the form

$$\frac{d^2\phi}{d\theta^2} = 0.$$

    (a) Solve the differential equation subject to the boundary conditions $\phi(\pi/4) = 30$ and $\phi(0) = 0$.

    (b) Find the complex potential $\mathbf{\Omega}(z)$.

    (c) Sketch the equipotential curves and the lines of force.

14. The steady-state temperature $\phi(r)$ between the two concentric circular conducting cylinders shown in Figure 3.4.12 satisfies Laplace's equation in polar coordinates in the form

$$r^2\frac{d^2\phi}{dr^2} + r\frac{d\phi}{dr} = 0.$$

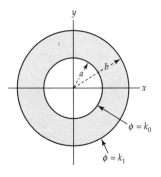

Figure 3.4.12 Figure for Problem 14

    (a) Show that a solution of the differential equation subject to the boundary conditions $\phi(a) = k_0$ and $\phi(b) = k_1$, where k_0 and k_1 are constant potentials, is given by $\phi(r) = A\log_e r + B$, where

$$A = \frac{k_0 - k_1}{\log_e(a/b)} \quad \text{and} \quad B = \frac{-k_0\log_e b + k_1\log_e a}{\log_e(a/b)}.$$

    [*Hint:* The differential equation is known as a Cauchy-Euler equation.]

    (b) Find the complex potential $\mathbf{\Omega}(z)$.

    (c) Sketch the isotherms and the lines of heat flux.

Focus on Concepts

15. Consider the function $f(z) = z + \dfrac{1}{z}$. Describe the level curve $v(x,y) = 0$.

16. The level curves of $u(x, y) = x^2 - y^2$ and $v(x, y) = 2xy$ discussed in Example 1 intersect at $z = 0$. Sketch the level curves that intersect at $z = 0$. Explain why these level curves are *not* orthogonal.

17. Reread the discussion on orthogonal families on page 149 that includes the proof that the tangent lines L_1 and L_2 are orthogonal. In the proof that concludes with (4), explain where the assumption $f'(z_0) \neq 0$ is used.

18. Suppose the two families of curves $u(x, y) = c_1$ and $v(x, y) = c_2$, are orthogonal trajectories in a domain D. Discuss: Is the function $f(z) = u(x, y) + iv(x, y)$ necessarily analytic in D?

Computer Lab Assignments

In Problems 19 and 20, use a CAS or graphing software to plot some representative curves in each of the orthogonal families $u(x, y) = c_1$ and $v(x, y) = c_2$ defined by the given analytic function first on different coordinate axes and then on the same set of coordinate axes.

19. $f(z) = \dfrac{z - 1}{z + 1}$

20. $f(z) = \sqrt{r}\left(\cos\dfrac{\theta}{2} + i\sin\dfrac{\theta}{2}\right), r > 0, -\pi < \theta < \pi$

21. The function $\Omega(z) = A\left(z + \dfrac{1}{z}\right)$, $A > 0$, is a complex potential of a two-dimensional fluid flow.

   (a) Assume $A = 1$. Determine the potential function $\phi(x, y)$ and stream function $\psi(x, y)$ of the flow.

   (b) Express the potential function ϕ and stream function ψ in terms of polar coordinates.

   (c) Use a CAS or graphing software to plot representative curves from each of the orthogonal families $\phi(r, \theta) = c_1$ and $\psi(r, \theta) = c_2$ on the same coordinate axes.

22. The function $\Omega(z) = \log_e\left|\dfrac{z + 1}{z - 1}\right| + i\text{Arg}\left(\dfrac{z + 1}{z - 1}\right)$ is a complex potential of a two-dimensional electrostatic field.

   (a) Show that the equipotential curves $\phi(x, y) = c_1$ and the lines of force $\psi(x, y) = c_1$ are, respectively

   $$(x - \coth c_1)^2 + y^2 = \text{csch}^2\, c_1 \quad \text{and} \quad x^2 + (y + \cot c_2)^2 = \csc^2 c_2.$$

   Observe that the equipotential curves and the lines of force are both families of circles.

   (b) The centers of the equipotential curves in part (a) are $(\coth c_1, 0)$. Approximately, where are the centers located when $c_1 \to \infty$? When $c_1 \to -\infty$? Where are the centers located when $c_1 \to 0^+$? When $c_1 \to 0^-$?

   (c) Verify that each circular line of force passes through $z = 1$ and through $z = -1$.

   (d) Use a CAS or graphing software to plot representative circles from each family on the same coordinate axes. If you use a CAS do not use the contour plot application.

CHAPTER 3 REVIEW QUIZ *Answers to selected odd-numbered problems begin on page ANS-12.*

In Problems 1–12, answer true or false. If the statement is false, justify your answer by either explaining why it is false or giving a counterexample; if the statement is true, justify your answer by either proving the statement or citing an appropriate result in this chapter.

1. If a complex function f is differentiable at point z, then f is analytic at z.

2. The function is $f(z) = \dfrac{y}{x^2 + y^2} + i\dfrac{x}{x^2 + y^2}$ differentiable for all $z \neq 0$.

3. The function $f(z) = z^2 + \bar{z}$ is nowhere analytic.

4. The function $f(z) = \cos y - i \sin y$ is nowhere differentiable.

5. There does not exist an analytic function $f(z) = u(x, y) + iv(x, y)$ for which $u(x, y) = y^3 + 5x$.

6. The function $u(x, y) = e^{4x} \cos 2y$ is the real part of an analytic function.

7. If $f(z) = e^x \cos y + i e^x \sin y$, then $f'(z) = f(z)$.

8. If $u(x, y)$ and $v(x, y)$ are harmonic functions in a domain D, then the function $f(z) = \left(\dfrac{\partial u}{\partial y} - \dfrac{\partial v}{\partial x} \right) + i \left(\dfrac{\partial u}{\partial x} + \dfrac{\partial v}{\partial y} \right)$ is analytic in D.

9. If g is an entire function, then $f(z) = \left(iz^2 + z \right) g(z)$ is necessarily an entire function.

10. The Cauchy-Riemann equations are necessary conditions for differentiability.

11. The Cauchy-Riemann equations can be satisfied at a point z, but the function $f(z) = u(x, y) + iv(x, y)$ can be nondifferentiable at z.

12. If the function $f(z) = u(x, y) + iv(x, y)$ is analytic at a point z, then necessarily the function $g(z) = v(x, y) - iu(x, y)$ is analytic at z.

In Problems 13–22, try to fill in the blanks without referring back to the text.

13. If $f(z) = \dfrac{1}{z^2 + 5iz - 4}$, then $f'(z) = $ _____.

14. The function $f(z) = \dfrac{1}{z^2 + 5iz - 4}$ is not analytic at _____.

15. The function $f(z) = (2 - x)^3 + i(y - 1)^3$ is differentiable at $z = $ _____.

16. For $f(z) = 2x^3 + 3iy^2$, $f'\left(x + ix^2 \right) = $ _____.

17. The function $f(z) = \dfrac{x - 1}{(x - 1)^2 + (y - 1)^2} - i\dfrac{y - 1}{(x - 1)^2 + (y - 1)^2}$ is analytic in a domain D not containing the point $z = 1 + i$. In D, $f'(z) = $ _____.

18. Find an analytic function $f(z) = \log_e \sqrt{x^2 + y^2} + i$ _____ in a domain D not containing the origin.

19. The function $f(z)$ is analytic in a domain D and $f(z) = c + iv(x, y)$, where c is a real constant. Then f is a _____ in D.

20. $\displaystyle\lim_{z \to 2i} \dfrac{z^5 - 4iz^4 - 4z^3 + z^2 - 4iz + 4}{5z^4 - 20iz^3 - 21z^2 - 4iz + 4} = $ _____.

21. $u(x, y) = c_1$ where $u(x, y) = e^{-x}(x \sin y - y \cos y)$ and $v(x, y) = c_2$ where $v(x, y) = $ _____ are orthogonal families.

22. The statement *"There exists a function f that is analytic for $\mathrm{Re}(z) \geq 1$ and is not analytic anywhere else"* is false because _____.

Elementary Functions

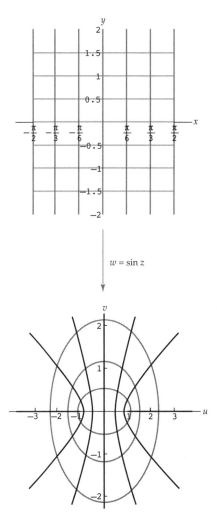

The mapping $w = \sin z$. See page 187.

Introduction In the last chapter we defined a class of functions that is of the most interest in complex analysis, the analytic functions. In this chapter we shall define and study a number of elementary complex analytic functions. In particular, we will investigate the complex exponential, logarithmic, power, trigonometric, hyperbolic, inverse trigonometric, and inverse hyperbolic functions. All of these functions will be shown to be analytic in a suitable domain and their derivatives will be found to agree with their real counterparts. We will also examine how these functions act as mappings of the complex plane. The set of elementary functions will be a useful source of examples that will be used for the remainder of this text.

4.1   Exponential and Logarithmic Functions

The real exponential and logarithmic functions play an important role in the study of real analysis and differential equations. In this section we define and study the complex analogues of these functions. In the first part of this section, we study the complex exponential function e^z, which has already been introduced in Sections 1.6 and 2.1. One concept that has not been discussed previously but will be addressed in this section is the exponential mapping $w = e^z$. In the second half of this section, we introduce the complex logarithm $\ln z$ to solve exponential equations of the form $e^w = z$. If x is a fixed positive *real* number, then there is a *single* solution to the equation $e^y = x$, namely the value $y = \log_e x$. However, we will see that when z is a fixed nonzero *complex* number there are *infinitely* many solutions to the equation $e^w = z$. Therefore, the complex logarithm $\ln z$ is a "multiple-valued function" in the interpretation of this term given in Section 2.4. The principal value of the complex logarithm will be defined to be a (single-valued) function that assigns to the complex input z one of the multiple values of $\ln z$. This principal value function will be shown to be an inverse function of the exponential function e^z defined on a suitably restricted domain of the complex plane. We conclude the section with a discussion of the analyticity of branches of the logarithm.

4.1.1 Complex Exponential Function

Exponential Function and its Derivative   We begin by repeating the definition of the complex exponential function given in Section 2.1.

Definition 4.1.1   Complex Exponential Function

The function e^z defined by

$$e^z = e^x \cos y + i e^x \sin y \tag{1}$$

is called the **complex exponential function**.

One reason why it is natural to call this function the *exponential* function was pointed out in Section 2.1. Namely, the function defined by (1) agrees with the real exponential function when z is real. That is, if z is real, then $z = x + 0i$, and Definition 4.1.1 gives:

$$e^{x+0i} = e^x \left(\cos 0 + i \sin 0 \right) = e^x (1 + i \cdot 0) = e^x. \tag{2}$$

The complex exponential function also shares important differential properties of the real exponential function. Recall that two important properties of the real exponential function are that e^x is differentiable everywhere and that $\dfrac{d}{dx} e^x = e^x$ for all x. The complex exponential function e^z has similar properties.

Theorem 4.1.1   Analyticity of e^z

The exponential function e^z is entire and its derivative is given by:

$$\frac{d}{dz} e^z = e^z. \tag{3}$$

Proof In order to establish that e^z is entire, we use the criterion for analyticity given in Theorem 3.2.2. We first note that the real and imaginary parts, $u(x,\ y) = e^x \cos y$ and $v(x,\ y) = e^x \sin y$, of e^z are continuous real functions and have continuous first-order partial derivatives for all $(x,\ y)$. In addition, the Cauchy-Riemann equations in u and v are easily verified:

$$\frac{\partial u}{\partial x} = e^x \cos y = \frac{\partial v}{\partial y} \quad \text{and} \quad \frac{\partial u}{\partial y} = -e^x \sin y = -\frac{\partial v}{\partial x}.$$

Therefore, the exponential function e^z is entire by Theorem 3.2.2. By (9) of Section 3.2, the derivative of e^z is:

$$\frac{d}{dz} e^z = \frac{\partial u}{\partial x} + i\frac{\partial v}{\partial x} = e^x \cos y + ie^x \sin y = e^z. \qquad \blacksquare$$

Using the fact that the real and imaginary parts of an analytic function are harmonic conjugates, we can also show that the *only* entire function f that agrees with the real exponential function e^x for real input and that satisfies the differential equation $f'(z) = f(z)$ is the complex exponential function e^z defined by (1). See Problem 50 in Exercises 4.1.

EXAMPLE 1   Derivatives of Exponential Functions

Find the derivative of each of the following functions:

(a) $iz^4\left(z^2 - e^z\right)$ and (b) $e^{z^2-(1+i)z+3}$.

Solution (a) Using (3) and the product rule (4) in Section 3.1:

$$\frac{d}{dz}\left[iz^4\left(z^2 - e^z\right)\right] = iz^4\left(2z - e^z\right) + 4iz^3\left(z^2 - e^z\right)$$

$$= 6iz^5 - iz^4 e^z - 4iz^3 e^z.$$

(b) Using (3) and the chain rule (6) in Section 3.1:

$$\frac{d}{dz}\left[e^{z^2-(1+i)z+3}\right] = e^{z^2-(1+i)z+3} \cdot (2z - 1 - i).$$

$$\blacksquare$$

Modulus, Argument, and Conjugate The modulus, argument, and conjugate of the exponential function are easily determined from (1). If we express the complex number $w = e^z$ in polar form:

$$w = e^x \cos y + ie^x \sin y = r\left(\cos\theta + i\sin\theta\right),$$

then we see that $r = e^x$ and $\theta = y + 2n\pi$, for $n = 0,\ \pm 1, \pm 2, \ldots$. Because r is the modulus and θ is an argument of w, we have:

$$|e^z| = e^x \tag{4}$$

and
$$\arg(e^z) = y + 2n\pi,\ n = 0,\ \pm 1, \pm 2, \ldots . \tag{5}$$

We know from calculus that $e^x > 0$ for all real x, and so it follows from (4) that $|e^z| > 0$. This implies that $e^z \neq 0$ for all complex z. Put another way, the point $w = 0$ is not in the range of the complex function $w = e^z$. Equation (4) does not, however, rule out the possibility that e^z is a negative real number. In fact, you should verify that if, say, $z = \pi i$, then $e^{\pi i}$ is real and $e^{\pi i} < 0$.

A formula for the conjugate of the complex exponential e^z is found using properties of the real cosine and sine functions. Since the real cosine function is even, we have $\cos y = \cos(-y)$ for all y, and since the real sine function is odd, we have $-\sin y = \sin(-y)$ for all y, and so:

$$\overline{e^z} = e^x \cos y - ie^x \sin y = e^x \cos(-y) + ie^x \sin(-y) = e^{x-iy} = e^{\bar{z}}.$$

Therefore, for all complex z, we have shown:

$$\overline{e^z} = e^{\bar{z}}. \tag{6}$$

Algebraic Properties   In Theorem 4.1.1 we proved that differentiating the complex exponential function is, in essence, the same as differentiating the real exponential function. These two functions also share the following algebraic properties.

Theorem 4.1.2   Algebraic Properties of e^z

If z_1 and z_2 are complex numbers, then

(i) $e^0 = 1$

(ii) $e^{z_1} e^{z_2} = e^{z_1 + z_2}$

(iii) $\dfrac{e^{z_1}}{e^{z_2}} = e^{z_1 - z_2}$

(iv) $\left(e^{z_1}\right)^n = e^{nz_1}$, $n = 0, \pm 1, \pm 2, \ldots$.

Proof of (i) and (ii) (i) The proof of property (i) follows from the observation that the complex exponential function agrees with the real exponential function for real input. That is, by (2), we have $e^{0+0i} = e^0$, and we know that, for the real exponential function, $e^0 = 1$.

(ii) Let $z_1 = x_1 + iy_1$ and $z_2 = x_2 + iy_2$. By Definition 4.1.1, we have:

$$\begin{aligned}
e^{z_1} e^{z_2} &= \left(e^{x_1} \cos y_1 + ie^{x_1} \sin y_1\right)\left(e^{x_2} \cos y_2 + ie^{x_2} \sin y_2\right) \\
&= e^{x_1+x_2}\left(\cos y_1 \cos y_2 - \sin y_1 \sin y_2\right) \\
&\quad + ie^{x_1+x_2}\left(\sin y_1 \cos y_2 + \cos y_1 \sin y_2\right).
\end{aligned}$$

Using the addition formulas for the real cosine and sine functions given on page 17 in Section 1.3, we can rewrite the foregoing expression as:

$$e^{z_1} e^{z_2} = e^{x_1+x_2} \cos(y_1 + y_2) + ie^{x_1+x_2} \sin(y_1 + y_2). \tag{7}$$

From (1), the right-hand side of (7) is $e^{z_1+z_2}$. Therefore, we have shown that $e^{z_1} e^{z_2} = e^{z_1+z_2}$. ❑

The proofs of Theorem 4.1.2(iii) and 4.1.2(iv) follow in a similar manner. See Problems 47 and 48 in Exercises 4.1.

Periodicity The most striking difference between the real and complex exponential functions is the periodicity of e^z. We say that a complex function f is **periodic** with period T if $f(z + T) = f(z)$ for all complex z. The real exponential function is not periodic, but the complex exponential function is because it is defined using the real cosine and sine functions, which are periodic. In particular, by (1) and Theorem 4.1.2(ii) we have $e^{z+2\pi i} = e^z e^{2\pi i} = e^z (\cos 2\pi + i \sin 2\pi)$. Since $\cos 2\pi = 1$ and $\sin 2\pi = 0$, this simplifies to:

$$e^{z+2\pi i} = e^z. \tag{8}$$

In summary, we have shown that:

> *The complex exponential function e^z is periodic with a pure imaginary period $2\pi i$.*

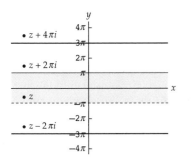

Figure 4.1.1 The fundamental region of e^z

Because (8) holds for all values of z we also have $e^{(z+2\pi i)+2\pi i} = e^{z+2\pi i}$. This fact combined with (8) implies that $e^{z+4\pi i} = e^z$. Now by repeating this process we find that $e^{z+2n\pi i} = e^z$ for $n = 0, \pm 1, \pm 2, \dots$. Thus, the complex exponential function is not one-to-one, and all values e^z are assumed in any infinite horizontal strip of width 2π in the z-plane. For example, all values of the function e^z are assumed in the set $-\infty < x < \infty$, $y_0 < y \leq y_0 + 2\pi$, where y_0 is a real constant. In Figure 4.1.1 we divide the complex plane into horizontal strips obtained by setting y_0 equal to any odd multiple of π. If the point z is in the infinite horizontal strip $-\infty < x < \infty$, $-\pi < y \leq \pi$, shown in color in Figure 4.1.1, then the values $f(z) = e^z$, $f(z+2\pi i) = e^{z+2\pi i}$, $f(z - 2\pi i) = e^{z-2\pi i}$, and so on are the same. The infinite horizontal strip defined by:

$$-\infty < x < \infty, \quad -\pi < y \leq \pi, \tag{9}$$

is called the **fundamental region** of the complex exponential function.

The Exponential Mapping Because all values of the complex exponential function e^z are assumed in the fundamental region defined by (9), the image of this region under the mapping $w = e^z$ is the same as the image of the entire complex plane. In order to determine the image of the fundamental region under $w = e^z$, we note that this region consists of the collection of vertical line segments $z(t) = a + it$, $-\pi < t \leq \pi$, where a is any real number. By (11) of Section 2.2, the image of the line segment $z(t) = a + it$, $-\pi < t \leq \pi$, under the exponential mapping is parametrized by $w(t) = e^{z(t)} = e^{a+it} = e^a e^{it}$, $-\pi < t \leq \pi$, and from (10) of Section 2.2 we see that $w(t)$ defines a circle centered at the origin with radius e^a. Because a can be any real number, the radius e^a of this circle can be any nonzero positive real number. Thus, the image of the fundamental region under the exponential mapping consists of the collection of all circles centered at the origin with nonzero radius. In other words, the image of the fundamental region $-\infty < x < \infty$, $-\pi < y \leq \pi$,

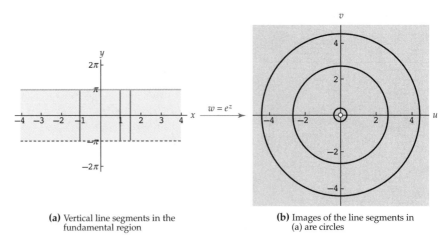

(a) Vertical line segments in the fundamental region

(b) Images of the line segments in (a) are circles

Figure 4.1.2 The image of the fundamental region under $w = e^z$

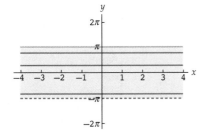

(a) Horizontal lines in the fundamental region

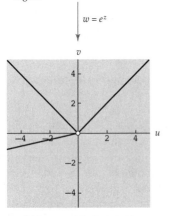

(b) Images of the lines in (a)

Figure 4.1.3 The mapping $w = e^z$

under $w = e^z$ is the set of all complex w with $w \neq 0$, or, equivalently, the set $|w| > 0$.* This agrees with our observation on pages 159–160 that the point $w = 0$ is not in the range of the complex exponential function.

The mapping $w = e^z$ of the fundamental region is shown in Figure 4.1.2. Each vertical line segment shown in color in Figure 4.1.2(a) is mapped onto a circle shown in black in Figure 4.1.2(b) by $w = e^z$. As the x-intercept of a vertical line segment increases, the radius of its image increases. Therefore, the leftmost line segment maps onto the innermost circle, the middle line segment maps onto the middle circle, and the rightmost line segment maps onto the outermost circle.

There is nothing particularly special about using vertical line segments to determine the image of the fundamental region under $w = e^z$. The image can also be found in the same manner by using, say, horizontal lines in the fundamental region. In order to see that this is so, consider the horizontal line $y = b$. This line can be parametrized by $z(t) = t + ib$, $-\infty < t < \infty$, and so its image under $w = e^z$ is given by $w(t) = e^{z(t)} = e^{t+ib} = e^t e^{ib}$, $-\infty < t < \infty$. Define a new parameter by $s = e^t$, and observe that $0 < s < \infty$ since $-\infty < t < \infty$. Using the parameter s, the image is given by $W(s) = e^{ib}s$, $0 < s < \infty$, which, by (8) of Section 2.2, is the set consisting of all points $w \neq 0$ in the ray emanating from the origin and containing the point $e^{ib} = \cos b + i \sin b$. The image can also be described by the equation $\arg(w) = b$. We represent this property of the complex exponential mapping in Figure 4.1.3. Each of the horizontal lines shown in color in Figure 4.1.3(a) is mapped onto a ray shown in black in Figure 4.1.3(b). As the y-intercept of a horizontal line increases, the angle the image ray makes with the positive u-axis increases. Therefore, the bottom-most line maps onto the ray in the third quadrant, the middle line maps onto the ray in the first quadrant, and the topmost line maps onto the ray in the second quadrant.

We now summarize these properties of exponential mapping.

---

*This set is sometimes called the punctured complex plane.

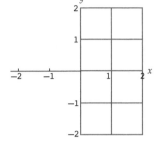

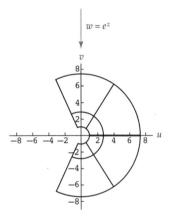

Exponential Mapping Properties

(*i*) $w = e^z$ maps the fundamental region $-\infty < x < \infty$, $-\pi < y \leq \pi$, onto the set $|w| > 0$.

(*ii*) $w = e^z$ maps the vertical line segment $x = a$, $-\pi < y \leq \pi$, onto the circle $|w| = e^a$.

(*iii*) $w = e^z$ maps the horizontal line $y = b$, $-\infty < x < \infty$, onto the ray $\arg(w) = b$.

(a) Figure for Example 2

$w = e^z$

(b) Image of the grid in (a)

Figure 4.1.4 The mapping $w = e^z$

EXAMPLE 2 **Exponential Mapping of a Grid**

Find the image of the grid shown in Figure 4.1.4(a) under $w = e^z$.

Solution The grid in Figure 4.1.4(a) consists of the vertical line segments $x = 0$, 1, and 2, $-2 \leq y \leq 2$, and the horizontal line segments $y = -2$, -1, 0, 1, and 2, $0 \leq x \leq 2$. Using the property (*ii*) of the exponential mapping, we have that the image of the vertical line segment $x = 0$, $-2 \leq y \leq 2$, is the circular arc $|w| = e^0 = 1$, $-2 \leq \arg(w) \leq 2$. In a similar manner, the segments $x = 1$ and $x = 2$, $-2 \leq y \leq 2$, map onto the arcs $|w| = e$ and $|w| = e^2$, $-2 \leq \arg(w) \leq 2$, respectively. By property (*iii*) of the exponential mapping, the horizontal line segment $y = 0$, $0 \leq x \leq 2$, maps onto the portion of the ray emanating from the origin defined by $\arg(w) = 0$, $1 \leq |w| \leq e^2$. This image is the line segment from 1 to e^2 on the u-axis. The remaining horizontal segments $y = -2$, -1, 1, and 2 map in the same way onto the segments defined by $\arg(w) = -2$, $\arg(w) = -1$, $\arg(w) = 1$, and $\arg(w) = 2$, $1 \leq |w| \leq e^2$, respectively. Therefore, the vertical line segments shown in color in Figure 4.1.4(a) map onto the circular arcs shown in black in Figure 4.1.4(b) with the line segment $x = a$ mapping onto the arc with radius e^a. In addition, the horizontal line segments shown in color in Figure 4.1.4(a) map onto the black line segments in Figure 4.1.4(b) with the line segment $y = b$ mapping onto the line segment making an angle of b radians with the positive u-axis. ☐

4.1.2 Complex Logarithmic Function

In real analysis, the natural logarithm function $\ln x$ is often defined as an inverse function of the real exponential function e^x. From this point on, we will use the alternative notation $\log_e x$ to represent the real exponential function. Because the real exponential function is one-to-one on its domain **R**, there is no ambiguity involved in defining this inverse function. The situation is very different in complex analysis because the complex exponential function e^z is not a one-to-one function on its domain **C**. In fact, given a fixed nonzero complex number z, the equation $e^w = z$ has infinitely many solutions. In order to see that this is so, assume that $w = u + iv$. If $e^w = z$, then $|e^w| = |z|$ and

Note: $\log_e x$ will be used to denote the real logarithm. ➡

$\arg(e^w) = \arg(z)$. From (4) and (5), it follows that $e^u = |z|$ and $v = \arg(z)$, or, equivalently, $u = \log_e |z|$ and $v = \arg(z)$. Therefore, given a nonzero complex number z we have shown that:

$$\text{If } e^w = z, \text{ then } w = \log_e |z| + i \arg(z). \tag{10}$$

Because there are infinitely many arguments of z, (10) gives infinitely many solutions w to the equation $e^w = z$. The set of values given by (10) defines a multiple-valued function $w = G(z)$, as described in Section 2.4, which is called the complex logarithm of z and denoted by $\ln z$. The following definition summarizes this discussion.

Definition 4.1.2   Complex Logarithm

The multiple-valued function $\ln z$ defined by:

$$\ln z = \log_e |z| + i \arg(z) \tag{11}$$

is called the **complex logarithm**.

Note: $\ln z$ will be used to denote the complex logarithm.

Hereafter, the notation $\ln z$ will always be used to denote the multiple-valued *complex* logarithm. By switching to exponential notation $z = re^{i\theta}$ in (11), we obtain the following alternative description of the complex logarithm:

$$\ln z = \log_e r + i(\theta + 2n\pi), \quad n = 0, \pm 1, \pm 2, \ldots. \tag{12}$$

From (10) we see that the complex logarithm can be used to find all solutions to the exponential equation $e^w = z$ when z is a nonzero complex number.

EXAMPLE 3   Solving Exponential Equations

Find all complex solutions to each of the following equations.
(a) $e^w = i$ (b) $e^w = 1 + i$ (c) $e^w = -2$

Solution For each equation $e^w = z$, the set of solutions is given by $w = \ln z$ where $\ln z$ is found using Definition 4.1.2.

(a) For $z = i$, we have $|z| = 1$ and $\arg(z) = \pi/2 + 2n\pi$. Thus, from (11) we obtain:

$$w = \ln i = \log_e 1 + i\left(\frac{\pi}{2} + 2n\pi\right).$$

Since $\log_e 1 = 0$, this simplifies to:

$$w = \frac{(4n + 1)\pi}{2} i, \quad n = 0, \pm 1, \pm 2 \ldots.$$

Therefore, each of the values:

$$w = \ldots, -\frac{7\pi}{2}i, -\frac{3\pi}{2}i, \frac{\pi}{2}i, \frac{5\pi}{2}i, \ldots$$

satisfies the equation $e^w = i$.

(b) For $z = 1+i$, we have $|z| = \sqrt{2}$ and $\arg(z) = \pi/4 + 2n\pi$. Thus, from (11) we obtain:

$$w = \ln(1+i) = \log_e \sqrt{2} + i\left(\frac{\pi}{4} + 2n\pi\right).$$

Because $\log_e \sqrt{2} = \frac{1}{2}\log_e 2$, this can be rewritten as:

$$w = \frac{1}{2}\log_e 2 + \frac{(8n+1)\pi}{4}i, \quad n = 0, \pm 1, \pm 2, \dots .$$

Each value of w is a solution to $e^w = 1+i$.

(c) Again we use (11). Since $z = -2$, we have $|z| = 2$ and $\arg(z) = \pi + 2n\pi$, and so:

$$w = \ln(-2) = \log_e 2 + i(\pi + 2n\pi).$$

That is,

$$w = \log_e 2 + (2n+1)\pi i, \quad n = 0, \pm 1, \pm 2, \dots .$$

Each value of w satisfies the equation $e^w = -2$.

❏

Logarithmic Identities Definition 4.1.2 can be used to prove that the complex logarithm satisfies the following identities, which are analogous to identities for the real logarithm.

Theorem 4.1.3   Algebraic Properties of ln z

If z_1 and z_2 are nonzero complex numbers and n is an integer, then

(i) $\ln(z_1 z_2) = \ln z_1 + \ln z_2$

(ii) $\ln\left(\dfrac{z_1}{z_2}\right) = \ln z_1 - \ln z_2$

(iii) $\ln z_1^n = n \ln z_1$.

Proof of (i) By Definition 4.1.2,

$$\ln z_1 + \ln z_2 = \log_e |z_1| + i\arg(z_1) + \log_e |z_2| + i\arg(z_2)$$
$$= \log_e |z_1| + \log_e |z_2| + i(\arg(z_1) + \arg(z_2)). \qquad (13)$$

Because the real logarithm has the property $\log_e a + \log_e b = \log_e(ab)$ for $a > 0$ and $b > 0$, we can write $\log_e |z_1 z_2| = \log_e |z_1| + \log_e |z_2|$. Moreover, from (8) of Section 1.3, we have $\arg(z_1) + \arg(z_2) = \arg(z_1 z_2)$. Therefore, (13) can be rewritten as:

$$\ln z_1 + \ln z_2 = \log_e |z_1 z_2| + i\arg(z_1 z_2) = \ln(z_1 z_2). \qquad ❏$$

Proofs of Theorems 4.1.3(ii) and 4.1.3(iii) are similar. See Problems 53 and 54 in Exercises 4.1.

Principal Value of a Complex Logarithm   It is interesting to note that the *complex* logarithm of a positive real number has infinitely many values. For example, the *complex* logarithm ln 5 is the set of values $1.6094 + 2n\pi i$, where n is any integer, whereas the *real* logarithm $\log_e 5$ has a single value: $\log_e 5 \approx 1.6094$. The unique value of ln 5 corresponding to $n = 0$ is the same as the value of the real logarithm $\log_e 5$. In general, this value of the complex logarithm is called the **principal value of the complex logarithm** since it is found by using the principal argument $\mathrm{Arg}(z)$ in place of the argument $\arg(z)$ in (11). We denote the principal value of the logarithm by the symbol $\mathrm{Ln}\, z$. Thus, the expression $f(z) = \mathrm{Ln}\, z$ defines a function, whereas, $F(z) = \ln z$ defines a multiple-valued function. We summarize this discussion in the following definition.

Notation used throughout this text ➡

Definition 4.1.3   **Principal Value of the Complex Logarithm**

The complex function $\mathrm{Ln}\, z$ defined by:

$$\mathrm{Ln}\, z = \log_e |z| + i\mathrm{Arg}(z) \tag{14}$$

is called the **principal value of the complex logarithm**.

We will use the terms *logarithmic function* and *logarithm* to refer to both the multiple-valued function $\ln z$ and the function $\mathrm{Ln}\, z$. In context, however, it should be clear to which of these we are referring. From (14), we see that the principal value of the complex logarithm can also be given by:

$$\mathrm{Ln}\, z = \log_e r + i\theta, \ -\pi < \theta \le \pi. \tag{15}$$

EXAMPLE 4   **Principal Value of the Complex Logarithm**

Compute the principal value of the complex logarithm $\mathrm{Ln}\, z$ for

(a) $z = i$ (b) $z = 1 + i$ (c) $z = -2$

Solution In each part we apply (14) of Definition 4.1.3.

(a) For $z = i$, we have $|z| = 1$ and $\mathrm{Arg}(z) = \pi/2$, and so:

$$\mathrm{Ln}\, i = \log_e 1 + \frac{\pi}{2}i.$$

However, since $\log_e 1 = 0$, this simplifies to:

$$\mathrm{Ln}\, i = \frac{\pi}{2}i.$$

(b) For $z = 1 + i$, we have $|z| = \sqrt{2}$ and $\mathrm{Arg}(z) = \pi/4$, and so:

$$\mathrm{Ln}(1 + i) = \log_e \sqrt{2} + \frac{\pi}{4}i.$$

Because $\log_e \sqrt{2} = \frac{1}{2}\log_e 2$, this can also be written as:

$$\mathrm{Ln}(1 + i) = \frac{1}{2}\log_e 2 + \frac{\pi}{4}i \approx 0.3466 + 0.7854i.$$

(c) For $z = -2$, we have $|z| = 2$ and $\text{Arg}(z) = \pi$, and so:

$$\text{Ln}(-2) = \log_e 2 + \pi i \approx 0.6931 + 3.1416i.$$

Observe that each of the values found in parts (a)–(c) could have also been found by setting $n = 0$ in the expressions for $\ln z$ from Example 3. ❏

Note: ➡ It is important to note that the identities for the complex logarithm in Theorem 4.1.3 are *not* necessarily satisfied by the principal value of the complex logarithm. For example, it is *not* true that $\text{Ln}(z_1 z_2) = \text{Ln}\, z_1 + \text{Ln}\, z_2$ for all complex numbers z_1 and z_2 (although it may be true for *some* complex numbers). See Problem 55 in Exercises 4.1.

Ln z as an Inverse Function

Because $\text{Ln}\, z$ is *one* of the values of the complex logarithm $\ln z$, it follows from (10) that:

$$e^{\text{Ln}\, z} = z \text{ for all } z \neq 0. \tag{16}$$

This suggests that the logarithmic function $\text{Ln}\, z$ is an inverse function of exponential function e^z. Because the complex exponential function is not one-to-one on its domain, this statement is not completely accurate. Rather, the relationship between these functions is similar to the relationship between the squaring function z^2 and the principal square root function $z^{1/2} = \sqrt{|z|}e^{i\text{Arg}(z)/2}$ defined by (7) in Section 2.4. The exponential function must first be restricted to a domain on which it is one-to-one in order to have a well-defined inverse function. In Problem 52 in Exercises 4.1, you will be asked to show that e^z is a one-to-one function on the fundamental region $-\infty < x < \infty$, $-\pi < y \leq \pi$, shown in Figure 4.1.1.

We now show that if the domain of e^z is restricted to the fundamental region, then the principal value of the complex logarithm $\text{Ln}\, z$ is its inverse function. To justify this claim, consider a point $z = x + iy$ in the fundamental region $-\infty < x < \infty$, $-\pi < y \leq \pi$. From (4) and (5), we have that $|e^z| = e^x$ and $\arg(e^z) = y + 2n\pi$, n an integer. Thus, y is an argument of e^z. Since z is in the fundamental region, we also have $-\pi < y \leq \pi$, and from this it follows that y is the principal argument of e^z. That is, $\text{Arg}(e^z) = y$. In addition, for the real logarithm we have $\log_e e^x = x$, and so from Definition 4.1.3 we obtain:

$$\begin{aligned}
\text{Ln}\, e^z &= \log_e |e^z| + i\text{Arg}(e^z) \\
&= \log_e e^x + iy \\
&= x + iy.
\end{aligned}$$

Thus, we have shown that:

$$\text{Ln}\, e^z = z \text{ if } -\infty < x < \infty \quad \text{and} \quad -\pi < y \leq \pi. \tag{17}$$

From (16) and (17), we conclude that $\text{Ln}\, z$ is the inverse function of e^z defined on the fundamental region. The following summarizes the relationship between these functions.

Ln z as an Inverse Function of e^z

If the complex exponential function $f(z) = e^z$ is defined on the fundamental region $-\infty < x < \infty$, $-\pi < y \leq \pi$, then f is one-to-one and the inverse function of f is the principal value of the complex logarithm $f^{-1}(z) = \text{Ln}\, z$.

Bear in mind that (16) holds for all nonzero complex numbers z, but (17) only holds if z is in the fundamental region. For example, for the point $z = 1 + \frac{3}{2}\pi i$, which is not in the fundamental region, we have:

$$\operatorname{Ln} e^{1+3\pi i/2} = 1 - \frac{1}{2}\pi i \neq 1 + \frac{3}{2}\pi i.$$

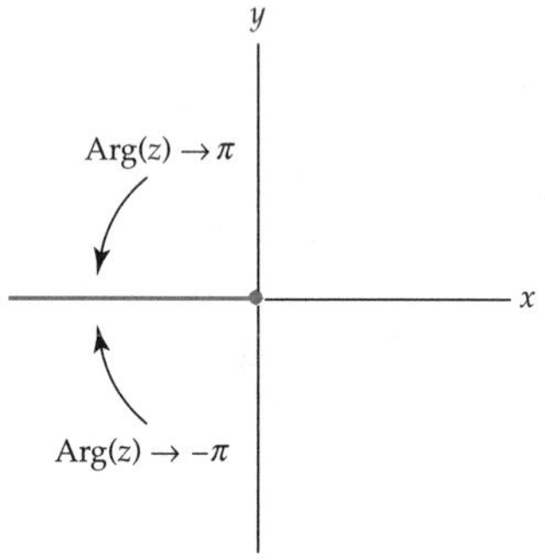

Figure 4.1.4 Ln z is discontinuous at $z = 0$ and on the negative real axis.

Analyticity The principal value of the complex logarithm Ln z is discontinuous at the point $z = 0$ since this function is not defined there. This function also turns out to be discontinuous at every point on the negative real axis. This is intuitively clear since the value of Ln z a point z near the negative x-axis in the second quadrant has imaginary part close to π, whereas the value of a nearby point in the third quadrant has imaginary part close to $-\pi$. See Figure 4.1.4. The function Ln z is, however, continuous on the set consisting of the complex plane excluding the nonpositive real axis. To see that this is so, we appeal to Theorem 2.6.3, which states that a complex function $f(z) = u(x, y) + iv(x, y)$ is continuous at a point $z = x + iy$ if and only if both u and v are continuous real functions at (x, y). By (14), the real and imaginary parts of Ln z are $u(x, y) = \log_e |z| = \log_e \sqrt{x^2 + y^2}$ and $v(x, y) = \operatorname{Arg}(z)$, respectively. From multivariable calculus we have that the function $u(x, y) = \log_e \sqrt{x^2 + y^2}$ is continuous at all points in the plane except $(0, 0)$ and from Problem 55 in Exercises 2.6 we have that the function $v(x, y) = \operatorname{Arg}(z)$ is continuous on the domain $|z| > 0$, $-\pi < \arg(z) < \pi$. Therefore, by Theorem 2.6.3, it follows that Ln z is a continuous function on the domain

$$|z| > 0, \quad -\pi < \arg(z) < \pi, \tag{18}$$

shown in gray in Figure 4.1.5. Put another way, the function f_1 defined by:

$$f_1(z) = \log_e r + i\theta \tag{19}$$

is continuous on the domain in (18) where $r = |z|$ and $\theta = \arg(z)$.

Since the function f_1 agrees with the principal value of the complex logarithm Ln z where they are both defined, it follows that f_1 assigns to the input z one of the values of the multiple-valued function $F(z) = \ln z$. Using the terminology of Section 2.6, we have shown that the function f_1 defined by (19) is a branch of the multiple-valued function $F(z) = \ln z$. (Recall that branches of a multiple-valued function F are denoted by f_1, f_2, and so on.) This branch is called the **principal branch of the complex logarithm**. The nonpositive real axis shown in color in Figure 4.1.5 is a **branch cut** for f_1 and the point $z = 0$ is a **branch point**. As the following theorem demonstrates, the branch f_1 is an analytic function on its domain.

Figure 4.1.5 Branch cut for f_1

Theorem 4.1.4   Analyticity of the Principal Branch of ln z

The principal branch f_1 of the complex logarithm defined by (19) is an analytic function and its derivative is given by:

$$f_1'(z) = \frac{1}{z}. \tag{20}$$

Proof We prove that f_1 is analytic by using the polar coordinate analogue to Theorem 3.2.2 of Section 3.2. Because f_1 is defined on the domain given in (18), if z is a point in this domain, then we can write $z = re^{i\theta}$ with $-\pi < \theta < \pi$. Since the real and imaginary parts of f_1 are $u(r, \theta) = \log_e r$ and $v(r, \theta) = \theta$, respectively, we find that:

$$\frac{\partial u}{\partial r} = \frac{1}{r}, \qquad \frac{\partial v}{\partial \theta} = 1,$$

$$\frac{\partial v}{\partial r} = 0, \quad \text{and} \quad \frac{\partial u}{\partial \theta} = 0.$$

Thus, u and v satisfy the Cauchy-Riemann equations in polar coordinates (10) in Section 3.2:

$$\frac{\partial u}{\partial r} = \frac{1}{r}\frac{\partial v}{\partial \theta} \quad \text{and} \quad \frac{\partial v}{\partial r} = -\frac{1}{r}\frac{\partial u}{\partial \theta}.$$

Because u, v, and the first partial derivatives of u and v are continuous at all points in the domain given in (18), it follows from Theorem 3.2.2 that f_1 is analytic in this domain. In addition, from (11) of Section 3.2, the derivative of f_1 is given by:

$$f_1'(z) = e^{-i\theta}\left(\frac{\partial u}{\partial r} + i\frac{\partial v}{\partial r}\right) = \frac{1}{re^{i\theta}} = \frac{1}{z}. \qquad \qquad \square$$

Because $f_1(z) = \text{Ln } z$ for each point z in the domain given in (18), it follows from Theorem 4.1.4 that $\text{Ln } z$ is differentiable in this domain, and that its derivative is given by f_1'. That is, if $|z| > 0$ and $-\pi < \arg(z) < \pi$ then:

$$\frac{d}{dz}\text{Ln } z = \frac{1}{z}. \tag{21}$$

EXAMPLE 5   Derivatives of Logarithmic Functions

Find the derivatives of the following functions in an appropriate domain:

(a) $z\text{Ln } z$ and   (b) $\text{Ln}(z + 1)$.

Solution (a) From the differentiation rules of Section 3.1 we have that the function $z\text{Ln } z$ is differentiable at all points where both of the functions z and $\text{Ln } z$ are differentiable. Because z is entire and $\text{Ln } z$ is differentiable on the domain given in (18), it follows that $z\text{Ln } z$ is differentiable on the domain defined by $|z| > 0$, $-\pi < \arg(z) < \pi$. In this domain the derivative is given by the product rule (4) of Section 3.1 and (21):

$$\frac{d}{dz}[z\text{Ln } z] = z \cdot \frac{1}{z} + 1 \cdot \text{Ln } z = 1 + \text{Ln } z.$$

(b) The function $\text{Ln}(z + 1)$ is a composition of the functions $\text{Ln } z$ and $z + 1$. Because the function $z+1$ is entire, it follows from the chain rule that $\text{Ln}(z+1)$ is differentiable at all points $w = z+1$ such that $|w| > 0$ and $-\pi < \arg(w) < \pi$. In other words, this function is differentiable at the point w whenever w does not lie on the nonpositive real axis. To determine the corresponding values of z for which $\text{Ln}(z + 1)$ is not differentiable, we first solve for z in terms of w to obtain $z = w - 1$. The equation $z = w - 1$ defines a linear mapping of the w-plane onto the z-plane given by translation by -1. Under this mapping the nonpositive real axis is mapped onto the ray emanating from $z = -1$ and

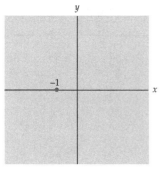

Figure 4.1.6 Ln$(z+1)$ is not differentiable on the ray shown in color.

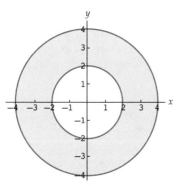

(a) The annulus $2 \leq |z| \leq 4$

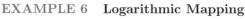

$w = \text{Ln } z$

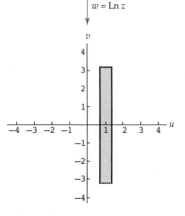

(b) The image of the annulus in (a)

Figure 4.1.7 The mapping $w = \text{Ln } z$

containing the point $z = -2$ shown in color in Figure 4.1.6. Thus, if the point $w = z + 1$ is on the nonpositive real axis, then the point z is on the ray shown in Figure 4.1.6. This implies that Ln$(z + 1)$ is differentiable at all points z that are not on this ray. For such points, the chain rule gives:

$$\frac{d}{dz}\text{Ln}(z+1) = \frac{1}{z+1} \cdot 1 = \frac{1}{z+1}.$$

❑

Logarithmic Mapping   The complex logarithmic mapping $w = \text{Ln } z$ can be understood in terms of the exponential mapping $w = e^z$ since these functions are inverses of each other. For example, because $w = e^z$ maps the fundamental region $-\infty < x < \infty$, $-\pi < y \leq \pi$, in the z-plane onto the set $|w| > 0$ in the w-plane, it follows that inverse mapping $w = \text{Ln } z$ maps the set $|z| > 0$ in the z-plane onto the region $-\infty < u < \infty$, $-\pi < v \leq \pi$, in the w-plane. Other properties of the exponential mapping can be similarly restated as properties of the logarithmic mapping. The following summarizes some of these properties.

Logarithmic Mapping Properties

(i) $w = \text{Ln } z$ maps the set $|z| > 0$ onto the region $-\infty < u < \infty$, $-\pi < v \leq \pi$.

(ii) $w = \text{Ln } z$ maps the circle $|z| = r$ onto the vertical line segment $u = \log_e r$, $-\pi < v \leq \pi$.

(iii) $w = \text{Ln } z$ maps the ray $\arg(z) = \theta$ onto the horizontal line $v = \theta$, $-\infty < u < \infty$.

EXAMPLE 6   Logarithmic Mapping

Find the image of the annulus $2 \leq |z| \leq 4$ under the logarithmic mapping $w = \text{Ln } z$.

Solution   From property (ii) of logarithmic mapping, the boundary circles $|z| = 2$ and $|z| = 4$ of the annulus map onto the vertical line segments $u = \log_e 2$ and $u = \log_e 4$, $-\pi < v \leq \pi$, respectively. In a similar manner, each circle $|z| = r$, $2 \leq r \leq 4$, maps onto a vertical line segment $u = \log_e r$, $-\pi < v \leq \pi$. Since the real logarithmic function is increasing on its domain, it follows that $u = \log_e r$ takes on all values in the interval $\log_e 2 \leq u \leq \log_e 4$ when $2 \leq r \leq 4$. Therefore, the image of the annulus $2 \leq |z| \leq 4$ shown in color in Figure 4.1.7(a) is the rectangular region $\log_e 2 \leq u \leq \log_e 4$, $-\pi < v \leq \pi$, shown in gray in Figure 4.1.7(b). ❑

Other Branches of ln z   The principal branch of the complex logarithm f_1 defined in (19) is just one of many possible branches of the multiple-valued function $F(z) = \ln z$. We can define other branches of F by simply changing the interval defining θ in (18) to a different interval of length 2π. For example,

$$f_2(z) = \log_e r + i\theta, \quad -\frac{\pi}{2} < \theta < \frac{3\pi}{2},$$

defines a branch of F whose branch cut is the nonpositive imaginary axis. You should verify that for the branch f_2 we have $f_2(1) = 0$, $f_2(2i) = \log_e 2 + \frac{1}{2}\pi i$, and $f_2(-1-i) = \frac{1}{2}\log_e 2 + \frac{5}{4}\pi i$.

In the same way that we proved that the principal branch f_1 of the complex logarithm is analytic, we can also show that any branch

$$f_k(z) = \log_e r + i\theta, \quad \theta_0 < \theta < \theta_0 + 2\pi,$$

of $F(z) = \ln z$ is analytic on its domain, and its derivative is given by:

$$f_k'(z) = \frac{1}{z}.$$

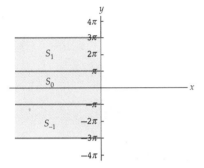

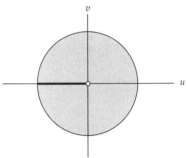

(a) Collection of half-infinite strips S_n

(b) The image of each strip S_n is the punctured unit disk

Figure 4.1.8 The mapping $w = e^z$

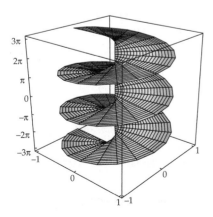

Figure 4.1.9 A Riemann surface for $w = e^z$

Remarks   *Comparison with Real Analysis*

(i) Although the complex exponential and logarithmic functions are similar to the real exponential and logarithmic functions in many ways, it is important to keep in mind their differences.

- The real exponential function is one-to-one, but the complex exponential is not.

- $\log_e x$ is a single-valued function, but $\ln z$ is multiple-valued.

- Many properties of real logarithms apply to the complex logarithm, such as $\ln(z_1 z_2) = \ln z_1 + \ln z_2$, but these properties don't always hold for the principal value $\mathrm{Ln}\, z$.

(ii) Since the complex exponential function is not one-to-one, we can use a Riemann surface, as described in the Remarks at the end of Section 2.4, to help visualize the mapping $w = e^z$. The Riemann surface that we construct will also help us to visualize the multiple-valued function $w = \ln z$. Consider the mapping $w = e^z$ on the half-plane $x \leq 0$. Each half-infinite strip S_n defined by $(2n-1)\pi < y \leq (2n+1)\pi$, $x \leq 0$, for $n = 0, \pm 1, \pm 2, \ldots$ is mapped onto the punctured unit disk $0 < |w| \leq 1$ shown in Figure 4.1.8(b) with the horizontal half-lines shown in color in Figure 4.1.8(a) mapping onto the segment $-1 \leq u < 0$ shown in black in Figure 4.1.8(b). Thus, $w = e^z$ describes an infinite-to-one covering of the punctured unit disk. To visualize this covering, we imagine there being a different image disk B_n for each half-infinite strip S_n. Now cut each disk B_n open along the segment $-1 \leq u < 0$. We construct a Riemann surface for $w = e^z$ by attaching, for each n, the cut disk B_n to the cut disk B_{n+1} along the edge that represents the image of the half-infinite line $y = (2n+1)\pi$. We place this surface in xyz-space so that for each z in the half-plane, the images $\ldots z_{-1}, z_0, z_1, \ldots$ of z in $B_{n-1}, B_0, B_1, \ldots$, respectively, lie directly above the point $w = e^z$ in the xy-plane. See Figure 4.1.9. By projecting the points of the Riemann surface vertically down onto the xy-plane we see the infinite-to-one nature of the mapping $w = e^z$. Conversely, the multiple-valued function $F(z) = \ln z$ may be visualized by considering all points in the Riemann surface lying directly above a point in the xy-plane. These infinitely many points on the Riemann surface correspond to the infinitely many values of $F(z)$ in the half-plane $u \leq 0$.

4.1.1 Complex Exponential Function

In Problems 1–4, find the derivative f' of the given function f.

1. $f(z) = z^2 e^{z+i}$

2. $f(z) = \dfrac{3e^{2z} - ie^{-z}}{z^3 - 1 + i}$

3. $f(z) = e^{iz} - e^{-iz}$

4. $f(z) = ie^{1/z}$

In Problems 5–8, write the given expression in terms of x and y.

5. $\left| e^{z^2 - z} \right|$

6. $\arg\left(e^{z - i/z} \right)$

7. $\arg\left(e^{i(z + \bar{z})} \right)$

8. $\overline{ie^z + 1}$

In Problems 9–12, express the given function f in the form $f(z) = u(x, y) + iv(x, y)$.

9. $f(z) = e^{-iz}$

10. $f(z) = e^{2\bar{z} + i}$

11. $f(z) = e^{z^2}$

12. $f(z) = e^{1/z}$

13. Use the sufficient conditions for differentiability from page 140 in Section 3.2 to determine where the function $f(z) = e^{2\bar{z} + i}$ is differentiable.

14. Use the sufficient conditions for differentiability from page 140 in Section 3.2 to determine where the function $f(z) = e^{z^2}$ is differentiable.

In Problems 15–20, find the image of the given set under the exponential mapping.

15. The line $y = -2$.

16. The line $x = 3$.

17. The infinite strip $1 < x \leq 2$.

18. The square with vertices at 0, 1, $1 + i$, and i.

19. The rectangle $0 \leq x \leq \log_e 2$, $-\pi/4 \leq y \leq \pi/2$.

20. The semi-infinite strip $-\infty < x \leq 0$, $0 \leq y \leq \pi$.

4.1.2 Complex Logarithmic Function

In Problems 21–26, find all complex values of the given logarithm.

21. $\ln(-5)$

22. $\ln(-ei)$

23. $\ln(-2 + 2i)$

24. $\ln(1 + i)$

25. $\ln\left(\sqrt{2} + \sqrt{6}i \right)$

26. $\ln\left(-\sqrt{3} + i \right)$

In Problems 27–32, write the principal value of the logarithm in the form $a + ib$.

27. $\mathrm{Ln}(6 - 6i)$

28. $\mathrm{Ln}(-e^2)$

29. $\mathrm{Ln}(-12 + 5i)$

30. $\mathrm{Ln}(3 - 4i)$

31. $\mathrm{Ln}\left[\left(1 + \sqrt{3}i \right)^5 \right]$

32. $\mathrm{Ln}\left[(1 + i)^4 \right]$

In Problems 33–36, find all complex values of z satisfying the given equation.

33. $e^z = 4i$

34. $e^{1/z} = -1$

35. $e^{z-1} = -ie^3$

36. $e^{2z} + e^z + 1 = 0$

In Problems 37–40, find a domain in which the given function f is differentiable; then find the derivative f'.

37. $f(z) = 3z^2 - e^{2iz} + i\text{Ln}\,z$ **38.** $f(z) = (z+1)\text{Ln}\,z$

39. $f(z) = \dfrac{\text{Ln}(2z - i)}{z^2 + 1}$ **40.** $f(z) = \text{Ln}(z^2 + 1)$

In Problems 41–46, find the image of the given set under the mapping $w = \text{Ln}\,z$.

41. The ray $\arg(z) = \pi/6$.

42. The positive y-axis.

43. The circle $|z| = 4$.

44. The region in the first quadrant bounded by the circles $|z| = 1$ and $|z| = e$.

45. The annulus $3 \le |z| \le 5$.

46. The region outside the unit circle $|z| = 1$ and between the rays $\arg(z) = \pi/4$ and $\arg(z) = 3\pi/4$.

Focus on Concepts

47. Use (1) to prove that $e^{z_1}/e^{z_2} = e^{z_1 - z_2}$.

48. Use (1) and de Moivre's formula to prove that $\left(e^{z_1}\right)^n = e^{nz_1}$, n an integer.

49. Determine where the complex function $e^{\bar{z}}$ is analytic.

50. In this problem, we will show that the complex exponential function defined by (1) is the only complex entire function f that agrees with the real exponential function e^x when z is real and that has the property $f'(z) = f(z)$ for all z.

  (a) Assume that $f(z) = u(x,\, y) + iv(x,\, y)$ is an entire complex function for which $f'(z) = f(z)$. Explain why u and v satisfy the differential equations:

  $$u_x(x,\, y) = u(x,\, y) \quad \text{and} \quad v_x(x,\, y) = v(x,\, y).$$

  (b) Show that $u(x,\, y) = a(y)e^x$ and $v(x,\, y) = b(y)e^x$ are solutions to the differential equations in (a).

  (c) Explain why the assumption that $f(z)$ agrees with the real exponential function for z real implies that $a(0) = 1$ and $b(0) = 0$.

  (d) Explain why the functions $a(y)$ and $b(y)$ satisfy the system of differential equations:

  $$a(y) - b'(y) = 0$$
  $$a'(y) + b(y) = 0.$$

  (e) Solve the system of differential equations in (d) subject to the initial conditions $a(0) = 1$ and $b(0) = 0$.

  (f) Use parts (a)–(e) to show that the complex exponential function defined by (1) is the only complex entire function $f(z)$ that agrees with the real exponential function when z is real and that has the property $f'(z) = f(z)$ for all z.

51. Describe the image of the line $y = x$ under the exponential function. [*Hint:* Find a polar expression $r(\theta)$ of the image.]

52. Prove that e^z is a one-to-one function on the fundamental region $-\infty < x < \infty$, $-\pi < y \le \pi$.

53. Prove that $\ln\left(\dfrac{z_1}{z_2}\right) = \ln z_1 - \ln z_2$ for all nonzero complex numbers z_1 and z_2.

54. Prove that $\ln z_1^n = n \ln z_1$ for all nonzero complex numbers z_1 and all integers n.

55. **(a)** Find two complex numbers z_1 and z_2 so that $\text{Ln}\,(z_1 z_2) \neq \text{Ln}\,z_1 + \text{Ln}\,z_2$.

   (b) Find two complex numbers z_1 and z_2 so that $\text{Ln}\,(z_1 z_2) = \text{Ln}\,z_1 + \text{Ln}\,z_2$.

   (c) What must be true about z_1 and z_2 if $\text{Ln}\,(z_1 z_2) = \text{Ln}\,z_1 + \text{Ln}\,z_2$?

56. Is $\text{Ln}\,z_1^n = n\text{Ln}\,z_1$ for all integers n and complex numbers z_1? Defend your position with a short proof or a counterexample.

Computer Lab Assignments

Most CASs have a built in function to evaluate $\text{Ln}\,z$. For example, in *Mathematica*, the syntax **Log[a + b I]** finds the principal value of the complex logarithm of $a + bi$. To find a numerical approximation of this value enter the expression **N[Log[a + b I]]**. For example, *Mathematica* indicates that **N[Log[2 + 3 I]]** is approximately $1.28247 + 0.982794i$.

In Problems 57–62, use a CAS to compute $\text{Ln}\,z$.

57. $z = -1 - i$ **58.** $z = 2 - 3i$

59. $z = 3 + \pi i$ **60.** $z = 13 + \sqrt{2}i$

61. $z = 4 + 10i$ **62.** $z = \dfrac{12 - i}{2 + 3i}$

In Problems 63–66, use a CAS to find one solution to the equation.

63. $e^{5z-i} = 12i$ **64.** $e^{iz} = 2 - 5i$

65. $3e^{(2+i)z} = 5 - i$ **66.** $ie^{z-2} = \pi$

4.2   Complex Powers

To The Instructor: In this section, we study complex powers z^α where α is a complex constant. This section can be omitted without significantly affecting the development of topics in later chapters.

In Section 2.4 we examined special power functions of the form z^n and $z^{1/n}$ for n an integer and $n \geq 2$. These functions generalized the usual squaring, cubing, square root, cube root, and so on, functions of elementary calculus. In this section we analyze the problem of raising a complex number to an arbitrary real or complex power. For simple integer powers this process is easily understood in terms of complex multiplication. For example, $(1 + i)^3 = (1 + i)(1 + i)(1 + i) = -2 + 2i$. However, there is no similar description for a complex power such as $(1 + i)^i$. In order to define such expressions, we will use the complex exponential and logarithmic functions from Section 4.1.

Complex Powers   Complex powers, such as $(1 + i)^i$ mentioned in the introduction, are defined in terms of the complex exponential and logarithmic functions. Recall from (10) in Section 4.1 that $z = e^{\ln z}$ for all nonzero complex numbers z. Thus, when n is an integer it follows from Theorem 4.1.2(iv) that z^n can be written as $z^n = \left(e^{\ln z}\right)^n = e^{n \ln z}$. This formula, which holds for integer exponents n, suggests the following formula used to define the complex power z^α for any complex exponent α.

Definition 4.2.1   Complex Powers

If α is a complex number and $z \neq 0$, then the **complex power** z^{α} is defined to be:

$$z^{\alpha} = e^{\alpha \ln z}. \tag{1}$$

In general, (1) gives an infinite set of values because the complex logarithm $\ln z$ is multiple-valued. When n is an integer, however, the expression in (1) is single-valued (in agreement with fact that z^n is a function when n is an integer). To see that this is so, we use Theorem 4.1.2(*ii*) to obtain:

$$z^n = e^{n \ln z} = e^{n[\log_e |z| + i \arg(z)]} = e^{n \log_e |z|} e^{n \arg(z) i}. \tag{2}$$

If $\theta = \text{Arg}(z)$, then $\arg(z) = \theta + 2k\pi$ where k is an integer and so

$$e^{n \arg(z) i} = e^{n(\theta + 2k\pi) i} = e^{n\theta i} e^{2nk\pi i}.$$

From Definition 4.1.1 we have that $e^{2nk\pi i} = \cos(2nk\pi) + i \sin(2nk\pi)$. Because n and k are integers, it follows that $2nk\pi$ is an even multiple of π, and so $\cos(2nk\pi) = 1$ and $\sin(2nk\pi) = 0$. Consequently, $e^{2nk\pi i} = 1$ and (2) can be rewritten as:

$$z^n = e^{n \log_e |z|} e^{n\text{Arg}(z) i}, \tag{3}$$

which is single-valued.

Although the previous discussion shows that (1) can define a single-valued function, you should bear in mind that, in general,

$$z^{\alpha} = e^{\alpha \ln z} \tag{4}$$

defines a multiple-valued function. We call the multiple-valued function given by (4) a **complex power function**.

EXAMPLE 1   Complex Powers

Find the values of the given complex power: (**a**) i^{2i} (**b**) $(1 + i)^i$.

Solution In each part, the values of z^{α} are found using (1).

(**a**) In part (a) of Example 3 in Section 4.1 we saw that:

$$\ln i = \frac{(4n + 1)\pi}{2} i.$$

Note: All values of i^{2i} are real. ➡

Thus, by identifying $z = i$ and $\alpha = 2i$ in (1) we obtain:

$$i^{2i} = e^{2i \ln i} = e^{2i[(4n+1)\pi i/2]} = e^{-(4n+1)\pi}$$

for $n = 0, \pm 1, \pm 2, \ldots$. The values of i^{2i} corresponding to, say, $n = -1$, 0, and 1 are 12391.6, 0.0432, and 1.507×10^{-7}, respectively.

(**b**) In part (b) of Example 3 in Section 4.1 we saw that:

$$\ln(1 + i) = \frac{1}{2} \log_e 2 + \frac{(8n + 1)\pi}{4} i$$

for $n = 0, \pm 1, \pm 2, \ldots$. Thus, by identifying $z = 1 + i$ and $\alpha = i$ in (1) we obtain:

$$(1+i)^i = e^{i\ln(1+i)} = e^{i[(\log_e 2)/2 + (8n+1)\pi i/4]},$$

or
$$(1+i)^i = e^{-(8n+1)\pi/4 + i(\log_e 2)/2},$$

for $n = 0, \pm 1, \pm 2, \ldots$.   ❏

Complex powers defined by (1) satisfy the following properties that are analogous to properties of real powers:

$$z^{\alpha_1} z^{\alpha_2} = z^{\alpha_1 + \alpha_2}, \quad \frac{z^{\alpha_1}}{z^{\alpha_2}} = z^{\alpha_1 - \alpha_2},$$

and
$$(z^\alpha)^n = z^{n\alpha} \quad \text{for} \quad n = 0, \pm 1, \pm 2, \ldots. \tag{5}$$

Each of these properties can be derived from Definition 4.2.1 and Theorem 4.1.2. For example, by Definition 4.2.1, we have $z^{\alpha_1} z^{\alpha_2} = e^{\alpha_1 \ln z} e^{\alpha_2 \ln z}$. This can be rewritten as $z^{\alpha_1} z^{\alpha_2} = e^{\alpha_1 \ln z + \alpha_2 \ln z} = e^{(\alpha_1 + \alpha_2)\ln z}$ by using Theorem 4.1.2(*ii*). Since $e^{(\alpha_1 + \alpha_2)\ln z} = z^{\alpha_1 + \alpha_2}$ by (1), we have shown that $z^{\alpha_1} z^{\alpha_2} = z^{\alpha_1 + \alpha_2}$.

Not all properties of real exponents have analogous properties for complex exponents. See the Remarks at the end of this section for an example.

Principal Value of a Complex Power   As pointed out, the complex power z^α given in (1) is, in general, multiple-valued because it is defined using the multiple-valued complex logarithm $\ln z$. We can assign a unique value to z^α by using the principal value of the complex logarithm $\text{Ln } z$ in place of $\ln z$. This particular value of the complex power is called the **principal value** of z^α. For example, since $\text{Ln } i = \pi i/2$, the principal value of i^{2i} is the value of i^{2i} corresponding to $n = 0$ in part (a) of Example 1. That is, the principal value of i^{2i} is $e^{-\pi} \approx 0.0432$. We summarize this discussion in the following definition.

Definition 4.2.2   **Principal Value of a Complex Power**

If α is a complex number and $z \neq 0$, then the function defined by:

$$z^\alpha = e^{\alpha \text{Ln} z} \tag{6}$$

is called the **principal value of the complex power** z^α.

Note ➡   The notation z^α will be used to denote both the multiple-valued power function $F(z) = z^\alpha$ of (4) and the **principal value power function** given by (6). In context it will be clear which of these two we are referring to.

EXAMPLE 2   **Principal Value of a Complex Power**

Find the principal value of each complex power: **(a)** $(-3)^{i/\pi}$ **(b)** $(2i)^{1-i}$.

Solution In each part, we use (6) to find the principal value of z^α.

(a) For $z = -3$, we have $|z| = 3$ and $\text{Arg}(-3) = \pi$, and so $\text{Ln}(-3) = \log_e 3 + i\pi$ by (14) in Section 4.1. Thus, by identifying $z = -3$ and $\alpha = i/\pi$ in (6), we obtain:

$$(-3)^{i/\pi} = e^{(i/\pi)\text{Ln}(-3)} = e^{(i/\pi)(\log_e 3 + i\pi)},$$

or

$$(-3)^{i/\pi} = e^{-1 + i(\log_e 3)/\pi}.$$

Definition 4.1.1 in Section 4.1 gives

$$e^{-1 + i(\log_e 3)/\pi} = e^{-1}\left(\cos\frac{\log_e 3}{\pi} + i\sin\frac{\log_e 3}{\pi}\right),$$

and so,

$$(-3)^{i/\pi} \approx 0.3456 + 0.1260i.$$

(b) For $z = 2i$, we have $|z| = 2$ and $\text{Arg}(z) = \pi/2$, and so $\text{Ln}\,2i = \log_e 2 + i\pi/2$ by (14) in Section 4.1. By identifying $z = 2i$ and $\alpha = 1 - i$ in (6) we obtain:

$$(2i)^{1-i} = e^{(1-i)\text{Ln}2i} = e^{(1-i)(\log_e 2 + i\pi/2)},$$

or

$$(2i)^{1-i} = e^{\log_e 2 + \pi/2 - i(\log_e 2 - \pi/2)}.$$

We approximate this value using Definition 4.1.1 in Section 4.1:

$$(2i)^{1-i} = e^{\log_e 2 + \pi/2}\left[\cos\left(\log_e 2 - \frac{\pi}{2}\right) - i\sin\left(\log_e 2 - \frac{\pi}{2}\right)\right]$$
$$\approx 6.1474 + 7.4008i. \qquad \square$$

Analyticity   In general, the principal value of a complex power z^α defined by (6) is not a continuous function on the complex plane because the function $\text{Ln}\,z$ is not continuous on the complex plane. However, since the function $e^{\alpha z}$ is continuous on the entire complex plane, and since the function $\text{Ln}\,z$ is continuous on the domain $|z| > 0$, $-\pi < \arg(z) < \pi$, it follows that z^α is continuous on the domain $|z| > 0$, $-\pi < \arg(z) < \pi$. Using polar coordinates $r = |z|$ and $\theta = \arg(z)$ we have found that the function defined by:

Recall: Branches of a multiple-valued ➡ function F are denoted by f_1, f_2 and so on.

$$f_1(z) = e^{\alpha(\log_e r + i\theta)}, \quad -\pi < \theta < \pi \qquad (7)$$

is a branch of the multiple-valued function $F(z) = z^\alpha = e^{\alpha \ln z}$. This particular branch is called the **principal branch** of the complex power z^α; its branch cut is the nonpositive real axis, and $z = 0$ is a branch point.

The branch f_1 defined by (7) agrees with the principal value z^α defined by (6) on the domain $|z| > 0$, $-\pi < \arg(z) < \pi$. Consequently, the derivative of f_1 can be found using the chain rule (6) of Section 3.1:

$$f_1'(z) = \frac{d}{dz}e^{\alpha\text{Ln}z} = e^{\alpha\text{Ln}z}\frac{d}{dz}[\alpha\text{Ln}\,z] = e^{\alpha\text{Ln}z}\frac{\alpha}{z}. \qquad (8)$$

Using the principal value $z^\alpha = e^{\alpha\text{Ln}z}$ we find that (8) simplifies to $f_1'(z) = \alpha z^\alpha / z = \alpha z^{\alpha-1}$. That is, on the domain $|z| > 0$, $-\pi < \arg(z) < \pi$, the principal value of the complex power z^α is differentiable and

$$\frac{d}{dz}z^\alpha = \alpha z^{\alpha-1}. \qquad (9)$$

This demonstrates that the power rule (7) of Section 3.1 holds for the principal value of a complex power in the stated domain.

Other branches of the multiple-valued function $F(z) = z^\alpha$ can be defined using the formula in (7) with a different interval of length 2π defining θ. For example, $f_2(z) = e^{\alpha(\log_e r + i\theta)}$, $-\pi/4 < \theta < 7\pi/4$, defines a branch of F whose branch cut is the ray $\arg(z) = -\pi/4$ together with the branch point $z = 0$.

EXAMPLE 3   Derivative of a Power Function

Find the derivative of the principal value z^i at the point $z = 1 + i$.

Solution   Because the point $z = 1 + i$ is in the domain $|z| > 0$, $-\pi < \arg(z) < \pi$, it follows from (9) that:

$$\frac{d}{dz} z^i = i\, z^{i-1},$$

and so,

$$\left. \frac{d}{dz} z^i \right|_{z=1+i} = \left. i\, z^{i-1} \right|_{z=1+i} = i\, (1+i)^{i-1}.$$

We can use (5) to rewrite this value as:

$$i\, (1+i)^{i-1} = i\, (1+i)^i\, (1+i)^{-1} = i\, (1+i)^i\, \frac{1}{1+i} = \frac{1+i}{2}\, (1+i)^i.$$

Moreover, from part (b) of Example 1 with $n = 0$, the principal value of $(1+i)^i$ is:

$$(1+i)^i = e^{-\pi/4 + i(\log_e 2)/2},$$

and so

$$\left. \frac{d}{dz} z^i \right|_{z=1+i} = \frac{1+i}{2}\, e^{-\pi/4 + i(\log_e 2)/2} \approx 0.1370 + 0.2919i. \qquad \square$$

Remarks   *Comparison with Real Analysis*

(i) As mentioned on page 176, there are some properties of real powers that are not satisfied by complex powers. One example of this is that for complex powers, $(z^{\alpha_1})^{\alpha_2} \neq z^{\alpha_1 \alpha_2}$ unless α_2 is an integer. See Problem 14 in Exercises 4.2.

(ii) As with complex logarithms, some properties that do hold for complex powers do not hold for principal values of complex powers. For example, using Definition 4.2.1 and Theorem 4.1.2, we can prove that $(z_1 z_2)^\alpha = z_1^\alpha z_2^\alpha$ for any nonzero complex numbers z_1 and z_2. However, this property does not hold for principal values of these complex powers. In particular, if $z_1 = -1$, $z_2 = i$, and $\alpha = i$, then from (6) we have that the principal value of $(-1 \cdot i)^i$ is $e^{i\operatorname{Ln}(-i)} = e^{\pi/2}$. On the other hand, the product of the principal values of $(-1)^i$ and i^i is $e^{i\operatorname{Ln}(-1)} e^{i\operatorname{Ln}i} = e^{-\pi} e^{-\pi/2} = e^{-3\pi/2}$.

EXERCISES 4.2   *Answers to selected odd-numbered problems begin on page ANS-12.*

In Problems 1–6, find all values of the given complex power.

1. $(-1)^{3i}$ **2.** $3^{2i/\pi}$

3. $(1+i)^{1-i}$ **4.** $\left(1+\sqrt{3}i\right)^{i}$

5. $(-i)^{i}$ **6.** $(ei)^{\sqrt{2}}$

In Problems 7–12, find the principal value of the given complex power.

7. $(-1)^{3i}$ **8.** $3^{2i/\pi}$

9. 2^{4i} **10.** $i^{i/\pi}$

11. $\left(1+\sqrt{3}i\right)^{3i}$ **12.** $(1+i)^{2-i}$

13. Verify that $\dfrac{z^{\alpha_1}}{z^{\alpha_2}} = z^{\alpha_1 - \alpha_2}$ for $z \neq 0$.

14. (a) Verify that $(z^{\alpha})^{n} = z^{n\alpha}$ for $z \neq 0$ and n an integer.

  (b) Find an example that illustrates that for $z \neq 0$ we can have $(z^{\alpha_1})^{\alpha_2} \neq z^{\alpha_1 \alpha_2}$.

In Problems 15–18, find the derivative of the given function at the given point. Let z^{α} represent the principal value of the complex power defined on the domain $|z| > 0$, $-\pi < \arg(z) < \pi$.

15. $z^{3/2}$; $z = 1 + i$ **16.** z^{2i}; $z = i$

17. z^{1+i}; $z = 1 + \sqrt{3}i$ **18.** $z^{\sqrt{2}}$; $z = -i$

Focus on Concepts

19. For any complex number $z \neq 0$, evaluate z^{0}.

20. If $\alpha = x + iy$ where $x = 0, \pm 1, \pm 2, \ldots$, then what can you say about 1^{α}?

21. Show that if $\alpha = 1/n$ where n is a positive integer, then the principal value of z^{α} is the same as the principal nth root of z.

22. (a) Show that if α is a rational number (that is, $\alpha = m/n$ where m and n are integers with no common factor), then z^{α} is finite-valued. That is, show that there are only finitely many values of z^{α}.

  (b) Show that if α is an irrational number (that is, not a rational number) or a complex number, then z^{α} is infinite-valued.

23. Which of the identities listed in (5) hold for the principal value of z^{α}?

24. A useful property of real numbers is $x^{a}y^{a} = (xy)^{a}$.

  (a) Does the property $z^{\alpha}w^{\alpha} = (zw)^{\alpha}$ hold for complex powers?

  (b) Does the property $z^{\alpha}w^{\alpha} = (zw)^{\alpha}$ hold for the principal value of a complex power?

Computer Lab Assignments

Most CASs have a built in function to find the principal value of a complex power. In *Mathematica*, the syntax **(a + b I)^(c + d I)** is used to accomplish this. To find a numerical approximation of this value in *Mathematica* you enter the expression **N[(a + b I)^(c + d I)]**. For example, *Mathematica* indicates that **N[(1 + 2 I)^(3 + 2 I)]** is approximately $0.2647 - 1.1922i$.

In Problems 25–30, use a CAS to find the principal value of the given complex power.

25. $(1 - 5i)^i$ **26.** 5^{5-2i}

27. $(2 - i)^{3+2i}$ **28.** $(1 - 4i)^{1+3i}$

29. $(1 + i)^{(1+i)^{1+i}}$ **30.** $(1 - 3i)^{1/4}$

4.3 Trigonometric and Hyperbolic Functions

In this section we define the complex trigonometric and hyperbolic functions. As with the complex functions e^z and $\mathrm{Ln}\, z$ defined in a previous section, these functions will agree with their real counterparts for real input. In addition, we will show that the complex trigonometric and hyperbolic functions have the same derivatives and satisfy many of the same identities as the real trigonometric and hyperbolic functions.

4.3.1 Complex Trigonometric Functions

If x is a real variable, then it follows from Definition 4.1.1 that:

$$e^{ix} = \cos x + i \sin x \quad \text{and} \quad e^{-ix} = \cos x - i \sin x. \tag{1}$$

By adding these equations and simplifying, we obtain an equation that relates the real cosine function with the complex exponential function:

$$\cos x = \frac{e^{ix} + e^{-ix}}{2}. \tag{2}$$

In a similar manner, if we subtract the two equations in (1), then we obtain an expression for the real sine function:

$$\sin x = \frac{e^{ix} - e^{-ix}}{2i}. \tag{3}$$

The formulas for the *real* cosine and sine functions given in (2) and (3) can be used to define the *complex* sine and cosine functions. Namely, we define these complex trigonometric functions by replacing the real variable x with the complex variable z in (2) and (3).

Definition 4.3.1 Complex Sine and Cosine Functions

The complex **sine** and **cosine** functions are defined by:

$$\sin z = \frac{e^{iz} - e^{-iz}}{2i} \quad \text{and} \quad \cos z = \frac{e^{iz} + e^{-iz}}{2}. \tag{4}$$

It follows from (2) and (3) that the complex sine and cosine functions defined by (4) agree with the real sine and cosine functions for real input. We next define the complex tangent, cotangent, secant, and cosecant functions using the complex sine and cosine:

$$\tan z = \frac{\sin z}{\cos z}, \quad \cot z = \frac{\cos z}{\sin z}, \quad \sec z = \frac{1}{\cos z}, \quad \text{and} \quad \csc z = \frac{1}{\sin z}. \tag{5}$$

These functions also agree with their real counterparts for real input.

EXAMPLE 1   **Values of Complex Trigonometric Functions**

In each part, express the value of the given trigonometric function in the form $a + ib$.

(a) $\cos i$ **(b)** $\sin(2+i)$ **(c)** $\tan(\pi - 2i)$

Solution For each expression we apply the appropriate formula from (4) or (5) and simplify.

(a) By (4),

$$\cos i = \frac{e^{i \cdot i} + e^{-i \cdot i}}{2} = \frac{e^{-1} + e}{2} \approx 1.5431.$$

(b) By (4),

$$\begin{aligned}
\sin(2+i) &= \frac{e^{i(2+i)} - e^{-i(2+i)}}{2i} \\
&= \frac{e^{-1+2i} - e^{1-2i}}{2i} \\
&= \frac{e^{-1}(\cos 2 + i \sin 2) - e(\cos(-2) + i \sin(-2))}{2i} \\
&\approx \frac{0.9781 + 2.8062i}{2i} \\
&\approx 1.4031 - 0.4891i.
\end{aligned}$$

(c) By the first entry in (5) together with (4) we have:

$$\begin{aligned}
\tan(\pi - 2i) &= \frac{\left(e^{i(\pi-2i)} - e^{-i(\pi-2i)}\right)/2i}{\left(e^{i(\pi-2i)} + e^{-i(\pi-2i)}\right)/2} = \frac{e^{i(\pi-2i)} - e^{-i(\pi-2i)}}{\left(e^{i(\pi-2i)} + e^{-i(\pi-2i)}\right)i} \\
&= -\frac{e^2 - e^{-2}}{e^2 + e^{-2}} i \approx -0.9640i.
\end{aligned}$$

❏

Identities Most of the familiar identities for real trigonometric functions hold for the complex trigonometric functions. This follows from Definition 4.3.1 and properties of the complex exponential function. We now list some of the more useful of the trigonometric identities. Each of the results in (6)–(10) is identical to its real analogue.

$$\sin(-z) = -\sin z \quad \cos(-z) = \cos z \tag{6}$$

$$\cos^2 z + \sin^2 z = 1 \tag{7}$$

$$\sin(z_1 \pm z_2) = \sin z_1 \cos z_2 \pm \cos z_1 \sin z_2 \tag{8}$$

$$\cos(z_1 \pm z_2) = \cos z_1 \cos z_2 \mp \sin z_1 \sin z_2 \tag{9}$$

Observe that the double-angle formulas:

$$\sin 2z = 2 \sin z \cos z \quad \cos 2z = \cos^2 z - \sin^2 z \tag{10}$$

follow directly from (8) and (9).

We will verify only identity (7). The other identities follow in a similar manner. See Problems 13 and 14 in Exercises 4.3. In order to verify (7), we note that by (4) and properties of the complex exponential function from Theorem 4.1.2, we have

$$\cos^2 z = \left(\frac{e^{iz} + e^{-iz}}{2}\right)^2 = \frac{e^{2iz} + 2 + e^{-2iz}}{4} \quad ,$$

and

$$\sin^2 z = \left(\frac{e^{iz} - e^{-iz}}{2i}\right)^2 = -\frac{e^{2iz} - 2 + e^{-2iz}}{4}.$$

Therefore,

$$\cos^2 z + \sin^2 z = \frac{e^{2iz} + 2 + e^{-2iz} - e^{2iz} + 2 - e^{-2iz}}{4} = 1.$$

Note ➡   It is important to recognize that some properties of the real trigonometric functions are *not* satisfied by their complex counterparts. For example, $|\sin x| \leq 1$ and $|\cos x| \leq 1$ for all real x, but, from Example 1 we have $|\cos i| > 1$ and $|\sin(2 + i)| > 1$ since $|\cos i| \approx 1.5431$ and $|\sin(2 + i)| \approx 1.4859$, so these inequalities, in general, are not satisfied for complex input.

Periodicity   In Section 4.1 we proved that the complex exponential function is periodic with a pure imaginary period of $2\pi i$. That is, we showed that $e^{z+2\pi i} = e^z$ for all complex z. Replacing z with iz in this equation we obtain $e^{iz+2\pi i} = e^{i(z+2\pi)} = e^{iz}$. Thus, e^{iz} is a periodic function with real period 2π. Similarly, we can show that $e^{-i(z+2\pi)} = e^{-iz}$ and so e^{-iz} is also a periodic function with a real period of 2π. Now from Definition 4.3.1 it follows that:

$$\sin(z + 2\pi) = \frac{e^{i(z+2\pi)} - e^{-i(z+2\pi)}}{2i} = \frac{e^{iz} - e^{-iz}}{2i} = \sin z.$$

A similar statement also holds for the complex cosine function. In summary, we have:

$$\sin(z + 2\pi) = \sin z \quad \text{and} \quad \cos(z + 2\pi) = \cos z \tag{11}$$

for all z. Put another way, (11) shows that the complex sine and cosine are periodic functions with a real period of 2π. The periodicity of the secant and cosecant functions follows immediately from (11) and (5). The identities $\sin(z + \pi) = -\sin z$ and $\cos(z + \pi) = -\cos z$ can be used to show that the complex tangent and cotangent are periodic with a real period of π. See Problems 51 and 52 in Exercises 4.3.

Trigonometric Equations   We now turn our attention to solving simple trigonometric equations. Because the complex sine and cosine functions are periodic, there are always infinitely many solutions to equations of the form $\sin z = w$ or $\cos z = w$. One approach to solving such equations is to use Definition 4.3.1 in conjunction with the quadratic formula. We demonstrate this method in the following example.

EXAMPLE 2   Solving Trigonometric Equations

Find all solutions to the equation $\sin z = 5$.

Solution   By Definition 4.3.1, the equation $\sin z = 5$ is equivalent to the equation

$$\frac{e^{iz} - e^{-iz}}{2i} = 5.$$

By multiplying this equation by e^{iz} and simplifying we obtain

$$e^{2iz} - 10ie^{iz} - 1 = 0.$$

This equation is quadratic in e^{iz}. That is,

$$e^{2iz} - 10ie^{iz} - 1 = \left(e^{iz}\right)^2 - 10i\left(e^{iz}\right) - 1 = 0.$$

Thus, it follows from the quadratic formula (3) of Section 1.6 that the solutions of $e^{2iz} - 10ie^{iz} - 1 = 0$ are given by

$$e^{iz} = \frac{10i + (-96)^{1/2}}{2} = 5i \pm 2\sqrt{6}i = \left(5 \pm 2\sqrt{6}\right)i. \tag{12}$$

In order to find the values of z satisfying (12), we solve the two exponential equations in (12) using the complex logarithm. If $e^{iz} = \left(5 + 2\sqrt{6}\right)i$, then $iz = \ln\left(5i + 2\sqrt{6}i\right)$ or $z = -i\ln\left[\left(5 + 2\sqrt{6}\right)i\right]$. Because $\left(5 + 2\sqrt{6}\right)i$ is a pure imaginary number and $5 + 2\sqrt{6} > 0$, we have $\arg\left[\left(5 + 2\sqrt{6}\right)i\right] = \frac{1}{2}\pi + 2n\pi$. Thus,

$$z = -i\log\left[\left(5 + 2\sqrt{6}\right)i\right] = -i\left[\log_e\left(5 + 2\sqrt{6}\right) + i\left(\frac{\pi}{2} + 2n\pi\right)\right]$$

or

$$z = \frac{(4n+1)\pi}{2} - i\log_e\left(5 + 2\sqrt{6}\right) \tag{13}$$

for $n = 0, \pm 1, \pm 2, \ldots$. In a similar manner, we find that if $e^{iz} = \left(5 - 2\sqrt{6}\right)i$, then $z = -i\ln\left[\left(5 - 2\sqrt{6}\right)i\right]$. Since $\left(5 - 2\sqrt{6}\right)i$ is a pure imaginary number and $5 - 2\sqrt{6} > 0$, it has an argument of $\pi/2$, and so:

$$z = -i\log\left[\left(5 - 2\sqrt{6}\right)i\right] = -i\left[\log_e\left(5 - 2\sqrt{6}\right) + i\left(\frac{\pi}{2} + 2n\pi\right)\right]$$

$$z = \frac{(4n+1)\pi}{2} - i\log_e\left(5 - 2\sqrt{6}\right) \tag{14}$$

or

for $n = 0, \pm 1, \pm 2, \ldots$. Therefore, we have shown that if $\sin z = 5$, then z is one of the values given in (13) or (14). ❏

Modulus The modulus of a complex trigonometric function can also be helpful in solving trigonometric equations. To find a formula in terms of x and y for the modulus of the sine and cosine functions, we first express these functions in terms of their real and imaginary parts. If we replace the symbol z with the symbol $x + iy$ in the expression for $\sin z$ in (4), then we obtain:

$$\sin z = \frac{e^{-y+ix} - e^{y-ix}}{2i} = \frac{e^{-y}(\cos x + i\sin x) - e^{y}(\cos x - i\sin x)}{2i}$$

$$= \sin x\left(\frac{e^{y} + e^{-y}}{2}\right) + i\cos x\left(\frac{e^{y} - e^{-y}}{2}\right). \tag{15}$$

Since the real hyperbolic sine and cosine functions are defined by $\sinh y = \dfrac{e^{y} - e^{-y}}{2}$ and $\cosh y = \dfrac{e^{y} + e^{-y}}{2}$ we can rewrite (15) as

$$\sin z = \sin x\cosh y + i\cos x\sinh y. \tag{16}$$

A similar computation enables us to express the complex cosine function in terms of its real and imaginary parts as:

$$\cos z = \cos x \cosh y - i \sin x \sinh y. \tag{17}$$

We now use (16) and (17) to derive formulas for the modulus of the complex sine and cosine functions. From (16) we have:

$$|\sin z| = \sqrt{\sin^2 x \cosh^2 y + \cos^2 x \sinh^2 y}.$$

This formula can be simplified using the identities $\cos^2 x + \sin^2 x = 1$ and $\cosh^2 y = 1 + \sinh^2 y$ for the real trigonometric and hyperbolic functions:

$$|\sin z| = \sqrt{\sin^2 x \left(1 + \sinh^2 y\right) + \cos^2 x \sinh^2 y}$$
$$= \sqrt{\sin^2 x + \left(\cos^2 x + \sin^2 x\right) \sinh^2 y},$$

or

$$|\sin z| = \sqrt{\sin^2 x + \sinh^2 y}. \tag{18}$$

After a similar computation we obtain the following expression for the modulus of the complex cosine function:

$$|\cos z| = \sqrt{\cos^2 x + \sinh^2 y}. \tag{19}$$

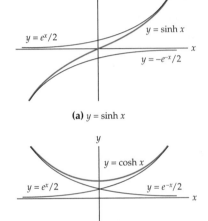

(a) $y = \sinh x$

(b) $y = \cosh x$

Figure 4.3.1 The real hyperbolic functions

You may recall from calculus that the real hyperbolic function $\sinh x$ is unbounded on the real line. See Figure 4.3.1(a). As a result of this fact, the expressions in (18) and (19) can be made arbitrarily large by choosing y to be arbitrarily large. Thus, the complex sine and cosine functions are unbounded on the complex plane. That is, there does not exist a real constant M so that $|\sin z| < M$ for all z in **C**, nor does there exist a real constant M so that $|\cos z| < M$ for all z in **C**. This, of course, is quite different from the situation for the real sine and cosine functions for which $|\sin x| \leq 1$ and $|\cos x| \leq 1$ for all real x.

Zeros   The formulas derived for the modulus of the complex sine and cosine are helpful in determining the zeros of these functions. Recall that the zeros of the real sine function occur at integer multiples of π, and that the zeros of the real cosine function occur at odd integer multiples of $\pi/2$. Since the complex sine and cosine functions agree with their real counterparts for real input, it follows that these zeros of the *real* sine and cosine functions are also zeros of the *complex* sine and cosine functions. It is a natural question to ask whether the complex sine and cosine functions have any additional zeros in the complex plane. One way of answering this question is by solving the equations $\sin z = 0$ and $\cos z = 0$ in the manner presented in Example 2. A different method involves recognizing that a complex number is equal to 0 if and only if its modulus is 0. Thus, solving the equation $\sin z = 0$ is equivalent to solving the equation $|\sin z| = 0$. Using (18) we see that if $|\sin z| = 0$, then $\sqrt{\sin^2 x + \sinh^2 y} = 0$, which is equivalent to:

$$\sin^2 x + \sinh^2 y = 0.$$

Since $\sin^2 x$ and $\sinh^2 y$ are both nonnegative real numbers, this last equation is satisfied if and only if $\sin x = 0$ and $\sinh y = 0$. As just noted, $\sin x = 0$ when

$x = n\pi$, $n = 0, \pm 1, \pm 2, \ldots$, and inspection of Figure 4.3.1(a) indicates that $\sinh y = 0$ only when $y = 0$. Therefore, the only solutions of the equation $\sin z = 0$ in the complex plane are the real numbers $z = n\pi$, $n = 0, \pm 1$, $\pm 2, \ldots$. That is, the zeros of the complex sine function are the same as the zeros of the real sine functions; there are no additional zeros of sine in the complex plane. This is not the same as the situation for polynomial functions where there are often additional zeros in the complex plane.

In essentially the same manner we can show that the only zeros of the complex cosine function are the real numbers $z = (2n+1)\pi/2$, $n = 0, \pm 1$, $\pm 2 \ldots$. See Problem 41 in Exercises 4.3. In summary we have:

All of the zeros of $\sin z$ and $\cos z$ are real.

$$\sin z = 0 \text{ if and only if } z = n\pi, \tag{20}$$

and

$$\cos z = 0 \text{ if and only if } z = \frac{(2n+1)\pi}{2} \tag{21}$$

for $n = 0, \pm 1, \pm 2, \ldots$.

Analyticity The derivatives of the complex sine and cosine functions are found using the chain rule (6) in Section 3.1. For the complex sine function we have:

$$\frac{d}{dz}\sin z = \frac{d}{dz}\left(\frac{e^{iz} - e^{-iz}}{2i}\right) = \frac{ie^{iz} + ie^{-iz}}{2i} = \frac{e^{iz} + e^{-iz}}{2},$$

or

$$\frac{d}{dz}\sin z = \cos z.$$

Since this derivative is defined for all complex z, $\sin z$ is an entire function. In a similar manner, we find

$$\frac{d}{dz}\cos z = -\sin z.$$

The derivatives of $\sin z$ and $\cos z$ can then be used to show that derivatives of all of the complex trigonometric functions are the same as derivatives of the real trigonometric functions. The derivatives of the six complex trigonometric functions are summarized in the following.

> ### Derivatives of Complex Trigonometric Functions
>
> $$\frac{d}{dz}\sin z = \cos z \qquad\qquad \frac{d}{dz}\cos z = -\sin z$$
> $$\frac{d}{dz}\tan z = \sec^2 z \qquad\qquad \frac{d}{dz}\cot z = -\csc^2 z$$
> $$\frac{d}{dz}\sec z = \sec z \tan z \qquad\qquad \frac{d}{dz}\csc z = -\csc z \cot z$$

The sine and cosine functions are entire, but the tangent, cotangent, secant, and cosecant functions are only analytic at those points where the denominator is nonzero. From (20) and (21), it then follows that the tangent and secant functions have singularities at $z = (2n+1)\pi/2$ for $n = 0, \pm 1$, $\pm 2 \ldots$, whereas the cotangent and cosecant functions have singularities at $z = n\pi$ for $n = 0, \pm 1, \pm 2 \ldots$.

Trigonometric Mapping   We will now discuss the complex mapping $w = \sin z$ of the z-plane onto the w-plane. Because $\sin z$ is periodic with a real period of 2π, this function takes on all values in any infinite vertical strip $x_0 < x \leq x_0 + 2\pi$, $-\infty < y < \infty$. In a manner similar to that used to study the exponential mapping $w = e^z$, this allows us to study the mapping $w = \sin z$ on the entire complex plane by analyzing it on any one of these strips. Consider, say, the strip $-\pi < x \leq \pi$, $-\infty < y < \infty$. Before we examine the complex mapping $w = \sin z$ on this strip, observe that $\sin z$ is not one-to-one on this region. For example, $z_1 = 0$ and $z_2 = \pi$ are in this region and $\sin 0 = \sin \pi = 0$. From the identity $\sin(-z + \pi) = \sin z$ it follows that the image of the strip $-\pi < x \leq -\pi/2$, $-\infty < y < \infty$, is the same as the image of the strip $\pi/2 < x \leq \pi$, $-\infty < y < \infty$, under $w = \sin z$. Therefore, we need only consider the mapping $w = \sin z$ on the region $-\pi/2 \leq x \leq \pi/2$, $-\infty < y < \infty$, to gain an understanding of this mapping on the entire z-plane. In Problem 45 of Exercises 4.3 you will be asked to show that the complex sine function is one-to-one on the domain $-\pi/2 < x < \pi/2$, $-\infty < y < \infty$.

EXAMPLE 3   The Mapping $w = \sin z$

Describe the image of the region $-\pi/2 \leq x \leq \pi/2$, $-\infty < y < \infty$, under the complex mapping $w = \sin z$.

Solution   Similar to the discussion on pages 161–162 of Section 4.1, one approach to this problem is to determine the image of vertical lines $x = a$ with $-\pi/2 \leq a \leq \pi/2$ under $w = \sin z$. Assume for the moment that $a \neq -\pi/2$, 0, or $\pi/2$. From (16) the image of the vertical line $x = a$ under $w = \sin z$ is given by:

$$u = \sin a \cosh y, \quad v = \cos a \sinh y, \quad -\infty < y < \infty. \tag{22}$$

We will eliminate the variable y in (22) to obtain a single Cartesian equation relating u and v. Since $-\pi/2 < a < \pi/2$ and $a \neq 0$, it follows that $\sin a \neq 0$ and $\cos a \neq 0$, and so from (22) we obtain $\cosh y = \dfrac{u}{\sin a}$ and $\sinh y = \dfrac{v}{\cos a}$. The identity $\cosh^2 y - \sinh^2 y = 1$ for real hyperbolic functions then gives the following equation:

$$\left(\frac{u}{\sin a}\right)^2 - \left(\frac{v}{\cos a}\right)^2 = 1. \tag{23}$$

The Cartesian equation in (23) is a hyperbola with vertices at $(\pm \sin a, \ 0)$ and slant asymptotes $v = \pm\left(\dfrac{\cos a}{\sin a}\right) u$. Because the point $(a, \ 0)$ is on the line $x = a$, the point $(\sin a, \ 0)$ must be on the image of the line. Therefore, the image of the vertical line $x = a$ with $-\pi/2 < a < \pi/2$ and $a \neq 0$ under $w = \sin z$ is the branch* of the hyperbola (23) that contains the point $(\sin a, 0)$. Because $\sin(-z) = -\sin z$ for all z, it also follows that the image of the line $x = -a$ is branch of the hyperbola (23) containing the point $(-\sin a, \ 0)$. We illustrate this mapping property of $w = \sin z$ in Figure 4.3.2, where the vertical lines shown in color in Figure 4.3.2(a) are mapped onto the hyperbolas shown in black in Figure 4.3.2(b). The line $x = \pi/3$ is mapped onto the branch of the hyperbola containing the point $\left(\frac{1}{2}\sqrt{3}, 0\right)$ and the line $x = \pi/6$ is mapped

---
*Do not confuse this term with "branch of a multiple-valued function."

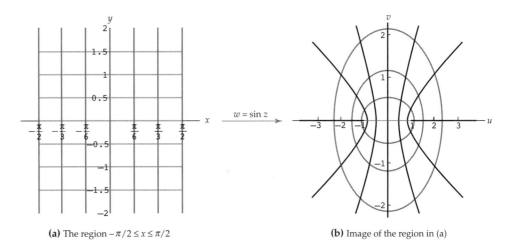

(a) The region $-\pi/2 \leq x \leq \pi/2$ (b) Image of the region in (a)

Figure 4.3.2 The mapping $w = \sin z$

onto the branch containing the point $\left(\frac{1}{2}, 0\right)$. Similarly, the line $x = -\pi/3$ is mapped onto the branch containing the point $\left(-\frac{1}{2}\sqrt{3}, 0\right)$ and the line $x = -\pi/6$ is mapped onto the branch containing the point $\left(-\frac{1}{2}, 0\right)$.

The images of the lines $x = -\pi/2$, $x = \pi/2$ and $x = 0$ cannot be found from (23). However, from (22) we see that the image of the line $x = -\pi/2$ is the set of points $u \leq -1$ on the negative real axis, that the image of the line $x = \pi/2$ is the set of points $u \geq 1$ on the positive real axis, and that the image of the line $x = 0$ is the imaginary axis $u = 0$. See Figure 4.3.2. In summary, we have shown that the image of the infinite vertical strip $-\pi/2 \leq x \leq \pi/2$, $-\infty < y < \infty$, under $w = \sin z$, is the entire w-plane. ☐

In Example 3 the image could also be found using horizontal line segments $y = b$, $-\pi/2 \leq x \leq \pi/2$, instead of vertical lines. In this case, the images are given by:

$$u = \sin x \cosh b, \quad v = \cos x \sinh b, \quad -\frac{\pi}{2} < x < \frac{\pi}{2}.$$

When $b \neq 0$, this set is also given by the Cartesian equation:

$$\left(\frac{u}{\cosh b}\right)^2 + \left(\frac{v}{\sinh b}\right)^2 = 1, \tag{24}$$

which is an ellipse with u-intercepts at $(\pm \cosh b, 0)$ and v-intercepts at $(0, \pm \sinh b)$. If $b > 0$, then the image of the line segment $y = b$ is the upper-half of the ellipse defined by (24) and the image of the line segment $y = -b$ is the bottom-half of the ellipse. Thus, the horizontal line segments shown in color in Figure 4.3.2(a) are mapped onto the ellipses shown in gray in Figure 4.3.2(b). The innermost pair of horizontal line segments are mapped onto the innermost ellipse, the middle pair of line segments are mapped onto the middle ellipse, and the outermost pair of line segments are mapped onto the outermost ellipse. As a final point, observe that if $b = 0$, then the image of the line segment $y = 0$, $-\pi/2 < x < \pi/2$, is the line segment $-1 \leq u \leq 1$, $v = 0$ on the real axis.

The mapping $w = \cos z$ can be analyzed in a similar manner, or, since $\cos z = \sin(z + \pi/2)$ from (9), we can view the mapping $w = \cos z$ as a composition of the translation $w = z + \pi/2$ and the mapping $w = \sin z$. See Problem 46 in Exercises 4.3.

4.3.2 Complex Hyperbolic Functions

The *real* hyperbolic sine and hyperbolic cosine functions are defined using the *real* exponential function as follows:

$$\sinh x = \frac{e^x - e^{-x}}{2} \quad \text{and} \quad \cosh x = \frac{e^x + e^{-x}}{2}.$$

The *complex* hyperbolic sine and cosine functions are defined in an analogous manner using the *complex* exponential function.

Definition 4.3.2   Complex Hyperbolic Sine and Cosine

The complex **hyperbolic sine** and **hyperbolic cosine** functions are defined by:

$$\sinh z = \frac{e^z - e^{-z}}{2} \quad \text{and} \quad \cosh z = \frac{e^z + e^{-z}}{2}. \tag{25}$$

Since the complex exponential function agrees with the real exponential function for real input, it follows from (25) that the complex hyperbolic functions agree with the real hyperbolic functions for real input. However, unlike the real hyperbolic functions whose graphs are shown in Figure 4.3.1, the complex hyperbolic functions are periodic and have infinitely many zeros. See Problem 50 in Exercises 4.3.

All zeros of $\sinh z$ and $\cosh z$ are purely imaginary.

The complex hyperbolic tangent, cotangent, secant, and cosecant are defined in terms of $\sinh z$ and $\cosh z$:

$$\tanh z = \frac{\sinh z}{\cosh z}, \quad \coth z = \frac{\cosh z}{\sinh z}, \quad \text{sech } z = \frac{1}{\cosh z}, \quad \text{and csch } z = \frac{1}{\sinh z}. \tag{26}$$

Observe that the hyperbolic sine and cosine functions are entire because the functions e^z and e^{-z} are entire. Moreover, from the chain rule (6) in Section 3.1, we have:

$$\frac{d}{dz} \sinh z = \frac{d}{dz} \left(\frac{e^z - e^{-z}}{2} \right) = \frac{e^z + e^{-z}}{2}$$

or
$$\frac{d}{dz} \sinh z = \cosh z.$$

A similar computation for $\cosh z$ yields

$$\frac{d}{dz} \cosh z = \sinh z.$$

Derivatives of the remaining four remaining hyperbolic functions can then be found using (26) and the quotient rule (5) in Section 3.1.

Derivatives of Complex Hyperbolic Functions

$$\frac{d}{dz}\sinh z = \cosh z \qquad\qquad \frac{d}{dz}\cosh z = \sinh z$$

$$\frac{d}{dz}\tanh z = \operatorname{sech}^2 z \qquad\qquad \frac{d}{dz}\coth z = -\operatorname{csch}^2 z$$

$$\frac{d}{dz}\operatorname{sech} z = -\operatorname{sech} z \tanh z \qquad\qquad \frac{d}{dz}\operatorname{csch} z = -\operatorname{csch} z \coth z$$

Relation To Sine and Cosine The real trigonometric and the real hyperbolic functions share many similar properties. For example,

$$\frac{d}{dx}\sin x = \cos x \quad \text{and} \quad \frac{d}{dx}\sinh x = \cosh x.$$

Aside from the similar notational appearance and the similarities of their respective Taylor series, there is no simple way to relate the real trigonometric and the real hyperbolic functions. When dealing with the *complex* trigonometric and hyperbolic functions, however, there is a simple and beautiful connection between the two. We derive this relationship by replacing z with iz in the definition of $\sinh z$ and then comparing the result with (4):

$$\sinh(iz) = \frac{e^{iz} - e^{-iz}}{2} = i\left(\frac{e^{iz} - e^{-iz}}{2i}\right) = i\sin z,$$

or

$$-i\sinh(iz) = \sin z.$$

In a similar manner, if we substitute iz for z in the expression for $\sin z$ and compare with (25), then we find that $\sinh z = -i\sin(iz)$. After repeating this process for $\cos z$ and $\cosh z$ we obtain the following important relationships between the complex trigonometric and hyperbolic functions:

$$\sin z = -i\sinh(iz) \qquad \text{and} \qquad \cos z = \cosh(iz) \qquad (27)$$

$$\sinh z = -i\sin(iz) \qquad \text{and} \qquad \cosh z = \cos(iz). \qquad (28)$$

Relations between the other trigonometric and hyperbolic functions can now be derived from (27) and (28). For example,

$$\tan(iz) = \frac{\sin(iz)}{\cos(iz)} = \frac{i\sinh z}{\cosh z} = i\tanh z.$$

We can also use (27) and (28) to derive hyperbolic identities from trigonometric identities. We next list some of the more commonly used hyperbolic identities. Each of the results in (29)–(32) is identical to its real analogue.

$$\sinh(-z) = -\sinh z \quad \cosh(-z) = \cosh z \qquad (29)$$

$$\cosh^2 z - \sinh^2 z = 1 \qquad (30)$$

$$\sinh(z_1 \pm z_2) = \sinh z_1 \cosh z_2 \pm \cosh z_1 \sinh z_2 \qquad (31)$$

$$\cosh(z_1 \pm z_2) = \cosh z_1 \cosh z_2 \pm \sinh z_1 \sinh z_2 \qquad (32)$$

In the following example we verify the addition formula given in (32). The other identities can be verified in a similar manner. See Problems 29 and 30 in Exercises 4.3.

EXAMPLE 4   **A Hyperbolic Identity**

Verify that $\cosh(z_1 + z_1) = \cosh z_1 \cosh z_2 + \sinh z_1 \sinh z_2$ for all complex z_1 and z_2.

Solution By (28), $\cosh(z_1 + z_2) = \cos(iz_1 + iz_2)$, and so by the trigonometric identity (9) and additional applications of (27) and (28), we obtain:

$$\cosh(z_1 + z_2) = \cos(iz_1 + iz_2)$$
$$= \cos iz_1 \cos iz_2 - \sin iz_1 \sin iz_2$$
$$= \cos iz_1 \cos iz_2 + (-i \sin iz_1)(-i \sin iz_2)$$
$$= \cosh z_1 \cosh z_2 + \sinh z_1 \sinh z_2.$$

The relations between the complex trigonometric and hyperbolic functions given in (27) and (28) also allow us determine the action of the hyperbolic functions as complex mappings. For example, because $\sinh z = -i \sin(iz)$, the complex mapping $w = \sinh z$ can be considered as the composition of the three complex mappings $w = iz$, $w = \sin z$, and $w = -iz$. See Problem 47 in Exercises 4.3.

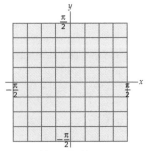

(a) The square S_0

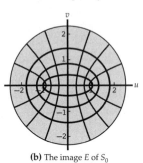

(b) The image E of S_0

Figure 4.3.3 The mapping $w = \sin z$

Remarks *Comparison with Real Analysis*

(*i*) In real analysis, the exponential function was just one of a number of apparently equally important elementary functions. In complex analysis, however, the complex exponential function assumes a much greater role. All of the complex elementary functions can be defined solely in terms of the complex exponential and logarithmic functions. A recurring theme throughout the study of complex analysis involves using the exponential and logarithmic functions to evaluate, differentiate, integrate, and map with elementary functions.

(*ii*) As functions of a real variable x, $\sinh x$ and $\cosh x$ are not periodic. In contrast, the complex functions $\sinh z$ and $\cosh z$ are periodic. See Problem 49 in Exercises 4.3. Moreover, $\cosh x$ has no zeros and $\sinh x$ has a single zero at $x = 0$. See Figure 4.3.1. The complex functions $\sinh z$ and $\cosh z$, on the other hand, both have infinitely many zeros. See Problem 50 in Exercises 4.3.

(*iii*) Since the complex sine function is periodic, the mapping $w = \sin z$ is not one-to-one on the complex plane. Constructing a Riemann surface, for this function, as described in the Remarks at the end of Section 2.4 and Section 4.1, will help us visualize the complex mapping $w = \sin z$. In order to construct a Riemann surface consider the mapping on the square S_0 defined by $-\pi/2 \le x \le \pi/2$, $-\pi/2 \le y \le \pi/2$. From Example 3, we find that the square S_0 shown in color in Figure 4.3.3(a) maps onto the elliptical region E shown in gray in Figure 4.3.3(b). Similarly, the adjacent square S_1 defined by $\pi/2 \le x \le 3\pi/2$, $-\pi/2 \le y \le \pi/2$, also maps onto E.

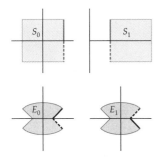

Figure 4.3.4 The cut elliptical regions E_0 and E_1

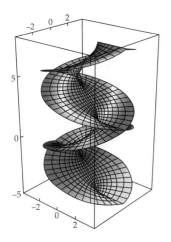

Figure 4.3.5 A Riemann surface for $w = \sin z$

A Riemann surface is constructed by starting with two copies of E, E_0, and E_1, representing the images of S_0 and S_1, respectively. We then cut E_0 and E_1 open along the line segments in the real axis from 1 to $\cosh(\pi/2)$ and from -1 to $-\cosh(\pi/2)$. As shown in Figure 4.3.4, the segment shown in color in the boundary of S_0 is mapped onto the segment shown in black in the boundary of E_0, while the dashed segment shown in color in the boundary of S_0 is mapped onto the dashed segment shown in black in the boundary of E_0. In a similar manner, the segments shown in color in the boundary of S_1 are mapped onto the segments shown in black in the boundary of E_1. Part of the Riemann surface consists of the two elliptical regions E_0 and E_1 with the segments shown in black glued together and the dashed segments glued together. To complete the Riemann surface, we take for every integer n an elliptical region E_n representing the image of the square S_n defined by $(2n-1)\pi/2 \leq x \leq (2n+1)\pi/2$, $-\pi/2 \leq y \leq \pi/2$. Each region E_n is cut open, as E_0 and E_1 were, and E_n is glued to E_{n+1} along their boundaries in a manner analogous to that used for E_0 and E_1. This Riemann surface, placed in xyz-space, is illustrated in Figure 4.3.5.

EXERCISES 4.3 *Answers to selected odd-numbered problems begin on page ANS-13.*

4.3.1 Complex Trigonometric Functions

In Problems 1–8, express the value of the given trigonometric function in the form $a + ib$.

1. $\sin(4i)$ **2.** $\cos(-3i)$

3. $\cos(2 - 4i)$ **4.** $\sin\left(\dfrac{\pi}{4} + i\right)$

5. $\tan(2i)$ **6.** $\cot(\pi + 2i)$

7. $\sec\left(\dfrac{\pi}{2} - i\right)$ **8.** $\csc(1 + i)$

In Problems 9–12, find all complex values z satisfying the given equation.

9. $\sin z = i$ **10.** $\cos z = 4$

11. $\sin z = \cos z$ **12.** $\cos z = i\sin z$

In Problems 13–16, verify the given trigonometric identity.

13. $\sin(-z) = -\sin z$ **14.** $\cos(z_1 + z_2) = \cos z_1 \cos z_2 - \sin z_1 \sin z_2$

15. $\overline{\cos z} = \cos \bar{z}$ **16.** $\sin\left(z - \dfrac{\pi}{2}\right) = -\cos z$

In Problems 17–20, find the derivative of the given function.

17. $\sin(z^2)$ **18.** $\cos(ie^z)$

19. $z \tan \dfrac{1}{z}$ **20.** $\sec(z^2 + (1 - i)z + i)$

4.3.2 Complex Hyperbolic Functions

In Problems 21–24, express the value of the given hyperbolic function in the form $a + ib$.

21. $\cosh(\pi i)$ **22.** $\sinh\left(\dfrac{\pi}{2}i\right)$

23. $\cosh\left(1 + \dfrac{\pi}{6}i\right)$ **24.** $\tanh(2 + 3i)$

In Problems 25–28, find all complex values z satisfying the given equation.

25. $\cosh z = i$ **26.** $\sinh z = -1$

27. $\sinh z = \cosh z$ **28.** $\sinh z = e^z$

In Problems 29–32, verify the given hyperbolic identity.

29. $\cosh^2 z - \sinh^2 z = 1$

30. $\sinh(z_1 + z_2) = \sinh z_1 \cosh z_2 + \cosh z_1 \sinh z_2$

31. $|\sinh z|^2 = \sinh^2 x + \sin^2 y$

32. $\operatorname{Im}(\cosh z) = \sinh x \sin y$

In Problems 33–36, find the derivative of the given function.

33. $\sin z \sinh z$ **34.** $\tanh z$

35. $\tanh(iz - 2)$ **36.** $\cosh\left(iz + e^{iz}\right)$

Focus on Concepts

37. Recall that Euler's formula states that $e^{i\theta} = \cos\theta + i\sin\theta$ for any real number θ. Prove that, in fact, $e^{iz} = \cos z + i\sin z$ for any complex number z.

38. Solve the equation $\sin z = \cosh 2$ by equating real and imaginary parts.

39. If $\sin z = a$ with $-1 \le a \le 1$, then what can you say about z? Justify your answer.

40. If $|\sin z| \le 1$, then what can you say about z? Justify your answer.

41. Show that all the zeros of $\cos z$ are $z = (2n+1)\pi/2$ for $n = 0, \pm1, \pm2, \dots$.

42. Find all z such that $|\tan z| = 1$.

43. Find the real and imaginary parts of the function $\sin \bar{z}$ and use them to show that this function is nowhere analytic.

44. Without calculating the partial derivatives, explain why $\sin x \cosh y$ and $\cos x \sinh y$ are harmonic functions in $\mathbf{C}$.

45. Prove that $\sin z$ is a one-to-one function on the domain $-\pi/2 < x < \pi/2$, $-\infty < y < \infty$.

46. Use the identity $\cos z = \sin\left(z + \frac{1}{2}\pi\right)$ to find the image of the region $-\pi \le x \le 0$ under the mapping $w = \cos z$. Describe the images of vertical and horizontal lines in the region.

47. Use the identity $\sinh z = -i\sin(iz)$ to find the image of the region $-\pi/2 \le y \le \pi/2$, $-\infty < x < \infty$, under the mapping $w = \sinh z$. Describe the images of vertical and horizontal lines in the region. [*Hint*: The identity implies that $w = \sinh z$ is a composition of linear mappings and the complex mapping $w = \sin z$.]

48. Find the image of the region defined by $-\pi/2 \le x \le \pi/2$, $y \ge 0$, under the mapping $w = (\sin z)^{1/4}$, where $z^{1/4}$ represents the principal fourth root function.

49. Find the period of each of the following complex functions.

  (a) $\cosh z$ **(b)** $\sinh z$ **(c)** $\tanh z$

50. Find all zeros of each of the following functions.

  (a) $\cosh z$ **(b)** $\sinh z$

51. Verify the following identities.

  (a) $\sin(z + \pi) = -\sin z$ **(b)** $\cos(z + \pi) = -\cos z$

52. Use the identities in Problem 51 to show that $\tan z$ is a periodic function with a real period of π.

4.4 Inverse Trigonometric and Hyperbolic Functions

To The Instructor: In this section, we introduce the inverse complex trigonometric and hyperbolic functions. This material can be skipped without affecting the development of topics in subsequent chapters.

The complex logarithmic function $\ln z$ was defined in Section 4.1 to solve equations of the form $e^w = z$. Because the complex exponential function is periodic, there are infinitely many solutions to such equations, and, consequently, $\ln z$ is necessarily a multiple-valued function. In this section we repeat this process for equations involving the complex trigonometric and hyperbolic functions. Because the complex trigonometric and hyperbolic functions are periodic, their inverse functions are multiple-valued. Furthermore, since the complex trigonometric and hyperbolic functions are defined in terms of the complex exponential function, their inverses will involve the complex logarithm.

Inverse Sine In (11) in Section 4.3 we found that the complex sine function is periodic with a real period of 2π. We also found that the sine function maps the complex plane onto the complex plane, that is, Range$(\sin z) = \mathbf{C}$. See Figure 4.3.2. These two properties imply that for any complex number z there exists infinitely many solutions w to the equation $\sin w = z$. An explicit formula for w is found by following the procedure used in Example 2 of Section 4.3. We begin by using Definition 4.3.1 to rewrite the equation $\sin w = z$ as:

$$\frac{e^{iw} - e^{-iw}}{2i} = z \quad \text{or} \quad e^{2iw} - 2ize^{iw} - 1 = 0.$$

Because $e^{2iw} - 2ize^{iw} - 1 = 0$ is a quadratic equation in e^{iw} we can then use the quadratic formula (3) in Section 1.6 to solve for e^{iw}:

$$e^{iw} = iz + \left(1 - z^2\right)^{1/2}. \tag{1}$$

Since we are using the quadratic formula, we should keep in mind that the expression $\left(1 - z^2\right)^{1/2}$ in (1) represents the *two* square roots of $1 - z^2$. Finally, we solve for w using the complex logarithm:

$$iw = \ln\left[iz + \left(1 - z^2\right)^{1/2}\right] \quad \text{or} \quad w = -i\ln\left[iz + \left(1 - z^2\right)^{1/2}\right]. \tag{2}$$

Each value of w obtained from the second equation in (2) satisfies the equation $\sin w = z$. Therefore, we call the multiple-valued function defined by the second equation in (2) the **inverse sine**. We summarize this discussion in the following definition.

Definition 4.4.1 Inverse Sine

The multiple-valued function $\sin^{-1} z$ defined by:

$$\sin^{-1} z = -i\ln\left[iz + \left(1 - z^2\right)^{1/2}\right] \tag{3}$$

is called the **inverse sine**.

At times, we will also call the inverse sine the **arcsine** and we will denote it by $\arcsin z$. It is clear from (3) that the inverse sine is multiple-valued since it is defined in terms of the complex logarithm $\ln z$. It is also worth repeating that the expression $\left(1 - z^2\right)^{1/2}$ in (3) represents the two square roots of $1 - z^2$.

EXAMPLE 1   Values of Inverse Sine

Find all values of $\sin^{-1} \sqrt{5}$.

Solution By setting $z = \sqrt{5}$ in (3) we obtain:

$$\sin^{-1} \sqrt{5} = -i \ln \left[i\sqrt{5} + \left(1 - \left(\sqrt{5} \right)^2 \right)^{1/2} \right]$$

$$= -i \ln \left[i\sqrt{5} + (-4)^{1/2} \right].$$

The two square roots $(-4)^{1/2}$ of -4 are found to be $\pm 2i$ using (4) in Section 1.4, and so:

$$\sin^{-1} \sqrt{5} = -i \ln \left[i\sqrt{5} \pm 2i \right] = -i \ln \left[\left(\sqrt{5} \pm 2 \right) i \right].$$

Because $\left(\sqrt{5} \pm 2 \right) i$ is a pure imaginary number with positive imaginary part (both $\sqrt{5} + 2$ and $\sqrt{5} - 2$ are positive), we have $\left| \left(\sqrt{5} \pm 2 \right) i \right| = \sqrt{5} \pm 2$ and $\arg \left[\left(\sqrt{5} \pm 2 \right) i \right] = \pi/2$. Thus, from (11) in Section 4.1 we have

$$\ln \left[\left(\sqrt{5} \pm 2 \right) i \right] = \log_e \left(\sqrt{5} \pm 2 \right) + i \left(\frac{\pi}{2} + 2n\pi \right)$$

for $n = 0, \pm 1, \pm 2, \ldots$. This expression can be simplified by observing that

$$\log_e \left(\sqrt{5} - 2 \right) = \log_e \frac{1}{\sqrt{5} + 2} = \log_e 1 - \log_e \left(\sqrt{5} + 2 \right) = 0 - \log_e \left(\sqrt{5} + 2 \right),$$

and so $\log_e \left(\sqrt{5} \pm 2 \right) = \pm \log_e \left(\sqrt{5} + 2 \right)$. Therefore,

$$-i \ln \left[\left(\sqrt{5} \pm 2 \right) i \right] = -i \left[\log_e \left(\sqrt{5} \pm 2 \right) + i \left(\frac{\pi}{2} + 2n\pi \right) \right]$$

$$= -i \left[\pm \log_e \left(\sqrt{5} + 2 \right) + i \frac{(4n+1)\pi}{2} \right],$$

and so

$$\sin^{-1} \sqrt{5} = \frac{(4n+1)\pi}{2} \pm i \log_e \left(\sqrt{5} + 2 \right)$$

for $n = 0, \pm 1, \pm 2, \ldots$.   ❑

Inverse Cosine and Tangent We can easily modify the procedure used on page 193 to solve the equations $\cos w = z$ and $\tan w = z$. This leads to definitions of the inverse cosine and the inverse tangent, which we now state.

Definition 4.4.2   **Inverse Cosine and Inverse Tangent**

The multiple-valued function $\cos^{-1} z$ defined by:

$$\cos^{-1} z = -i \ln \left[z + i \left(1 - z^2 \right)^{1/2} \right] \qquad (4)$$

is called the **inverse cosine**. The multiple-valued function $\tan^{-1} z$ defined by:

$$\tan^{-1} z = \frac{i}{2} \ln \left(\frac{i+z}{i-z} \right) \qquad (5)$$

is called the **inverse tangent**.

Both the inverse cosine and inverse tangent are multiple-valued functions since they are defined in terms of the complex logarithm $\ln z$. As with the inverse sine, the expression $\left(1 - z^2 \right)^{1/2}$ in (4) represents the two square roots of the complex number $1 - z^2$. Every value of $w = \cos^{-1} z$ satisfies the equation $\cos w = z$, and, similarly, every value of $w = \tan^{-1} z$ satisfies the equation $\tan w = z$.

Branches and Analyticity   The inverse sine and inverse cosine are multiple-valued functions that can be made single-valued by specifying a single value of the square root to use for the expression $\left(1 - z^2 \right)^{1/2}$ and a single value of the complex logarithm to use in (3) or (4). The inverse tangent, on the other hand, can be made single-valued by just specifying a single value of $\ln z$ to use. For example, we can define a function f_1 that gives a value of the inverse sine by using the principal square root and the principal value of the complex logarithm in (3). If, say, $z = \sqrt{5}$, then the principal square root of $1 - \left(\sqrt{5} \right)^2 = -4$ is $2i$, and

$$\text{Ln} \left(i\sqrt{5} + 2i \right) = \log_e \left(\sqrt{5} + 2 \right) + \pi i/2.$$

Identifying these values in (3) gives:

$$f_1 \left(\sqrt{5} \right) = \frac{\pi}{2} - i \log_e \left(\sqrt{5} + 2 \right) \approx 1.5708 - 1.4436i.$$

Thus, we see that the value of the function f_1 at $z = \sqrt{5}$ is the value of $\sin^{-1} \sqrt{5}$ associated to $n = 0$ and the square root $2i$ in Example 1.

A branch of a multiple-valued inverse trigonometric function may be obtained by choosing a branch of the square root function and a branch of the complex logarithm. Determining the domain of a branch defined in this manner can be quite involved. Because this is an elementary text, we will not discuss this topic further. On the other hand, the derivatives of branches of the multiple-valued inverse trigonometric functions are easily found using implicit differentiation. To see that this is so, suppose that f_1 is a branch of the multiple-valued function $F(z) = \sin^{-1} z$. If $w = f_1(z)$, then we know that $z = \sin w$. By differentiating both sides of this last equation with respect to z and applying the chain rule (6) in Section 3.1, we obtain:

$$1 = \cos w \cdot \frac{dw}{dz} \quad \text{or} \quad \frac{dw}{dz} = \frac{1}{\cos w}. \qquad (6)$$

Now, from the trigonometric identity $\cos^2 w + \sin^2 w = 1$, we have $\cos w = \left(1 - \sin^2 w\right)^{1/2}$, and since $z = \sin w$, this may be written as $\cos w = \left(1 - z^2\right)^{1/2}$. Therefore, after substituting this expression for $\cos w$ in (6) we obtain the following result:

$$f_1'(z) = \frac{dw}{dz} = \frac{1}{\left(1 - z^2\right)^{1/2}}.$$

If we let $\sin^{-1} z$ denote the branch f_1, then this formula may be restated in a less cumbersome manner as:

$$\frac{d}{dz}\sin^{-1} z = \frac{1}{\left(1 - z^2\right)^{1/2}}.$$

We must be careful, however, to use the same branch of the square root function that defined $\sin^{-1} z$ when finding values of its derivative.

In a similar manner, derivatives of branches of the inverse cosine and the inverse tangent can be found. In the following formulas, the symbols $\sin^{-1} z$, $\cos^{-1} z$, and $\tan^{-1} z$ represent branches of the corresponding multiple-valued functions. These formulas for the derivatives hold only on the domains of these branches.

Derivatives of Branches $\sin^{-1} z$, $\cos^{-1} z$, and $\tan^{-1} z$

$$\frac{d}{dz}\sin^{-1} z = \frac{1}{\left(1 - z^2\right)^{1/2}} \qquad (7)$$

$$\frac{d}{dz}\cos^{-1} z = \frac{-1}{\left(1 - z^2\right)^{1/2}} \qquad (8)$$

$$\frac{d}{dz}\tan^{-1} z = \frac{1}{1 + z^2} \qquad (9)$$

When finding the value of a derivative with (7) or (8), we must use the same square root as is used to define the branch. These formulas are similar to those for the derivatives of the real inverse trigonometric functions. The difference between the real and complex formulas is the specific choice of a branch of the square root function needed for (7) and (8).

EXAMPLE 2   **Derivative of a Branch of Inverse Sine**

Let $\sin^{-1} z$ represent a branch of the inverse sine obtained by using the principal branches of the square root and the logarithm defined by (7) of Section 4.2 and (19) of Section 4.1, respectively. Find the derivative of this branch at $z = i$.

Solution   We note in passing that this branch is differentiable at $z = i$ because $1 - i^2 = 2$ is not on the branch cut of the principal branch of the square root function, and because $i(i) + \left(1 - i^2\right)^{1/2} = -1 + \sqrt{2}$ is not on the branch cut of the principal branch of the complex logarithm. Thus, by (7) we have:

$$\frac{d}{dz} \sin^{-1} z \bigg|_{z=i} = \frac{1}{(1-z^2)^{1/2}} \bigg|_{z=i} = \frac{1}{(1-i^2)^{1/2}} = \frac{1}{2^{1/2}}.$$

Using the principal branch of the square root, we obtain $2^{1/2} = \sqrt{2}$. Therefore, the derivative is $1/\sqrt{2}$ or $\frac{1}{2}\sqrt{2}$. ❑

Observe that the branch of the inverse sine used in Example 2 is not defined at, say, $z = \sqrt{5}$ because the point $1 - \left(\sqrt{5}\right)^2 = -4$ is on the branch cut of the principal branch of the square root. We can define a different branch of the inverse sine that is defined at this point. For example, consider the branch $f_2(z) = \sqrt{r}e^{i\theta/2}$, $0 < \theta < 2\pi$, of the square root function. Because $-4 = 4e^{i\pi}$, we have that $f_2(-4) = 2i$. You should verify that if we define $\sin^{-1} z$ to be the branch of inverse sine obtained using the branch f_2 of the square root and the principal branch of the logarithm, then:

$$\sin^{-1}\sqrt{5} = \frac{1}{2}\pi - i\log_e\left(\sqrt{5} + 2\right) \quad \text{and} \quad \frac{d}{dz}\sin^{-1} z \bigg|_{z=\sqrt{5}} = -\frac{i}{2}.$$

Inverse Hyperbolic Functions The foregoing discussion of inverse trigonometric functions can be repeated for hyperbolic functions. This leads to the definition of the inverse hyperbolic functions stated below. Once again these inverses are defined in terms of the complex logarithm because the hyperbolic functions are defined in terms of the complex exponential.

Definition 4.4.3   Inverse Hyperbolic Sine, Cosine, and Tangent

The multiple-valued functions $\sinh^{-1} z$, $\cosh^{-1} z$, and $\tanh^{-1} z$, defined by:

$$\sinh^{-1} z = \ln\left[z + \left(z^2 + 1\right)^{1/2}\right], \tag{10}$$

$$\cosh^{-1} z = \ln\left[z + \left(z^2 - 1\right)^{1/2}\right], \tag{11}$$

and
$$\tanh^{-1} z = \frac{1}{2}\ln\left(\frac{1+z}{1-z}\right) \tag{12}$$

are called the **inverse hyperbolic sine,** the **inverse hyperbolic cosine,** and the **inverse hyperbolic tangent,** respectively.

The expressions (10)–(12) in Definition 4.4.3 allow us to solve equations involving the complex hyperbolic functions. In particular, if $w = \sinh^{-1} z$, then $\sinh w = z$; if $w = \cosh^{-1} z$, then $\cosh w = z$; and if $w = \tanh^{-1} z$, then $\tanh w = z$.

Branches of the inverse hyperbolic functions are defined by choosing branches of the square root and complex logarithm, or, in the case of the inverse hyperbolic tangent, just choosing a branch of the complex logarithm. The derivative of a branch can be found using implicit differentiation. The following result gives formulas for the derivatives of branches of the inverse hyperbolic functions. In these formulas, the symbols $\sinh^{-1} z$, $\cosh^{-1} z$, and $\tanh^{-1} z$ represent branches of the corresponding inverse hyperbolic multiple-valued functions.

Derivatives of Branches $\sinh^{-1} z$, $\cosh^{-1} z$, *and* $\tanh^{-1} z$

$$\frac{d}{dz}\sinh^{-1} z = \frac{1}{\left(z^2+1\right)^{1/2}} \tag{13}$$

$$\frac{d}{dz}\cosh^{-1} z = \frac{1}{\left(z^2-1\right)^{1/2}} \tag{14}$$

$$\frac{d}{dz}\tanh^{-1} z = \frac{1}{1-z^2} \tag{15}$$

As with the inverse trigonometric functions, we should take care to be consistent in our use of branches when evaluating derivatives. Formulas (13)–(15) for the derivatives of branches of the *complex* inverse hyperbolic functions are the same as the analogous formulas for the derivatives of the *real* inverse hyperbolic functions except for the choice of branch required in (13) and (14).

EXAMPLE 3   Inverse Hyperbolic Cosine

Let $\cosh^{-1} z$ represent the branch of the inverse hyperbolic cosine obtained by using the branch $f_2(z) = \sqrt{r}e^{i\theta/2}$, $0 < \theta < 2\pi$, of the square root and the principal branch of the complex logarithm. Find the following values.

(a) $\cosh^{-1}\dfrac{\sqrt{2}}{2}$ (b) $\dfrac{d}{dz}\cosh^{-1} z\Big|_{z=\sqrt{2}/2}$

Solution (a) In order to find $\cosh^{-1}\left(\frac{1}{2}\sqrt{2}\right)$, we use (11) with $z = \frac{1}{2}\sqrt{2}$ and the stated branches of the square root and logarithm. When $z = \frac{1}{2}\sqrt{2}$, we have that $z^2 - 1 = -\frac{1}{2}$. Since $-\frac{1}{2}$ has exponential form $\frac{1}{2}e^{i\pi}$, the square root given by the branch f_2 is:

$$f_2\left(\frac{1}{2}e^{i\pi}\right) = \sqrt{\frac{1}{2}}e^{i\pi/2} = \frac{1}{\sqrt{2}}i = \frac{\sqrt{2}}{2}i.$$

The value of our branch of the inverse cosine is then given by:

$$\cosh^{-1}\frac{\sqrt{2}}{2} = \ln\left[z + \left(z^2-1\right)^{1/2}\right] = \ln\left[\frac{\sqrt{2}}{2} + \frac{\sqrt{2}}{2}i\right],$$

where we take the value of the principal branch of the logarithm. Because $\left|\frac{1}{2}\sqrt{2} + \frac{1}{2}\sqrt{2}i\right| = 1$ and $\mathrm{Arg}\left(\frac{1}{2}\sqrt{2} + \frac{1}{2}\sqrt{2}i\right) = \frac{1}{4}\pi$, the principal branch of the logarithm is $\log_e 1 + i\left(\frac{1}{4}\pi\right) = \frac{1}{4}\pi i$. Therefore,

$$\cosh^{-1}\frac{\sqrt{2}}{2} = \frac{\pi}{4}i.$$

(b) From (14) we have:

$$\frac{d}{dz}\cosh^{-1} z\Big|_{z=\sqrt{2}/2} = \frac{1}{\left[\left(\sqrt{2}/2\right)^2 - 1\right]^{1/2}} = \frac{1}{\left(-1/2\right)^{1/2}}.$$

After using f_2 to find the square root in this expression we obtain:

$$\frac{d}{dz}\cosh^{-1}z\bigg|_{z=\sqrt{2}/2}=\frac{1}{\sqrt{2}i/2}=-\sqrt{2}i.$$

❑

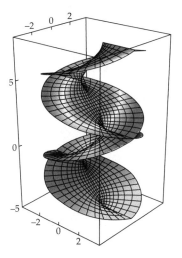

Figure 4.4.1 A Riemann surface for
$w = \sin^{-1} z$

Remarks

The multiple-valued function $F(z) = \sin^{-1} z$ can be visualized using the Riemann surface constructed for $\sin z$ in the Remarks in Section 4.3 and shown in Figure 4.4.1. In order to see the image of a point z_0 under the multiple-valued mapping $w = \sin^{-1} z$, we imagine that z_0 is lying in the xy-plane in Figure 4.4.1. We then consider all points on the Riemann surface lying directly over z_0. Each of these points on the surface corresponds to a unique point in one of the squares S_n described in the Remarks in Section 4.3. Thus, this infinite set of points in the Riemann surface represents the infinitely many images of z_0 under $w = \sin^{-1} z$.

EXERCISES 4.4   *Answers to selected odd-numbered problems begin on page ANS-13.*

In Problems 1–10, find all values of the given quantity.

1. $\cos^{-1} i$

2. $\sin^{-1} 1$

3. $\sin^{-1} \sqrt{2}$

4. $\cos^{-1} \dfrac{5}{3}$

5. $\tan^{-1} 1$

6. $\tan^{-1} 2i$

7. $\sinh^{-1} i$

8. $\cosh^{-1} \dfrac{1}{2}$

9. $\tanh^{-1} (1 + 2i)$

10. $\tanh^{-1} \left(\sqrt{2}i\right)$

In Problems 11–16, use the stated branch of the multiple-valued function $z^{1/2}$ and principal branch of $\ln z$ to (**a**) find the value of the inverse trigonometric or hyperbolic function at the given point and (**b**) compute the value of the derivative of the function at the given point.

11. $\sin^{-1} z$, $z = \frac{1}{2}i$; use the principal branch of $z^{1/2}$

12. $\cos^{-1} z$, $z = \frac{5}{3}$; use the branch $\sqrt{r}e^{i\theta/2}$, $0 < \theta < 2\pi$, of $z^{1/2}$

13. $\tan^{-1} z$, $z = 1 + i$

14. $\sinh^{-1} z$, $z = 0$; use the principal branch of $z^{1/2}$

15. $\cosh^{-1} z$, $z = -i$; use the branch $\sqrt{r}e^{i\theta/2}$, $-2\pi < \theta < 0$, of $z^{1/2}$

16. $\tanh^{-1} z$, $z = 3i$

Focus on Concepts

17. Derive formula (4) for $\cos^{-1} z$ by modifying the procedure used to derive the formula for arcsine on page 193.

18. Derive formula (10) for $\sinh^{-1} z$ by modifying the procedure used to derive the formula for arcsine on page 193.

19. Use implicit differentiation to derive formula (8) for the derivative of a branch of the inverse cosine.

20. Use implicit differentiation to derive formula (12) for the derivative of a branch of the inverse hyperbolic tangent.

21. (a) Prove that $\sin z$ is one-to-one on the domain defined by the inequalities $-\pi/2 < x < \pi/2, -\infty < y < \infty$.

    (b) Which square root and which branch of the logarithm should be used so that the mapping $w = \sin^{-1} z$ takes the half-plane $\text{Im}(z) > 0$ onto the region $-\pi/2 < u < \pi/2$, $v > 0$, that is, so that $w = \sin^{-1} z$ is the inverse mapping of the mapping in part (a)?

22. Prove the following identities.

    (a) $\sin^{-1}\left[\left(1 - z^2\right)^{1/2}\right] = \cos^{-1}\left(\pm z\right)$

    (b) $\sin^{-1} z + \cos^{-1} z = \frac{1}{2}(4n + 1)\pi$, $n = 0, \pm 1, \pm 2, \ldots$

4.5   Applications

In Section 3.4 we saw the important role that harmonic functions play in the fields of electrostatics, fluid flow, gravitation, and heat flow. It is often the case that in order to solve an applied problem in one of these fields we need to find a function $\phi(x,\ y)$, which is harmonic in a domain D and which takes on specified values on the boundary of D. In this section we will see that mapping by analytic functions can often help solve these types of problems.

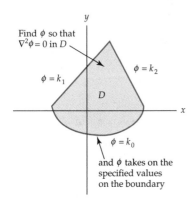

Find ϕ so that $\nabla^2\phi = 0$ in D

$\phi = k_1$

$\phi = k_2$

D

$\phi = k_0$

and ϕ takes on the specified values on the boundary

Figure 4.5.1 Dirichlet problem

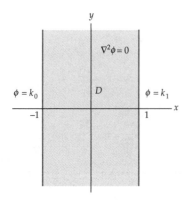

$\nabla^2\phi = 0$

$\phi = k_0$ D $\phi = k_1$

Figure 4.5.2 Dirichlet problem from Example 2 of Section 3.4

Dirichlet Problems   Suppose that D is a domain in the complex plane. Recall from Section 3.3 that a real-valued function ϕ of two real variables x and y is called harmonic in D if ϕ has continuous first and second-order partial derivatives and if ϕ satisfies Laplace's equation $\nabla^2\phi = 0$,

or
$$\frac{\partial^2\phi}{\partial x^2} + \frac{\partial^2\phi}{\partial y^2} = 0. \tag{1}$$

In Section 3.4 we defined a **Dirichlet problem** to be the problem of finding a function $\phi(x,\ y)$ that is harmonic in D and that takes on specified values on the boundary of D. See Figure 4.5.1. The specifications of the values of the function ϕ on the boundary of D are called **boundary conditions**. For example, consider the problem:

$$Solve: \frac{\partial^2\phi}{\partial x^2} + \frac{\partial^2\phi}{\partial y^2} = 0, \quad -1 < x < 1, -\infty < y < \infty$$

$$Subject\ to: \phi(-1, y) = k_0, \quad \phi(1, y) = k_1, \quad -\infty < y < \infty,$$

where k_0 and k_1 are real constants. This is a Dirichlet problem in the domain D bounded by the vertical lines $x = -1$ and $x = 1$. See Figure 4.5.2. In Example 2 in Section 3.4 we used elementary techniques from differential equations to find the solution

$$\phi(x, y) = \frac{k_1 - k_0}{2}x + \frac{k_1 + k_0}{2} \tag{2}$$

of this particular Dirichlet problem. You should reread this example in Section 3.4 to remind yourself of how this solution was found.

Harmonic Functions and Analytic Mappings In part, the Dirichlet problem represented in Figure 4.5.2 was relatively easy to solve because of the simple shape of the domain D. The techniques used to solve this type of Dirichlet problem do not, in general, apply to Dirichlet problems in a more complicated domain. A function f that is analytic in a domain D and that maps D onto a domain D' is called an **analytic mapping** of D onto D'. It is often the case that a Dirichlet problem in a complicated domain D can be solved by finding an analytic mapping of D onto a domain D' in which the associated, or *transformed*, Dirichlet problem is easier to solve. This technique will be presented briefly here and discussed in greater detail in Chapter 7. The key to this method for solving Dirichlet problems is the following theorem, which shows that Laplace's equation is invariant under an analytic mapping.

Theorem 4.5.1   Harmonic Function under an Analytic Mapping

Let $f(z) = u(x, y) + iv(x, y)$ be an analytic mapping of a domain D in the z-plane onto a domain D' in the w-plane. If the function $\Phi(u, v)$ is harmonic in D', then the function $\phi(x, y) = \Phi(u(x, y), v(x, y))$ is harmonic in D.

Proof In order to prove that the function $\phi(x, y)$ is harmonic in D, we must show that $\phi(x, y)$ satisfies Laplace's equation (1) in D. We begin by finding the partial derivatives of $\phi(x, y)$ with respect to x. Since

$$\phi(x, y) = \Phi(u(x, y), v(x, y)),$$

the chain rule of partial differentiation gives:

$$\frac{\partial \phi}{\partial x} = \frac{\partial \Phi}{\partial u}\frac{\partial u}{\partial x} + \frac{\partial \Phi}{\partial v}\frac{\partial v}{\partial x}.$$

A second application of the chain rule combined with the product rule gives the second partial derivative with respect to x:

$$\frac{\partial^2 \phi}{\partial x^2} = \left(\frac{\partial^2 \Phi}{\partial u^2}\frac{\partial u}{\partial x} + \frac{\partial^2 \Phi}{\partial v \partial u}\frac{\partial v}{\partial x}\right)\frac{\partial u}{\partial x} + \frac{\partial \Phi}{\partial u}\frac{\partial^2 u}{\partial x^2}$$
$$+ \left(\frac{\partial^2 \Phi}{\partial u \partial v}\frac{\partial u}{\partial x} + \frac{\partial^2 \Phi}{\partial v^2}\frac{\partial v}{\partial x}\right)\frac{\partial v}{\partial x} + \frac{\partial \Phi}{\partial v}\frac{\partial^2 v}{\partial x^2}. \tag{3}$$

In a similar manner we find the second partial derivative with respect to y:

$$\frac{\partial^2 \phi}{\partial y^2} = \left(\frac{\partial^2 \Phi}{\partial u^2}\frac{\partial u}{\partial y} + \frac{\partial^2 \Phi}{\partial v \partial u}\frac{\partial v}{\partial y}\right)\frac{\partial u}{\partial y} + \frac{\partial \Phi}{\partial u}\frac{\partial^2 u}{\partial y^2}$$
$$+ \left(\frac{\partial^2 \Phi}{\partial u \partial v}\frac{\partial u}{\partial y} + \frac{\partial^2 \Phi}{\partial v^2}\frac{\partial v}{\partial y}\right)\frac{\partial v}{\partial y} + \frac{\partial \Phi}{\partial v}\frac{\partial^2 v}{\partial y^2}. \tag{4}$$

By adding equations (3) and (4) we obtain:

$$\nabla^2 \phi = \frac{\partial^2 \Phi}{\partial u^2}\left(\left(\frac{\partial u}{\partial x}\right)^2 + \left(\frac{\partial u}{\partial y}\right)^2\right) + \frac{\partial^2 \Phi}{\partial v^2}\left(\left(\frac{\partial v}{\partial x}\right)^2 + \left(\frac{\partial v}{\partial y}\right)^2\right)$$
$$+ \frac{\partial \Phi}{\partial u}\left(\frac{\partial^2 u}{\partial x^2} + \frac{\partial^2 u}{\partial y^2}\right) + \frac{\partial \Phi}{\partial v}\left(\frac{\partial^2 v}{\partial x^2} + \frac{\partial^2 v}{\partial y^2}\right) \tag{5}$$
$$+ \left(\frac{\partial^2 \Phi}{\partial v \partial u} + \frac{\partial^2 \Phi}{\partial u \partial v}\right)\left(\frac{\partial u}{\partial x}\frac{\partial v}{\partial x} + \frac{\partial u}{\partial y}\frac{\partial v}{\partial y}\right).$$

Because f is an analytic function in D, we know from Theorem 3.2.1 that the Cauchy-Riemann equations are satisfied by $\dfrac{\partial u}{\partial x} = \dfrac{\partial v}{\partial y}$ and $\dfrac{\partial u}{\partial y} = -\dfrac{\partial v}{\partial x}$. Moreover, from Theorem 3.3.1 we have that u and v are harmonic conjugates in D, and so $\dfrac{\partial^2 u}{\partial x^2} + \dfrac{\partial^2 u}{\partial y^2} = 0$ and $\dfrac{\partial^2 v}{\partial x^2} + \dfrac{\partial^2 v}{\partial y^2} = 0$. Thus, (5) becomes:

$$\nabla^2 \phi = \frac{\partial^2 \Phi}{\partial u^2} \left(\left(\frac{\partial u}{\partial x} \right)^2 + \left(\frac{\partial v}{\partial x} \right)^2 \right) + \frac{\partial^2 \Phi}{\partial v^2} \left(\left(\frac{\partial v}{\partial x} \right)^2 + \left(\frac{\partial u}{\partial x} \right)^2 \right)$$

$$= \nabla^2 \Phi \left(\left(\frac{\partial u}{\partial x} \right)^2 + \left(\frac{\partial v}{\partial x} \right)^2 \right).$$

Using (9) in Section 3.2, we see that $\left(\dfrac{\partial u}{\partial x} \right)^2 + \left(\dfrac{\partial v}{\partial x} \right)^2 = |f'(z)|^2$, and so this equation for $\nabla^2 \phi$ simplifies to the following:

$$\nabla^2 \phi = \nabla^2 \Phi \cdot |f'(z)|^2 \tag{6}$$

Since $\Phi(u,\, v)$ is harmonic in D', $\nabla^2 \Phi = 0$, and so (6) becomes

$$\nabla^2 \phi = 0 \cdot |f'(z)|^2 = 0. \tag{7}$$

Finally, from (7) we conclude that $\phi(x,\, y)$ satisfies Laplace's equation in D. Therefore, the function $\phi(x,\, y)$ is harmonic in D.   ❏

A Method to Solve Dirichlet Problems   We now present a method for solving Dirichlet problems using Theorem 4.5.1. Let D be a domain whose boundary consists of the curves $C_1,\, C_2,\, \dots,\, C_n$. Suppose that we wish to find a function $\phi(x,\, y)$ that is harmonic in D and that takes on the values $k_1,\, k_2,\, \dots,\, k_n$ on the boundary curves $C_1,\, C_2,\, \dots,\, C_n$, respectively. Our method for solving such a problem consists of the following four steps.

Steps for Solving a Dirichlet Problem

1. *Find an analytic function $f(z) = u(x,\, y) + iv(x,\, y)$ that maps the domain D in the z-plane onto a simpler domain D' in the w-plane and that maps the boundary curves $C_1,\, C_2,\, \dots,\, C_n$ onto the curves $C_1',\, C_2',\, \dots,\, C_n'$, respectively.*

2. *Transform the boundary conditions on $C_1,\, C_2,\, \dots,\, C_n$ to boundary conditions on $C_1',\, C_2',\, \dots,\, C_n'$.*

3. *Solve this new (and easier) Dirichlet problem in D' to obtain a harmonic function $\Phi(u,\, v)$.*

4. *Substitute the real and imaginary parts $u(x,\, y)$ and $v(x,\, y)$ of f for the variables u and v in $\Phi(u,\, v)$. By Theorem 4.5.1, the function $\phi(x,\, y) = \Phi(u(x,\, y),\, v(x,\, y))$ is a solution to the Dirichlet problem in D.*

We illustrate the general idea of these steps in Figure 4.5.3.

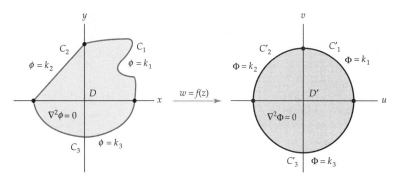

Figure 4.5.3 Transforming a Dirichlet problem

EXAMPLE 1 Using Mappings to Solve a Dirichlet Problem

Let D be the domain in the z-plane bounded by the lines $y = x$ and $y = x + 2$ shown in color in Figure 4.5.4. Find a function $\phi(x, y)$ that is harmonic in D and satisfies the boundary conditions $\phi(x, x + 2) = -2$ and $\phi(x, x) = 3$.

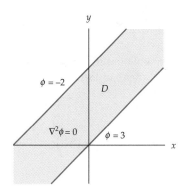

Figure 4.5.4 Figure for Example 1

Solution We will solve this problem using the four steps given on page 202.

Step 1 Inspection of the domain D in Figure 4.5.4 suggests that we take D' to be a domain bounded by the lines $u = -1$ and $u = 1$ in which a solution of the associated Dirichlet problem is given by (2).

Our first step is to find an analytic mapping from D onto D'. In order to do so, we first rotate the region D through $\pi/4$ radians counterclockwise about the origin. Under this rotation, the boundary lines $y = x + 2$ and $y = x$ are mapped onto the vertical lines $u = -\sqrt{2}$ and $u = 0$, respectively. If we next magnify this domain by a factor of $\sqrt{2}$, we obtain a domain bounded by the lines $u = -2$ and $u = 0$. Finally, we translate this image by 1 in order to obtain a domain bounded by the lines $u = 1$ and $u = -1$ as desired. Recall from Section 2.3 that rotation through $\pi/4$ radians about the origin is given by the mapping $R(z) = e^{i\pi/4}$, magnification by $\sqrt{2}$ is given by $M(z) = \sqrt{2}z$, and translation by 1 is given by mapping $T(z) = z + 1$. Therefore, the domain D is mapped onto the domain D' by the composition

$$f(z) = T(M(R(z))) = \sqrt{2}e^{i\pi/4}z + 1 = (1 + i)z + 1.$$

Since the function f is a linear function, it is entire, and so we have completed Step 1.

Step 2 We now transform the boundary conditions on D to boundary conditions on D'. In order to do so, we must find the images under $w = f(z)$ of the boundary lines $y = x$ and $y = x + 2$ of D. By replacing the symbol z with $x + iy$, we can express the mapping $w = (1 + i)z + 1$ as:

$$w = (1 + i)(x + iy) + 1 = x - y + 1 + (x + y)i. \tag{8}$$

From (8) we find that the image of the boundary line $y = x + 2$ is the set of points:

$$w = u + iv = x - (x + 2) + 1 + (x + (x + 2))i = -1 + 2(x + 1)i$$

which is the line $u = -1$. In a similar manner, we also find that the image of the boundary line $y = x$ is the set of points:

$$w = u + iv = x - (x) + 1 + (x + (x))i = 1 + 2xi$$

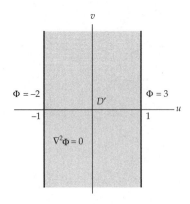

Figure 4.5.5 The transformed Dirichlet problem for Example 1

which is the line $u = 1$. Therefore, the boundary condition $\phi(x,\ x+2) = -2$ is transformed to the boundary condition $\Phi(-1,\ v) = -2$, and the boundary condition $\phi(x,\ x) = 3$ is transformed to the boundary condition $\Phi(1,\ v) = 3$. See Figure 4.5.5.

Step 3 A solution of the Dirichlet problem in D' is given by (2) with x and y replaced by u and v, and with $k_0 = -2$ and $k_1 = 3$:

$$\Phi(u,v) = \frac{3-(-2)}{2}u + \frac{-2+3}{2} = \frac{5}{2}u + \frac{1}{2}.$$

Step 4 The final step in our solution is to substitute the real and imaginary parts of f into Φ for the variables u and v to obtain the desired solution ϕ. From (8) we see that the real and imaginary parts of f are:

$$u(x,y) = x - y + 1 \quad \text{and} \quad v(x,y) = x + y,$$

respectively, and so the function:

$$\phi(x,y) = \Phi(u(x,y), v(x,y)) = \frac{5}{2}(x - y + 1) + \frac{1}{2} = \frac{5}{2}x - \frac{5}{2}y + 3 \quad (9)$$

is a solution of the Dirichlet problem in D. You are encouraged to verify by direct calculation that the function ϕ given in (9) satisfies Laplace's equation and the boundary conditions $\phi(x,\ x) = 3$ and $\phi(x,\ x+2) = -2$. ❑

In Section 3.4, we saw that if ϕ is harmonic in a domain D and if ψ is a harmonic conjugate of ϕ in D, then the **complex potential function** $\mathbf{\Omega}(z)$ given by:

$$\mathbf{\Omega}(z) = \phi(x,y) + i\psi(x,y)$$

is an analytic function in D. The level curves of ϕ and ψ form an orthogonal family of curves as defined in Section 3.4, and the physical meanings of these curves are summarized in Table 3.4.1 on page 152. For example, if the function ϕ in Example 1 represents the electrostatic potential between two infinitely long conducting plates, then the level curves $\phi(x,y) = \frac{5}{2}x - \frac{5}{2}y + 3 = c_1$ represent equipotential curves. These equipotential curves, which are lines with slope 1, are shown in color in Figure 4.5.6.

It is a straightforward computation to find a harmonic conjugate $\Psi(u,v) = \frac{5}{2}v$ of $\Phi(u,v) = \frac{5}{2}u + \frac{1}{2}$ and form the complex potential function $\mathbf{\Omega}(w) = \frac{5}{2}w + \frac{1}{2}$ of the Dirichlet problem in D' for Example 1. Consequently, a complex potential function for the Dirichlet problem in D is found by replacing the symbol w with the analytic mapping $w = (1+i)z + 1$:

$$\mathbf{\Omega}(z) = \frac{5}{2}[(1+i)z + 1] + \frac{1}{2} = \left(\frac{5}{2} + \frac{5}{2}i\right)z + 3.$$

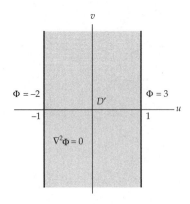

Figure 4.5.6 Equipotential curves and lines of force for Example 1

Therefore, a harmonic conjugate of ϕ is $\psi(x,y) = \text{Im}(\mathbf{\Omega}(z)) = \frac{5}{2}x + \frac{5}{2}y$. If ϕ is electrostatic potential, then the level curves of $\psi(x,y) = \frac{5}{2}x + \frac{5}{2}y = c_2$ represent lines of force. These lines of force, which are line segments with slope -1, are shown in black in Figure 4.5.6.

The method used in Example 1 can be generalized to solve a Dirichlet problem in any domain D bounded by two parallel lines. The key to solving such a problem is finding an appropriate linear function that maps the boundary lines of D onto the boundary lines of the domain shown in Figure 4.5.2. See Problems 1–4 in Exercises 4.5.

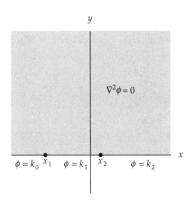

Figure 4.5.7 A Dirichlet problem in the half-plane $y > 0$

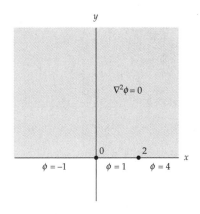

Figure 4.5.8 A Dirichlet problem in the half-plane $y > 0$

Dirichlet Problem in a Half-Plane   Let D be the upper half-plane $y > 0$, and let $x_1 < x_2 < \ldots < x_n$ be n distinct points on the real axis (which is the boundary of D). For many applications, it is useful to know a solution ϕ of the Dirichlet problem in D that satisfies the boundary conditions $\phi(x, 0) = k_0$ for $x < x_1$, $\phi(x, 0) = k_1$ for $x_1 < x < x_2$, $\phi(x, 0) = k_2$ for $x_2 < x < x_3$, ..., $\phi(x, 0) = k_n$ for $x_n < x$. See Figure 4.5.7. If $z = x + iy$, then a solution of this Dirichlet problem given by:

$$\phi(x, y) = k_n + \frac{1}{\pi} \sum_{j=1}^{n} (k_{j-1} - k_j) \operatorname{Arg}(z - x_j). \tag{10}$$

The derivation of this solution will be discussed in Section 7.4. As an application of (10), consider the Dirichlet problem:

$$Solve: \frac{\partial^2 \phi}{\partial x^2} + \frac{\partial^2 \phi}{\partial y^2} = 0, \ -\infty < x < \infty, \ y > 0.$$

$$Subject \ to: \ \phi(x, 0) = \begin{cases} -1, & -\infty < x < 0 \\ 1, & 0 < x < 2 \\ 4, & 2 < x < \infty. \end{cases}$$

See Figure 4.5.8. A solution of this problem is given by (10) with $x_1 = 0$, $x_2 = 2$, $k_0 = -1$, $k_1 = 1$, and $k_2 = 4$:

$$\phi(x, y) = 4 - \frac{2}{\pi} \operatorname{Arg}(z) - \frac{3}{\pi} \operatorname{Arg}(z - 2). \tag{11}$$

We will now directly verify that the function $\phi(x, y)$ in (11) is a solution of this Dirichlet problem. To see that ϕ is harmonic in the domain $y > 0$, we note that ϕ is the real part of the function

$$\mathbf{\Omega}(z) = 4 + (2i/\pi)\operatorname{Ln}(z) + (3i/\pi)\operatorname{Ln}(z - 2).$$

Since $\mathbf{\Omega}$ is analytic in the domain $y > 0$, it follows that ϕ is harmonic in the domain $y > 0$. We next verify that ϕ satisfies the boundary conditions shown in Figure 4.5.8. If $-\infty < x < 0$ and $y = 0$, then $z = x + iy$ is on the negative real axis and so $\operatorname{Arg}(z) = \pi$. In this case, we also have that $z - 2$ is on the negative real axis and so $\operatorname{Arg}(z - 2) = \pi$ as well. Substituting these values in (11) yields:

$$\phi(x, 0) = 4 - \frac{2}{\pi}\pi - \frac{3}{\pi}\pi = -1.$$

On the other hand, if $0 < x < 2$ and $y = 0$, then z is on the positive real axis, while $z - 2$ is on the negative real axis. Thus, $\operatorname{Arg}(z) = 0$ and $\operatorname{Arg}(z - 2) = \pi$. After substituting these values in (11) we see that:

$$\phi(x, 0) = 4 - \frac{2}{\pi}0 - \frac{3}{\pi}\pi = 1.$$

Finally, if $2 < x < \infty$ and $y = 0$, then z and $z - 2$ are on the positive real axis, and so $\operatorname{Arg}(z) = \operatorname{Arg}(z - 2) = 0$. Therefore,

$$\phi(x, 0) = 4 - \frac{2}{\pi}0 - \frac{3}{\pi}0 = 4.$$

Therefore, we have shown that the function ϕ in (11) is a solution of the Dirichlet problem shown Figure 4.5.8 as claimed.

EXAMPLE 2   A Heat Flow Application

Find the steady-state temperature $\phi(x, y)$ in the vertical semi-infinite strip shown in color in Figure 4.5.9. That is, solve the Dirichlet problem in the domain D defined by $-\pi/2 < x < \pi/2$, $y > 0$, where the boundary conditions are:

$$\phi(-\pi/2, y) = 40, \quad \phi(\pi/2, y) = 10, \quad y > 0$$

$$\phi(x, 0) = \begin{cases} 20, & -\pi/2 < x < 0 \\ 50, & 0 < x < \pi/2. \end{cases}$$

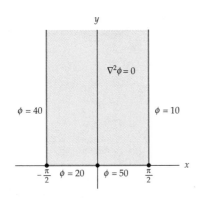

Figure 4.5.9 Figure for Example 2

Solution   From Section 3.4, the steady-state temperature ϕ must satisfy Laplace's equation (1) in D. We proceed as in Example 1.

Step 1   In Section 4.3, we saw that the mapping $w = \sin z$ takes the domain D onto the upper-half plane D' given by $v > 0$. See Example 3 in Section 4.3. Because $\sin z$ is an entire function, $w = \sin z$ is an analytic mapping of D onto D'.

Step 2   From Example 3 in Section 4.3 we have that $w = \sin z$ maps:

(i) the half-line $x = -\pi/2$, $y > 0$, onto the half-line $v = 0$, $u < -1$,

(ii) the segment $y = 0$, $-\pi/2 < x < 0$ onto the segment $v = 0$, $-1 < u < 0$,

(iii) the segment $y = 0$, $0 < x < \pi/2$ onto the segment $v = 0$, $0 < u < 1$, and

(iv) the half-line $x = \pi/2$, $y > 0$, onto the half-line $v = 0$, $u > 1$.

This transforms the Dirichlet problem in the domain D shown in color in Figure 4.5.9 onto the Dirichlet problem in the half-plane $v > 0$ shown in gray in Figure 4.5.10. That is, the transformed Dirichlet problem is:

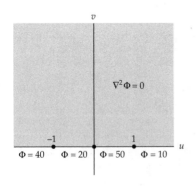

Figure 4.5.10 Transformed Dirichlet problem for Example 2

$$Solve: \frac{\partial^2 \Phi}{\partial u^2} + \frac{\partial^2 \Phi}{\partial v^2} = 0$$

$$Subject\ to:\ \phi(u, 0) = \begin{cases} 40, & -\infty < u < -1 \\ 20, & -1 < u < 0 \\ 50, & 0 < u < 1 \\ 10, & 1 < u < \infty. \end{cases}$$

Step 3   A solution of the transformed Dirichlet problem in Step 2 is given by (10) with the symbols x, y, and z replaced by u, v, and w, respectively. Setting $k_0 = 40$, $k_1 = 20$, $k_2 = 50$, $k_3 = 10$, $u_1 = -1$, $u_2 = 0$, and $u_3 = 1$, we obtain:

$$\Phi(u, v) = 10 + \frac{20}{\pi} \text{Arg}(w + 1) - \frac{30}{\pi} \text{Arg}(w) + \frac{40}{\pi} \text{Arg}(w - 1). \tag{12}$$

Step 4 A solution ϕ of the Dirichlet problem in the domain D is found by replacing the variables u and v in (12) with the real and imaginary parts of the analytic function $f(z) = \sin z$. This is equivalent to replacing w with $\sin z$ in (12):

$$\phi(x,y) = 10 + \frac{20}{\pi}\text{Arg}(\sin z + 1) - \frac{30}{\pi}\text{Arg}(\sin z) + \frac{40}{\pi}\text{Arg}(\sin z - 1). \quad (13)$$

If desired, the function ϕ can be written in terms of x and y, provided that we are careful with our use of the real arctangent function. In particular, *if the values of the arctangent are chosen to lie between* 0 *and* π, then the function ϕ in (13) can be written as:

$$\phi(x,y) = 10 + \frac{20}{\pi}\arctan\left(\frac{\cos x \sinh y}{\sin x \cosh y + 1}\right) - \frac{30}{\pi}\arctan\left(\frac{\cos x \sinh y}{\sin x \cosh y}\right)$$
$$+ \frac{40}{\pi}\arctan\left(\frac{\cos x \sin y}{\sin x \cosh y - 1}\right).$$

❑

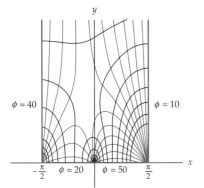

$\phi = 40$ $\phi = 10$

$-\frac{\pi}{2}$ $\phi = 20$ $\phi = 50$ $\frac{\pi}{2}$

Figure 4.5.11 The isotherms and lines of heat flux for Example 2

Observe that the function

$$\mathbf{\Omega}(z) = 10 - \frac{20i}{\pi}\text{Ln}(\sin z + 1) + \frac{30i}{\pi}\text{Ln}(\sin z) - \frac{40i}{\pi}\text{Ln}(\sin z - 1)$$

is analytic in the domain D given by $-\pi/2 < x < \pi/2$, $y > 0$, and shown in color in Figure 4.5.9. Since the real part of $\mathbf{\Omega}(z)$ is the function ϕ given by (13), the imaginary part ψ of $\mathbf{\Omega}(z)$ is a harmonic conjugate of ϕ. Therefore, $\mathbf{\Omega}(z)$ is a complex potential function of the function ϕ in Example 2. In heat flow problems, the level curves of the steady-state temperature ϕ are called isotherms, whereas the level curves of its harmonic conjugate ψ are called lines of heat flux. In Figure 4.5.11 we have sketched the level curves for the heat flow problem in Example 2. The isotherms are the curves shown in color and lines of heat flux are the curves shown in black.

EXERCISES 4.5   *Answers to selected odd-numbered problems begin on page ANS-13.*

In Problems 1–4, (**a**) use a linear mapping and (2) to find the electrostatic potential $\phi(x, y)$ in the domain D that satisfies the given boundary conditions, (**b**) find a complex potential function $\mathbf{\Omega}(z)$ for $\phi(x, y)$, and (**c**) sketch the equipotential curves and the lines of force.

1. The domain D is bounded by the lines $x = 2$ and $x = 7$, and the boundary conditions are $\phi(2, y) = 3$ and $\phi(7, y) = -2$.

2. The domain D is bounded by the lines $y = 0$ and $y = 3$, and the boundary conditions are $\phi(x, 0) = 1$ and $\phi(x, 3) = 2$.

3. The domain D is bounded by the lines $y = \sqrt{3}x$ and $y = \sqrt{3}x + 4$, and the boundary conditions are $\phi\left(x, \sqrt{3}x\right) = 10$ and $\phi\left(x, \sqrt{3}x + 4\right) = 5$.

4. The domain D is bounded by the lines $y = x + 2$ and $y = x + 4$, and the boundary conditions are $\phi(x, x + 2) = -4$ and $\phi(x, x + 4) = 5$.

In Problems 5–8, (**a**) use the analytic mapping $w = \sin z$ and, if necessary, linear mappings together with (10) to find the steady-state temperature $\phi(x, y)$ in the domain D that satisfies the given boundary conditions, and (**b**) find a complex potential function $\boldsymbol{\Omega}(z)$ for $\phi(x,\ y)$.

5. The domain D is given by $\pi/2 < x < 3\pi/2$, $y > 0$, and the boundary conditions are $\phi\,(\pi/2,\ y) = 20$, $\phi\,(x,\ 0) = -13$, and $\phi\,(3\pi/2,\ y) = 12$.

6. The domain D is bounded by $-3 < x < 3$, $y > 1$, and the boundary conditions are $\phi\,(-3,\ y) = 1$, $\phi\,(x,\ 1) = 3$, and $\phi\,(3,\ y) = 5$.

7. The domain D is bounded by $-\pi/2 < y < \pi/2$, $x > 0$, and the boundary conditions are $\phi\,(x, -\pi/2) = 15$, $\phi(0,\ y) = 32$, and $\phi\,(x,\ \pi/2) = 23$.

8. The domain D is bounded by the lines $y = x + 2$, $y = x - 2$, and $y = -x$. In D the points $z = x + iy$ satisfy $y \geq -x$. The boundary conditions are $\phi\,(x,\ x + 2) = 10$, $\phi\,(x,\ -x) = 7$, and $\phi\,(x,\ x - 2) = 5$.

Focus on Concepts

9. Use the analytic mapping $w = z^2$ and (10) to solve the Dirichlet problem shown in Figure 4.5.12. Find a complex potential function $\boldsymbol{\Omega}(z)$ for $\phi(x,y)$.

10. Use the analytic mapping $w = z^4$ and (10) to solve the Dirichlet problem shown in Figure 4.5.13. Find a complex potential function $\boldsymbol{\Omega}(z)$ for $\phi(x,\ y)$.

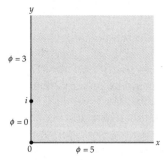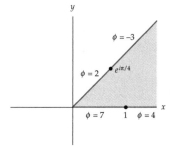

Figure 4.5.12 Figure for Problem 9   **Figure 4.5.13** Figure for Problem 10

11. Use the analytic mapping $w = \sin^{-1} z$ and (2) to solve the Dirichlet problem shown in Figure 4.5.14. Find the complex potential function $\boldsymbol{\Omega}(z)$ for $\phi(x,\ y)$.

12. Use the analytic mapping $w = z + \dfrac{1}{z}$ and (10) to solve the Dirichlet problem shown in Figure 4.5.15. Find a complex potential function $\boldsymbol{\Omega}(z)$ for $\phi(x,y)$.

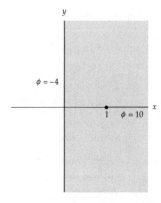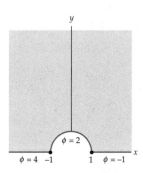

Figure 4.5.14 Figure for Problem 11   **Figure 4.5.15** Figure for Problem 12

Computer Lab Assignments

In Problems 13–16, use a CAS to plot the isotherms and lines of heat flux for the given heat flow.

13. The heat flow in Problem 5.

14. The heat flow in Problem 6.

15. The heat flow in Problem 7.

16. The heat flow in Problem 8.

In Problems 17–20, use a CAS to plot the level curves $\phi = c_1$ and $\psi = c_2$ of the given complex potential function $\Omega(z)$.

17. $\Omega(z)$ is the complex potential function in Problem 9.

18. $\Omega(z)$ is the complex potential function in Problem 10.

19. $\Omega(z)$ is the complex potential function in Problem 11.

20. $\Omega(z)$ is the complex potential function in Problem 12.

CHAPTER 4 REVIEW QUIZ

Answers to selected odd-numbered problems begin on page ANS-13.

In Problems 1–20, answer true or false. If the statement is false, justify your answer by either explaining why it is false or giving a counterexample; if the statement is true, justify your answer by either proving the statement or citing an appropriate result in this chapter.

1. If $|e^z| = 1$, then z is a pure imaginary number.

2. $\operatorname{Re}(e^z) = \cos y$.

3. The mapping $w = e^z$ takes vertical lines in the z-plane onto horizontal lines in the w-plane.

4. There are infinitely many solutions z to the equation $e^z = w$.

5. $\ln i = \frac{1}{2}\pi i$.

6. $\operatorname{Im}(\ln z) = \arg(z)$.

7. For all nonzero complex z, $e^{\operatorname{Ln} z} = z$.

8. If w_1 and w_2 are two values of $\ln z$, then $\operatorname{Re}(w_1) = \operatorname{Re}(w_2)$.

9. $\operatorname{Ln}\dfrac{1}{z} = -\operatorname{Ln} z$ for all nonzero z.

10. For all nonzero complex numbers, $\operatorname{Ln}(z_1 z_2) = \operatorname{Ln} z_1 + \operatorname{Ln} z_2$.

11. $\operatorname{Ln} z$ is an entire function.

12. The principal value of i^{i+1} is $e^{-\pi/2+i}$.

13. The complex power z^α is always multiple-valued.

14. $\cos z$ is a periodic function with a period of 2π.

15. There are complex z such that $|\sin z| > 1$.

16. $\tan z$ has singularities at $z = (2n+1)\,\pi/2$, for $n = 0, \pm 1, \pm 2, \ldots$.

17. $\cosh z = \cos(iz)$.

18. $z = \frac{1}{2}\pi i$ is a zero of $\cosh z$.

19. The function $\sin \bar{z}$ is nowhere analytic.

20. Every branch of $\tan^{-1} z$ is entire.

In Problems 21–40, try to fill in the blanks without referring back to the text.

21. The real and imaginary parts of e^z are $u(x,\,y) = $ _____ and $v(x,\,y) = $ _____ .

22. The domain of $\operatorname{Ln} z$ is _____ , and its range is _____ .

23. $\operatorname{Ln}\left(\sqrt{3}+i\right) = $ _____ .

24. The complex exponential function e^z is periodic with a period of _____ .

25. If $e^{iz} = 2$, then $z = $ _____ .

26. $\operatorname{Ln}\left(e^{1-\pi i}\right) = $ _____ .

27. $\operatorname{Ln} z$ is discontinuous on _____ .

28. The line segment $x = a$, $-\pi < y \le \pi$, is mapped onto _____ by the mapping $w = e^z$.

29. $\ln\left(1+i\right) = $ _____ .

30. If $\ln z$ is pure imaginary, then $|z| = $ _____ .

31. $z_1 = 1$ and $z_2 = $ _____ are two real numbers for which the principal value $z^i = 1$.

32. The principal value of i^i is _____ .

33. On the domain $|z| > 0$, $-\pi < \arg(z) < \pi$, the derivative of the principal value of z^α is _____ .

34. The complex sine function is defined by $\sin z = $ _____ .

35. $\cos\left(4i\right) = $ _____ .

36. The semi-infinite vertical strip $-\pi/2 \le x \le \pi/2$, $y \ge 0$, is mapped onto _____ by $w = \sin z$.

37. The real and imaginary parts of $\sin z$ are _____ and _____ , respectively.

38. The complex sine and hyperbolic sine functions are related by the formulas $\sin(iz) = $ _____ and $\sinh(iz) = $ _____ .

39. $\tanh^{-1} z$ is not defined for $z = $ _____ .

40. In order to compute a specific value of $\sin^{-1} z$ you need to choose a branch of _____ and a branch of _____ .

Integration in the Complex Plane

Chapter Outline

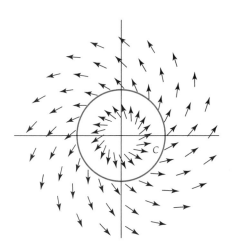

Normalized velocity vector field for $f(z) = (1 + i)z$. See page 264.

Introduction To define an integral of a complex function f, we start with a complex function f defined along some curve C or contour in the complex plane. We shall see in this section that the definition of a complex integral, its properties, and method of evaluation are quite similar to those of a real line integral in the Cartesian plane.

5.1   Real Integrals

Definite Integral   It is likely that you have retained at least two associations from your study of elementary calculus: the derivative with slope, and the definite integral with area. But as the derivative $f'(x)$ of a real function $y = f(x)$ has other uses besides finding slopes of tangent lines, so too the value of a definite integral $\int_a^b f(x)\,dx$ need not be area "under a curve." Recall, if $F(x)$ is an antiderivative of a continuous function f, that is, F is a function for which $F'(x) = f(x)$, then the definite integral of f on the interval $[a, b]$ is the number

$$\int_a^b f(x)\,dx = F(x)\big|_a^b = F(b) - F(a). \tag{1}$$

For example, $\int_{-1}^2 x^2\,dx = \frac{1}{3}x^3\big|_{-1}^2 = \frac{8}{3} - \left(-\frac{1}{3}\right) = 3$. Bear in mind that the **fundamental theorem of calculus**, just given in (1), is a method of *evaluating* $\int_a^b f(x)\,dx$; it is not the *definition* of $\int_a^b f(x)\,dx$.

In the discussion that follows we present the definitions of two types of **real integrals**. We begin with the five steps leading to the definition of the definite (or Riemann) integral of a function f; we follow it with the definition of line integrals in the Cartesian plane. Both definitions rest on the limit concept.

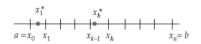

Figure 5.1.1 Partition of $[a, b]$ with x_k^* in each subinterval $[x_{k-1},\ x_k]$

Steps Leading to the Definition of the Definite Integral

1. *Let f be a function of a single variable x defined at all points in a closed interval $[a, b]$.*

2. *Let P be a partition:*

$$a = x_0 < x_1 < x_2 < \cdots < x_{n-1} < x_n = b$$

   of $[a, b]$ into n subintervals $[x_{k-1},\ x_k]$ of length $\Delta x_k = x_k - x_{k-1}$. See Figure 5.1.1.

3. *Let $\|P\|$ be the **norm** of the partition P of $[a, b]$, that is, the length of the longest subinterval.*

4. *Choose a number x_k^* in each subinterval $[x_{k-1},\ x_k]$ of $[a, b]$. See Figure 5.1.1.*

5. *Form n products $f(x_k^*)\Delta x_k$, $k = 1, 2, \ldots, n$, and then sum these products:*

$$\sum_{k=1}^{n} f(x_k^*)\,\Delta x_k.$$

Definition 5.1.1   Definite Integral

The **definite integral** of f on $[a, b]$ is

$$\int_a^b f(x)\,dx = \lim_{\|P\|\to 0} \sum_{k=1}^{n} f(x_k^*)\,\Delta x_k. \qquad (2)$$

Whenever the limit in (2) exists we say that f is **integrable** on the interval $[a, b]$ or that the definite integral of f **exists**. It can be proved that if f is continuous on $[a, b]$, then the integral defined in (2) exists.

The notion of the definite integral $\int_a^b f(x)\,dx$, that is, *integration of a real function $f(x)$ over an interval on the x-axis from $x = a$ to $x = b$* can be generalized to *integration of a real multivariable function $G(x, y)$ on a curve C from point A to point B* in the Cartesian plane. To this end we need to introduce some terminology about curves.

Terminology   Suppose a curve C in the plane is parametrized by a set of equations $x = x(t)$, $y = y(t)$, $a \le t \le b$, where $x(t)$ and $y(t)$ are continuous real functions. Let the **initial** and **terminal points** of C, that is, $(x(a), y(a))$ and $(x(b), y(b))$, be denoted by the symbols A and B, respectively. We say that:

(i) C is a **smooth curve** if x' and y' are continuous on the closed interval $[a, b]$ and not simultaneously zero on the open interval (a, b).

(ii) C is a **piecewise smooth curve** if it consists of a finite number of smooth curves $C_1, C_2, \ldots, C_n$ joined end to end, that is, the terminal point of one curve C_k coinciding with the initial point of the next curve C_{k+1}.

(iii) C is a **simple curve** if the curve C does not cross itself except possibly at $t = a$ and $t = b$.

(iv) C is a **closed curve** if $A = B$.

(v) C is a **simple closed curve** if the curve C does not cross itself and $A = B$; that is, C is simple and closed.

Figure 5.1.2 illustrates each type of curve defined in (i)–(v).

Line Integrals in the Plane   The following five steps lead to the definition of three **line integrals**[*] in the plane and are analogous to the five steps given prior to the definition of the definite integral.

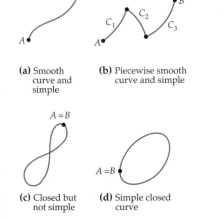

(a) Smooth curve and simple

(b) Piecewise smooth curve and simple

(c) Closed but not simple

(d) Simple closed curve

Figure 5.1.2 Types of curves in the plane

---

[*]An unfortunate choice of names. **Curve integrals** would be more appropriate.

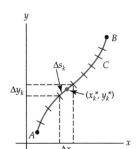

Figure 5.1.3 Partition of curve C into n subarcs induced by a partition P of the parameter interval $[a,\ b]$

Steps Leading to the Definition of Line Integrals

1. *Let G be a function of two real variables x and y defined at all points on a smooth curve C that lies in some region of the xy-plane. Let C be defined by the parametrization $x = x(t)$, $y = y(t)$, $a \le t \le b$.*

2. *Let P be a partition of the parameter interval $[a,\ b]$ into n subintervals $[t_{k-1},\ t_k]$ of length $\Delta t_k = t_k - t_{k-1}$:*

$$a = t_0 < t_1 < t_2 < \cdots < t_{n-1} < t_n = b.$$

   The partition P induces a partition of the curve C into n subarcs of length Δs_k. Let the projection of each subarc onto the x- and y-axes have lengths Δx_k and Δy_k, respectively. See Figure 5.1.3.

3. *Let $\|P\|$ be the norm of the partition P of $[a,\ b]$, that is, the length of the longest subinterval.*

4. *Choose a point $(x_k^*,\ y_k^*)$ on each subarc of C. See Figure 5.1.3.*

5. *Form n products $G(x_k^*, y_k^*)\Delta x_k$, $G(x_k^*, y_k^*)\Delta y_k$, $G(x_k^*, y_k^*)\Delta s_k$, $k = 1,\ 2, \ldots,\ n$, and then sum these products*

$$\sum_{k=1}^{n} G(x_k^*,\ y_k^*)\Delta x_k, \quad \sum_{k=1}^{n} G(x_k^*,\ y_k^*)\Delta y_k, \quad \text{and} \quad \sum_{k=1}^{n} G(x_k^*,\ y_k^*)\Delta s_k.$$

Definition 5.1.2   Line Integrals in the Plane

(i) The **line integral of G along C with respect to x** is

$$\int_C G(x, y)\, dx = \lim_{\|P\| \to 0} \sum_{k=1}^{n} G(x_k^*,\ y_k^*)\,\Delta x_k. \qquad (3)$$

(ii) The **line integral of G along C with respect to y** is

$$\int_C G(x, y)\, dy = \lim_{\|P\| \to 0} \sum_{k=1}^{n} G(x_k^*,\ y_k^*)\,\Delta y_k. \qquad (4)$$

(iii) The **line integral of G along C with respect to arc length s** is

$$\int_C G(x, y)\, ds = \lim_{\|P\| \to 0} \sum_{k=1}^{n} G(x_k^*,\ y_k^*)\,\Delta s_k. \qquad (5)$$

It can be proved that if G is continuous on C, then the three types of line integrals defined in (3), (4), and (5) exist. We shall assume continuity of G as matter of course. The curve C is referred to as the **path** of integration.

Method of Evaluation–C Defined Parametrically The line integrals in Definition 5.1.2 can be evaluated in two ways, depending on whether the curve C is defined by a pair of parametric equations or by an explicit function. Either way, *the basic idea is to convert a line integral to a* *definite integral* in a single variable. If C is smooth curve parametrized by $x = x(t)$, $y = y(t)$, $a \le t \le b$, then replace x and y in the integral by the functions $x(t)$ and $y(t)$, and the appropriate differential dx, dy, or ds by

$$dx = x'(t)\, dt, \ dy = y'(t)\, dt, \quad \text{or} \quad ds = \sqrt{[x'(t)]^2 + [y'(t)]^2}\, dt.$$

The term $ds = \sqrt{[x'(t)]^2 + [y'(t)]^2}\, dt$ is called the differential of the arc length. In this manner each of the line integrals in Definition 5.1.2 becomes a definite integral in which the variable of integration is the parameter t. That is,

$$\int_C G(x, y)\, dx = \int_a^b G\left(x(t), y(t)\right) x'(t)\, dt, \tag{6}$$

$$\int_C G(x, y)\, dy = \int_a^b G\left(x(t), y(t)\right) y'(t)\, dt, \tag{7}$$

$$\int_C G(x, y)\, ds = \int_a^b G\left(x(t), y(t)\right) \sqrt{[x'(t)]^2 + [y'(t)]^2}\, dt. \tag{8}$$

EXAMPLE 1 C Defined Parametrically

Evaluate (a) $\int_C xy^2 dx$, (b) $\int_C xy^2 dy$, and (c) $\int_C xy^2 ds$, where the path of integration C is the quarter circle defined by $x = 4\cos t$, $y = 4\sin t$, $0 \le t \le \pi/2$.

Solution The path C of integration is shown in color in Figure 5.1.4. In each of the three given line integrals, x is replaced by $4\cos t$ and y is replaced by $4\sin t$.

(a) Since $dx = -4\sin t\, dt$, we have from (6):

$$\int_C xy^2 dx = \int_0^{\pi/2} (4\cos t)(4\sin t)^2 (-4\sin t\, dt)$$

$$= -256 \int_0^{\pi/2} \sin^3 t \cos t\, dt = -256 \left[\frac{1}{4}\sin^4 t\right]_0^{\pi/2} = -64.$$

(b) Since $dy = 4\cos t\, dt$, we have from (7):

$$\int_C xy^2 dy = \int_0^{\pi/2} (4\cos t)(4\sin t)^2 (4\cos t\, dt)$$

$$= 256 \int_0^{\pi/2} \sin^2 t \cos^2 t\, dt$$

$$= 256 \int_0^{\pi/2} \frac{1}{4}\sin^2 2t\, dt$$

$$= 64 \int_0^{\pi/2} \frac{1}{2}(1 - \cos 4t)\, dt = 32\left[t - \frac{1}{4}\sin 4t\right]_0^{\pi/2} = 16\pi.$$

Note in this integration we have used two trigonometric identities: $\sin 2\theta = 2\sin\theta\cos\theta$ and $\sin^2\theta = \frac{1}{2}(1 - \cos 2\theta)$.

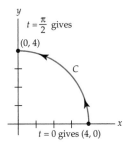

$t = \frac{\pi}{2}$ gives
$(0, 4)$
C
$t = 0$ gives $(4, 0)$

Figure 5.1.4 Path C of integration

(c) Since $ds = \sqrt{16\left(\sin^2 t + \cos^2 t\right)}\,dt = 4\,dt$, it follows from (8):

$$\int_C xy^2\,ds = \int_0^{\pi/2}(4\cos t)(4\sin t)^2(4\,dt)$$

$$= 256\int_0^{\pi/2}\sin^2 t\cos t\,dt = 256\left[\frac{1}{3}\sin^3 t\right]_0^{\pi/2} = \frac{256}{3}.$$

❑

Method of Evaluation–C Defined by a Function   If the path of integration C is the graph of an explicit function $y = f(x)$, $a \le x \le b$, then we can use x as a parameter. In this situation, the differential of y is $dy = f'(x)\,dx$, and the differential of arc length is $ds = \sqrt{1 + [f'(x)]^2}\,dx$. After substituting, the three line integrals of Definition 5.1.2 become, in turn, the definite integrals:

$$\int_C G(x,y)\,dx = \int_a^b G\left(x, f(x)\right)dx, \tag{9}$$

$$\int_C G(x,y)\,dy = \int_a^b G\left(x, f(x)\right)f'(x)\,dx, \tag{10}$$

$$\int_C G(x,y)\,ds = \int_a^b G\left(x, f(x)\right)\sqrt{1 + [f'(x)]^2}\,dx. \tag{11}$$

A line integral along a *piecewise smooth* curve C is defined as the sum of the integrals over the various smooth curves whose union comprises C. For example, to evaluate $\int_C G(x,\,y)\,ds$ when C is composed of two smooth curves C_1 and C_2, we begin by writing

$$\int_C G(x,y)\,ds = \int_{C_1} G(x,y)\,ds + \int_{C_2} G(x,y)\,ds. \tag{12}$$

The integrals $\int_{C_1} G(x,\,y)\,ds$ and $\int_{C_2} G(x,\,y)\,ds$ are then evaluated in the manner given in (8) or (11).

EXAMPLE 2 C Defined by an Explicit Function

Evaluate $\int_C xy\,dx + x^2\,dy$, where C is the graph of $y = x^3$, $-1 \le x \le 2$.

Solution   The curve C is illustrated in Figure 5.1.5 and is defined by the explicit function $y = x^3$. Hence we can use x as the parameter. Using the differential $dy = 3x^2\,dx$, we apply (9) and (10):

$$\int_C xy\,dx + x^2\,dy = \int_{-1}^2 x\left(x^3\right)dx + x^2\left(3x^2\,dx\right)$$

$$= \int_{-1}^2 4x^4\,dx = \frac{4}{5}x^5\bigg|_{-1}^2 = \frac{132}{5}.$$

❑

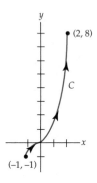

Figure 5.1.5 Graph of $y = x^3$ on the interval $-1 \le x \le 2$

Notation In many applications, line integrals appear as a sum $\int_C P(x,y)\,dx + \int_C Q(x,y)\,dy$. It is common practice to write this sum as one integral without parentheses as

$$\int_C P(x,y)\,dx + Q(x,y)\,dy \qquad \text{or simply} \qquad \int_C P\,dx + Q\,dy. \qquad (13)$$

A line integral along a closed curve C is usually denoted by

$$\oint_C P\,dx + Q\,dy.$$

EXAMPLE 3 C is a Closed Curve

Evaluate $\oint_C x\,dx$, where C is the circle defined by $x = \cos t$, $y = \sin t$, $0 \le t \le 2\pi$.

Solution The differential of $x = \cos t$ is $dx = -\sin t\,dt$, and so from (6),

$$\oint_C x\,dx = \int_0^{2\pi} \cos t\,(-\sin t\,dt) = \frac{1}{2}\cos^2 t\,\Big|_0^{2\pi} = \frac{1}{2}[1-1] = 0.$$

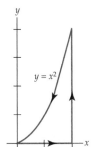

Figure 5.1.6 Piecewise smooth path of integration

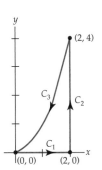

Figure 5.1.7 C consists of the union of C_1, C_2, and C_3.

EXAMPLE 4 C is a Closed Curve

Evaluate $\oint_C y^2\,dx - x^2\,dy$, where C is the closed curve shown in Figure 5.1.6.

Solution Since C is piecewise smooth, we proceed as illustrated in (12); namely, the given integral is expressed as a sum of integrals. Symbolically, we write $\oint_C = \int_{C_1} + \int_{C_2} + \int_{C_3}$, where the C_1, C_2, and C_3 are the curves labeled in Figure 5.1.7. On C_1, we use x as a parameter. Since $y = 0$, $dy = 0$; therefore

$$\int_{C_1} y^2\,dx - x^2\,dy = \int_0^2 0\,dx - x^2(0) = 0.$$

On C_2, we use y as a parameter. From $x = 2$, $dx = 0$, we have

$$\int_{C_2} y^2\,dx - x^2\,dy = \int_0^4 y^2(0) - 4\,dy = -\int_0^4 4\,dy = -16.$$

On C_3, we again use x as a parameter. From $y = x^2$, we get $dy = 2x\,dx$ and so

$$\int_{C_3} y^2\,dx - x^2\,dy = \int_2^0 \left(x^2\right)^2 dx - x^2\,(2x\,dx)$$

$$= \int_2^0 \left(x^4 - 2x^3\right) dx = \left(\frac{1}{5}x^5 - \frac{1}{2}x^4\right)\Big|_2^0 = \frac{8}{5}.$$

Hence, $$\oint_C y^2\,dx - x^2\,dy = \int_{C_1} + \int_{C_2} + \int_{C_3} = 0 + (-16) + \frac{8}{5} = -\frac{72}{5}.$$

Orientation of a Curve

Orientation of a Curve   In definite integration we normally assume that the interval of integration is $a \le x \le b$ and the symbol $\int_a^b f(x)\,dx$ indicates that we are integrating in the positive direction on the x-axis. Integration in the opposite direction, from $x = b$ to $x = a$, results in the negative of the original integral:

$$\int_b^a f(x)\,dx = -\int_a^b f(x)\,dx. \tag{14}$$

Line integrals possess a property similar to (14), but first we have to introduce the notion of **orientation** of the path C. If C is not a closed curve, then we say the **positive direction** on C, or that C has **positive orientation**, if we traverse C from its initial point A to its terminal point B. In other words, if $x = x(t)$, $y = y(t)$, $a \le t \le b$ are parametric equations for C, then the positive direction on C corresponds to increasing values of the parameter t. If C is traversed in the sense opposite to that of the positive orientation, then C is said to have **negative orientation**. If C has an orientation (positive or negative), then the **opposite curve**, the curve with the opposite orientation, will be denoted by the symbol $-C$. In Figure 5.1.8 if we assume that A and B are the initial and terminal points of the curve C, respectively, then the arrows on curve C indicate that we are traversing the curve from its initial point to its terminal point, and so C has positive orientation. The curve to the right of C that is labeled $-C$ then has negative orientation. Finally, if $-C$ denotes the curve having the opposite orientation of C, then the analogue of (14) for line integrals is

$$\int_{-C} P\,dx + Q\,dy = -\int_C P\,dx + Q\,dy, \tag{15}$$

or, equivalently

$$\int_{-C} P\,dx + Q\,dy + \int_C P\,dx + Q\,dy = 0. \tag{16}$$

For example, in part (a) of Example 1 we saw that $\int_C xy^2\,dx = -64$; we conclude from (15) that $\int_{-C} xy^2\,dx = 64$.

Note ➡   It is important to be aware that a line integral is independent of the parametrization of the curve C, provided C is given the same orientation by all sets of parametric equations defining the curve. See Problem 33 in Exercises 5.1.

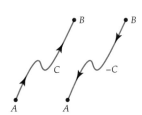

Figure 5.1.8 Curve C and its opposite $-C$

EXERCISES 5.1   *Answers to selected odd-numbered problems begin on page ANS-14.*

In Problems 1–10, evaluate the definite integral. If necessary, review the techniques of integration in your calculus text.

1. $\displaystyle\int_{-1}^3 x(x-1)(x+2)\,dx$

2. $\displaystyle\int_{-1}^0 t^2\,dt + \int_0^2 x^2\,dx + \int_2^3 u^2\,du$

3. $\displaystyle\int_{1/2}^1 \sin 2\pi x\,dx$

4. $\displaystyle\int_0^{\pi/8} \sec^2 2x\,dx$

5. $\displaystyle\int_0^4 \frac{dx}{2x+1}$

6. $\displaystyle\int_{\ln 2}^{\ln 3} e^{-x}\,dx$

7. $\displaystyle\int_2^4 xe^{-x/2}\,dx$

8. $\displaystyle\int_1^e \ln x\,dx$

9. $\displaystyle\int_2^4 \frac{dx}{x^2 - 6x + 5}$

10. $\displaystyle\int_2^4 \frac{2x-1}{(x+3)^2}\,dx$

In Problems 11–14, evaluate the line integrals $\int_C G(x,\,y)\,dx$, $\int_C G(x,\,y)\,dy$, and $\int_C G(x,\,y)\,ds$ on the indicated curve C.

11. $G(x,\,y) = 2xy$; $x = 5\cos t$, $y = 5\sin t$, $0 \le t \le \pi/4$

12. $G(x,\,y) = x^3 + 2xy^2 + 2x$; $x = 2t$, $y = t^2$, $0 \le t \le 1$

13. $G(x,\,y) = 3x^2 + 6y^2$; $y = 2x + 1$, $-1 \le x \le 0$

14. $G(x,\,y) = x^2/y^3$; $2y = 3x^{3/2}$, $1 \le t \le 8$

In Problems 15–18, evaluate $\int_C (2x + y)\,dx + xy\,dy$ on the given curve from $(-1,\,2)$ to $(2,\,5)$.

15. $y = x + 3$ **16.** $y = x^2 + 1$

17.

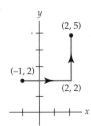

18.

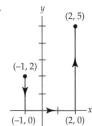

Figure 5.1.9 Figure for Problem 17   **Figure 5.1.10** Figure for Problem 18

In Problems 19–22, evaluate $\displaystyle\int_C y\,dx + x\,dy$ on the given curve from $(0,\,0)$ to $(1,\,1)$.

19. $y = x^2$ **20.** $y = x$

21. C consists of the line segments from $(0,\,0)$ to $(0,\,1)$ and from $(0,\,1)$ to $(1,\,1)$.

22. C consists of the line segments from $(0,\,0)$ to $(1,\,0)$ and from $(1,\,0)$ to $(1,\,1)$.

23. Evaluate $\displaystyle\int_C \left(6x^2 + 2y^2\right) dx + 4xy\,dy$, where C is given by $x = \sqrt{t}$, $y = t$, $4 \le t \le 9$.

24. Evaluate $\displaystyle\int_C -y^2\,dx + xy\,dy$, where C is given by $x = 2t$, $y = t^3$, $0 \le t \le 2$.

25. Evaluate $\displaystyle\int_C 2x^3 y\,dx + (3x + y)\,dy$, where C is given by $x = y^2$ from $(1,\,-1)$ to $(1,\,1)$.

26. Evaluate $\displaystyle\int_C 4x\,dx + 2y\,dy$, where C is given by $x = y^3 + 1$ from $(0,\,-1)$ to $(9,\,2)$.

In Problems 27 and 28, evaluate $\displaystyle\oint_C \left(x^2 + y^2\right) dx - 2xy\,dy$ on the given closed curve.

27.

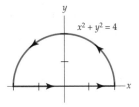

Figure 5.1.11 Figure for Problem 27

28.

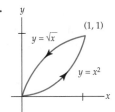

Figure 5.1.12 Figure for Problem 28

In Problems 29 and 30, evaluate $\oint_C x^2 y^3\,dx - xy^2\,dy$ on the given closed curve.

29.

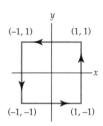

30.

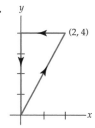

Figure 5.1.13 Figure for Problem 29   Figure 5.1.14 Figure for Problem 30

31. Evaluate $\oint_C \left(x^2 - y^2\right)\,ds$, where C is given by $x = 5\cos t$, $y = 5\sin t$, $0 \le t \le 2\pi$.

32. Evaluate $\int_{-C} y\,dx - x\,dy$, where C is given by $x = 2\cos t$, $y = 3\sin t$, $0 \le t \le \pi$.

33. Verify that the line integral $\int_C y^2\,dx + xy\,dy$ has the same value on C for each of the following parametrizations:

$$C : x = 2t + 1,\ y = 4t + 2,\ 0 \le t \le 1$$
$$C : x = t^2,\ y = 2t^2,\ 1 \le t \le \sqrt{3}$$
$$C : x = \ln t,\ y = 2\ln t,\ e \le t \le e^3.$$

34. Consider the three curves between $(0, 0)$ and $(2, 4)$:

$$C : x = t,\ y = 2t,\ 0 \le t \le 2$$
$$C : x = t,\ y = t^2,\ 0 \le t \le 2$$
$$C : x = 2t - 4,\ y = 4t - 8,\ 2 \le t \le 3.$$

Show that $\int_{C_1} xy\,ds = \int_{C_3} xy\,ds$, but $\int_{C_1} xy\,ds \ne \int_{C_2} xy\,ds$. Explain.

35. If $\rho(x, y)$ is the density of a wire (mass per unit length), then the mass of the wire is $m = \int_C \rho(x, y)\,ds$. Find the mass of a wire having the shape of a semicircle $x = 1 + \cos t$, $y = \sin t$, $0 \le t \le \pi$, if the density at a point P is directly proportional to the distance from the y-axis.

36. The coordinates of the center of mass of a wire with variable density are given by $\bar{x} = M_y/m$, $\bar{y} = M_x/m$ where

$$m = \int_C \rho(x, y)\,ds, \quad M_x = \int_C y\rho(x, y)\,ds, \quad M_y = \int_C x\rho(x, y)\,ds.$$

Find the center of mass of the wire in Problem 35.

5.2 Complex Integrals

In the preceding section we reviewed two types of real integrals. We saw that the definition of the definite integral starts with a real function $y = f(x)$ that is defined on an interval on the x-axis. Because a planar curve is the two-dimensional analogue of an interval, we then generalized the definition of $\int_a^b f(x)\,dx$ to integrals of real functions of two variables defined on a curve C in the Cartesian plane. We shall see in this section that a complex integral is defined in a manner that is quite similar to that of a line integral in the Cartesian plane.

Since curves play a big part in the definition of a complex integral, we begin with a brief review of how curves are represented in the complex plane.

Curves Revisited   Suppose the continuous real-valued functions $x = x(t)$, $y = y(t)$, $a \le t \le b$, are parametric equations of a curve C in the complex plane. If we use these equations as the real and imaginary parts in $z = x + iy$, we saw in Section 2.2 that we can describe the points z on C by means of a complex-valued function of a real variable t called a **parametrization** of C:

$$z(t) = x(t) + iy(t), \ a \leq t \leq b. \tag{1}$$

For example, the parametric equations $x = \cos t$, $y = \sin t$, $0 \leq t \leq 2\pi$, describe a unit circle centered at the origin. A parametrization of this circle is $z(t) = \cos t + i \sin t$, or $z(t) = e^{it}, 0 \leq t \leq 2\pi$. See (6)–(10) in Section 2.2.

The point $z(a) = x(a) + iy(a)$ or $A = (x(a), \ y(a))$ is called the **initial point** of C and $z(b) = x(b) + iy(b)$ or $B = (x(b), \ y(b))$ is its **terminal point**. We also saw in Section 2.7 that $z(t) = x(t) + iy(t)$ could also be interpreted as a two-dimensional vector function. Consequently, $z(a)$ and $z(b)$ can be interpreted as position vectors. As t varies from $t = a$ to $t = b$ we can envision the curve C being traced out by the moving arrowhead of $z(t)$. See Figure 5.2.1.

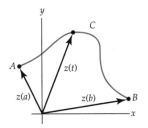

Figure 5.2.1 $z(t) = x(t) + iy(t)$ as a position vector

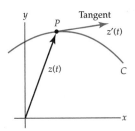

Figure 5.2.2 $z'(t) = x'(t) + iy'(t)$ as a tangent vector

Figure 5.2.3 Curve C is not smooth since it has a cusp.

Contours The notions of curves in the complex plane that are smooth, piecewise smooth, simple, closed, and simple closed are easily formulated in terms of the vector function (1). Suppose the derivative of (1) is $z'(t) = x'(t) + iy'(t)$. We say a curve C in the complex plane is **smooth** if $z'(t)$ is continuous and never zero in the interval $a \leq t \leq b$. As shown Figure 5.2.2, since the vector $z'(t)$ is not zero at any point P on C, the vector $z'(t)$ is tangent to C at P. In other words, a smooth curve has a continuously turning tangent; put yet another way, a smooth curve can have no sharp corners or cusps. See Figure 5.2.3. A **piecewise smooth curve** C has a continuously turning tangent, except possibly at the points where the component smooth curves $C_1, C_2, \ldots, C_n$ are joined together. A curve C in the complex plane is said to be a **simple** if $z(t_1) \neq z(t_2)$ for $t_1 \neq t_2$, except possibly for $t = a$ and $t = b$. C is a **closed curve** if $z(a) = z(b)$. C is a **simple closed curve** if $z(t_1) \neq z(t_2)$ for $t_1 \neq t_2$ and $z(a) = z(b)$. In complex analysis, a piecewise smooth curve C is called a **contour** or **path**.

Just as we did in the preceding section, we define the **positive direction** on a contour C to be the direction on the curve corresponding to increasing values of the parameter t. It is also said that the curve C has **positive orientation**. In the case of a simple closed curve C, the positive direction roughly corresponds to the counterclockwise direction or the direction that a person must walk on C in order to keep the interior of C to the left. For example, the circle $z(t) = e^{it}$, $0 \leq t \leq 2\pi$, has positive orientation. See Figure 5.2.4. The **negative direction** on a contour C is the direction opposite the positive direction. If C has an orientation, the **opposite curve**, that is, a curve with opposite orientation, is denoted by $-C$. On a simple closed curve, the negative direction corresponds to the clockwise direction.

(a) Positive direction

(b) Positive direction

Figure 5.2.4 Interior of each curve is to the left.

Complex Integral An integral of a function f of a complex variable z that is defined on a contour C is denoted by $\int_C f(z) \, dz$ and is called a **complex integral**. The following list of assumptions is a prelude to the definition of a complex integral. For the sake of comparison, you are encouraged to review the lists given prior to Definitions 5.1.1 and 5.1.2. Also, look over Figure 5.2.5 as you read this new list.

Steps Leading to the Definition of the Complex Integral

1. *Let f be a function of a complex variable z defined at all points on a smooth curve C that lies in some region of the complex plane. Let C be defined by the parametrization $z(t) = x(t) + iy(t)$, $a \le t \le b$.*

2. *Let P be a partition of the parameter interval $[a, b]$ into n subintervals $[t_{k-1}, t_k]$ of length $\Delta t_k = t_k - t_{k-1}$:*

$$a = t_0 < t_1 < t_2 < \cdots < t_{n-1} < t_n = b.$$

   The partition P induces a partition of the curve C into n subarcs whose initial and terminal points are the pairs of numbers

   $$z_0 = x(t_0) + iy(t_0), \qquad z_1 = x(t_1) + iy(t_1),$$
   $$z_1 = x(t_1) + iy(t_1), \qquad z_2 = x(t_2) + iy(t_2),$$
   $$\vdots \qquad\qquad\qquad \vdots$$
   $$z_{n-1} = x(t_{n-1}) + iy(t_{n-1}), \qquad z_n = x(t_n) + iy(t_n).$$

   Let $\Delta z_k = z_k - z_{k-1}$, $k = 1, 2, \dots, n$. See Figure 5.2.5.

3. *Let $\|P\|$ be the norm of the partition P of $[a, b]$, that is, the length of the longest subinterval.*

4. *Choose a point $z_k^* = x_k^* + iy_k^*$ on each subarc of C. See Figure 5.2.5.*

5. *Form n products $f(z_k^*)\Delta z_k$, $k = 1, 2, \dots, n$, and then sum these products:*

$$\sum_{k=1}^{n} f(z_k^*)\Delta z_k.$$

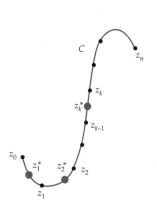

Figure 5.2.5 Partition of curve C into n subarcs is induced by a partition P of the parameter interval $[a, b]$.

Definition 5.2.1   Complex Integral

The **complex integral** of f on C is

$$\int_C f(z)\, dz = \lim_{\|P\| \to 0} \sum_{k=1}^{n} f(z_k^*)\Delta z_k. \tag{2}$$

If the limit in (2) exists, then f is said to be **integrable** on C. The limit exists whenever f is continuous at all points on C and C is either smooth or piecewise smooth. Consequently we shall, hereafter, assume these conditions as a matter of course. Moreover, we will use the notation $\oint_C f(z)\, dz$ to represent a complex integral around a *positively oriented* closed curve C. When it is important to distinguish the direction of integration around a closed curve, we will employ the notations

General assumptions throughout this text ➡

$$\oint_C f(z)\, dz \qquad \text{and} \qquad \oint_C f(z)\, dz$$

to denote integration in the positive and negative directions, respectively.

    The point having been made that the definitions of real integrals discussed in Section 5.1 and Definition 5.2.1 are formally the same, we shall from now *Note* ➡ on refer to a complex integral $\int_C f(z)\,dz$ by its more common name, **contour integral**.

Complex-Valued Function of a Real Variable

Before turning to the properties of contour integrals and the all-important question of how to evaluate a contour integral, we need to digress briefly to enlarge upon the concept of a **complex-valued function of a real variable** introduced in the Remarks in Section 2.1. As already mentioned, a parametrization of a curve C of the form given in (1) is a case in point. Let's consider another simple example. If t represents a real variable, then the output of the function $f(t) = (2t + i)^2$ is a complex number. For $t = 2$,

$$f(2) = (4 + i)^2 = 16 + 8i + i^2 = 15 + 8i.$$

In general, if f_1 and f_2 are real-valued functions of a real variable t (that is, real functions), then $f(t) = f_1(t) + if_2(t)$ is a complex-valued function of a real variable t. What we are really interested in at the moment is the definite integral $\int_a^b f(t)\,dt$, in other words, integration of complex-valued function $f(t) = f_1(t) + if_2(t)$ of real variable t carried out over a real interval. Continuing with the specific function $f(t) = (2t + i)^2$ it seems logical to write on, say, the interval $0 \le t \le 1$,

$$\int_0^1 (2t + i)^2\,dt = \int_0^1 \left(4t^2 - 1 + 4ti\right)dt = \int_0^1 (4t^2 - 1)dt + i\int_0^1 4t\,dt. \qquad (3)$$

The integrals $\int_0^1 (4t^2 - 1)dt$ and $\int_0^1 4t\,dt$ in (3) are real, and so one would be inclined to call them the real and imaginary parts of $\int_0^1 (2t + i)^2 dt$. Each of these real integrals can be evaluated using the fundamental theorem of calculus ((1) of Section 5.1):

$$\int_0^1 (4t^2 - 1)dt = \left(\frac{4}{3}t^3 - t\right)\Bigg|_0^1 = \frac{1}{3} \quad \text{and} \quad \int_0^1 4t\,dt = 2t^2\Big|_0^1 = 2.$$

Thus (3) becomes $\int_0^1 (2t + i)^2 dt = \frac{1}{3} + 2i$.

    Since the preceding integration seems very ordinary and routine, we give the following generalization. If f_1 and f_2 are real-valued functions of a real variable t continuous on a common interval $a \le t \le b$, then we *define* the integral of the complex-valued function $f(t) = f_1(t) + if_2(t)$ on $a \le t \le b$ in terms of the definite integrals of the real and imaginary parts of f:

$$\int_a^b f(t)\,dt = \int_a^b f_1(t)\,dt + i\int_a^b f_2(t)\,dt. \qquad (4)$$

The continuity of f_1 and f_2 on $[a,\, b]$ guarantees that both $\int_a^b f_1(t)\,dt$ and $\int_a^b f_2(t)\,dt$ exist.

    All of the following familiar properties of integrals can be proved directly from the definition given in (4). If $f(t) = f_1(t) + if_2(t)$ and

$g(t) = g_1(t) + ig_2(t)$ are complex-valued functions of a real variable t continuous on an interval $a \leq t \leq b$, then

$$\int_a^b k\, f(t)\, dt = k \int_a^b f(t)\, dt, \quad k \text{ a complex constant}, \tag{5}$$

$$\int_a^b (f(t) + g(t))\, dt = \int_a^b f(t)\, dt + \int_a^b g(t)\, dt, \tag{6}$$

$$\int_a^b f(t)\, dt = \int_a^c f(t)\, dt + \int_c^b f(t)\, dt, \tag{7}$$

$$\int_b^a f(t)\, dt = -\int_a^b f(t)\, dt. \tag{8}$$

These properties are important in the evaluation of contour integrals. They will often be used without mention.

In (7) we choose to assume that the real number c is in the interval $[a, b]$.

We now resume our discussion of contour integrals.

Evaluation of Contour Integrals To facilitate the discussion on how to evaluate a contour integral $\int_C f(z)\, dz$, let us write (2) in an abbreviated form. If we use $u + iv$ for $f(z)$, $\Delta x + i\Delta y$ for Δz, lim for $\lim_{||P|| \to 0}$, $\sum$ for $\sum_{k=1}^n$ and then suppress all subscripts, (2) becomes

$$\int_C f(z)dz = \lim \sum (u + iv)(\Delta x + i\Delta y)$$
$$= \lim \left[\sum (u\Delta x - v\Delta y) + i \sum (v\Delta x + u\Delta y) \right].$$

The interpretation of the last line is

$$\int_C f(z)\, dz = \int_C u\, dx - v\, dy + i \int_C v\, dx + u\, dy. \tag{9}$$

See Definition 5.1.2. In other words, the real and imaginary parts of a contour integral $\int_C f(z)\, dz$ are a pair of real line integrals $\int_C u\, dx - v\, dy$ and $\int_C v\, dx + u\, dy$. Now if $x = x(t)$, $y = y(t)$, $a \leq t \leq b$ are parametric equations of C, then $dx = x'(t)\, dt$, $dy = y'(t)\, dt$. By replacing the symbols x, y, dx, and dy by $x(t)$, $y(t)$, $x'(t)\, dt$, and $y'(t)\, dt$, respectively, the right side of (9) becomes

$$\overbrace{\int_a^b [u(x(t), y(t))\, x'(t) - v(x(t), y(t))\, y'(t)]\, dt}^{\int_C u dx - v dy}$$
$$+ i \underbrace{\int_a^b [v(x(t), y(t))\, x'(t) + u(x(t), y(t))\, y'(t)]\, dt}_{\int_C v dx + u dy}. \tag{10}$$

If we use the complex-valued function (1) to describe the contour C, then (10) is the same as $\int_a^b f(z(t))\, z'(t)\, dt$ when the integrand

$$f(z(t))\, z'(t) = [u(x(t), y(t)) + iv(x(t), y(t))]\, [x'(t) + iy'(t)]$$

is multiplied out and $\int_a^b f(z(t))\, z'(t)\, dt$ is expressed in terms of its real and imaginary parts. Thus we arrive at a practical means of evaluating a contour integral.

> **Theorem 5.2.1** **Evaluation of a Contour Integral**
>
> If f is continuous on a smooth curve C given by the parametrization $z(t) = x(t) + iy(t)$, $a \le t \le b$, then
>
> $$\int_C f(z)\, dz = \int_a^b f(z(t))\, z'(t)\, dt. \tag{11}$$

The foregoing results in (10) and (11) bear repeating—this time in somewhat different words. Suppose $z(t) = x(t) + iy(t)$ and $z'(t) = x'(t) + iy'(t)$. Then the integrand $f(z(t))\, z'(t)$ is a complex-valued function of a real variable t. Hence the integral $\int_a^b f(z(t))\, z'(t)\, dt$ is evaluated in the manner defined in (4). The next example illustrates the method.

EXAMPLE 1   **Evaluating a Contour Integral**

Evaluate $\int_C \bar{z}\, dz$, where C is parametrized by $z(t) = 3t + it^2$, $-1 \le t \le 4$.

Solution With the identification $f(z) = \bar{z}$ we have $f(z(t)) = \overline{3t + it^2} = 3t - it^2$. Also, $z'(t) = 3 + 2it$, and so by (11) the integral is

$$\int_C \bar{z}\, dz = \int_{-1}^4 (3t - it^2)(3 + 2it)\, dt = \int_{-1}^4 \left[2t^3 + 9t + 3t^2 i \right] dt.$$

Now in view of (4), the last integral is the same as

$$\int_C \bar{z}\, dz = \int_{-1}^4 (2t^3 + 9t)\, dt + i \int_{-1}^4 3t^2\, dt$$

$$= \left(\frac{1}{2} t^4 + \frac{9}{2} t^2 \right) \Big|_{-1}^4 + it^3 \Big|_{-1}^4 = 195 + 65i.$$

❑

EXAMPLE 2   **Evaluating a Contour Integral**

Evaluate $\oint_C \dfrac{1}{z}\, dz$, where C is the circle $x = \cos t$, $y = \sin t$, $0 \le t \le 2\pi$.

Solution In this case $z(t) = \cos t + i \sin t = e^{it}$, $z'(t) = ie^{it}$, and $f(z(t)) = \dfrac{1}{z(t)} = e^{-it}$. Hence,

$$\oint_C \frac{1}{z}\, dz = \int_0^{2\pi} (e^{-it})\, ie^{it}\, dt = i \int_0^{2\pi} dt = 2\pi i.$$

❑

As discussed in Section 5.1, for some curves the real variable x itself can be used as the parameter. For example, to evaluate $\int_C (8x^2 - iy)\, dz$

on the line segment $y = 5x$, $0 \leq x \leq 2$, we write $z = x + 5xi$, $dz = (1 + 5i)dx$, $\int_C (8x^2 - iy)\,dz = \int_0^2 (8x^2 - 5ix)(1 + 5i)\,dx$, and then integrate in the usual manner:

$$\int_C (8x^2 - iy)\,dz = (1 + 5i)\int_0^2 (8x^2 - 5ix)\,dx$$

$$= (1 + 5i)\left[\frac{8}{3}x^3\right]_0^2 - (1 + 5i)i\left[\frac{5}{2}x^2\right]_0^2 = \frac{214}{3} + \frac{290}{3}i.$$

In general, if x and y are related by means of a continuous real function $y = f(x)$, then the corresponding curve C in the complex plane can be parametrized by $z(x) = x + if(x)$. Equivalently, we can let $x = t$ so that a set of parametric equations for C is $x = t$, $y = f(t)$.

Properties   The following properties of contour integrals are analogous to the properties of real line integrals as well as the properties listed in (5)–(8).

Theorem 5.2.2   Properties of Contour Integrals

Suppose the functions f and g are continuous in a domain D, and C is a smooth curve lying entirely in D. Then

(*i*) $\int_C kf(z)\,dz = k\int_C f(z)\,dz$, k a complex constant.

(*ii*) $\int_C [f(z) + g(z)]\,dz = \int_C f(z)\,dz + \int_C g(z)\,dz$.

(*iii*) $\int_C f(z)\,dz = \int_{C_1} f(z)\,dz + \int_{C_2} f(z)\,dz$, where C consists of the smooth curves C_1 and C_2 joined end to end.

(*iv*) $\int_{-C} f(z)\,dz = -\int_C f(z)\,dz$, where $-C$ denotes the curve having the opposite orientation of C.

The four parts of Theorem 5.2.2 also hold if C is a piecewise smooth curve in D.

EXAMPLE 3   *C* Is a Piecewise Smooth Curve

Evaluate $\int_C (x^2 + iy^2)\,dz$, where C is the contour shown in Figure 5.2.6.

Solution   In view of Theorem 5.2.2(*iii*) we write

$$\int_C (x^2 + iy^2)\,dz = \int_{C_1} (x^2 + iy^2)\,dz + \int_{C_2} (x^2 + iy^2)\,dz.$$

Since the curve C_1 is defined by $y = x$, $0 \leq x \leq 1$, it makes sense to use x as a parameter. Therefore, $z(x) = x + ix$, $z'(x) = 1 + i$, $f(z) = x^2 + iy^2$, $f(z(x)) = x^2 + ix^2$, and

$$\int_{C_1} (x^2 + iy^2)\,dz = \int_0^1 \overbrace{(x^2 + ix^2)}^{(1+i)x^2}(1 + i)\,dx$$

$$= (1 + i)^2 \int_0^1 x^2\,dx = \frac{(1 + i)^2}{3} = \frac{2}{3}i. \tag{12}$$

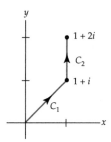

Figure 5.2.6 Contour C is piecewise smooth.

The curve C_2 is defined by $x = 1$, $1 \leq y \leq 2$. If we use y as a parameter, then $z(y) = 1 + iy$, $z'(y) = i$, $f(z(y)) = 1 + iy^2$, and

$$\int_{C_2} (x^2 + iy^2)\, dz = \int_1^2 (1 + iy^2) i\, dy = -\int_1^2 y^2\, dy + i \int_1^2 dy = -\frac{7}{3} + i. \quad (13)$$

Combining (10) and (13) gives $\int_C (x^2 + iy^2)\, dz = \frac{2}{3}i + \left(-\frac{7}{3} + i\right) = -\frac{7}{3} + \frac{5}{3}i.$ ❏

There are times in the application of complex integration that it is useful to find an upper bound for the modulus or absolute value of a contour integral. In the next theorem we use the fact that the length of a plane curve is $L = \int_a^b \sqrt{[x'(t)]^2 + [y'(t)]^2}\, dt$. But if $z'(t) = x'(t) + iy'(t)$, then $|z'(t)| = \sqrt{[x'(t)]^2 + [y'(t)]^2}$ and, consequently, $L = \int_a^b |z'(t)|\, dt$.

> **Theorem 5.2.3   A Bounding Theorem**
>
> If f is continuous on a smooth curve C and if $|f(z)| \leq M$ for all z on C, then $\left|\int_C f(z)\, dz\right| \leq ML$, where L is the length of C.

Proof It follows from the form of the triangle inequality given in (11) of Section 1.2 that

$$\left|\sum_{k=1}^n f(z_k^*)\Delta z_k\right| \leq \sum_{k=1}^n |f(z_k^*)|\, |\Delta z_k| \leq M \sum_{k=1}^n |\Delta z_k|. \quad (14)$$

Because $|\Delta z_k| = \sqrt{(\Delta x_k)^2 + (\Delta y_k)^2}$, we can interpret $|\Delta z_k|$ as the length of the chord joining the points z_k and z_{k-1} on C. Moreover, since the sum of the lengths of the chords cannot be greater than the length L of C, the inequality (14) continues as $|\sum_{k=1}^n f(z_k^*)\Delta z_k| \leq ML$. Finally, the continuity of f guarantees that $\int_C f(z)\, dz$ exists, and so if we let $\|P\| \to 0$, the last inequality yields $\left|\int_C f(z)\, dz\right| \leq ML$. ❏

Theorem 5.2.3 is used often in the theory of complex integration and is sometimes referred to as the **ML-inequality**. It follows from the discussion on page 112 of Section 2.6 that since f is continuous on the contour C, the bound M for the values $f(z)$ in Theorem 5.2.3 will always exist.

EXAMPLE 4   A Bound for a Contour Integral

Find an upper bound for the absolute value of $\oint_C \dfrac{e^z}{z+1}\, dz$ where C is the circle $|z| = 4$.

Solution First, the length L (circumference) of the circle of radius 4 is 8π. Next, from the inequality (7) of Section 1.2, it follows for all points z on the circle that $|z+1| \geq |z| - 1 = 4 - 1 = 3$. Thus

$$\left|\frac{e^z}{z+1}\right| \leq \frac{|e^z|}{|z| - 1} = \frac{|e^z|}{3}. \quad (15)$$

In addition, $|e^z| = |e^x(\cos y + i \sin y)| = e^x$. For points on the circle $|z| = 4$, the maximum that $x = \text{Re}(z)$ can be is 4, and so (15) yields

$$\left| \frac{e^z}{z+1} \right| \leq \frac{e^4}{3}.$$

From the *ML*-inequality (Theorem 5.2.3) we have

$$\left| \oint_C \frac{e^z}{z+1} \, dz \right| \leq \frac{8\pi e^4}{3}.$$

❏

Remarks

There is no unique parametrization for a contour C. You should verify that

$$z(t) = e^{it} = \cos t + i \sin t, \ 0 \leq t \leq 2\pi$$
$$z(t) = e^{2\pi it} = \cos 2\pi t + i \sin 2\pi t, \ 0 \leq t \leq 1$$
$$z(t) = e^{\pi it/2} = \cos \frac{\pi t}{2} + i \sin \frac{\pi t}{2}, \ 0 \leq t \leq 4$$

are all parametrizations, oriented in the positive direction, for the unit circle $|z| = 1$.

EXERCISES 5.2   *Answers to selected odd-numbered problems begin on page ANS-14.*

In Problems 1–16, evaluate the given integral along the indicated contour.

1. $\displaystyle\int_C (z+3)\, dz$, where C is $x = 2t$, $y = 4t - 1$, $1 \leq t \leq 3$

2. $\displaystyle\int_C (2\bar{z} - z)\, dz$, where C is $x = -t$, $y = t^2 + 2$, $0 \leq t \leq 2$

3. $\displaystyle\int_C z^2 \, dz$, where C is $z(t) = 3t + 2it$, $-2 \leq t \leq 2$

4. $\displaystyle\int_C (3z^2 - 2z)\, dz$, where C is $z(t) = t + it^2$, $0 \leq t \leq 1$

5. $\displaystyle\int_C \frac{z+1}{z} \, dz$, where C is the right half of the circle $|z| = 1$ from $z = -i$ to $z = i$

6. $\displaystyle\int_C |z|^2 \, dz$, where C is $x = t^2$, $y = 1/t$, $1 \leq t \leq 2$

7. $\displaystyle\oint_C \text{Re}(z)\, dz$, where C is the circle $|z| = 1$

8. $\displaystyle\oint_C \left(\frac{1}{(z+i)^3} - \frac{5}{z+i} + 8 \right) dz$, where C is the circle $|z + i| = 1$

9. $\displaystyle\int_C (x^2 + iy^3)\, dz$, where C is the straight line from $z = 1$ to $z = i$

10. $\displaystyle\int_C (x^2 - iy^3)\, dz$, where C is the lower half of the circle $|z| = 1$ from $z = -1$ to $z = 1$

11. $\int_C e^z\,dz$, where C is the polygonal path consisting of the line segments from $z=0$ to $z=2$ and from $z=2$ to $z=1+\pi i$

12. $\int_C \sin z\,dz$, where C is the polygonal path consisting of the line segments from $z=0$ to $z=1$ and from $z=1$ to $z=1+i$

13. $\int_C \operatorname{Im}(z-i)\,dz$, where C is the polygonal path consisting of the circular arc along $|z|=1$ from $z=1$ to $z=i$ and the line segment from $z=i$ to $z=-1$

14. $\int_C dz$, where C is the left half of the ellipse $\frac{1}{36}x^2+\frac{1}{4}y^2=1$ from $z=2i$ to $z=-2i$

15. $\oint_C ze^z\,dz$, where C is the square with vertices $z=0$, $z=1$, $z=1+i$, and $z=i$

16. $\int_C f(z)\,dz$, where $f(z)=\begin{cases}2, & x<0 \\ 6x, & x>0\end{cases}$ and C is the parabola $y=x^2$ from $z=-1+i$ to $z=1+i$

In Problems 17–20, evaluate the given integral along the contour C given in Figure 5.2.7.

17. $\oint_C x\,dz$

18. $\oint_C (2z-1)\,dz$

19. $\oint_C z^2\,dz$

20. $\oint_C \bar{z}^2\,dz$

In Problems 21–24, evaluate $\int_C (z^2-z+2)\,dz$ from i to 1 along the contour C given in the figures.

21.

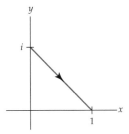

Figure 5.2.8 Figure for Problem 21

22.

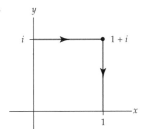

Figure 5.2.9 Figure for Problem 22

23.

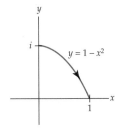

Figure 5.2.10 Figure for Problem 23

24.

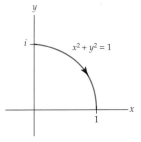

Figure 5.2.11 Figure for Problem 24

In Problems 25–28, find an upper bound for the absolute value of the given integral along the indicated contour.

25. $\oint_C \dfrac{e^z}{z^2+1}\,dz$, where C is the circle $|z|=5$

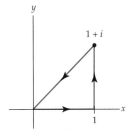

Figure 5.2.7 Figure for Problems 17–20

26. $\displaystyle\int_C \frac{1}{z^2 - 2i}\, dz$, where C is the right half of the circle $|z| = 6$ from $z = -6i$ to $z = 6i$

27. $\displaystyle\int_C (z^2 + 4)\, dz$, where C is the line segment from $z = 0$ to $z = 1 + i$

28. $\displaystyle\int_C \frac{1}{z^3}\, dz$, where C is one-quarter of the circle $|z| = 4$ from $z = 4i$ to $z = 4$

> **Focus on Concepts**

29. (a) Use Definition 5.2.1 to show for any smooth curve C between z_0 and z_n that $\int_C dz = z_n - z_0$.

   (b) Use the result in part (a) to verify your answer to Problem 14.

   (c) What is $\oint_C dz$ when C is a simple closed curve?

30. Use Definition 5.2.1 to show for any smooth curve C between z_0 and z_n that $\int_C z\, dz = \frac{1}{2}(z_n^2 - z_0^2)$. [*Hint*: The integral exists. So choose $z_k^* = z_k$ and $z_k^* = z_{k-1}$.]

31. Use the results of Problems 29 and 30 to evaluate $\int_C (6z + 4)\, dz$ where C is:

   (a) The straight line from $1 + i$ to $2 + 3i$.

   (b) The closed contour $x^4 + y^4 = 4$.

32. Find an upper bound for the absolute value of the integral $\displaystyle\int_C \frac{1}{z^2 + 1}\, dz$, where the contour C is the line segment from $z = 3$ to $z = 3 + i$. Use the fact that $|z^2 + 1| = |z - i|\,|z + i|$ where $|z - i|$ and $|z + i|$ represent, respectively, the distances from i and $-i$ to points z on C.

33. Find an upper bound for the absolute value of the integral $\displaystyle\oint_C \frac{z^2}{(z^2 + 9)^2}\, dz$, where C is a circle of radius $R > 3$ centered at the origin. Use the fact that $|z^2 + 9| = |z - 3i|\,|z + 3i|$ and apply the triangle inequality (10) of Section 1.2 to each factor.

34. Find an upper bound for the absolute value of the integral $\int_C \text{Ln}(z + 3)\, dz$, where the contour C is the line segment from $z = 3i$ to $z = 4 + 3i$.

5.3   Cauchy-Goursat Theorem

In this section we shall concentrate on contour integrals, where the contour C is a simple closed curve with a positive (counterclockwise) orientation. Specifically, we shall see that when f is analytic in a special kind of domain D, the value of the contour integral $\oint_C f(z)\, dz$ is the same for *any* simple closed curve C that lies entirely within D. This theorem, called the **Cauchy-Goursat theorem**, is one of *the* fundamental results in complex analysis.

   Preliminary to discussing the Cauchy-Goursat theorem and some of its ramifications, we need to distinguish two kinds of domains in the complex plain: simply connected and multiply connected.

Simply and Multiply Connected Domains   Recall from Section 1.5 that a domain is an open connected set in the complex plane. We say that a domain D is **simply connected** if every simple closed contour C lying entirely in D can be shrunk to a point without leaving D. See Figure 5.3.1. In other words, if we draw any simple closed contour C so that it lies entirely

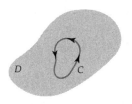

Figure 5.3.1 Simply connected domain D

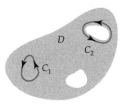

Figure 5.3.2 Multiply connected domain D

Green's theorem expresses a real line integral as a double integral ➥

within a simply connected domain, then C encloses only points of the domain D. Expressed yet another way, a simply connected domain has no "holes" in it. The entire complex plane is an example of a simply connected domain; the annulus defined by $1 < |z| < 2$ is not simply connected. (Why?) A domain that is not simply connected is called a **multiply connected domain**; that is, a multiply connected domain has "holes" in it. Note in Figure 5.3.2 that if the curve C_2 enclosing the "hole" were shrunk to a point, the curve would have to leave D eventually. We call a domain with one "hole" **doubly connected**, a domain with two "holes" **triply connected**, and so on. The open disk defined by $|z| < 2$ is a simply connected domain; the open circular annulus defined by $1 < |z| < 2$ is a doubly connected domain.

Cauchy's Theorem In 1825 the French mathematician Louis-Augustin Cauchy proved one the most important theorems in complex analysis.

> ### Cauchy's Theorem
> *Suppose that a function f is analytic in a simply connected domain D and that f' is continuous in D. Then for every simple closed contour C in D, $\oint_C f(z)\,dz = 0$.* $\qquad$ (1)

Cauchy's Proof of (1) The proof of this theorem is an immediate consequence of Green's theorem in the plane and the Cauchy-Riemann equations. Recall from calculus that if C is a positively oriented, piecewise smooth, simple closed curve forming the boundary of a region R within D, and if the real-valued functions $P(x,\,y)$ and $Q(x,\,y)$ along with their first-order partial derivatives are continuous on a domain that contains C and R, then

$$\oint_C P\,dx + Q\,dy = \iint_R \left(\frac{\partial Q}{\partial x} - \frac{\partial P}{\partial y} \right) dA. \qquad (2)$$

Now in the statement (1) we have assumed that f' is continuous throughout the domain D. As a consequence, the real and imaginary parts of $f(z) = u + iv$ and their first partial derivatives are continuous throughout D. By (9) of Section 5.2 we write $\oint_C f(z)\,dz$ in terms of real line integrals and apply Green's theorem (2) to each line integral:

$$\oint_C f(z)\,dz = \oint_C u(x,y)\,dx - v(x,y)\,dy + i\oint_C v(x,y)\,dx + u(x,y)\,dy$$

$$= \iint_R \left(-\frac{\partial v}{\partial x} - \frac{\partial u}{\partial y} \right) dA + i\iint_R \left(\frac{\partial u}{\partial x} - \frac{\partial v}{\partial y} \right) dA. \qquad (3)$$

Because f is analytic in D, the real functions u and v satisfy the Cauchy-Riemann equations, $\partial u/\partial x = \partial v/\partial y$ and $\partial u/\partial y = -\partial v/\partial x$, at every point in D. Using the Cauchy-Riemann equations to replace $\partial u/\partial y$ and $\partial u/\partial x$ in (3) shows that

$$\oint_C f(z)\,dz = \iint_R \left(-\frac{\partial v}{\partial x} + \frac{\partial v}{\partial x} \right) dA + i\iint_R \left(\frac{\partial v}{\partial y} - \frac{\partial v}{\partial y} \right) dA$$

$$= \iint_R (0)\,dA + i\iint_R (0)\,dA = 0.$$

This completes the proof. ❏

In 1883 the French mathematician Edouard Goursat proved that the assumption of continuity of f' is not necessary to reach the conclusion of

Cauchy's theorem. The resulting modified version of Cauchy's theorem is known today as the **Cauchy-Goursat theorem**. As one might expect, with fewer hypotheses, the proof of this version of Cauchy's theorem is more complicated than the one just presented. A form of the proof devised by Goursat is outlined in Appendix II.

Theorem 5.3.1   Cauchy-Goursat Theorem

Suppose that a function f is analytic in a simply connected domain D. Then for every simple closed contour C in D, $\oint_C f(z)\,dz = 0$.

Since the interior of a simple closed contour is a simply connected domain, the Cauchy-Goursat theorem can be stated in the slightly more practical manner:

If f is analytic at all points within and on a simple closed contour C, then $\oint_C f(z)\,dz = 0$.   (4)

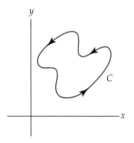

Figure 5.3.3 Contour for Example 1

EXAMPLE 1   Applying the Cauchy-Goursat Theorem

Evaluate $\oint_C e^z\,dz$, where the contour C is shown in Figure 5.3.3.

Solution The function $f(z) = e^z$ is entire and consequently is analytic at all points within and on the simple closed contour C. It follows from the form of the Cauchy-Goursat theorem given in (4) that $\oint_C e^z\,dz = 0$ ❏

The point of Example 1 is that $\oint_C e^z\,dz = 0$ for *any* simple closed contour in the complex plane. Indeed, it follows that for any simple closed contour C and any entire function f, such as $f(z) = \sin z$, $f(z) = \cos z$, and $p(z) = a_n z^n + a_{n-1} z^n + \cdots + a_1 z + a_0$, $n = 0, 1, 2, \ldots$, that

$$\oint_C \sin z\,dz = 0, \quad \oint_C \cos z\,dz = 0, \quad \oint_C p(z)\,dz = 0,$$

and so on.

EXAMPLE 2   Applying the Cauchy-Goursat Theorem

Evaluate $\oint_C \dfrac{dz}{z^2}$, where the contour C is the ellipse $(x-2)^2 + \frac{1}{4}(y-5)^2 = 1$.

Solution The rational function $f(z) = 1/z^2$ is analytic everywhere except at $z = 0$. But $z = 0$ is not a point interior to or on the simple closed elliptical contour C. Thus, from (4) we have that $\oint_C \dfrac{dz}{z^2} = 0$.   ❏

Cauchy-Goursat Theorem for Multiply Connected

Domains If f is analytic in a multiply connected domain D then we cannot conclude that $\oint_C f(z)\,dz = 0$ for every simple closed contour C in D. To begin, suppose that D is a doubly connected domain and C and C_1 are simple closed

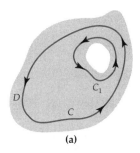

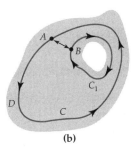

Figure 5.3.4 Doubly connected domain D

contours such that C_1 surrounds the "hole" in the domain and is interior to C. See Figure 5.3.4(a). Suppose, also, that f is analytic on each contour and at each point interior to C but exterior to C_1. By introducing the crosscut AB shown in Figure 5.3.4(b), the region bounded between the curves is now simply connected. From (iv) of Theorem 5.2.2, the integral from A to B has the opposite value of the integral from B to A, and so from (4) we have

$$\oint_C f(z)\,dz + \int_{AB} f(z)\,dz + \int_{-AB} f(z)\,dz + \oint_{C_1} f(z)\,dz = 0$$

or

$$\oint_C f(z)\,dz = \oint_{C_1} f(z)\,dz. \tag{5}$$

The last result is sometimes called the **principle of deformation of contours** since we can think of the contour C_1 as a continuous deformation of the contour C. Under this deformation of contours, if the region between C and C_1 contains only points where f is analytic, then the value of the integral does not change. In other words, (5) allows us to evaluate an integral over a complicated simple closed contour C by replacing C with a contour C_1 that is more convenient.

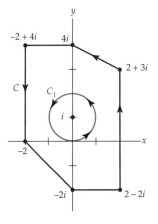

Figure 5.3.5 We use the simpler contour C_1 in Example 3.

EXAMPLE 3   Applying Deformation of Contours

Evaluate $\oint_C \dfrac{dz}{z-i}$, where C is the contour shown in black in Figure 5.3.5.

Solution   Note that $f(z) = \dfrac{1}{z-i}$ is analytic on the multiply connected domain consisting of the complex plane excluding the point $z = i$. In view of (5), we choose the more convenient circular contour C_1 drawn in color in the figure. By taking the radius of the circle to be $r = 1$, we are guaranteed that C_1 lies within C. In other words, C_1 is the circle $|z-i| = 1$, which from (10) of Section 2.2 can be parametrized by $z = i + e^{it}$, $0 \le t \le 2\pi$. From $z - i = e^{it}$ and $dz = ie^{it}\,dt$ we obtain

$$\oint_C \frac{dz}{z-i} = \oint_{C_1} \frac{dz}{z-i} = \int_0^{2\pi} \frac{ie^{it}}{e^{it}}\,dt = i\int_0^{2\pi} dt = 2\pi i.$$

$\square$

The result obtained in Example 3 can be generalized. Using the principle of deformation of contours (5) and proceeding as in the example, it can be shown that if z_0 is any constant complex number interior to *any* simple closed contour C, then for n an integer we have

$$\oint_C \frac{dz}{(z-z_0)^n} = \begin{cases} 2\pi i & n = 1 \\ 0, & n \ne 1. \end{cases} \tag{6}$$

For example, $1/(z-z_0)^{-3} = (z-z_0)^3$ is a polynomial. $\Rightarrow$

The fact that the integral in (6) is zero when $n \ne 1$ follows only partially from the Cauchy-Goursat theorem. When n is zero or a negative integer, $1/(z-z_0)^n$ is a *polynomial* and therefore entire. Theorem 5.3.1 and the discussion following Example 1 then indicate that $\oint_C dz/(z-z_0)^n = 0$. It is left as an exercise to show that the integral is still zero when n is a positive integer different from 1. See Problem 24 in Exercises 5.3.

Analyticity of the function f at all points within and on a simple closed contour C is *sufficient* to guarantee that $\oint_C f(z)\,dz = 0$. However, the result in (6) emphasizes that analyticity is not *necessary*; in other words, it can happen that $\oint_C f(z)\,dz = 0$ without f being analytic within C. For instance, if C in Example 2 is the circle $|z| = 1$, then (6), with the identifications $n = 2$ and $z_0 = 0$, immediately gives $\oint_C \dfrac{dz}{z^2} = 0$. Note that $f(z) = 1/z^2$ is not analytic at $z = 0$ within C.

EXAMPLE 4   Applying Formula (6)

Evaluate $\oint_C \dfrac{5z+7}{z^2+2z-3}dz$, where C is circle $|z-2| = 2$.

Solution Since the denominator factors as $z^2 + 2z - 3 = (z-1)(z+3)$ the integrand fails to be analytic at $z = 1$ and $z = -3$. Of these two points, only $z = 1$ lies within the contour C, which is a circle centered at $z = 2$ of radius $r = 2$. Now by partial fractions

$$\frac{5z+7}{z^2+2z-3} = \frac{3}{z-1} + \frac{2}{z+3}$$

and so

$$\oint_C \frac{5z+7}{z^2+2z-3}\,dz = 3\oint_C \frac{1}{z-1}\,dz + 2\oint_C \frac{1}{z+3}\,dz. \tag{7}$$

In view of the result given in (6), the first integral in (7) has the value $2\pi i$, whereas the value of the second integral is 0 by the Cauchy-Goursat theorem. Hence, (7) becomes

$$\oint_C \frac{5z+7}{z^2+2z-3}\,dz = 3(2\pi i) + 2(0) = 6\pi i.$$

□

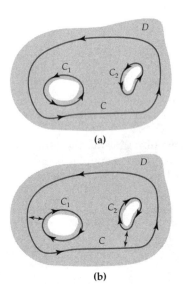

Figure 5.3.6 Triply connected domain D

If C, C_1, and C_2 are simple closed contours as shown in Figure 5.3.6(a) and if f is analytic on each of the three contours as well as at each point interior to C but exterior to both C_1 and C_2, then by introducing crosscuts between C_1 and C and between C_2 and C, as illustrated in Figure 5.3.6(b), it follows from Theorem 5.3.1 that

$$\oint_C f(z)\,dz + \oint_{C_1} f(z)\,dz + \oint_{C_2} f(z)\,dz = 0$$

and so

$$\oint_C f(z)\,dz = \oint_{C_1} f(z)\,dz + \oint_{C_2} f(z)\,dz. \tag{7}$$

The next theorem summarizes the general result for a multiply connected domain with n "holes."

> **Theorem 5.3.2 Cauchy-Goursat Theorem for Multiply Connected Domains**
>
> Suppose $C, C_1, \ldots, C_n$ are simple closed curves with a positive orientation such that $C_1, C_2, \ldots, C_n$ are interior to C but the regions interior to each C_k, $k = 1, 2, \ldots, n$, have no points in common. If f is analytic on each contour and at each point interior to C but exterior to all the C_k, $k = 1, 2, \ldots, n$, then
>
> $$\oint_C f(z)\,dz = \sum_{k=1}^{n} \oint_{C_k} f(z)\,dz. \tag{8}$$

EXAMPLE 5   Applying Theorem 5.3.2

Evaluate $\oint_C \dfrac{dz}{z^2 + 1}$, where C is the circle $|z| = 4$.

Solution In this case the denominator of the integrand factors as $z^2 + 1 = (z - i)(z + i)$. Consequently, the integrand $1/(z^2 + 1)$ is not analytic at $z = i$ and at $z = -i$. Both of these points lie within the contour C. Using partial fraction decomposition once more, we have

$$\frac{1}{z^2 + 1} = \frac{1}{2i}\frac{1}{z - i} - \frac{1}{2i}\frac{1}{z + i}$$

and

$$\oint_C \frac{dz}{z^2 + 1} = \frac{1}{2i}\oint_C \left[\frac{1}{z - i} - \frac{1}{z + i}\right] dz.$$

We now surround the points $z = i$ and $z = -i$ by circular contours C_1 and C_2, respectively, that lie entirely within C. Specifically, the choice $|z - i| = \frac{1}{2}$ for C_1 and $|z + i| = \frac{1}{2}$ for C_2 will suffice. See Figure 5.3.7. From Theorem 5.3.2 we can write

$$\oint_C \frac{dz}{z^2 + 1} = \frac{1}{2i}\oint_{C_1}\left[\frac{1}{z - i} - \frac{1}{z + i}\right]dz + \frac{1}{2i}\oint_{C_2}\left[\frac{1}{z - i} - \frac{1}{z + i}\right]dz$$

$$= \frac{1}{2i}\oint_{C_1}\frac{1}{z - i}dz - \frac{1}{2i}\oint_{C_1}\frac{1}{z + i}dz + \frac{1}{2i}\oint_{C_2}\frac{1}{z - i}dz - \frac{1}{2i}\oint_{C_2}\frac{1}{z + i}dz. \tag{9}$$

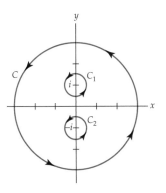

Figure 5.3.7 Contour for Example 5

Because $1/(z + i)$ is analytic on C_1 and at each point in its interior and because $1/(z - i)$ is analytic on C_2 and at each point in its interior, it follows from (4) that the second and third integrals in (9) are zero. Moreover, it follows from (6), with $n = 1$, that

$$\oint_{C_1} \frac{dz}{z - i} = 2\pi i \qquad \text{and} \qquad \oint_{C_2} \frac{dz}{z + i} = 2\pi i.$$

Thus (9) becomes

$$\oint_C \frac{dz}{z^2 + 1} = \pi - \pi = 0.$$

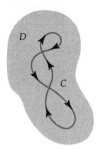

Figure 5.3.8 Contour C is closed but not simple.

Remarks

Throughout the foregoing discussion we assumed that C was a simple closed contour, in other words, C did not intersect itself. Although we shall not give the proof, it can be shown that the Cauchy-Goursat theorem is valid for any closed contour C in a simply connected domain D. As shown in Figure 5.3.8, the contour C is closed but not simple. Nevertheless, if f is analytic in D, then $\oint_C f(z)\,dz = 0$. See Problem 23 in Exercises 5.3.

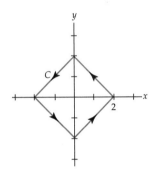

Figure 5.3.9 Figure for Problem 9

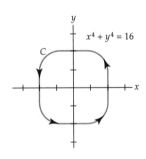

Figure 5.3.10 Figure for Problem 10

EXERCISES 5.3   *Answers to selected odd-numbered problems begin on page ANS-14.*

In Problems 1–8, show that $\oint_C f(z)\,dz = 0$, where f is the given function and C is the unit circle $|z| = 1$.

1. $f(z) = z^3 - 1 + 3i$

2. $f(z) = z^2 + \dfrac{1}{z-4}$

3. $f(z) = \dfrac{z}{2z+3}$

4. $f(z) = \dfrac{z-3}{z^2+2z+2}$

5. $f(z) = \dfrac{\sin z}{(z^2-25)(z^2+9)}$

6. $f(z) = \dfrac{e^z}{2z^2+11z+15}$

7. $f(z) = \tan z$

8. $f(z) = \dfrac{z^2-9}{\cosh z}$

9. Evaluate $\oint_C \dfrac{1}{z}\,dz$, where C is the contour shown in Figure 5.3.9.

10. Evaluate $\oint_C \dfrac{5}{z+1+i}\,dz$, where C is the contour shown in Figure 5.3.10.

In Problems 11–22, use any of the results in this section to evaluate the given integral along the indicated closed contour(s).

11. $\oint_C \left(z + \dfrac{1}{z}\right) dz;\ |z| = 2$

12. $\oint_C \left(z + \dfrac{1}{z^2}\right) dz;\ |z| = 2$

13. $\oint_C \dfrac{z}{z^2-\pi^2}\,dz;\ |z| = 3$

14. $\oint_C \dfrac{10}{(z+i)^4}\,dz;\ |z+i| = 1$

15. $\oint_C \dfrac{2z+1}{z^2+z}\,dz;$ (a) $|z| = \frac{1}{2}$, (b) $|z| = 2$, (c) $|z-3i| = 1$

16. $\oint_C \dfrac{2z}{z^2+3}\,dz;$ (a) $|z| = 1$, (b) $|z-2i| = 1$, (c) $|z| = 4$

17. $\oint_C \dfrac{-3z+2}{z^2-8z+12}\,dz;$ (a) $|z-5| = 2$, (b) $|z| = 9$

18. $\oint_C \left(\dfrac{3}{z+2} - \dfrac{1}{z-2i}\right) dz;$ (a) $|z| = 5$, (b) $|z-2i| = \frac{1}{2}$

19. $\oint_C \dfrac{z-1}{z(z-i)(z-3i)}\,dz;\ |z-i| = \frac{1}{2}$

20. $\oint_C \dfrac{1}{z^3+2iz^2}\,dz;\ |z| = 1$

21. $\oint_C \mathrm{Ln}(z+10)\,dz;\ |z| = 2$

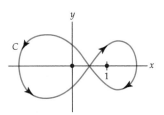

Figure 5.3.11 Figure for Problem 23

22. $\displaystyle\oint_C \left[\frac{5}{(z-2)^3} + \frac{3}{(z-2)^2} - \frac{10}{z-2} + 7\csc z\right] dz; \; |z-2| = \frac{1}{2}$

23. Evaluate $\displaystyle\oint_C \frac{8z-3}{z^2-z}\, dz$, where C is the "figure-eight" contour shown in Figure 5.3.11. [*Hint*: Express C as the union of two closed curves C_1 and C_2.]

24. Suppose z_0 is any constant complex number interior to any simple closed curve contour C. Show that for a positive integer n,

$$\oint_C \frac{dz}{(z-z_0)^n} = \begin{cases} 2\pi i, & n = 1 \\ 0, & n > 1. \end{cases}$$

In Problems 25 and 26, evaluate the given contour integral by any means.

25. $\displaystyle\oint_C \left(\frac{e^z}{z+3} - 3\bar{z}\right) dz$, where C is the unit circle $|z| = 1$

26. $\displaystyle\oint_C \left(z^3 + z^2 + \mathrm{Re}(z)\right) dz$, where C is the triangle with vertices $z = 0$, $z = 1+2i$, and $z = 1$

Focus on Concepts

27. Explain why $\displaystyle\oint_C f(z)\,dz = 0$ for each of the following functions and C is any simple closed contour in the complex plane.

    (a) $f(z) = (5iz^4 - 4z^2 + 2 - 6i)^9$ **(b)** $f(z) = (z^2 - 3iz)e^{5z}$

    (c) $f(z) = \dfrac{\sin z}{e^{z^2}}$ **(d)** $f(z) = z\cos^2 z$

28. Describe contours C for which we are guaranteed that $\displaystyle\oint_C f(z)\,dz = 0$ for each of the following functions.

    (a) $f(z) = \dfrac{1}{z^3 + z}$ **(b)** $f(z) = \csc z$

    (c) $f(z) = \dfrac{1}{1 - e^z}$ **(d)** $f(z) = \mathrm{Ln}\, z$

29. Explain why the integral in Problem 25 is the same as

$$\oint_C \left(\frac{e^z}{z+3} - \frac{3}{z}\right) dz$$

and why, in view of (6), this form makes the integral slightly easier to evaluate.

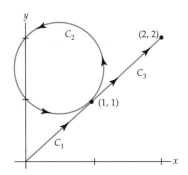

Figure 5.3.12 Figure for Problem 30

30. Evaluate $\int_C e^z\, dz$ from $z = 0$ to $z = 2+2i$ on the contour C shown in Figure 5.3.12 that consists of the line $y = x$ and a circle tangent to the line at $(1, 1)$.

31. From Example 1 we know the value of $\oint_C e^z dz$ for any simple closed contour C in the complex plane. In particular, use $|z| = 1$ as C and the parametrization $z = e^{i\theta}$, $0 \leq \theta \leq 2\pi$, to discover the values of the real integrals

$$\int_0^{2\pi} e^{\cos\theta}\sin(\theta + \sin\theta)d\theta \quad \text{and} \quad \int_0^{2\pi} e^{\cos\theta}\cos(\theta + \sin\theta)d\theta.$$

32. Let n be a positive integer. Use a complex contour integral to show that

$$\int_0^{2\pi} e^{\sin n\theta}\cos\left(\theta - \cos n\theta\right) d\theta = 0 \quad \text{and} \quad \int_0^{2\pi} e^{\sin n\theta}\sin\left(\theta - \cos n\theta\right) d\theta = 0.$$

[*Hint*: Modify the integrand from Problem 31.]

5.4   Independence of Path

In Section 5.1 we saw that when a real function f possesses an elementary antiderivative, that is, a function F for which $F'(x) = f(x)$, a definite integral can be evaluated by the fundamental theorem of calculus:

$$\int_a^b f(x)\,dx = F(b) - F(a). \tag{1}$$

Note that $\int_a^b f(x)\,dx$ depends only on the numbers a and b at the initial and terminal points of the interval of integration. In contrast, the value of a real line integral $\int_C P\,dx + Q\,dy$ generally depends on the curve C. However, there exist integrals $\int_C P\,dx + Q\,dy$ whose value depends only on the initial point A and terminal point B of the curve C, and not on C itself. In this case we say that the line integral is **independent of the path**. For example, $\int_C y\,dx + x\,dy$ is independent of the path. See Problems 19–22 in Exercises 5.1. A line integral that is independent of the path can be evaluated in a manner similar to (1). It seems natural then to ask:

Is there a complex version of the fundamental theorem of calculus?
Can a contour integral $\int_C f(z)\,dz$ be independent of the path?

In this section we will see that the answer to both of these questions is yes.

Path Independence   The definition of **path independence** for a contour integral $\int_C f(z)\,dz$ is essentially the same as for a real line integral $\int_C P\,dx + Q\,dy$.

Definition 5.4.1   Independence of the Path

Let z_0 and z_1 be points in a domain D. A contour integral $\int_C f(z)\,dz$ is said to be **independent of the path** if its value is the same for all contours C in D with initial point z_0 and terminal point z_1.

At the end of the preceding section we noted that the Cauchy-Goursat theorem also holds for closed contours, not just simple closed contours, in a simply connected domain D. Now suppose, as shown in Figure 5.4.1, that C and C_1 are two contours lying entirely in a simply connected domain D and both with initial point z_0 and terminal point z_1. Note that C joined with the opposite curve $-C_1$ forms a closed contour. Thus, if f is analytic in D, it follows from the Cauchy-Goursat theorem that

$$\int_C f(z)\,dz + \int_{-C_1} f(z)\,dz = 0. \tag{2}$$

But (2) is equivalent to

$$\int_C f(z)\,dz = \int_{C_1} f(z)\,dz. \tag{3}$$

The result in (3) is also an example of the principle of deformation of contours introduced in (5) of Section 5.3. We summarize the last result as a theorem.

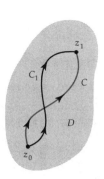

Figure 5.4.1 If f is analytic in D, integrals on C and C_1 are equal.

> **Theorem 5.4.1 Analyticity Implies Path Independence**
>
> Suppose that a function f is analytic in a simply connected domain D and C is any contour in D. Then $\int_C f(z)\,dz$ is independent of the path C.

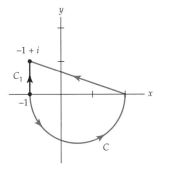

Figure 5.4.2 Contour for Example 1

EXAMPLE 1 Choosing a Different Path

Evaluate $\int_C 2z\,dz$, where C is the contour shown in color in Figure 5.4.2.

Solution Since the function $f(z) = 2z$ is entire, we can, in view of Theorem 5.4.1, replace the piecewise smooth path C by any convenient contour C_1 joining $z_0 = -1$ and $z_1 = -1 + i$. Specifically, if we choose the contour C_1 to be the vertical line segment $x = -1$, $0 \le y \le 1$, shown in black in Figure 5.4.2, then $z = -1 + iy$ and $dz = i\,dy$. Therefore,

$$\int_C 2z\,dz = \int_{C_1} 2z\,dz = -2\int_0^1 y\,dy - 2i\int_0^1 dy = -1 - 2i.$$

$\square$

A contour integral $\int_C f(z)\,dz$ that is independent of the path C is usually written $\int_{z_0}^{z_1} f(z)\,dz$, where z_0 and z_1 are the initial and terminal points of C. Hence, in Example 1 we can write $\int_{-1}^{-1+i} 2z\,dz$.

There is an easier way to evaluate the contour integral in Example 1, but before proceeding we need another definition.

> **Definition 5.4.2 Antiderivative**
>
> Suppose that a function f is continuous on a domain D. If there exists a function F such that $F'(z) = f(z)$ for each z in D, then F is called an **antiderivative** of f.

For example, the function $F(z) = -\cos z$ is an antiderivative of $f(z) = \sin z$ since $F'(z) = \sin z$. As in calculus of a real variable, the most general antiderivative, or **indefinite integral,** of a function $f(z)$ is written $\int f(z)\,dz = F(z) + C$, where $F'(z) = f(z)$ and C is some complex constant. For example, $\int \sin z\,dz = -\cos z + C$.

Since an antiderivative F of a function f has a derivative at each point in a domain D, it is necessarily analytic and hence continuous at each point in D.

Recall, differentiability implies continuity.

We are now in a position to prove the complex analogue of (1).

> **Theorem 5.4.2 Fundamental Theorem for Contour Integrals**
>
> Suppose that a function f is continuous on a domain D and F is an antiderivative of f in D. Then for any contour C in D with initial point z_0 and terminal point z_1,
>
> $$\int_C f(z)\,dz = F(z_1) - F(z_0). \qquad (4)$$

Proof We will prove (4) in the case when C is a smooth curve parametrized by $z = z(t)$, $a \leq t \leq b$. The initial and terminal points on C are then $z(a) = z_0$ and $z(b) = z_1$. Using (11) of Section 5.2 and the fact that $F'(z) = f(z)$ for each z in D, we then have

$$\int_C f(z)\, dz = \int_a^b f(z(t)) z'(t)\, dt = \int_a^b F'(z(t)) z'(t)\, dt$$

$$= \int_a^b \frac{d}{dt} F(z(t))\, dt \quad \leftarrow \text{chain rule}$$

$$= F(z(t)) \big|_a^b$$

$$= F(z(b)) - F(z(a)) = F(z_1) - F(z_0). \qquad \square$$

EXAMPLE 2   Applying Theorem 5.4.2

In Example 1 we saw that the integral $\int_C 2z\, dz$, where C is shown in Figure 5.4.2, is independent of the path. Now since the $f(z) = 2z$ is an entire function, it is continuous. Moreover, $F(z) = z^2$ is an antiderivative of f since $F'(z) = 2z = f(z)$. Hence, by (4) of Theorem 5.4.2 we have

$$\int_{-1}^{-1+i} 2z\, dz = z^2 \big|_{-1}^{-1+i} = (-1+i)^2 - (-1)^2 = -1 - 2i.$$

$\square$

EXAMPLE 3   Applying Theorem 5.4.2

Evaluate $\int_C \cos z\, dz$, where C is any contour with initial point $z_0 = 0$ and terminal point $z_1 = 2 + i$.

Solution $F(z) = \sin z$ is an antiderivative of $f(z) = \cos z$ since $F'(z) = \cos z = f(z)$. Therefore, from (4) we have

$$\int_C \cos z\, dz = \int_0^{2+i} \cos z\, dz = \sin z \big|_0^{2+i} = \sin(2+i) - \sin 0 = \sin(2+i).$$

If we desire a complex number of the form $a + ib$ for an answer, we can use $\sin(2 + i) \approx 1.4031 - 0.4891i$ (see part (b) of Example 1 in Section 4.3). Hence,

$$\int_C \cos z\, dz = \int_0^{2+i} \cos z\, dz \approx 1.4031 - 0.4891i.$$

$\square$

Some Conclusions We can draw several immediate conclusions from Theorem 5.4.2. First, observe that if the contour C is closed, then $z_0 = z_1$ and, consequently,

$$\oint_C f(z)\, dz = 0. \tag{5}$$

Next, since the value of $\int_C f(z)\,dz$ depends only on the points z_0 and z_1, this value is the same for any contour C in D connecting these points. In other words:

If a continuous function f has an antiderivative F in D, then $\int_C f(z)\,dz$ is independent of the path. (6)

In addition, we have the following sufficient condition for the existence of an antiderivative.

If f is continuous and $\int_C f(z)\,dz$ is independent of the path C in a domain D, then f has an antiderivative everywhere in D. (7)

The last statement is important and deserves a proof.

Proof of (7) Assume f is continuous and $\int_C f(z)\,dz$ is independent of the path in a domain D and that F is a function defined by $F(z) = \int_{z_0}^{z} f(s)\,ds$, where s denotes a complex variable, z_0 is a fixed point in D, and z represents any point in D. We wish to show that $F'(z) = f(z)$, that is,

Important ➡

$$F(z) = \int_{z_0}^{z} f(s)\,ds \tag{8}$$

is an antiderivative of f in D. Now,

$$F(z + \Delta z) - F(z) = \int_{z_0}^{z+\Delta z} f(s)\,ds - \int_{z_0}^{z} f(s)\,ds = \int_{z}^{z+\Delta z} f(s)\,ds. \tag{9}$$

Because D is a domain, we can choose Δz so that $z + \Delta z$ is in D. Moreover, z and $z + \Delta z$ can be joined by a straight segment as shown in Figure 5.4.3. This is the contour we use in the last integral in (9). With z fixed, we can write

$$f(z)\Delta z = f(z) \int_{z}^{z+\Delta z} ds = \int_{z}^{z+\Delta z} f(z)\,ds$$

or

$$f(z) = \frac{1}{\Delta z} \int_{z}^{z+\Delta z} f(z)\,ds. \tag{10}$$

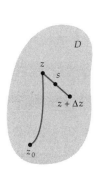

Figure 5.4.3 Contour used in proof of (7)

From (9) and (10) we have

$$\frac{F(z + \Delta z) - F(z)}{\Delta z} - f(z) = \frac{1}{\Delta z} \int_{z}^{z+\Delta z} [f(s) - f(z)]\,ds.$$

Now f is continuous at the point z. This means for any $\varepsilon > 0$ there exists a $\delta > 0$ so that $|f(s) - f(z)| < \varepsilon$ whenever $|s - z| < \delta$. Consequently, if we choose Δz so that $|\Delta z| < \delta$, it follows from the *ML*-inequality of Section 5.2 that

$$\left| \frac{F(z + \Delta z) - F(z)}{\Delta z} - f(z) \right| = \left| \frac{1}{\Delta z} \int_{z}^{z+\Delta z} [f(s) - f(z)]\,ds \right|$$

$$= \left| \frac{1}{\Delta z} \right| \left| \int_{z}^{z+\Delta z} [f(s) - f(z)]\,ds \right| \leq \left| \frac{1}{\Delta z} \right| \varepsilon |\Delta z| = \varepsilon.$$

Hence, we have shown that

$$\lim_{\Delta z \to 0} \frac{F(z + \Delta z) - F(z)}{\Delta z} = f(z) \quad \text{or} \quad F'(z) = f(z).$$ ❑

If f is an analytic function in a simply connected domain D, it is necessarily continuous throughout D. This fact, when put together with the results in Theorem 5.4.1 and (7), leads to a theorem which states that an analytic function possesses an analytic antiderivative.

Theorem 5.4.3   Existence of an Antiderivative

Suppose that a function f is analytic in a simply connected domain D. Then f has an antiderivative in D; that is, there exists a function F such that $F'(z) = f(z)$ for all z in D.

Be careful when using Ln z as an antiderivative of $1/z$. ➡

In (21) of Section 4.1 we saw for $|z| > 0$, $-\pi < \arg(z) < \pi$, that $1/z$ is the derivative of Ln z. This means that under some circumstances Ln z is an antiderivative of $1/z$. But care must be exercised in using this result. For example, suppose D is the entire complex plane without the origin. The function $1/z$ is analytic in this *multiply connected* domain. If C is any simple closed contour containing the origin, it does *not* follow from (5) that $\oint_C dz/z = 0$. In fact, from (6) of Section 5.3 with $n = 1$ and the identification of $z_0 = 0$, we see that

$$\oint_C \frac{1}{z}\, dz = 2\pi i.$$

In this case, Ln z is not an antiderivative of $1/z$ in D since Ln z is not analytic in D. Recall, Ln z fails to be analytic on the nonpositive real axis which is the branch cut of the principal branch $f_1(z)$ of the logarithm. See page 168 in Section 4.1.

EXAMPLE 4   Using the Logarithmic Function

Evaluate $\displaystyle\int_C \frac{1}{z}\, dz$, where C is the contour shown in Figure 5.4.4.

Solution   Suppose that D is the simply connected domain defined by $x > 0$, $y > 0$, in other words, D is the first quadrant in the z-plane. In this case, Ln z *is* an antiderivative of $1/z$ since both these functions are analytic in D. Hence by (4),

$$\int_3^{2i} \frac{1}{z}\, dz = \text{Ln } z \Big|_3^{2i} = \text{Ln } 2i - \text{Ln } 3.$$

From (14) of Section 4.1,

$$\text{Ln } 2i = \log_e 2 + \frac{\pi}{2}i \qquad \text{and} \qquad \text{Ln } 3 = \log_e 3$$

$\log_e 2 - \log_e 3 = \log_e \frac{2}{3}$ ➡   and so,

$$\int_3^{2i} \frac{1}{z}\, dz = \log_e \frac{2}{3} + \frac{\pi}{2}i \approx -0.4055 + 1.5708i.$$ ❑

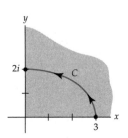

Figure 5.4.4 Contour for Example 4

EXAMPLE 5   **Using an Antiderivative of $z^{-1/2}$**

Evaluate $\displaystyle\int_C \frac{1}{z^{1/2}}\, dz$, where C is the line segment between $z_0 = i$ and $z_1 = 9$.

Solution   Throughout, we take $f_1(z) = z^{1/2}$ to be the principal branch of the square root function. In the domain $|z| > 0$, $-\pi < \arg(z) < \pi$, the function $f_1(z) = 1/z^{1/2} = z^{-1/2}$ is analytic and possesses the antiderivative $F(z) = 2z^{1/2}$ (see (9) in Section 4.2). Hence,

See Problem 5 in Exercise 1.4

$$\int_i^9 \frac{1}{z^{1/2}}\,dz = 2z^{1/2}\bigg|_i^9 = 2\left[3 - \overbrace{\left(\frac{\sqrt{2}}{2} + i\frac{\sqrt{2}}{2}\right)}\right]$$

$$= (6 - \sqrt{2}) - i\sqrt{2}.$$

☐

Remarks   *Comparison with Real Analysis*

(*i*) In the study of techniques of integration in calculus you learned that indefinite integrals of certain kinds of products could be evaluated by integration by parts:

$$\int f(x)g'(x)dx = f(x)g(x) - \int g(x)f'(x)dx. \tag{11}$$

You undoubtedly have used (11) in the more compact form $\int u\,dv = uv - \int v\,du$. Formula (11) carries over to complex analysis. Suppose f and g are analytic in a simply connected domain D. Then

$$\int f(z)g'(z)dz = f(z)g(z) - \int g(z)f'(z)dz. \tag{12}$$

In addition, if z_0 and z_1 are the initial and terminal points of a contour C lying entirely in D, then

$$\int_{z_0}^{z_1} f(z)g'(z)\,dz = f(z)g(z)\bigg|_{z_0}^{z_1} - \int_{z_0}^{z_1} g(z)f'(z)\,dz. \tag{13}$$

These results can be proved in a straightforward manner using Theorem 5.4.2 on the function $\dfrac{d}{dz}fg$. See Problems 21–24 in Exercises 5.4.

(*ii*) If f is a real function continuous on the closed interval $[a, b]$, then there exists a number c in the open interval (a, b) such that

$$\int_a^b f(x)\,dx = f(c)(b - a). \tag{14}$$

The result in (14) is known as the **mean-value theorem** for definite integrals. If f is a complex function analytic in a simply connected domain D, it is continuous at every point on a contour C in D with initial point z_0 and terminal point z_1. One might expect a result parallel to (14) for an integral $\int_{z_0}^{z_1} f(z)\,dz$. However, there is no such complex counterpart.

EXERCISES 5.4   *Answers to selected odd-numbered problems begin on page ANS-14.*

In Problems 1 and 2, evaluate the given integral, where the contour C is given in the figure, (**a**) by using an alternative path of integration and (**b**) by using Theorem 5.4.2.

1. $\displaystyle\int_C (4z - 1)\, dz$

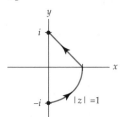

2. $\displaystyle\int_C e^z\, dz$

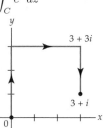

Figure 5.4.5 Figure for Problem 1   **Figure 5.4.6** Figure for Problem 2

In Problems 3 and 4, evaluate the given integral along the indicated contour C.

3. $\displaystyle\int_C 2z\, dz$, where C is $z(t) = 2t^3 + i(t^4 - 4t^3 + 2)$, $-1 \le t \le 1$

4. $\displaystyle\int_C 2z\, dz$, where C is $z(t) = 2\cos^3 \pi t - i\sin^2 \dfrac{\pi}{4} t$, $0 \le t \le 2$

In Problems 5–20, use Theorem 5.4.2 to evaluate the given integral. Write each answer in the form $a + ib$.

5. $\displaystyle\int_0^{3+i} z^2\, dz$

6. $\displaystyle\int_{-2i}^{1} (3z^2 - 4z + 5i)\, dz$

7. $\displaystyle\int_{1-i}^{1+i} z^3\, dz$

8. $\displaystyle\int_{-3i}^{2i} (z^3 - z)\, dz$

9. $\displaystyle\int_{-i/2}^{1-i} (2z + 1)^2\, dz$

10. $\displaystyle\int_1^{i} (iz + 1)^3\, dz$

11. $\displaystyle\int_{i/2}^{i} e^{\pi z}\, dz$

12. $\displaystyle\int_{1-i}^{1+2i} z e^{z^2}\, dz$

13. $\displaystyle\int_{\pi}^{\pi+2i} \sin \dfrac{z}{2}\, dz$

14. $\displaystyle\int_{1-2i}^{\pi i} \cos z\, dz$

15. $\displaystyle\int_{\pi i}^{2\pi i} \cosh z\, dz$

16. $\displaystyle\int_{i}^{1+(\pi/2)i} \sinh 3z\, dz$

17. $\displaystyle\int_C \dfrac{1}{z}\, dz$, C is the arc of the circle $z = 4e^{it}$, $-\pi/2 \le t \le \pi/2$

18. $\displaystyle\int_C \dfrac{1}{z}\, dz$, C is the line segment between $1 + i$ and $4 + 4i$

19. $\displaystyle\int_{-4i}^{4i} \dfrac{1}{z^2}\, dz$, C is any contour not passing through the origin

20. $\displaystyle\int_{1-i}^{1+\sqrt{3}i} \left(\dfrac{1}{z} + \dfrac{1}{z^2}\right)\, dz$, C is any contour in the right half-plane $\operatorname{Re}(z) > 0$

In Problems 21–24, use integration by parts (13) to evaluate the given integral. Write each answer in the form $a + ib$.

21. $\displaystyle\int_{\pi}^{i} e^z \cos z\, dz$

22. $\displaystyle\int_0^{i} z \sin z\, dz$

23. $\displaystyle\int_{i}^{1+i} z e^z\, dz$

24. $\displaystyle\int_0^{\pi i} z^2 e^z\, dz$

In Problems 25 and 26, use Theorem 5.4.2 to evaluate the given integral. In each integral $z^{1/2}$ is the principal branch of the square root function. Write each answer in the form $a + ib$.

25. $\displaystyle\int_C \frac{1}{4z^{1/2}}\,dz$, C is the arc of the circle $z = 4e^{it}$, $-\pi/2 \le t \le \pi/2$

26. $\displaystyle\int_C 3z^{1/2}\,dz$, C is the line segment between $z_0 = 1$ and $z_1 = 9i$

Focus on Concepts

27. Find an antiderivative of $f(z) = \sin z^2$. Do not think profound thoughts.

28. Give a domain D over which $f(z) = z(z+1)^{1/2}$ is analytic. Then find an antiderivative of f in D.

29. Let $\alpha = a + ib$ be a complex constant.

(a) Use Theorem 5.4.2 to evaluate $\displaystyle\int_{x_0}^{x} e^{\alpha z}\,dz$ where x_0 and x are real values.

(b) Explain how part (a) and Theorem 5.2.1 with the parametrization $z(t) = t$, $x_0 \le t \le x$ can be used to derive the (real) integral formula

$$\int e^{ax} \cos bx\,dx = \frac{e^{ax}(a\cos bx + b\sin bx)}{a^2 + b^2} + C.$$

30. Find $\int e^{ax} \sin bx\,dx$. [*Hint:* See Problem 29.]

5.5 Cauchy's Integral Formulas and Their Consequences

In the last two sections we saw the importance of the Cauchy-Goursat theorem in the evaluation of contour integrals. In this section we are going to examine several more consequences of the Cauchy-Goursat theorem. Unquestionably, the most significant of these is the following result:

The value of a analytic function f at any point z_0 in a simply connected domain can be represented by a contour integral.

After establishing this proposition we shall use it to further show that:

An analytic function f in a simply connected domain possesses derivatives of all orders.

The ramifications of these two results alone will keep us busy not only for the remainder of this section but in the next chapter as well.

5.5.1 Cauchy's Two Integral Formulas

First Formula If f is analytic in a simply connected domain D and z_0 is any point in D, the quotient $f(z)/(z - z_0)$ is not defined at z_0 and hence is *not* analytic in D. Therefore, we *cannot* conclude that the integral of $f(z)/(z - z_0)$ around a simple closed contour C that contains z_0 is zero by the Cauchy-Goursat theorem. Indeed, as we shall now see, the integral of

$f(z)/(z - z_0)$ around C has the value $2\pi i f(z_0)$. The first of two remarkable formulas is known simply as the **Cauchy integral formula**.

> **Theorem 5.5.1   Cauchy's Integral Formula**
>
> Suppose that f is analytic in a simply connected domain D and C is any simple closed contour lying entirely within D. Then for any point z_0 within C,
>
> $$f(z_0) = \frac{1}{2\pi i} \oint_C \frac{f(z)}{z - z_0}\, dz. \tag{1}$$

Proof Let D be a simply connected domain, C a simple closed contour in D, and z_0 an interior point of C. In addition, let C_1 be a circle centered at z_0 with radius small enough so that C_1 lies within the interior of C. By the principle of deformation of contours, (5) of Section 5.3, we can write

$$\oint_C \frac{f(z)}{z - z_0}\, dz = \oint_{C_1} \frac{f(z)}{z - z_0}\, dz. \tag{2}$$

We wish to show that the value of the integral on the right is $2\pi i f(z_0)$. To this end we add and subtract the constant $f(z_0)$ in the numerator of the integrand,

$$\oint_{C_1} \frac{f(z)}{z - z_0}\, dz = \oint_{C_1} \frac{f(z_0) - f(z_0) + f(z)}{z - z_0}\, dz.$$

$$= f(z_0) \oint_{C_1} \frac{1}{z - z_0}\, dz + \oint_{C_1} \frac{f(z) - f(z_0)}{z - z_0}\, dz. \tag{3}$$

From (6) of Section 5.3 we know that

$$\oint_{C_1} \frac{1}{z - z_0}\, dz = 2\pi i \tag{4}$$

and so (3) becomes

$$\oint_{C_1} \frac{f(z)}{z - z_0}\, dz = 2\pi i\, f(z_0) + \oint_{C_1} \frac{f(z) - f(z_0)}{z - z_0}\, dz. \tag{5}$$

Since f is continuous at z_0, we know that for any arbitrarily small $\varepsilon > 0$ there exists a $\delta > 0$ such that $|f(z) - f(z_0)| < \varepsilon$ whenever $|z - z_0| < \delta$. In particular, if we choose the circle C_1 to be $|z - z_0| = \frac{1}{2}\delta < \delta$, then by the ML-inequality (Theorem 5.2.3) the absolute value of the integral on the right side of the equality in (5) satisfies

$$\left| \oint_{C_1} \frac{f(z) - f(z_0)}{z - z_0}\, dz \right| \leq \frac{\varepsilon}{\delta/2} 2\pi \left(\frac{\delta}{2} \right) = 2\pi\varepsilon.$$

In other words, the absolute value of the integral can be made arbitrarily small by taking the radius of the circle C_1 to be sufficiently small. This can happen only if the integral is 0. Thus (5) is $\oint_{C_1} \frac{f(z)}{z - z_0}\, dz = 2\pi i\, f(z_0)$. The theorem is proved by dividing both sides of the last result by $2\pi i$. ❏

Note ➡ Because the symbol z represents a point on the contour C, (1) indicates that *the values of an analytic function f at points z_0 inside a simple closed contour C are determined by the values of f on the contour C.*

Cauchy's integral formula (1) can be used to evaluate contour integrals. Since we often work problems without a simply connected domain explicitly defined, a more practical restatement of Theorem 5.5.1 is:

> *If f is analytic at all points within and on a simple closed contour C, and z_0 is any point interior to C, then* $f(z_0) = \dfrac{1}{2\pi i} \displaystyle\oint_C \dfrac{f(z)}{z - z_0}\, dz.$

EXAMPLE 1   Using Cauchy's Integral Formula

Evaluate $\displaystyle\oint_C \dfrac{z^2 - 4z + 4}{z + i}\, dz$, where C is the circle $|z| = 2$.

Solution First, we identify $f(z) = z^2 - 4z + 4$ and $z_0 = -i$ as a point within the circle C. Next, we observe that f is analytic at all points within and on the contour C. Thus, by the Cauchy integral formula (1) we obtain

$$\oint_C \frac{z^2 - 4z + 4}{z + i}\, dz = 2\pi i f(-i) = 2\pi i(3 + 4i) = \pi(-8 + 6i).$$

❑

EXAMPLE 2   Using Cauchy's Integral Formula

Evaluate $\displaystyle\oint_C \dfrac{z}{z^2 + 9}\, dz$, where C is the circle $|z - 2i| = 4$.

Solution By factoring the denominator as $z^2 + 9 = (z - 3i)(z + 3i)$ we see that $3i$ is the only point within the closed contour C at which the integrand fails to be analytic. See Figure 5.5.1. Then by rewriting the integrand as

$$\frac{z}{z^2 + 9} = \frac{\overbrace{\dfrac{z}{z + 3i}}^{f(z)}}{z - 3i},$$

we can identify $f(z) = z/(z + 3i)$. The function f is analytic at all points within and on the contour C. Hence, from Cauchy's integral formula (1) we have

$$\oint_C \frac{z}{z^2 + 9}\, dz = \int_C \frac{\dfrac{z}{z + 3i}}{z - 3i}\, dz = 2\pi i\, f(3i) = 2\pi i\frac{3i}{6i} = \pi i.$$

❑

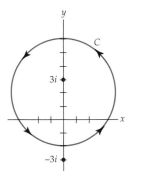

Figure 5.5.1 Contour for Example 2

Second Formula We shall now build on Theorem 5.5.1 by using it to prove that the values of the derivatives $f^{(n)}(z_0), n = 1, 2, 3, \ldots$ of an analytic function are also given by a integral formula. This second integral formula is similar to (1) and is known by the name **Cauchy's integral formula for derivatives**.

> **Theorem 5.5.2   Cauchy's Integral Formula for Derivatives**
>
> Suppose that f is analytic in a simply connected domain D and C is any simple closed contour lying entirely within D. Then for any point z_0 within C,
>
> $$f^{(n)}(z_0) = \frac{n!}{2\pi i}\oint_C \frac{f(z)}{(z-z_0)^{n+1}}\,dz. \qquad (6)$$

Partial Proof   We will prove (6) only for the case $n = 1$. The remainder of the proof can be completed using the principle of mathematical induction. We begin with the definition of the derivative and (1):

$$f'(z_0) = \lim_{\Delta z \to 0} \frac{f(z_0 + \Delta z) - f(z_0)}{\Delta z}$$

$$= \lim_{\Delta z \to 0} \frac{1}{2\pi i\,\Delta z}\left[\oint_C \frac{f(z)}{z-(z_0+\Delta z)}\,dz - \oint_C \frac{f(z)}{z-z_0}\,dz\right]$$

$$= \lim_{\Delta z \to 0} \frac{1}{2\pi i}\oint_C \frac{f(z)}{(z-z_0-\Delta z)(z-z_0)}\,dz.$$

Before continuing, let us set out some preliminaries. Continuity of f on the contour C guarantees that f is bounded (see page 112 of Section 2.6), that is, there exists a real number M such that $|f(z)| \le M$ for all points z on C. In addition, let L be the length of C and let δ denote the shortest distance between points on C and the point z_0. Thus for all points z on C we have

$$|z - z_0| \ge \delta \qquad \text{or} \qquad \frac{1}{|z-z_0|^2} \le \frac{1}{\delta^2}.$$

Furthermore, if we choose $|\Delta z| \le \tfrac{1}{2}\delta$, then by (10) of Section 1.2,

$$|z - z_0 - \Delta z| \ge \big|\,|z-z_0| - |\Delta z|\,\big| \ge \delta - |\Delta z| \ge \tfrac{1}{2}\delta$$

and so,
$$\frac{1}{|z-z_0-\Delta z|} \le \frac{2}{\delta}.$$

Now,

$$\left|\oint_C \frac{f(z)}{(z-z_0)^2}\,dz - \oint_C \frac{f(z)}{(z-z_0-\Delta z)(z-z_0)}\,dz\right|$$

$$= \left|\oint_C \frac{-\Delta z\,f(z)}{(z-z_0-\Delta z)(z-z_0)^2}\,dz\right| \le \frac{2ML\,|\Delta z|}{\delta^3}.$$

Because the last expression approaches zero as $\Delta z \to 0$, we have shown that

$$f'(z_0) = \lim_{\Delta z \to 0} \frac{f(z_0+\Delta z) - f(z_0)}{\Delta z} = \frac{1}{2\pi i}\oint_C \frac{f(z)}{(z-z_0)^2}\,dz,$$

which is (6) for $n = 1$.   ❏

Like (1), formula (6) can be used to evaluate integrals.

EXAMPLE 3   Using Cauchy's Integral Formula for Derivatives

Evaluate $\oint_C \dfrac{z+1}{z^4 + 2iz^3}\, dz$, where C is the circle $|z| = 1$.

Solution Inspection of the integrand shows that it is not analytic at $z = 0$ and $z = -2i$, but only $z = 0$ lies within the closed contour. By writing the integrand as

$$\frac{z+1}{z^4 + 2iz^3} = \frac{\dfrac{z+1}{z+2i}}{z^3}$$

we can identify, $z_0 = 0$, $n = 2$, and $f(z) = (z+1)/(z+2i)$. The quotient rule gives $f''(z) = (2 - 4i)/(z + 2i)^3$ and so $f''(0) = (2i - 1)/4i$. Hence from (6) we find

$$\oint_C \frac{z+1}{z^4 + 4z^3}\, dz = \frac{2\pi i}{2!} f''(0) = -\frac{\pi}{4} + \frac{\pi}{2}i.$$

❏

EXAMPLE 4   Using Cauchy's Integral Formula for Derivatives

Evaluate $\displaystyle\int_C \frac{z^3 + 3}{z(z - i)^2}\, dz$, where C is the figure-eight contour shown in Figure 5.5.2.

Solution Although C is not a simple closed contour, we can think of it as the union of two simple closed contours C_1 and C_2 as indicated in Figure 5.5.2. Since the arrows on C_1 flow clockwise or in the negative direction, the opposite curve $-C_1$ has positive orientation. Hence, we write

$$\int_C \frac{z^3 + 3}{z(z - i)^2}\, dz = \int_{C_1} \frac{z^3 + 3}{z(z - i)^2}\, dz + \int_{C_2} \frac{z^3 + 3}{z(z - i)^2}\, dz$$

$$= -\oint_{-C_1} \frac{\dfrac{z^3 + 3}{(z - i)^2}}{z}\, dz + \oint_{C_2} \frac{\dfrac{z^3 + 3}{z}}{(z - i)^2}\, dz = -I_1 + I_2,$$

and we are in a position to use both formulas (1) and (6).

To evaluate I_1 we identify $z_0 = 0$, $f(z) = (z^3 + 3)/(z - i)^2$, and $f(0) = -3$. By (1) it follows that

$$I_1 = \oint_{-C_1} \frac{\dfrac{z^3 + 3}{(z - i)^2}}{z}\, dz = 2\pi i\, f(0) = 2\pi i(-3) = -6\pi i.$$

To evaluate I_2 we now identify $z_0 = i$, $n = 1$, $f(z) = (z^3 + 3)/z$, $f'(z) = (2z^3 - 3)/z^2$, and $f'(i) = 3 + 2i$. From (6) we obtain

$$I_2 = \oint_{C_2} \frac{\dfrac{z^3 + 3}{z}}{(z - i)^2}\, dz = \frac{2\pi i}{1!} f'(i) = 2\pi i(3 + 2i) = -4\pi + 6\pi i.$$

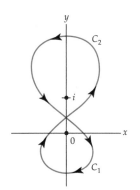

Figure 5.5.2 Contour for Example 4

Finally, we get

$$\int_C \frac{z^3 + 3}{z(z-i)^2}\, dz = -I_1 + I_2 = 6\pi i + (-4\pi + 6\pi i) = -4\pi + 12\pi i.$$

❏

5.5.2 Some Consequences of the Integral Formulas

An immediate and important corollary to Theorem 5.5.2 is summarized next.

Theorem 5.5.3   Derivative of an Analytic Function Is Analytic

Suppose that f is analytic in a simply connected domain D. Then f possesses derivatives of all orders at every point z in D. The derivatives f', f'', f''', ... are analytic functions in D.

If a function $f(z) = u(x,\ y) + iv(x,\ y)$ is analytic in a simply connected domain D, we have just seen its derivatives of all orders exist at any point z in D and so f', f'', f''', ... are continuous. From

$$f'(z) = \frac{\partial u}{\partial x} + i\frac{\partial v}{\partial x} = \frac{\partial v}{\partial y} - i\frac{\partial u}{\partial y}$$

$$f''(z) = \frac{\partial^2 u}{\partial x^2} + i\frac{\partial^2 v}{\partial x^2} = \frac{\partial^2 v}{\partial y \partial x} - i\frac{\partial^2 u}{\partial y \partial x}$$

$$\vdots$$

we can also conclude that the real functions u and v have continuous partial derivatives of all orders at a point of analyticity.

Cauchy's Inequality   We next present an inequality derived from the Cauchy integral formula for derivatives.

Theorem 5.5.4   Cauchy's Inequality

Suppose that f is analytic in a simply connected domain D and C is a circle defined by $|z - z_0| = r$ that lies entirely in D. If $|f(z)| \le M$ for all points z on C, then

$$\left| f^{(n)}(z_0) \right| \le \frac{n!M}{r^n}. \tag{7}$$

Proof   From the hypothesis,

$$\left| \frac{f(z)}{(z-z_0)^{n+1}} \right| = \frac{|f(z)|}{r^{n+1}} \le \frac{M}{r^{n+1}}.$$

Thus, from (6) and the *ML*-inequality (Theorem 5.2.3), we have

$$\left| f^{(n)}(z_0) \right| = \left| \frac{n!}{2\pi i} \oint_C \frac{f(z)}{(z-z_0)^{n+1}}\, dz \right| \le \frac{n!}{2\pi} \frac{M}{r^{n+1}} 2\pi r = \frac{n!M}{r^n}.$$

❏

The number M in Theorem 5.5.4 depends on the circle $|z - z_0| = r$. But notice in (7) that if $n = 0$, then $M \geq |f(z_0)|$ for *any* circle C centered at z_0 as long as C lies within D. In other words, an upper bound M of $|f(z)|$ on C cannot be smaller than $|f(z_0)|$.

Liouville's Theorem   Cauchy's inequality (7) is a key ingredient in the proof of the next result. Although it bears the name "Liouville's theorem," it probably was first proved by Cauchy. The gist of the theorem is that an entire function f, one that is analytic for all z, cannot be bounded unless f itself is a constant.

Theorem 5.5.5   Liouville's Theorem

The only bounded entire functions are constants.

Proof   Suppose f is an entire function and is bounded, that is, $|f(z)| \leq M$ for all z. Then for any point z_0, (7) gives $|f'(z_0)| \leq M/r$. By making r arbitrarily large we can make $|f'(z_0)|$ as small as we wish. This means $f'(z_0) = 0$ for all points z_0 in the complex plane. Hence, by Theorem 3.2.3(ii), f must be a constant.   ❑

Fundamental Theorem of Algebra   Theorem 5.5.5 enables us to establish a result usually learned—but never proved—in elementary algebra.

Theorem 5.5.6   Fundamental Theorem of Algebra

If $p(z)$ is a nonconstant polynomial, then the equation $p(z) = 0$ has at least one root.

Proof   Let us suppose that the polynomial $p(z) = a_n z^n + a_{n-1} z^{n-1} + \cdots + a_1 z + a_0$, $n > 0$, is not 0 for any complex number z. This implies that the reciprocal of p, $f(z) = 1/p(z)$, is an entire function. Now

$$|f(z)| = \frac{1}{|a_n z^n + a_{n-1} z^{n-1} + \cdots + a_1 z + a_0|}$$

$$= \frac{1}{|z|^n |a_n + (a_{n-1}/z + \cdots + a_1/z^{n-1} + a_0/z^n)|}.$$

Choose a real number $M > 1$ such that $M > |2na_j/a_n|$ for $j = 0, 1, \ldots, n-1$. Then for $|z| > M$ we have $|a_j/z^{n-j}| < |a_n|/2n$, and so

$$\left| a_{n-1}/z + \cdots + a_1/z^{n-1} + a_0/z \right| \leq |a_n|/2n + \cdots + |a_n|/2n = |a_n|/2.$$

From inequality (8) of Section 1.2 we then have:

$$|f(z)| \leq \frac{1}{|z|^n \left| |a_n| - |a_{n-1}/z + \cdots + a_1/z^{n-1} + a_0/z^n| \right|}$$

$$\leq \frac{1}{M^n \left| |a_n| - |a_n|/2 \right|} = \frac{2}{M^n |a_n|}.$$

Thus, we see that f is bounded on the exterior of the disk $|z| \leq M$. However, since f is continuous it must also be bounded on the disk $|z| \leq M$ (see

Section 2.6). We conclude that the function f is a bounded entire function. It then follows from Liouville's theorem that f is a constant, and therefore p is a constant. But this is a contradiction to our underlying assumption that p was *not* a constant polynomial. We conclude that there must exist at least one number z for which $p(z) = 0$. ❏

It is left as an exercise to show, using Theorem 5.5.6, that if $p(z)$ is a nonconstant polynomial of degree n, then $p(z) = 0$ has exactly n roots (counting multiple roots). See Problem 29 in Exercises 5.5.

Morera's Theorem   The proof of the next theorem enshrined the name of the Italian mathematician Giacinto Morera forever in texts on complex analysis. Morera's theorem, which gives a sufficient condition for analyticity, is often taken to be the converse of the Cauchy-Goursat theorem. For its proof we return to Theorem 5.5.3.

Theorem 5.5.7   Morera's Theorem

If f is continuous in a simply connected domain D and if $\oint_C f(z)\,dz = 0$ for every closed contour C in D, then f is analytic in D.

Proof   By the hypotheses of continuity of f and $\oint_C f(z)\,dz = 0$ for every closed contour C in D, we conclude that $\int_C f(z)\,dz$ is independent of the path. In the proof of (7) of Section 5.4 we then saw that the function F defined by $F(z) = \int_{z_0}^{z} f(s)\,ds$ (where s denotes a complex variable, z_0 is a fixed point in D, and z represents any point in D) is an antiderivative of f; that is, $F'(z) = f(z)$. Hence, F is analytic in D. In addition, $F'(z)$ is analytic in view of Theorem 5.5.3. Since $f(z) = F'(z)$, we see that f is analytic in D. ❏

An alternative proof of this last result is outlined in Problem 31 in Exercises 5.5.

We could go on at length stating more and more results whose proofs rest on a foundation of theory that includes the Cauchy-Goursat theorem and the Cauchy integral formulas. But we shall stop after one more theorem.

In Section 2.6 we saw that if a function f is continuous on a closed and bounded region R, then f is bounded; that is, there is some constant M such that $|f(z)| \le M$ for z in R. If the boundary of R is a simple closed curve C, then the next theorem, which we present without proof, tells us that $|f(z)|$ assumes its maximum value at some point z on the boundary C.

Theorem 5.5.8   Maximum Modulus Theorem

Suppose that f is analytic and nonconstant on a closed region R bounded by a simple closed curve C. Then the modulus $|f(z)|$ attains its maximum on C.

If the stipulation that $f(z) \ne 0$ for all z in R is added to the hypotheses of Theorem 5.5.8, then the modulus $|f(z)|$ also attains its *minimum* on C. See Problems 27 and 33 in Exercises 5.5.

EXAMPLE 5   Maximum Modulus

Find the maximum modulus of $f(z) = 2z + 5i$ on the closed circular region defined by $|z| \leq 2$.

Solution From (2) of Section 1.2 we know that $|z|^2 = z\bar{z}$. By replacing the symbol z by $2z + 5i$ we have

$$|2z + 5i|^2 = (2z + 5i)\overline{(2z + 5i)} = (2z + 5i)(2\bar{z} - 5i) = 4z\bar{z} - 10i(z - \bar{z}) + 25. \qquad (8)$$

But from (6) of Section 1.1, $\bar{z} - z = 2i \, \text{Im}(z)$, and so (8) is

$$|2z + 5i|^2 = 4|z|^2 + 20 \, \text{Im}(z) + 25. \qquad (9)$$

Because f is a polynomial, it is analytic on the region defined by $|z| \leq 2$. By Theorem 5.5.8, $\max_{|z| \leq 2} |2z + 5i|$ occurs on the boundary $|z| = 2$. Therefore, if $|z| = 2$, then (9) yields

$$|2z + 5i| = \sqrt{41 + 20 \, \text{Im}(z)}. \qquad (10)$$

The last expression attains its maximum when $\text{Im}(z)$ attains its maximum on $|z| = 2$, namely, at the point $z = 2i$. Thus, $\max_{|z| \leq 2} |2z + 5i| = \sqrt{81} = 9$. ❑

Note in Example 5 that $f(z) = 0$ only at $z = -\frac{5}{2}i$ and that this point is outside the region defined by $|z| \leq 2$. Hence we can conclude that (10) attains its minimum when $\text{Im}(z)$ attains its minimum on $|z| = 2$ at $z = -2i$. As a result, $\min_{|z| \leq 2} |2z + 5i| = \sqrt{1} = 1$.

Remarks *Comparison with Real Analysis*

(i) As a consequence of Theorem 5.5.3, if f is differentiable at all points in some domain, then the derivatives f', f'', f''', ... all exist on the domain. In real analysis this need not be true. For example, the function $f(x) = x^{5/3}$ is differentiable on the entire real line. However, $f''(x) = \frac{10}{9}x^{-1/3}$ does not exist at $x = 0$.

(ii) In real analysis there are many infinitely differentiable functions that are bounded on a closed interval $[a, b]$, that is, $|f(x)| \leq M$ for all $a \leq x \leq b$. For example, $\sin x$ is infinitely differentiable and bounded on $[-\pi, \pi]$ since $|\sin x| \leq 1$ for all x. The absolute value $|\sin x|$, however, does not attain its maximum on the boundary of the closed interval since $|\sin(-\pi)| = |\sin \pi| = 0$. Compare this fact with the Maximum Modulus Theorem 5.5.8.

EXERCISES 5.5   *Answers to selected odd-numbered problems begin on page ANS-15.*

5.5.1 Cauchy's Integral Formulas

In Problems 1–22, use Theorems 5.5.1 and 5.5.2, when appropriate, to evaluate the given integral along the indicated closed contour(s).

1. $\oint_C \dfrac{4}{z - 3i}\, dz;\ |z| = 5$

2. $\oint_C \dfrac{z^2}{(z - 3i)^2}\, dz;\ |z| = 5$

3. $\oint_C \dfrac{e^z}{z - \pi i}\, dz;\ |z| = 4$

4. $\oint_C \dfrac{1 + e^z}{z}\, dz;\ |z| = 1$

5. $\oint_C \dfrac{z^2 - 3z + 4i}{z + 2i}\, dz;\ |z| = 3$ **6.** $\oint_C \dfrac{\cos z}{3z - \pi}\, dz;\ |z| = 1.1$

7. $\oint_C \dfrac{z^2}{z^2 + 4}\, dz;\ $ **(a)** $|z - i| = 2,\ $ **(b)** $|z + 2i| = 1$

8. $\oint_C \dfrac{z^2 + 3z + 2i}{z^2 + 3z - 4}\, dz;\ $ **(a)** $|z| = 2,\ $ **(b)** $|z + 5| = \frac{3}{2}$

9. $\oint_C \dfrac{z^2 + 4}{z^2 - 5iz - 4}\, dz;\ |z - 3i| = 1.3$ **10.** $\oint_C \dfrac{\sin z}{z^2 + \pi^2}\, dz;\ |z - 2i| = 2$

11. $\oint_C \dfrac{e^{z^2}}{(z - i)^3}\, dz;\ |z - i| = 1$ **12.** $\oint_C \dfrac{z}{(z + i)^4}\, dz;\ |z| = 2$

13. $\oint_C \dfrac{\cos 2z}{z^5}\, dz;\ |z| = 1$ **14.** $\oint_C \dfrac{e^{-z} \sin z}{z^3}\, dz;\ |z - 1| = 3$

15. $\oint_C \dfrac{2z + 5}{z^2 - 2z}\, dz;\ $ **(a)** $|z| = \frac{1}{2},\ $ **(b)** $|z + 1| = 2\ $ **(c)** $|z - 3| = 2,\ $ **(d)** $|z + 2i| = 1$

16. $\oint_C \dfrac{z}{(z - 1)(z - 2)}\, dz;\ $ **(a)** $|z| = \frac{1}{2},\ $ **(b)** $|z + 1| = 1\ $ **(c)** $|z - 1| = \frac{1}{2},\ $ **(d)** $|z| = 4$

17. $\oint_C \dfrac{z + 2}{z^2(z - 1 - i)}\, dz;\ $ **(a)** $|z| = 1,\ $ **(b)** $|z - 1 - i| = 1$

18. $\oint_C \dfrac{1}{z^3(z - 4)}\, dz;\ $ **(a)** $|z| = 1,\ $ **(b)** $|z - 2| = 1$

19. $\oint_C \left(\dfrac{e^{2iz}}{z^4} - \dfrac{z^4}{(z - i)^3} \right) dz;\ |z| = 6$

20. $\oint_C \left(\dfrac{\cosh z}{(z - \pi)^3} - \dfrac{\sin^2 z}{(2z - \pi)^3} \right) dz;\ |z| = 3$

21. $\oint_C \dfrac{1}{z^3(z - 1)^2}\, dz;\ |z - 2| = 5$

22. $\oint_C \dfrac{1}{z^2(z^2 + 1)}\, dz;\ |z - i| = \frac{3}{2}$

In Problems 23 and 24, evaluate the given integral, where C is the figure-eight contour in the figure.

23. $\oint_C \dfrac{3z + 1}{z(z - 2)^2}\, dz$ **24.** $\oint_C \dfrac{e^{iz}}{(z^2 + 1)^2}\, dz$

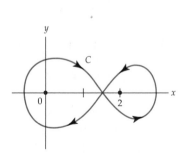

Figure 5.5.3 Figure for Problem 23

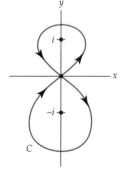

Figure 5.5.4 Figure for Problem 24

5.5.2 Some Consequences of the Integral Formulas

In Problems 25 and 26, proceed as in Example 5 to find the maximum modulus of the given function on indicated closed circular region.

25. $f(z) = -iz + i$; $|z| \leq 5$ **26.** $f(z) = z^2 + 4z$; $|z| \leq 1$

27. Suppose the boundary C of the closed circular region R defined by $|z| \leq 1$ is parametrized by $x = \cos t$, $y = \sin t$, $0 \leq t \leq 2\pi$. By considering $|f(z(t))|$, find the maximum modulus and the minimum modulus of the given analytic function f and the points z on C that give these values.

   (a) $f(z) = (iz + 3)^2$

   (b) $f(z) = (z - 2 - 2\sqrt{3}\,i)^2$

   (c) $f(z) = -2iz^2 + 5$

[*Hint*: In parts (b) and (c), it may help to recall from calculus how to find the relative extrema of a real-valued function of real variable t.]

Focus on Concepts

28. (*Cauchy's Integral Formula*) Suppose f is analytic within and on a circle C of radius r with center at z_0. Use (1) to obtain

$$f(z_0) = \frac{1}{2\pi} \int_0^{2\pi} f(z_0 + re^{i\theta})\, d\theta.$$

This result is known as **Gauss' mean-value theorem** and shows that the value of f at the center z_0 of the circle is the average of all the values of f on the circumference of C.

29. (*Fundamental Theorem of Algebra*) Suppose

$$p(z) = a_n z^n + a_{n-1} z^{n-1} + \cdots + a_1 z + a_0$$

is a polynomial of degree $n > 1$ and that z_1 is number such that $p(z_1) = 0$. Then

   (a) Show that $p(z) = p(z) - p(z_1) = a_n(z^n - z_1^n) + a_{n-1}(z^{n-1} - z_1^{n-1}) + \cdots + a_1(z - z_1)$.

   (b) Use the result in part (a) to show that $p(z) = (z - z_1)q(z)$, where q is a polynomial of degree $n - 1$.

   (c) Use the result in part (b) to give a sound explanation why the equation $p(z) = 0$ has n roots.

30. Use Problem 29 to factor the polynomial

$$p(z) = z^3 + (3 - 4i)z^2 - (15 + 4i)z - 1 + 12i.$$

Do not use technology.

31. (*Morera's Theorem*) Sometimes in the proof of Theorem 5.5.7 continuity of f' in D is also assumed. If this is the case, then (3) of Section 5.1 and Green's theorem can be used to write $\oint_C f(z)\, dz$ as

$$\oint_C f(z)\, dz = \oint_C u\, dx - v\, dy + i \oint_C v\, dx + u\, dy$$

$$= \iint_R \left(-\frac{\partial v}{\partial x} - \frac{\partial u}{\partial y} \right) dA + i \iint_R \left(\frac{\partial u}{\partial x} - \frac{\partial v}{\partial y} \right) dA,$$

where R denotes the region bounded by C. Supply the next step(s) in the proof and state the conclusion.

32. (*Maximum Modulus Theorem*) Critique the following reasoning:

Consider the function $f(z) = z^2 + 5z - 1$ defined the closed circular region defined by $|z| \leq 1$. It follows from the triangle inequality, (10) of Section 1.2, that

$$|z^2 + 5z - 1| \leq |z|^2 + 5|z| + |-1|.$$

Since the maximum modulus of f occurs on $|z| = 1$, the inequality shows that the maximum modulus of $f(z) = z^2 + 5z - 1$ over the region is 7.

33. In this problem we will start you out in the proof of the **minimum modulus theorem**.

> *If f is analytic on a closed region R bounded by a simple closed curve C and $f(z) \neq 0$ for all z in R, then the modulus $|f(z)|$ attains its minimum on C.*

Define the function $g(z) = 1/f(z)$, reread Theorem 5.5.8, and then complete the proof of the theorem.

34. Suppose $f(z) = z + 1 - i$ is defined over the triangular region R that has vertices i, 1, and $1 + i$. Discuss how the concept of *distance* from the point $-1 + i$ can be used to find the points on the boundary of R for which $|f(z)|$ attains its maximum value and its minimum value.

5.6   Applications

In Section 1.2 we introduced the notion that a complex number could be interpreted as a two-dimensional vector. Because of this, we saw in Section 2.7 that a two-dimensional vector field $\mathbf{F}(x, y) = P(x, y)\mathbf{i} + Q(x, y)\mathbf{j}$ could be represented by means of a complex-valued function f by taking the components P and Q of $\mathbf{F}$ to be the real and imaginary parts of f; that is, $f(z) = P(x, y) + iQ(x, y)$ is a vector whose initial point is z. In this section we will explore the use of this complex representation of the vector field $\mathbf{F}(x, y)$ in the context of analyzing certain aspects of fluid flow. Because the vector field consists of vectors representing velocities at various points in the flow, $\mathbf{F}(x, y)$ or $f(z)$ is called a **velocity field**. The magnitude $\|\mathbf{F}\|$ of $\mathbf{F}$, or the modulus $|f(z)|$ of the complex representation f, is called **speed**.

It will assumed throughout this section that every domain D is simply connected.

Irrotational Vector Field   Throughout this section we consider only a two-dimensional or **planar flow** of a fluid (see Sections 2.7 and 3.4). This assumption permits us to analyze a single sheet of fluid flowing across a domain D in the plane. Suppose that $\mathbf{F}(x, y) = P(x, y)\mathbf{i} + Q(x, y)\mathbf{j}$ represents a *steady-state* velocity field of this fluid flow in D. In other words, the velocity of the fluid at a point in the sheet depends only on its position (x, y) and not on time t. In the study of fluids, if curl $\mathbf{F} = \mathbf{0}$, then the fluid flow is said to be **irrotational**. If a paddle device, such as shown in Figure 5.6.1, is inserted in a flowing fluid, then the curl of its velocity field $\mathbf{F}$ is a measure of the tendency of the fluid to turn the device about its vertical axis (imagine this vertical axis pointing straight out of the page). The flows illustrated in Figures 5.6.1(a) and 5.6.1(b) are irrotational because the paddle device is not turning. The word "irrotational" is somewhat misleading because, as is seen in Figure 5.6.1(b), it does *not* mean that the fluid does not rotate. Rather, if curl $\mathbf{F} = \mathbf{0}$, then the flow of the fluid is free of turbulence

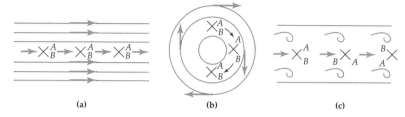

Figure 5.6.1 Three fluid flows

in the form of vortices or whirlpools that would cause the paddle to turn. In the case of Figure 5.6.1(c), the flow is rotational; notice the vortices and that the paddle device is depicted as turning.

The divergence of the vector field $\mathbf{F}(x, y) = P(x, y)\mathbf{i} + Q(x, y)\mathbf{j}$ is a measure of the rate of change of the density of the fluid at a point. If div $\mathbf{F} = 0$, the fluid is said to be **incompressible**; that is, an incompressible fluid is homogeneous (constant density) throughout the domain D. In a simply connected domain D, an incompressible flow has the special property that the amount of fluid in the interior of any simple closed contour C is independent of time. The rate at which fluid enters the interior of C matches the rate at which it leaves, and consequently there can be no fluid sources or sinks at points in D. In electromagnetic theory, if $\mathbf{F}$ represents a vector field for which div $\mathbf{F} = 0$, then $\mathbf{F}$ is said to be **solenoidal**. Let us assume that P and Q are continuous and have continuous partial derivatives in D. Then from vector calculus div $\mathbf{F}$ (or $\nabla \cdot \mathbf{F}$) is a scalar function and curl $\mathbf{F}$ (or $\nabla \times \mathbf{F}$) is a vector:

$$\operatorname{div} \mathbf{F} = \frac{\partial P}{\partial x} + \frac{\partial Q}{\partial y} \qquad \text{and} \qquad \operatorname{curl} \mathbf{F} = \left(\frac{\partial Q}{\partial x} - \frac{\partial P}{\partial y} \right) \mathbf{k}. \qquad (1)$$

In the case of an **ideal fluid**, that is, an *incompressible* nonviscous fluid whose planar flow is *irrotational*, we see from (1) that div $\mathbf{F} = 0$ and curl $\mathbf{F} = \mathbf{0}$ yield the simultaneous equations

$$\frac{\partial P}{\partial x} = -\frac{\partial Q}{\partial y} \qquad \text{and} \qquad \frac{\partial P}{\partial y} = \frac{\partial Q}{\partial x}. \qquad (2)$$

The system of partial differential equations in (2) is reminiscent of the Cauchy-Riemann equations—a criterion for analyticity presented in Theorem 3.2.2 of Section 3.2. If the vector field $\mathbf{F}(x, y) = P(x, y)\mathbf{i} + Q(x, y)\mathbf{j}$ is represented by the complex function $f(z) = P(x, y) + iQ(x, y)$, then it turns out that (2) implies that the conjugate of f, that is, $g(z) = \overline{f(z)} = P(x, y) - iQ(x, y)$, is an analytic function in D.

Theorem 5.6.1   Vector Fields and Analyticity

Suppose the functions u, v, P, and Q are continuous and have continuous first partial derivatives in a domain D.

(i) If $\mathbf{F}(x, y) = P(x, y)\mathbf{i} + Q(x, y)\mathbf{j}$ is a vector field for which div $\mathbf{F} = 0$ and curl $\mathbf{F} = \mathbf{0}$ in D, and if $f(z) = P(x, y) + iQ(x, y)$ is the complex representation of $\mathbf{F}$, then the function $g(z) = \overline{f(z)} = P(x, y) - iQ(x, y)$ is analytic in D.

(ii) Conversely, if $g(z) = u(x, y) + iv(x, y)$ is analytic in D, then the function $f(z) = \overline{g(z)} = u(x, y) - iv(x, y)$ is the complex representation of a vector field $\mathbf{F}(x, y) = P(x, y)\mathbf{i} + Q(x, y)\mathbf{j}$ for which div $\mathbf{F} = 0$ and curl $\mathbf{F} = \mathbf{0}$ in D.

Proof (*i*) If we let $u(x, y)$ and $v(x, y)$ denote the real and imaginary parts of $g(z) = \overline{f(z)} = P(x, y) - iQ(x, y)$, then $P = u$ and $Q = -v$. Because div **F** $= 0$ and curl **F** $= \mathbf{0}$, the equations in (2) become, respectively,

$$\frac{\partial u}{\partial x} = -\frac{\partial(-v)}{\partial y} \quad \text{and} \quad \frac{\partial u}{\partial y} = \frac{\partial(-v)}{\partial x}.$$

That is,

$$\frac{\partial u}{\partial x} = \frac{\partial v}{\partial y} \quad \text{and} \quad \frac{\partial u}{\partial y} = -\frac{\partial v}{\partial x}. \tag{3}$$

The equations in (3) are the usual Cauchy-Riemann equations, and so by Theorem 3.2.2 we conclude that $g(z) = \overline{f(z)} = P(x, y) - iQ(x, y)$ is analytic in D.

(*ii*) We now let $P(x, y)$ and $Q(x, y)$ denote the real and imaginary parts of $f(z) = \overline{g(z)} = u(x, y) - iv(x, y)$. Since $u = P$ and $v = -Q$, the Cauchy-Riemann equations become

$$\frac{\partial P}{\partial x} = \frac{\partial(-Q)}{\partial y} = -\frac{\partial Q}{\partial y} \quad \text{and} \quad \frac{\partial P}{\partial y} = -\frac{\partial(-Q)}{\partial x} = \frac{\partial Q}{\partial x}. \tag{4}$$

These are the equations in (2) and so div **F** $= 0$ and curl **F** $= \mathbf{0}$.   ❏

EXAMPLE 1   Vector Field Gives an Analytic Function

The two-dimensional vector field

$$\mathbf{F}(x, y) = \frac{K}{2\pi}\left[\frac{y - y_0}{(x - x_0)^2 + (y - y_0)^2}\mathbf{i} - \frac{x - x_0}{(x - x_0)^2 + (y - y_0)^2}\mathbf{j}\right], K > 0,$$

can be interpreted as the velocity field of the flow of an ideal fluid in a domain D of the xy-plane not containing (x_0, y_0). It is easily verified that the fluid is incompressible (div **F** $= 0$) and irrotational (curl **F** $= \mathbf{0}$) in D. The complex representation of **F** is

$$f(z) = \frac{K}{2\pi}\left[\frac{y - y_0}{(x - x_0)^2 + (y - y_0)^2} - i\frac{x - x_0}{(x - x_0)^2 + (y - y_0)^2}\right].$$

Indeed, by rewriting f as

$$f(z) = \frac{K}{2\pi i}\left[\frac{x - x_0}{(x - x_0)^2 + (y - y_0)^2} + i\frac{y - y_0}{(x - x_0)^2 + (y - y_0)^2}\right]$$

and using $z_0 = x_0 + iy_0$ and $z = x + iy$, you should be able to recognize $f(z)$ is the same as

$$f(z) = \frac{K}{2\pi i}\frac{z - z_0}{|z - z_0|^2} \quad \text{or} \quad f(z) = \frac{K}{2\pi i}\frac{1}{\bar{z} - \bar{z}_0}.$$

Hence, from Theorem 5.6.1(*i*), the complex function $g(z) = \overline{f(z)}$ is a rational function,

$$g(z) = \overline{f(z)} = -\frac{K}{2\pi i}\frac{1}{z - z_0}, \ K > 0,$$

and is analytic in a domain D of the z-plane not containing z_0.   ❏

Any analytic function $g(z)$ can be interpreted as a complex representation of the velocity field $\mathbf{F}$ of a planar fluid flow. *But in view of Theorem 5.6.1(ii),* it is the function f defined as the conjugate of g, $f(z) = \overline{g(z)}$, that is a complex representation of a velocity field $\mathbf{F}(x, y) = P(x, y)\mathbf{i} + Q(x, y)\mathbf{j}$ of the planar flow of an *ideal fluid* in some domain D of the plane.

EXAMPLE 2   Analytic Function Gives a Vector Field

The polynomial function $g(z) = kz = k(x + iy), k > 0$, is analytic in any domain D of the complex plane. From Theorem 5.6.1(ii), $f(z) = \overline{g(z)} = k\bar{z} = kx - iky$ is the complex representation of a velocity field $\mathbf{F}$ of an ideal fluid in D. With the identifications $P(x, y) = kx$ and $Q(x, y) = -ky$, we have $\mathbf{F}(x, y) = k(x\mathbf{i} - y\mathbf{j})$. A quick inspection of (2) verifies that div $\mathbf{F} = 0$ and curl $\mathbf{F} = \mathbf{0}$.   ❑

Streamlines Revisited We can now tie up a few loose ends between Section 2.7 and Section 3.4. In Section 2.7, we saw that if $\mathbf{F}(x, y) = P(x, y)\mathbf{i} + Q(x, y)\mathbf{j}$ or $f(z) = P(x, y) + iQ(x, y)$ represented the velocity field of *any* planar fluid flow, then the actual path $z(t) = x(t) + iy(t)$ of a particle (such as a small cork) placed in the flow must satisfy the system of first-order differential equations:

$$\begin{aligned} \frac{dx}{dt} &= P(x, y) \\ \frac{dy}{dt} &= Q(x, y). \end{aligned} \tag{5}$$

The family of all solutions of (5) were called **streamlines** of the flow.

In Section 3.4 we saw that in the case of a planar flow of an ideal fluid, the velocity vector $\mathbf{F}$ could be represented by the gradient of a real-valued function ϕ called a **velocity potential**. The level curves $\phi(x, y) = c_1$ were called **equipotential curves**. Most importantly, the function ϕ is a solution of Laplace's equation in some domain D and so is harmonic in D. After finding the harmonic conjugate $\psi(x, y)$ of ϕ, we formed a function

$$\mathbf{\Omega}(z) = \phi(x, y) + i\psi(x, y), \tag{6}$$

called the **complex velocity potential,** which is analytic in D. In that context, we called $\psi(x, y)$ a **stream function** and its level curves $\psi(x, y) = c_2$ **streamlines**.

We shall now demonstrate that we are talking about the same thing in (5) and (6). Let us assume that $\mathbf{F}(x, y) = P(x, y)\mathbf{i} + Q(x, y)\mathbf{j}$ is the velocity field of the flow of an ideal fluid in some domain D. Let's start by reviewing the content of the paragraph containing (6). Because $\mathbf{F}(x, y) = P(x, y)\mathbf{i} + Q(x, y)\mathbf{j}$ is a gradient field, there exists a scalar function ϕ such that

$$\mathbf{F}(x, y) = \nabla\phi = \frac{\partial\phi}{\partial x}\mathbf{i} + \frac{\partial\phi}{\partial y}\mathbf{j}, \tag{7}$$

and so $$\frac{\partial\phi}{\partial x} = P(x, y) \quad \text{and} \quad \frac{\partial\phi}{\partial y} = Q(x, y). \tag{8}$$

Because ϕ is harmonic in D, we call it a potential function for $\mathbf{F}$. We then find its harmonic conjugate ψ and use it to form the complex velocity potential

$\mathbf{\Omega}(z) = \phi + i\psi$. Since $\mathbf{\Omega}(z)$ is analytic in D, we can use the Cauchy-Riemann equations

$$\frac{\partial \phi}{\partial x} = \frac{\partial \psi}{\partial y} \quad \text{and} \quad \frac{\partial \phi}{\partial y} = -\frac{\partial \psi}{\partial x} \tag{9}$$

to rewrite (8) as

$$\frac{\partial \psi}{\partial y} = P(x, y) \quad \text{and} \quad \frac{\partial \psi}{\partial x} = -Q(x, y). \tag{10}$$

Now let us re-examine the system of differential equations in (5). If we divide the second equation in the system by the first, we obtain a single first-order differential equation $dy/dx = Q(x, y)/P(x, y)$ or

$$-Q(x, y)dx + P(x, y)dy = 0. \tag{11}$$

Then by (2), we see that P and Q are related by $\partial P/\partial x = \partial(-Q)/\partial y = -\partial Q/\partial y$. This last condition proves that (11) is an **exact** first-order differential equation. More to the point, the equations in (10) show us that (11) is the same as

$$\frac{\partial \psi}{\partial x}dx + \frac{\partial \psi}{\partial y}dy = 0. \tag{12}$$

Note ➡   If you have not had a course in differential equations, then the foregoing manipulations may not impress you. But those readers with some knowledge of that subject should recognize the result in (12) establishes that (11) is equivalent to the exact differential $d(\psi(x, y)) = 0$. Integrating this last equation shows that all solutions of (5) satisfy $\psi(x, y) = c_2$. In other words, the streamlines of the velocity field $\mathbf{F}(x, y) = P(x, y)\mathbf{i} + Q(x, y)\mathbf{j}$ obtained from (5) are the same as the level curves of the harmonic conjugate ψ of ϕ in (6).

Complex Potential Revisited

Assume that $\mathbf{F}(x, y) = P(x, y)\mathbf{i} + Q(x, y)\mathbf{j}$ is the velocity field of the flow of an ideal fluid in some domain D of the plane and that $\mathbf{\Omega}(z) = \phi(x, y) + i\psi(x, y)$ is the complex velocity potential of the flow. We know from Theorem 5.6.1(i) that from the complex representation $f(z) = P(x, y) + iQ(x, y)$ of $\mathbf{F}$ we can construct an analytic function $g(z) = \overline{f(z)} = P(x, y) - iQ(x, y)$. The two analytic functions g and $\mathbf{\Omega}$ are related. To see why this is so, we first write the gradient vector $\mathbf{F}$ in (7) in equivalent complex notation as

$$f(z) = \frac{\partial \phi}{\partial x} + i\frac{\partial \phi}{\partial y}. \tag{13}$$

Now by (9) of Section 3.2, the derivative of the analytic function $\mathbf{\Omega}(z) = \phi(x, y) + i\psi(x, y)$ is the analytic function

$$\mathbf{\Omega}'(z) = \frac{\partial \phi}{\partial x} + i\frac{\partial \psi}{\partial x}. \tag{14}$$

We now replace $\partial \psi/\partial x$ in (14) using the second of the Cauchy-Riemann equations in (9):

$$\mathbf{\Omega}'(z) = \frac{\partial \phi}{\partial x} - i\frac{\partial \phi}{\partial y}. \tag{15}$$

By comparing (13) and (15) we see immediately that $\overline{f(z)} = \Omega'(z)$ and, consequently,

$$g(z) = \Omega'(z). \tag{16}$$

The conjugate of this analytic function, $\overline{g(z)} = \overline{\overline{f(z)}} = f(z)$, is the complex representation of the vector field $\mathbf{F}$ whose complex potential is $\Omega(z)$. In symbols,

$$f(z) = \overline{\Omega'(z)}. \tag{17}$$

Because $f(z)$ is a complex representation of velocity vector field, the quantity $\overline{\Omega'(z)}$ in (17) is sometimes referred to as the **complex velocity**.

You may legitimately ask: Is (17) merely interesting or is it useful? Answer: Useful. Here is one practical observation: Any function analytic in some domain D can be regarded as a complex potential for the planar flow of an ideal fluid.

EXAMPLE 3 Complex Potentials

(a) The analytic function $\Omega(z) = \frac{1}{2}kz^2$, $k > 0$, is a complex potential for the flow in Example 2. By (16), the derivative $g(z) = \Omega'(z) = kz$ is an analytic function. By (17), the conjugate $f(z) = \overline{\Omega'(z)}$ or $f(z) = k\bar{z} = kx - iky$ is the complex representation of the velocity vector field $\mathbf{F}$ of the flow of an ideal fluid in some domain D of the plane. The complex potential of $\mathbf{F}$ is $\Omega(z)$. Since $\Omega(z) = \frac{1}{2}k(x^2 - y^2 + 2xyi)$, we see that the streamlines of the flow are $xy = c_2$.

(b) The complex function $\Omega(z) = Az$, $A > 0$, is a complex potential for a very simple but important type of flow. From

$$\Omega'(z) = A \quad \text{and} \quad \overline{\Omega'(z)} = A,$$

we see that $\Omega(z)$ is the complex potential of vector field $\mathbf{F}$ whose complex representation is $f(z) = A$. Because the speed $|f| = A$ is constant at every point, we say that the velocity field $\mathbf{F}(x, y) = A\mathbf{i}$ is a *uniform* flow. In other words, in a domain D such as the upper half-plane, a particle in the fluid moves with a constant speed. From $\Omega(z) = Az = Ax + iAy$, we see that a path of a moving particle, a streamline for the flow, is a horizontal line from the family defined by $y = c_2$. Notice that the boundary of the domain D, $y = 0$, is itself a streamline. See Figure 5.6.2. ❏

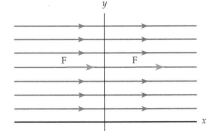

Figure 5.6.2 Uniform flow

Circulation and Net Flux Given a simple closed curve C oriented counterclockwise in the plane and a complex function f that represents the velocity field of a planar fluid flow, we can ask the following two questions:

(i) To what degree does the fluid tend to flow around the curve C?

(ii) What is the net difference between the rates at which fluid enters and leaves the region bounded by the curve C?

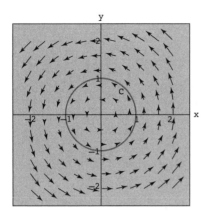

Figure 5.6.3 Positive circulation and zero net flux

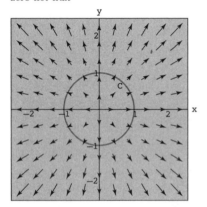

Figure 5.6.4 Zero circulation and positive net flux

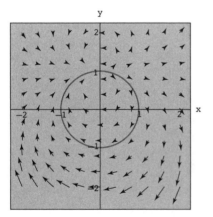

Figure 5.6.5 Velocity field for part (a) of Example 4

The quantities considered in questions (i) and (ii) are called the **circulation** around C and the **net flux** across C, respectively. A precise definition of circulation and flux depends on the use of a contour integral involving the complex representation f and will be given shortly. In the meantime, we can decide whether the circulation or net flux is positive, negative, or 0 by graphing the velocity vector field f of the flow. As with arguments of complex numbers, we consider the counterclockwise direction of a flow as the "positive" direction. Thus, a flow will have a positive circulation around C if the fluid tends to flow counterclockwise around C. Similarly, a negative circulation means the fluid tends to flow clockwise around C, and a 0 circulation means that the flow is perpendicular to C. For example, in Figure 5.6.3, the circulation is positive since the fluid tends to flow counterclockwise around C, whereas the circulation in Figure 5.6.4 is 0 since the flow is perpendicular to the curve C. In a similar manner, we consider a positive net flux to mean that fluid leaves the region bounded by the curve C at a greater rate than it enters. This indicates the presence of a **source** within C, that is, a point at which fluid is produced. Conversely, a negative net flux indicates that fluid enters the region bounded by C at a greater rate than it leaves, and this indicates the presence of a **sink** inside C, that is, a point at which fluid disappears. If the net flux is 0, then the fluid enters and leaves C at the same rate. In Figure 5.6.3, the flow is tangent to the circle C. Thus, no fluid crosses C, and this implies that the net flux across C is 0. On the other hand, in Figure 5.6.4, the net flux across C is positive because the flow appears to be only leaving the region bounded by C.

EXAMPLE 4   Circulation and Net Flux of a Flow

Let C be the unit circle $|z| = 1$ in the complex plane. For each flow f determine graphically whether the circulation around C is positive, negative, or 0. Also, determine whether the net flux across C is positive, negative, or 0: **(a)** $f(z) = (z - i)^2$ **(b)** $f(z) = 1/z$.

Solution In each part, we have used a computer algebra system to plot the velocity vector field f and the curve C.

(a) The velocity field $f(z) = (z - i)^2$ is given in Figure 5.6.5. Because the vector field f shows that the fluid flows clockwise about C, we conclude that the circulation is negative. Moreover, since it appears that no fluid crosses the curve C, the net flux is 0.

(b) The velocity field $f(z) = 1/z$ given in Figure 5.6.6 indicates that the fluid flows an equal amount counterclockwise as it does clockwise about C. This suggests that the circulation is 0. In addition, because the same amount of fluid appears to enter the region bounded by C as exits the region bounded by C, we also have that the net flux is 0.   ❑

Circulation and Net Flux Revisited Let $\mathbf{T}$ denote the unit tangent vector to a positively oriented, simple closed contour C. If $\mathbf{F}(x, y) = P(x, y)\mathbf{i} + Q(x, y)\mathbf{j}$ represents a velocity vector field of a two-dimensional fluid flow, we define the **circulation** of $\mathbf{F}$ along C as the real line integral $\oint_C \mathbf{F} \cdot d\mathbf{r}$, where $d\mathbf{r} = dx\mathbf{i} + dy\mathbf{j}$. Since $d\mathbf{r}/dt = (d\mathbf{r}/ds)(ds/dt)$, where

$d\mathbf{r}/ds = (dx/ds)\,\mathbf{i} + (dy/ds)\,\mathbf{j} = \mathbf{T}$ is a unit tangent to C, the line integral can be written in terms of the tangential component of the velocity vector $\mathbf{F}$; that is,

$$\text{circulation} = \oint_C \mathbf{F} \cdot \mathbf{T}\,ds. \tag{18}$$

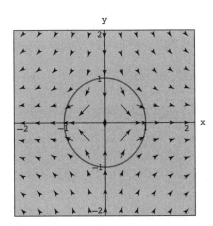

As we have already discussed, the circulation of $\mathbf{F}$ is a measure of the amount by which the fluid tends to turn the curve C by rotating, or circulating, around it. If $\mathbf{F}$ is perpendicular to $\mathbf{T}$ for every (x, y) on C, then $\oint_C \mathbf{F} \cdot \mathbf{T}\,ds = 0$ and the curve does not move at all. On the other hand, $\oint_C \mathbf{F} \cdot \mathbf{T}\,ds > 0$ and $\oint_C \mathbf{F} \cdot \mathbf{T}\,ds < 0$ mean that the fluid tends to rotate C in the counterclockwise and clockwise directions, respectively. See Figure 5.6.7.

Figure 5.6.6 Velocity field for part (b) of Example 4

Now if $\mathbf{N} = (dy/ds)\,\mathbf{i} - (dx/ds)\,\mathbf{j}$ denotes the normal unit vector to a positively oriented, simple closed contour C, and if $\mathbf{F}(x, y) = P(x, y)\mathbf{i} + Q(x, y)\mathbf{j}$ again represents a velocity field of a two-dimensional fluid flow, we define the **net flux** of $\mathbf{F}$ as the real line integral of the normal component of the velocity vector $\mathbf{F}$:

$$\text{net flux} = \oint_C \mathbf{F} \cdot \mathbf{N}\,ds. \tag{19}$$

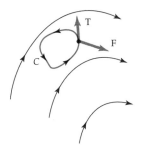

Specifically, (19) defines a *net rate* at which the fluid is crossing the curve C in the direction of the normal $\mathbf{N}$ and is measured in units of area per unit time. In other words, the net flux across C is the difference between the rate at which fluid enters and the rate at which fluid leaves the region bounded by C. A nonzero value of $\oint_C \mathbf{F} \cdot \mathbf{N}\,ds$ indicates the presence of sources or sinks for the fluid inside the curve C.

Figure 5.6.7 Flow of fluid tends to turn C.

Now if $f(z) = P(x, y) + iQ(x, y)$ is the complex representation of the velocity field $\mathbf{F}$ of a fluid, the line integrals (18) and (19) can be computed simultaneously by evaluating the single contour integral $\oint_C \overline{f(z)}\,dz$. To see how this is done, first observe:

$$\oint_C \mathbf{F} \cdot \mathbf{T}\,ds = \oint_C (P\,\mathbf{i} + Q\,\mathbf{j}) \cdot \left(\frac{dx}{ds}\mathbf{i} + \frac{dy}{ds}\mathbf{j}\right)\,ds = \oint_C P\,dx + Q\,dy$$

$$\oint_C \mathbf{F} \cdot \mathbf{N}\,ds = \oint_C (P\,\mathbf{i} + Q\,\mathbf{j}) \cdot \left(\frac{dy}{ds}\mathbf{i} - \frac{dx}{ds}\mathbf{j}\right)\,ds = \oint_C P\,dy - Q\,dx,$$

and then:

$$\oint_C \overline{f(z)}\,dz = \oint_C (P - iQ)\,(dx + i\,dy)$$

$$= \left(\oint_C P\,dx + Q\,dy\right) + i\left(\oint_C P\,dy - Q\,dx\right)$$

$$= \left(\oint_C \mathbf{F} \cdot \mathbf{T}\,ds\right) + i\left(\oint_C \mathbf{F} \cdot \mathbf{N}\,ds\right). \tag{20}$$

Equation (20) shows that (18) and (19) can be found by computing $\oint_C \overline{f(z)}\,dz$ and identifying the real and imaginary parts of the result. That is,

$$\text{circulation} = \text{Re}\left(\oint_C \overline{f(z)}\,dz\right) \quad \text{and} \quad \text{net flux} = \text{Im}\left(\oint_C \overline{f(z)}\,dz\right). \tag{21}$$

EXAMPLE 5   Circulation and Net Flux

Use (21) to compute the circulation and the net flux for the flow and curve C in part (a) of Example 4.

Solution From part (a) of Example 4, the flow is $f(z) = (z-i)^2$ and C is the circle $|z| = 1$. Then $\overline{f(z)} = \left(\overline{z-i}\right)^2 = (\bar{z}+i)^2 = \bar{z}^2 + 2i\bar{z} - 1$, and C is parametrized by $z(t) = e^{it}$, $0 \le t \le 2\pi$. Using $dz = ie^{it}dt$ and the integration method in (11) of Section 5.2, we have:

$$
\begin{aligned}
\oint_C \overline{f(z)}\,dz &= \int_0^{2\pi} \left(e^{-2it} + 2ie^{-it} - 1\right)ie^{it}dt \\
&= i\int_0^{2\pi} \left(e^{-it} + 2i - e^{it}\right)dt \\
&= \left[-e^{-it} - 2t - e^{it}\right]_0^{2\pi} = -4\pi + 0i.
\end{aligned}
\tag{22}
$$

In the last computation we used $e^{-2\pi i} = e^{2\pi i} = e^0 = 1$. Now, by comparing the results obtained in (22) to (21), we see that the circulation around C is -4π and the net flux across C is 0. The negative circulation and zero net flux are consistent with our geometric analysis in Figure 5.6.5 for the flow f in part (a) of Example 4.   ❏

The analysis of the flow f in part (b) in Example 4 is left as a exercise. See Problem 25 in Exercises 5.6.

EXAMPLE 6   Circulation and Net Flux

Suppose the velocity field of a fluid flow is $f(z) = (1+i)z$. Compute the circulation and net flux across C, where C is the unit circle $|z| = 1$.

Solution Since $\overline{f(z)} = (1-i)\bar{z}$ and $z(t) = e^{it}$, $0 \le t \le 2\pi$, we have from (21)

$$
\oint_C \overline{f(z)}\,dz = \int_0^{2\pi}(1-i)e^{-it}ie^{it}dt = (1+i)\int_0^{2\pi}dt = 2\pi + 2\pi i.
$$

Thus the circulation around C is 2π and net flux across C is also 2π.   ❏

EXAMPLE 7   Applying the Cauchy-Goursat Theorem

Suppose the velocity field of a fluid flow is $f(z) = \overline{\cos z}$. Compute the circulation and net flux across C, where C is the square with vertices $z = 1$, $z = i$, $z = -1$, and $z = -i$.

Solution We must compute $\oint_C \overline{f(z)}\,dz = \oint_C \overline{\cos z}\,dz = \oint_C \cos z\,dz$, and then take the real and imaginary parts of the integral to find the circulation and net flux, respectively. But since the function $\cos z$ is analytic everywhere, we immediately have $\oint_C \cos z\,dz = 0$ by the Cauchy-Goursat theorem. The circulation and net flux are therefore both 0.

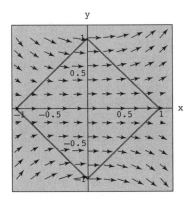

Figure 5.6.8 Velocity field for Example 7

The velocity field $f(z) = \overline{\cos z}$ and the contour C are shown in Figure 5.6.8. The results just obtained for the circulation and net flux are consistent with our earlier discussion in Example 4 about the geometry of flows. ❑

EXAMPLE 8 Circulation and Net Flux

The complex function $f(z) = k/(\overline{z} - \overline{z}_0)$ where $k = a + ib$ and z_0 are complex constants, gives rise to a flow in the domain defined by $z \neq z_0$. If C is a simple closed contour with z_0 in its interior, then the Cauchy integral formula, (1) of Section 5.5, gives

$$\oint_C \overline{f(z)}\, dz = \oint_C \frac{a - ib}{z - z_0}\, dz = 2\pi i(a - ib) = 2\pi b + 2\pi ai.$$

From (21) we see that the circulation around C is $\mathrm{Re}(2\pi b + 2\pi ai) = 2\pi b$ and the net flux across C is $\mathrm{Im}(2\pi b + 2\pi ai) = 2\pi a$. ❑

Note in Example 8, if z_0 were an exterior point of the region bounded by C, then it would follow from the Cauchy-Goursat theorem that both the circulation and the next flux are zero. Moreover, when the constant k is real ($a \neq 0$, $b = 0$), the circulation around C is 0, but the net flux across C is $2\pi k$. From our earlier discussion in this section, it follows that complex number z_0 is a source for the flow when $k > 0$ and a sink when $k < 0$. Velocity fields corresponding to these two cases are shown in Figure 5.6.9. The flow illustrated in Figure 5.6.4 is of the type shown in Figure 5.6.9(a).

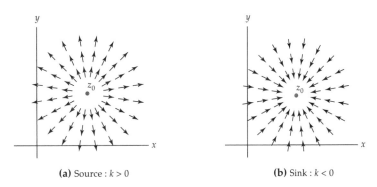

(a) Source : $k > 0$ **(b)** Sink : $k < 0$

Figure 5.6.9 Two normalized velocity fields

EXERCISES 5.6   *Answers to selected odd-numbered problems begin on page ANS-15.*

In Problems 1–4, for the given velocity field $\mathbf{F}(x, y)$, verify that div $\mathbf{F} = 0$ and curl $\mathbf{F} = \mathbf{0}$ in an appropriate domain D.

1. $\mathbf{F}(x, y) = (\cos\theta_0)\mathbf{i} + (\sin\theta_0)\mathbf{j}$, θ_0 a constant

2. $\mathbf{F}(x, y) = -y\mathbf{i} - x\mathbf{j}$

3. $\mathbf{F}(x, y) = 2x\mathbf{i} + (3 - 2y)\mathbf{j}$

4. $\mathbf{F}(x, y) = \dfrac{x}{x^2 + y^2}\mathbf{i} + \dfrac{y}{x^2 + y^2}\mathbf{j}$

In Problems 5–8, give the complex representation $f(z)$ of the velocity field $\mathbf{F}(x, y)$. Express the function $g(z) = \overline{f(z)}$ in terms of the symbol z and verify that $g(z)$ is an analytic function in an appropriate domain D.

5. $\mathbf{F}(x, y)$ in Problem 1

6. $\mathbf{F}(x, y)$ in Problem 2

7. $\mathbf{F}(x, y)$ in Problem 3

8. $\mathbf{F}(x, y)$ in Problem 4

In Problems 9–12, find the velocity field $\mathbf{F}(x, y)$ of the flow of an ideal fluid determined by the given analytic function $g(z)$.

9. $g(z) = (1 + i)z^2$

10. $g(z) = \sin z$

11. $g(z) = e^x \cos y + i e^x \sin y$

12. $g(z) = x^3 - 3xy^2 + i\left(3x^2 y - y^3\right)$

In Problems 13–16, find a complex velocity potential $\mathbf{\Omega}(z)$ of the complex representation $f(z)$ of the indicated velocity field $\mathbf{F}(x, y)$. Verify your answer using (17). Describe the equipotential lines and the streamlines.

13. $\mathbf{F}(x, y)$ in Problem 1

14. $\mathbf{F}(x, y)$ in Problem 2

15. $\mathbf{F}(x, y)$ in Problem 3

16. $\mathbf{F}(x, y)$ in Problem 4

In Problems 17 and 18, the given analytic function $\mathbf{\Omega}(z)$ is a complex velocity potential for the flow of an ideal fluid. Find the velocity field $\mathbf{F}(x, y)$ of the flow.

17. $\mathbf{\Omega}(z) = \frac{1}{3}iz^3$

18. $\mathbf{\Omega}(z) = \frac{1}{4}z^4 + z$

19. Show that

$$\mathbf{F}(x, y) = A\left[\left(1 - \frac{x^2 - y^2}{(x^2 + y^2)^2}\right)\mathbf{i} - \frac{2xy}{(x^2 + y^2)^2}\mathbf{j}\right], \ A > 0,$$

is a velocity field for an ideal fluid in any domain D not containing the origin.

20. Verify that the analytic function $\mathbf{\Omega}(z) = A\left(z + \dfrac{1}{z}\right)$ is a complex velocity potential for the flow whose velocity field $\mathbf{F}(x, y)$ is in Problem 19.

21. (a) Consider the velocity field in Problem 19. Describe the field $\mathbf{F}(x, y)$ at a point (x, y) far from the origin.

(b) For the complex velocity potential in Problem 20, how does the observation that $\mathbf{\Omega}(z) \to Az$ as $|z|$ increases verify your answer to part (a)?

22. A **stagnation point** in a fluid flow is a point at which the velocity field $\mathbf{F}(x, y) = \mathbf{0}$. Find the stagnation points for:

(a) the flow in Example 3(a).

(b) the flow in Problem 19.

23. For any two real numbers k and x_1, the function $\mathbf{\Omega}(z) = k\text{Ln}(z - x_1)$ is analytic in the upper half-plane and therefore is complex potential for the flow of an ideal fluid. The real number x_1 is a sink when $k < 0$ and a source for the flow when $k > 0$.

(a) Show that the streamlines are rays emanating from x_1.

(b) Show that the complex representation $f(z)$ of the velocity field $\mathbf{F}(x, y)$ of the flow is

$$f(z) = k\frac{z - x_1}{|z - x_1|^2}$$

and conclude that the flow is directed toward x_1 precisely when $k < 0$.

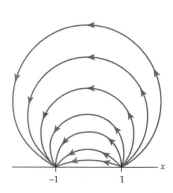

Figure 5.6.10 Figure for Problem 24

24. The complex potential $\mathbf{\Omega}(z) = k\text{Ln}(z - 1) - k\text{Ln}(z + 1)$, $k > 0$, determines the flow of an ideal fluid on the upper half-plane $y > 0$ with a single source at $z = 1$ and a single sink at $z = -1$. Show that the streamlines are the family of circles $x^2 + (y - c_2)^2 = 1 + c_2^2$. See Figure 5.6.10.

In Problems 25–30, compute the circulation and net flux for the given flow and the indicated closed contour C.

25. $f(z) = \dfrac{1}{z}$; where C is the circle $|z| = 1$

26. $f(z) = 2z$; where C is the circle $|z| = 1$

27. $f(z) = \dfrac{1}{z-1}$; where C is the circle $|z - 1| = 2$

28. $f(z) = \bar{z}$; where C is the square with vertices $z = 0,\ z = 1,\ z = 1+i,\ z = i$

29. $\mathbf{F}(x,\ y) = (4x + 3y)\mathbf{i} + (2x - y)\mathbf{j}$, where C is the circle $x^2 + y^2 = 4$

30. $\mathbf{F}(x,\ y) = (x + 2y)\mathbf{i} + (x - y)\mathbf{j}$, where C is the square with vertices $z = 0,\ z = 1+i,\ z = 2i,\ z = -1+i$

Focus on Concepts

31. Suppose $f(z) = P(x,\ y) + iQ(x,\ y)$ is a complex representation of a velocity field $\mathbf{F}$ of the flow of an ideal fluid on a simply connected domain D of the complex plane. Assume P and Q have continuous partial derivatives throughout D. If C is any simple closed curve C lying within D, show that the circulation around C and the net flux across C are zero.

32. The flow described by the velocity field $f(z) = (a + ib)/\bar{z}$ is said to have a **vortex** at $z = 0$. The geometric nature of the streamlines depends on the choice of a and b.

(a) Show that if $z(t) = x(t) + iy(t)$ is the path of a particle in the flow, then

$$\frac{dx}{dt} = \frac{ax - by}{x^2 + y^2}$$

$$\frac{dy}{dt} = \frac{bx + ay}{x^2 + y^2}.$$

(b) Rectangular and polar coordinates are related by $r^2 = x^2 + y^2$, $\tan\theta = y/x$.

Use these equations to show that

$$\frac{dr}{dt} = \frac{1}{r}\left(x\frac{dx}{dt} + y\frac{dy}{dt}\right),\ \frac{d\theta}{dt} = \frac{1}{r^2}\left(-y\frac{dx}{dt} + x\frac{dy}{dt}\right).$$

(c) Use the equations in parts (a) and (b) to establish that

$$\frac{dr}{dt} = \frac{a}{r},\ \frac{d\theta}{dt} = \frac{b}{r^2}.$$

(d) Use the equations in part (c) to conclude that the streamlines of the flow are logarithmic spirals $r = ce^{a\theta/b}$, $b \neq 0$. Use a graphing utility to verify that a particle traverses a path in a counterclockwise direction if and only if $a < 0$, and in a clockwise direction if and only if $b < 0$. Which of these directions corresponds to motion spiraling into the vortex?

CHAPTER 5 REVIEW QUIZ *Answers to selected odd-numbered problems begin on page ANS-15.*

In Problems 1–20, answer true or false. If the statement is false, justify your answer by either explaining why it is false or giving a counterexample; if the statement is true, justify your answer by either proving the statement or citing an appropriate result in this chapter.

1. If $z(t)$, $a \leq t \leq b$, is a parametrization of a contour C and $z(a) = z(b)$, then C is a simple closed contour.

2. The real line integral $\int_C (x^2 + y^2)\, dx + 2xy\, dy$, where C is given by $y = x^3$ from $(0,\,0)$ to $(1,\,1)$, has the same value on the curve $y = x^6$ from $(0,\,0)$ to $(1,\,1)$.

3. The sector defined by $-\pi/6 < \arg(z) < \pi/6$ is a simply connected domain.

4. If f is analytic at z_0, then f''' necessarily exists at z_0.

5. If f is analytic within and on a simple closed contour C and z_0 is any point within C, then the value of $f(z_0)$ is determined by the values of $f(z)$ on C.

6. If f is analytic on a simple closed contour C, then $\oint_C f(z)\, dz = 0$.

7. If f is continuous in a domain D and has an antiderivative F in D, then an integral $\int_C f(z)\, dz$ has the same value on all contours C in D between the initial point z_0 and terminal point z_1.

8. If $\oint_C f(z)\, dz = 0$ for every simple closed contour C, then f is analytic within and on C.

9. The value of $\displaystyle\int_C \frac{z-2}{z}\, dz$ is the same for any path C in the right half-plane $\mathrm{Re}(z) > 0$ between $z = 1 + i$ and $z = 10 + 8i$.

10. If g is entire, then $\displaystyle\oint_C \frac{g(z)}{z-i}\, dz = \oint_{C_1} \frac{g(z)}{z-i}\, dz$, where C is the circle $|z| = 3$ and C_1 is the ellipse $x^2 + \frac{1}{9}y^2 = 1$.

11. $\displaystyle\oint_C \frac{1}{(z-z_0)(z-z_1)}\, dz = 0$ for every simple closed contour C that encloses the points z_0 and z_1.

12. If f is analytic within and on the simple closed contour C and z_0 is a point within C, then $\displaystyle\oint_C \frac{f'(z)}{z-z_0}\, dz = \oint_C \frac{f(z)}{(z-z_0)^2}\, dz$.

13. $\oint_C \mathrm{Re}\,(z)\, dz$ is independent of the path C between $z_0 = 0$ and $z_1 = 1 + i$.

14. $\int_C \left(4z^3 - 2z + 1\right) dz = \int_{-2}^{2} \left(4x^3 - 2x + 1\right) dx$, where the contour C is comprised of segments C_1 and C_2 shown in Figure 5.R.1.

15. $\int_{C_1} z^n\, dz = \int_{C_2} z^n\, dz$ for all integers n, where C_1 is $z(t) = e^{it}$, $0 \leq t \leq 2\pi$ and C_2 is $z(t) = Re^{it}$, $R > 1$, $0 \leq t \leq 2\pi$.

16. If f is continuous on the contour C, then $\int_C f(z)\, dz + \int_{-C} f(z)\, dz = 0$.

17. On any contour C with initial point $z_0 = -i$ and terminal point $z_1 = i$ that lies in a simply connected domain D not containing the origin or the negative real axis, $\displaystyle\int_{-i}^{i} \frac{1}{z}\, dz = \mathrm{Ln}(i) - \mathrm{Ln}(-i) = \pi i$.

18. $\displaystyle\oint_C \frac{1}{z^2 + 1}\, dz = 0$, where C is the ellipse $x^2 + \frac{1}{4}y^2 = 1$.

19. If $p(z)$ is a polynomial in z then the function $f(z) = 1/p(z)$ can be never be an entire function.

20. The function $f(z) = \cos z$ is entire and not a constant and so must be unbounded.

In Problems 21–40, try to fill in the blanks without referring back to the text.

21. $z(t) = e^{it^2}$, $0 \leq t \leq \sqrt{2\pi}$, is a parametrization for a _____ .

22. $z(t) = z_0 + e^{it}$, $0 \leq t \leq 2\pi$, is a parametrization for a _____ .

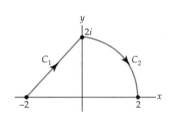

Figure 5.R.1 Figure for Problem 14

23. The difference between $z_1(t) = e^{it}$, $0 \le t \le 2\pi$ and $z_2(t) = e^{i(2\pi - t)}$, $0 \le t \le 2\pi$ is _____ .

24. $\oint_C (2y + x - 6ix^2)\, dz =$ _____ , where C is the triangle with vertices 0, i, $1 + i$, traversed counterclockwise.

25. If f is a polynomial function and C is a simple closed contour, then $\oint_C f(z)\, dz =$ _____ .

26. $\int_C z\, \text{Im}(z)\, dz =$ _____ , where C is given by $z(t) = 2t + t^2 i$, $0 \le t \le 1$.

27. $\int_C |z|^2\, dz =$ _____ , where C is the line segment for $1 - i$ to $1 + i$.

28. $\oint_C (\bar{z})^n\, dz =$ _____ , where n is an integer and C is $z(t) = e^{it}$, $0 \le t \le 2\pi$.

29. $\int_C \sin\frac{z}{2}\, dz =$ _____ , where C is given by $z(t) = 2i + 4e^{it}$, $0 \le t \le \pi/2$.

30. $\oint_C \sec z\, dz =$ _____ , where C is $|z| = 1$.

31. $\oint_C \frac{1}{z(z-1)}\, dz =$ _____ , where C is $|z - 1| = \frac{1}{2}$.

32. If $f(z) = \oint_C \frac{\xi^2 + 6\xi - 2}{\xi - z}\, d\xi$, where C is $|z| = 3$, then $f(1 + i) =$ _____ .

33. If $f(z) = z^3 + e^z$ and C is a contour $z = 8e^{it}$, $0 \le t \le 2\pi$, then $\oint_C \frac{f(z)}{(z + \pi i)^3}\, dz =$ _____ .

34. If $|f(z)| \le 2$ on the circle $|z| = 3$, then $\left| \oint_C f(z)\, dz \right| \le$ _____ .

35. If n is a positive integer and C is the contour $|z| = 2$, then $\oint_C z^{-n} e^z\, dz =$ _____ .

36. On $|z| = 1$, the contour integral $\oint_C \frac{\cos z}{z^n}\, dz$ equals _____ for $n = 1$, equals _____ for $n = 2$, and equals _____ for $n = 3$.

37. $\oint_C z^n\, dz = \begin{cases} 0, & \text{if } n \text{\underline{\hspace{1cm}}} \\ 2\pi i, & \text{if } n \text{\underline{\hspace{1cm}}} \end{cases}$, where n is an integer and C is the circle $|z| = 1$.

38. The value of the integral $\oint_C \frac{z}{z + i}\, dz$ on the contour C shown in Figure 5.R.2 is _____ .

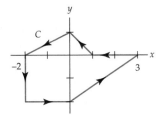

Figure 5.R.2 Figure for Problem 38

39. The value of the integral $\displaystyle\int_C (2z+1)\,dz$ on the contour C, comprised of line segments $C_1, C_2, \ldots, C_{11}$ shown in Figure 5.R.3 is _____ .

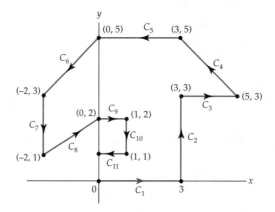

Figure 5.R.3 Figure for Problem 39

40. The value of the integral $\displaystyle\oint_C \frac{e^z}{z^2(z-\pi i)}\,dz$ on the closed contour C shown in Figure 5.R.4 is _____ .

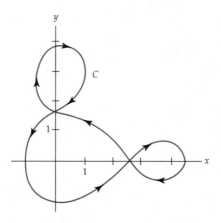

Figure 5.R.4 Figure for Problem 40

Series and Residues

Chapter Outline

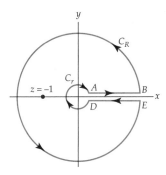

Special contour used in evaluating a real integral. See page 326.

Introduction   Cauchy's integral formula for derivatives indicates that if a function f is analytic at a point z_0, then it possesses derivatives of all orders at that point. As a consequence of this result we shall see that f can always be expanded in a power series centered at that point. On the other hand, if f fails to be analytic at z_0, we may still be able to expand it in a different kind of series known as a *Laurent series*. The notion of Laurent series leads to the concept of a *residue*, and this, in turn, leads to yet another way of evaluating complex and, in some instances, real integrals.

6.1   Sequences and Series

Much of the theory of complex sequences and series is analogous to that encountered in real calculus. In this section we explore the definitions of convergence and divergence for complex sequences and complex infinite series. In addition, we give some tests for convergence of infinite series. You are urged to pay particular attention to what is said about *geometric series* since this type of series will be important in the later sections of this chapter.

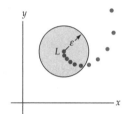

Figure 6.1.1 If $\{z_n\}$ converges to L, all but a finite number of terms are in every ε-neighborhood of L.

Sequences   A complex **sequence** $\{z_n\}$ is a function whose domain is the set of positive integers and whose range is a subset of the complex numbers **C**. In other words, to each integer $n = 1, 2, 3, \ldots$ we assign a single complex number z_n. For example, the sequence $\{1 + i^n\}$ is

$$1 + i, \quad 0, \quad 1 - i, \quad 2, \quad 1 + i, \ldots .$$
$$\uparrow \qquad \uparrow \qquad \uparrow \qquad \uparrow \qquad \uparrow \qquad\qquad (1)$$
$$n = 1, \ n = 2, \ n = 3, \ n = 4, \ n = 5, \ \ldots$$

If $\lim\limits_{n \to \infty} z_n = L$, then we say the sequence $\{z_n\}$ is **convergent.** In other words, $\{z_n\}$ converges to the number L if for each positive real number ε a positive integer N can be found such that $|z_n - L| < \varepsilon$ whenever $n > N$. Since $|z_n - L|$ is distance, the terms z_n of a sequence that converges to L can be made arbitrarily close to L. Put another way, when a sequence $\{z_n\}$ converges to L, then all but a finite number of the terms of the sequence are within every ε-neighborhood of L. See Figure 6.1.1. A sequence that is not convergent is said to be **divergent**.

The sequence $\{1 + i^n\}$ illustrated in (1) is divergent since the general term $z_n = 1 + i^n$ does not approach a fixed complex number as $n \to \infty$. Indeed, you should verify that the first four terms of this sequence repeat endlessly as n increases.

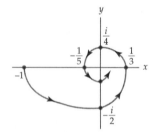

Figure 6.1.2 The terms of the sequence $\{i^{n+1}/n\}$ spiral in toward 0.

EXAMPLE 1   A Convergent Sequence

The sequence $\left\{ \dfrac{i^{n+1}}{n} \right\}$ converges since $\lim\limits_{n \to \infty} \dfrac{i^{n+1}}{n} = 0$. As we see from

$$-1, -\frac{i}{2}, \frac{1}{3}, \frac{i}{4}, -\frac{1}{5}, \cdots ,$$

and Figure 6.1.2, the terms of the sequence, marked by colored dots in the figure, spiral in toward the point $z = 0$ as n increases.   ❑

The following theorem for sequences is the analogue of Theorem 2.6.1 in Section 2.6.

Theorem 6.1.1   Criterion for Convergence

A sequence $\{z_n\}$ converges to a complex number $L = a + ib$ if and only if $\text{Re}(z_n)$ converges to $\text{Re}(L) = a$ and $\text{Im}(z_n)$ converges to $\text{Im}(L) = b$.

EXAMPLE 2 Illustrating Theorem 6.1.1

Consider the sequence $\left\{ \dfrac{3 + ni}{n + 2ni} \right\}$. From

$$z_n = \frac{3 + ni}{n + 2ni} = \frac{(3 + ni)(n - 2ni)}{n^2 + 4n^2} = \frac{2n^2 + 3n}{5n^2} + i\frac{n^2 - 6n}{5n^2},$$

we see that

$$\mathrm{Re}(z_n) = \frac{2n^2 + 3n}{5n^2} = \frac{2}{5} + \frac{3}{5n} \to \frac{2}{5}$$

and

$$\mathrm{Im}(z_n) = \frac{n^2 - 6n}{5n^2} = \frac{1}{5} - \frac{6}{5n} \to \frac{1}{5}$$

as $n \to \infty$. From Theorem 6.1.1, the last results are sufficient for us to conclude that the given sequence converges to $a + ib = \frac{2}{5} + \frac{1}{5}i$. ❑

Series An **infinite series** or **series** of complex numbers

$$\sum_{k=1}^{\infty} z_k = z_1 + z_2 + z_3 + \cdots + z_n + \cdots$$

is **convergent** if the sequence of partial sums $\{S_n\}$, where

$$S_n = z_1 + z_2 + z_3 + \cdots + z_n$$

converges. If $S_n \to L$ as $n \to \infty$, we say that the series converges to L or that the **sum** of the series is L.

Geometric Series A **geometric series** is any series of the form

$$\sum_{k=1}^{\infty} az^{k-1} = a + az + az^2 + \cdots + az^{n-1} + \cdots. \tag{2}$$

For (2), the nth term of the sequence of partial sums is

$$S_n = a + az + az^2 + \cdots + az^{n-1}. \tag{3}$$

When an infinite series is a geometric series, it is always possible to find a formula for S_n. To demonstrate why this is so, we multiply S_n in (3) by z,

$$zS_n = az + az^2 + az^3 + \cdots + az^n,$$

and subtract this result from S_n. All terms cancel except the first term in S_n and the last term in zS_n:

$$
\begin{aligned}
S_n - zS_n &= \left(a + az + az^2 + \cdots + az^{n-1} \right) \\
&\quad - \left(az + az^2 + az^3 + \cdots + az^{n-1} + az^n \right) \\
&= a - az^n
\end{aligned}
$$

or $(1-z)S_n = a(1-z^n)$. Solving the last equation for S_n gives us

$$S_n = \frac{a(1-z^n)}{1-z}. \tag{4}$$

Now $z^n \to 0$ as $n \to \infty$ whenever $|z| < 1$, and so $S_n \to a/(1-z)$. In other words, for $|z| < 1$ the sum of a geometric series (2) is $a/(1-z)$:

$$\frac{a}{1-z} = a + az + az^2 + \cdots + az^{n-1} + \cdots. \tag{5}$$

A geometric series (2) diverges when $|z| \geq 1$.

Special Geometric Series   We next present several immediate deductions from (4) and (5) that will be particularly helpful in the two sections that follow. If we set $a = 1$, the equality in (5) is

You should remember (6) and (7).   ➡

$$\frac{1}{1-z} = 1 + z + z^2 + z^3 + \cdots. \tag{6}$$

If we then replace the symbol z by $-z$ in (6), we get a similar result

$$\frac{1}{1+z} = 1 - z + z^2 - z^3 + \cdots. \tag{7}$$

Like (5), the equality in (7) is valid for $|z| < 1$ since $|-z| = |z|$. Now with $a = 1$, (4) gives us the sum of the first n terms of the series in (6):

$$\frac{1-z^n}{1-z} = 1 + z + z^2 + z^3 + \cdots + z^{n-1}.$$

If we rewrite the left side of the above equation as $\dfrac{1-z^n}{1-z} = \dfrac{1}{1-z} + \dfrac{-z^n}{1-z}$, we obtain an alternative form

$$\frac{1}{1-z} = 1 + z + z^2 + z^3 + \cdots + z^{n-1} + \frac{z^n}{1-z} \tag{8}$$

that will be put to use in proving the two principal theorems of this chapter.

EXAMPLE 3   Convergent Geometric Series

The infinite series

$$\sum_{k=1}^{\infty} \frac{(1+2i)^k}{5^k} = \frac{1+2i}{5} + \frac{(1+2i)^2}{5^2} + \frac{(1+2i)^3}{5^3} + \cdots$$

is a geometric series. It has the form given in (2) with $a = \frac{1}{5}(1+2i)$ and $z = \frac{1}{5}(1+2i)$. Since $|z| = \sqrt{5}/5 < 1$, the series is convergent and its sum is given by (5):

$$\sum_{k=1}^{\infty} \frac{(1+2i)^k}{5^k} = \frac{\dfrac{1+2i}{5}}{1 - \dfrac{1+2i}{5}} = \frac{1+2i}{4-2i} = \frac{1}{2}i.$$

❑

We turn now to some important theorems about convergence and divergence of an infinite series. You should have seen similar theorems in a course in elementary calculus.

Theorem 6.1.2   A Necessary Condition for Convergence

If $\sum_{k=1}^{\infty} z_k$ converges, then $\lim_{n \to \infty} z_n = 0$.

Proof Let L denote the sum of the series. Then $S_n \to L$ and $S_{n-1} \to L$ as $n \to \infty$. By taking the limit of both sides of $S_n - S_{n-1} = z_n$ as $n \to \infty$ we obtain the desired conclusion.   ❏

A Test for Divergence The contrapositive* of the proposition in Theorem 6.1.2 is the familiar nth term test for divergence of an infinite series.

Theorem 6.1.3   The nth Term Test for Divergence

If $\lim_{n \to \infty} z_n \neq 0$, then $\sum_{k=1}^{\infty} z_k$ diverges.

For example, the series $\sum_{k=1}^{\infty} (ik + 5)/k$ diverges since $z_n = (in + 5)/n \to i \neq 0$ as $n \to \infty$. The geometric series (2) diverges if $|z| \geq 1$ because even in the case when $\lim_{n \to \infty} |z^n|$ exists, the limit is not zero.

Definition 6.1.1   Absolute and Conditional Convergence

An infinite series $\sum_{k=1}^{\infty} z_k$ is said to be **absolutely convergent** if $\sum_{k=1}^{\infty} |z_k|$ converges. An infinite series $\sum_{k=1}^{\infty} z_k$ is said to be **conditionally convergent** if it converges but $\sum_{k=1}^{\infty} |z_k|$ diverges.

In elementary calculus a real series of the form $\sum_{k=1}^{\infty} \dfrac{1}{k^p}$ is called a **p-series** and converges for $p > 1$ and diverges for $p \leq 1$. We use this well-known result in the next example.

EXAMPLE 4   Absolute Convergence

The series $\sum_{k=1}^{\infty} \dfrac{i^k}{k^2}$ is absolutely convergent since the series $\sum_{k=1}^{\infty} \left| \dfrac{i^k}{k^2} \right|$ is the same as the real convergent p-series $\sum_{k=1}^{\infty} \dfrac{1}{k^2}$. Here we identify $p = 2 > 1$.   ❏

As in real calculus:

Absolute convergence implies convergence.

---

*See the footnote on page 139.

See Problem 47 in Exercises 6.1. We are able to conclude that the series in Example 4,

$$\sum_{k=1}^{\infty} \frac{i^k}{k^2} = i - \frac{1}{2^2} - \frac{i}{3^2} + \cdots$$

converges because it is was shown to be absolutely convergent.

Tests for Convergence Two of the most frequently used tests for convergence of infinite series are given in the next theorems.

Theorem 6.1.4   Ratio Test

Suppose $\sum_{k=1}^{\infty} z_k$ is a series of nonzero complex terms such that

$$\lim_{n \to \infty} \left| \frac{z_{n+1}}{z_n} \right| = L. \tag{9}$$

(*i*) If $L < 1$, then the series converges absolutely.

(*ii*) If $L > 1$ or $L = \infty$, then the series diverges.

(*iii*) If $L = 1$, the test is inconclusive.

Theorem 6.1.5   Root Test

Suppose $\sum_{k=1}^{\infty} z_k$ is a series of complex terms such that

$$\lim_{n \to \infty} \sqrt[n]{|z_n|} = L. \tag{10}$$

(*i*) If $L < 1$, then the series converges absolutely.

(*ii*) If $L > 1$ or $L = \infty$, then the series diverges.

(*iii*) If $L = 1$, the test is inconclusive.

Proofs of the ratio and root tests are similar to those of their real counterparts. See Problem 48 in Exercises 6.1. We are interested primarily in applying the tests in Theorems 6.1.4 and 6.1.5 to power series.

Power Series The notion of a power series is important in the study of analytic functions. An infinite series of the form

$$\sum_{k=0}^{\infty} a_k (z - z_0)^k = a_0 + a_1(z - z_0) + a_2(z - z_0)^2 + \cdots, \tag{11}$$

where the coefficients a_k are complex constants, is called a **power series** in $z - z_0$. The power series (11) is said to be **centered at z_0**; the complex point z_0 is referred to as the **center** of the series. In (11) it is also convenient to define $(z - z_0)^0 = 1$ even when $z = z_0$.

Circle of Convergence Every complex power series (11) has a **radius of convergence**. Analogous to the concept of an interval of convergence for real power series, a complex power series (11) has a **circle of convergence**, which is the circle centered at z_0 of largest radius $R > 0$ for which (11) converges at every point *within* the circle $|z - z_0| = R$. A power series converges absolutely at all points z within its circle of convergence, that is, for all z satisfying $|z - z_0| < R$, and diverges at all points z exterior to the circle, that is, for all z satisfying $|z - z_0| > R$. The radius of convergence can be:

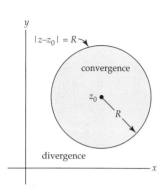

y

$|z-z_0| = R$

convergence

z_0

R

divergence

x

Figure 6.1.3 No general statement concerning convergence at points on the circle $|z - z_0| = R$ can be made.

(i) $R = 0$ (in which case (11) converges only at its center $z = z_0$),

(ii) R a finite positive number (in which case (11) converges at all interior points of the circle $|z - z_0| = R$), or

(iii) $R = \infty$ (in which case (11) converges for all z).

A power series may converge at some, all, or at none of the points *on* the actual circle of convergence. See Figure 6.1.3 and the next example.

EXAMPLE 5 Circle of Convergence

Consider the power series $\sum\limits_{k=1}^{\infty} \dfrac{z^{k+1}}{k}$. By the ratio test (9),

$$\lim_{n \to \infty} \left| \frac{\dfrac{z^{n+2}}{n+1}}{\dfrac{z^{n+1}}{n}} \right| = \lim_{n \to \infty} \frac{n}{n+1} |z| = |z|.$$

Thus the series converges absolutely for $|z| < 1$. The circle of convergence is $|z| = 1$ and the radius of convergence is $R = 1$. Note that on the circle of convergence $|z| = 1$, the series does not converge absolutely since $\sum_{k=1}^{\infty} \dfrac{1}{k}$ is the well-known divergent harmonic series. Bear in mind *this does not say* that the series diverges on the circle of convergence. In fact, at $z = -1$, $\sum_{k=1}^{\infty} \dfrac{(-1)^{k+1}}{k}$ is the convergent alternating harmonic series. Indeed, it can be shown that the series converges at all points on the circle $|z| = 1$ *except* at $z = 1$. ❑

It should be clear from Theorem 6.1.4 and Example 5 that for a power series $\sum_{k=0}^{\infty} a_k (z - z_0)^k$, the limit (9) depends only on the coefficients a_k. Thus, if

$$(i) \quad \lim_{n \to \infty} \left| \frac{a_{n+1}}{a_n} \right| = L \neq 0, \text{ the radius of convergence is } R = \frac{1}{L}; \qquad (12)$$

$$(ii) \quad \lim_{n \to \infty} \left| \frac{a_{n+1}}{a_n} \right| = 0, \text{ the radius of convergence is } R = \infty; \qquad (13)$$

$$(iii) \quad \lim_{n \to \infty} \left| \frac{a_{n+1}}{a_n} \right| = \infty, \text{ the radius of convergence is } R = 0. \qquad (14)$$

Similar conclusions can be made for the root test (10) by utilizing

$$\lim_{n \to \infty} \sqrt[n]{|a_n|}. \qquad (15)$$

For example, if $\lim_{n\to\infty} \sqrt[n]{|a_n|} = L \neq 0$, then $R = 1/L$.

EXAMPLE 6 Radius of Convergence

Consider the power series $\displaystyle\sum_{k=1}^{\infty} \frac{(-1)^{k+1}}{k!}(z-1-i)^k$. With the identification $a_n = (-1)^{n+1}/n!$ we have

$$\lim_{n\to\infty} \left| \frac{\dfrac{(-1)^{n+2}}{(n+1)!}}{\dfrac{(-1)^{n+1}}{n!}} \right| = \lim_{n\to\infty} \frac{1}{n+1} = 0.$$

Hence by (13) the radius of convergence is ∞; the power series with center $z_0 = 1 + i$ converges absolutely for all z, that is, for $|z - 1 - i| < \infty$.   ❑

EXAMPLE 7 Radius of Convergence

Consider the power series $\displaystyle\sum_{k=1}^{\infty} \left(\frac{6k+1}{2k+5}\right)^k (z-2i)^k$. With $a_n = \left(\dfrac{6n+1}{2n+5}\right)^n$, the root test in the form (15) gives

$$\lim_{n\to\infty} \sqrt[n]{|a_n|} = \lim_{n\to\infty} \frac{6n+1}{2n+5} = 3.$$

By reasoning similar to that leading to (12), we conclude that the radius of convergence of the series is $R = \frac{1}{3}$. The circle of convergence is $|z - 2i| = \frac{1}{3}$; the power series converges absolutely for $|z - 2i| < \frac{1}{3}$.   ❑

The Arithmetic of Power Series On occasion it may be to our advantage to perform certain *arithmetic operations* on one or more power series. Although it would take us too far afield to delve into properties of power series in a formal manner (stating and proving theorems), it will be helpful at points in this chapter to know what we can (or cannot) do to power series. So here are some facts.

- A power series $\sum_{k=0}^{\infty} a_k(z-z_0)^k$ can be multiplied by a nonzero complex constant c without affecting its convergence or divergence.

- A power series $\sum_{k=0}^{\infty} a_k(z-z_0)^k$ converges absolutely within its circle of convergence. As a consequence, within the circle of convergence the terms of the series can be rearranged and the rearranged series has the same sum L as the original series.

- Two power series $\sum_{k=0}^{\infty} a_k(z-z_0)^k$ and $\sum_{k=0}^{\infty} b_k(z-z_0)^k$ can be added and subtracted by adding or subtracting like terms. In symbols:

$$\sum_{k=0}^{\infty} a_k(z-z_0)^k \pm \sum_{k=0}^{\infty} b_k(z-z_0)^k = \sum_{k=0}^{\infty} (a_k \pm b_k)(z-z_0)^k.$$

If both series have the same nonzero radius R of convergence, the radius of convergence of $\sum_{k=0}^{\infty}(a_k \pm b_k)(z - z_0)^k$ is R. It should make intuitive sense that if one series has radius of convergence $r > 0$ and the other has radius of convergence $R > 0$, where $r \neq R$, then the radius of convergence of $\sum_{k=0}^{\infty}(a_k \pm b_k)(z - z_0)^k$ is the smaller of the two numbers r and R.

- Two power series can (with care) be multiplied and divided.

Remarks

(i) If $z_n = a_n + ib_n$ then the nth term of the sequence of partial sums for $\sum_{k=1}^{\infty} z_k$ can be written $S_n = \sum_{k=1}^{n}(a_k + ib_k) = \sum_{k=1}^{n} a_k + i\sum_{k=1}^{n} b_k$. Analogous to Theorem 6.1.2, $\sum_{k=1}^{\infty} z_k$ converges to a number $L = a + ib$ if and only if $\operatorname{Re}(S_n) = \sum_{k=1}^{n} a_k$ converges to a and $\operatorname{Im}(S_n) = \sum_{k=1}^{n} b_k$ converges to b. See Problem 35 in Exercises 6.1.

(ii) When written in terms of summation notation, a geometric series may not be immediately recognizable as equivalent to (2). In summation notation a geometric series need not start at $k = 1$ nor does the general term have to appear precisely as az^{k-1}. At first glance $\sum_{k=3}^{\infty} 40\dfrac{i^{k+2}}{2^{k-1}}$ does not appear to match the general form $\sum_{k=1}^{\infty} az^{k-1}$ of a geometric series. However, by writing out three terms,

$$\sum_{k=3}^{\infty} 40\frac{i^{k+2}}{2^{k-1}} = \overbrace{40\frac{i^5}{2^2}}^{a} + \overbrace{40\frac{i^6}{2^3}}^{az} + \overbrace{40\frac{i^7}{2^4}}^{az^2} + \cdots$$

we are able make the identifications $a = 40\left(i^5/2^2\right)$ and $z = i/2$ on the right-hand side of the equality. Since $|z| = \frac{1}{2} < 1$ the sum of the series is given by (5):

$$\sum_{k=3}^{\infty} 40\frac{i^{k+2}}{2^{k-1}} = \frac{40\dfrac{i^5}{2^2}}{1 - \dfrac{i}{2}} = -4 + 8i.$$

(iii) Although we have not proved it, it bears repeating: A power series $\sum_{k=0}^{\infty} a_k(z - z_0)^k$, $z \neq z_0$, *always* possesses a radius of convergence R that is either positive, 0, or ∞. We have seen in the discussion prior to Example 6 that the ratio and root tests lead to

$$\frac{1}{R} = \lim_{n \to \infty}\left|\frac{a_{n+1}}{a_n}\right| \quad \text{and} \quad \frac{1}{R} = \lim_{n \to \infty} \sqrt[n]{|a_n|}$$

assuming the appropriate limit exists. Since these formulas depend only on the coefficients, it is easy to make up examples where neither $\lim_{n \to \infty}|a_{n+1}/a_n|$ nor $\lim_{n \to \infty}|a_n|^{1/n}$ exist. What is R if neither of these limits exist? See Problems 45 and 46 in Exercises 6.1.

EXERCISES 6.1   *Answers to selected odd-numbered problems begin on page ANS-16.*

In Problems 1–4, write out the first five terms of the given sequence.

1. $\{5i^n\}$

2. $\{2 + (-i)^n\}$

3. $\{1 + e^{n\pi i}\}$

4. $\{(1+i)^n\}$ [*Hint:* Write in polar form.]

In Problems 5–10, determine whether the given sequence converges or diverges.

5. $\left\{\dfrac{3ni + 2}{n + ni}\right\}$

6. $\left\{\dfrac{ni + 2^n}{3ni + 5^n}\right\}$

7. $\left\{\dfrac{(ni+2)^2}{n^2 i}\right\}$

8. $\left\{\dfrac{n(1+i^n)}{n+1}\right\}$

9. $\left\{\dfrac{n + i^n}{\sqrt{n}}\right\}$

10. $\left\{e^{1/n} + 2\,(\tan^{-1} n)i\right\}$

In Problems 11 and 12, show that the given sequence $\{z_n\}$ converges to a complex number L by computing $\lim_{n\to\infty} \mathrm{Re}(z_n)$ and $\lim_{n\to\infty} \mathrm{Im}(z_n)$.

11. $\left\{\dfrac{4n + 3ni}{2n + i}\right\}$

12. $\left\{\left(\dfrac{1+i}{4}\right)^n\right\}$

In Problems 13 and 14, use the sequence of partial sums to show that the given series is convergent.

13. $\displaystyle\sum_{k=1}^{\infty}\left[\dfrac{1}{k + 2i} - \dfrac{1}{k+1+2i}\right]$

14. $\displaystyle\sum_{k=1}^{\infty}\dfrac{i}{k(k+1)}$

In Problems 15–20, determine whether the given geometric series is convergent or divergent. If convergent, find its sum.

15. $\displaystyle\sum_{k=0}^{\infty}(1-i)^k$

16. $\displaystyle\sum_{k=1}^{\infty}4i\left(\tfrac{1}{3}\right)^{k-1}$

17. $\displaystyle\sum_{k=1}^{\infty}\left(\dfrac{i}{2}\right)^k$

18. $\displaystyle\sum_{k=0}^{\infty}\tfrac{1}{2}i^k$

19. $\displaystyle\sum_{k=0}^{\infty}3\left(\dfrac{2}{1+2i}\right)^k$

20. $\displaystyle\sum_{k=2}^{\infty}\dfrac{i^k}{(1+i)^{k-1}}$

In Problems 21–30, find the circle and radius of convergence of the given power series.

21. $\displaystyle\sum_{k=0}^{\infty}\dfrac{1}{(1-2i)^{k+1}}(z-2i)^k$

22. $\displaystyle\sum_{k=1}^{\infty}\dfrac{1}{k}\left(\dfrac{i}{1+i}\right)z^k$

23. $\displaystyle\sum_{k=1}^{\infty}\dfrac{(-1)^k}{k2^k}(z-1-i)^k$

24. $\displaystyle\sum_{k=1}^{\infty}\dfrac{1}{k^2(3+4i)^k}(z+3i)^k$

25. $\displaystyle\sum_{k=0}^{\infty}(1+3i)^k(z-i)^k$

26. $\displaystyle\sum_{k=1}^{\infty}\dfrac{z^k}{k^k}$

27. $\displaystyle\sum_{k=0}^{\infty}\dfrac{(z-4-3i)^k}{5^{2k}}$

28. $\displaystyle\sum_{k=0}^{\infty}(-1)^k\left(\dfrac{1+2i}{2}\right)^k(z+2i)^k$

29. $\displaystyle\sum_{k=0}^{\infty}\dfrac{(2k)!}{(k+2)(k!)^2}(z-i)^{2k}$

30. $\displaystyle\sum_{k=0}^{\infty}\dfrac{k!}{(2k)^k}z^{3k}$

31. Show that the power series $\displaystyle\sum_{k=1}^{\infty}\dfrac{(z-i)^k}{k2^k}$ is not absolutely convergent on its circle of convergence. Determine at least one point on the circle of convergence at which the power series is convergent.

32. Show that the power series $\displaystyle\sum_{k=1}^{\infty}\dfrac{z^k}{k^2}$ converges at every point on its circle of convergence.

33. Reread Theorem 6.1.3. What conclusion can be drawn, if any, when $\lim_{n\to\infty}|z_n| \neq 0$?

34. Show that the power series $\sum_{k=1}^{\infty} kz^k$ diverges at every point on its circle of convergence.

Focus on Concepts

35. By considering the series $\sum_{k=0}^{\infty} r^k e^{ik\theta}$, $0 < r < 1$, show that

$$\sum_{k=0}^{\infty} r^k \cos k\theta = \frac{1 - r\cos\theta}{1 - 2r\cos\theta + r^2} \quad \text{and} \quad \sum_{k=0}^{\infty} r^k \sin k\theta = \frac{r\sin\theta}{1 - 2r\cos\theta + r^2}.$$

36. Suppose $\{z_n + w_n\}$ converges. Discuss: Does it follow that at least one of the sequences $\{z_n\}$ or $\{w_n\}$ converges?

37. A sequence $\{z_n\}$ is said to be **bounded** if the set S consisting of its terms is a bounded set (see Section 1.5). The sequence in Example 1 is bounded.

   (a) Prove that the sequence in Example 2 is bounded.

   (b) Give another example of a sequence consisting of complex terms that is bounded.

   (c) Give an example of a sequence consisting of complex terms that is unbounded.

38. Discuss: Is every convergent sequence $\{z_n\}$ bounded? (See Problem 37.)

   Is every bounded sequence convergent? Defend your answers with sound mathematics.

39. Does the sequence $\left\{ i^{1/n} \right\}$, where $i^{1/n}$ denotes the principal nth root of i, converge?

40. We saw that the equality in (6) was valid for $|z| < 1$. Show that

$$\frac{1}{1-z} = -z^{-1} - z^{-2} - z^{-3} - \cdots$$

   and give the values of z for which the equality is valid.

41. Consider (6) with the symbol z replaced by e^{iz}:

$$\frac{1}{1 - e^{iz}} = e^{iz} + e^{2iz} + e^{3iz} + \cdots .$$

   Give the region in the complex plane for which the foregoing series converges.

42. Sketch the region in the complex plane for which $\sum_{k=0}^{\infty} \left(\frac{z-1}{z+2} \right)^k$ converges.

43. Consider the power series $\sum_{k=0}^{\infty} a_k(z - 1 + 2i)^k$. Discuss: Can the series converge at $-3 + i$ and diverge at $5 - 3i$?

44. Use a sketch in the complex plane that illustrates the validity of each of the following theorems:

   (i) If a power series centered at z_0 converges at $z_1 \neq z_0$, then the series converges for every z for which $|z - z_0| < |z_1 - z_0|$.

   (ii) If a power series centered at z_0 diverges at z_2, then the series diverges for every z for which $|z - z_0| > |z_2 - z_0|$.

45. Consider the power series $f(z) = \sum_{k=0}^{\infty} a_k z^k$ where

$$a_k = \begin{cases} 2^k, & k = 0, 2, 4, \ldots \\ \dfrac{1}{7^k} & k = 1, 3, 5, \ldots \end{cases}$$

(a) Show that neither $\lim\limits_{n\to\infty}|a_{n+1}/a_n|$ nor $\lim_{n\to\infty}|a_n|^{1/n}$ exist.

(b) Find the radius of convergence of each power series:

$$f_1(z) = \sum_{k=0}^{\infty} 2^{2k} z^{2k} \qquad \text{and} \qquad f_2(z) = \sum_{k=0}^{\infty} \frac{1}{7^{2k+1}} z^{2k+1}.$$

(c) Verify that $f(z) = f_1(z) + f_2(z)$. Discuss: How can the radius of convergence R for the original power series be found from the foregoing observation? What is R?

46. Proceed as in Problem 45 to find the radius of convergence R for the power series

$$1 + 3z + z^2 + 27z^3 + z^4 + 243z^5 + z^6 + \cdots .$$

47. In this problem you are guided through the proof of the proposition:

If the series $\sum_{k=1}^{\infty} z_k$ converges absolutely, then the series converges.

Proof We begin with the hypothesis that $\sum_{k=1}^{\infty}|z_k|$ converges. If $z_k = a_k + ib_k$, then $\sum_{k=1}^{\infty}|a_k| \le \sum_{k=1}^{\infty}\sqrt{a_k^2 + b_k^2} = \sum_{k=1}^{\infty}|z_k|$.

(a) First, explain why the foregoing inequality is true. Second, explain why this inequality shows that the series $\sum_{k=1}^{\infty}|a_k|$ converges.

(b) Explain how your reasoning in part (a) also shows that $\sum_{k=1}^{\infty}|b_k|$ converges.

(c) Explain how parts (a) and (b) show that $\sum_{k=1}^{\infty} z_k$ converges.

48. In this problem you are guided through the proof of part (i) of the ratio test (9).

Proof We begin with the hypothesis that $\lim_{n\to\infty}|z_{n+1}/z_n| = L < 1$.

(a) First, explain why there is an integer N and a positive real number $r < 1$ such that $|z_{n+1}/z_n| < r$ whenever $n > N$. Second, explain why this implies that $|z_{N+j}| < r^j|z_N|$ for $j = 0, 1, 2, \ldots$.

(b) Use the comparison test from real analysis and part (a) to show that $\sum_{j=0}^{\infty}|z_{N+j}|$ converges.

(c) Explain why part (b) shows that $\sum_{k=0}^{\infty} z_k$ converges absolutely.

6.2   Taylor Series

The correspondence between a complex number z within the circle of convergence and the number to which the series $\sum_{k=0}^{\infty} a_k(z - z_0)^k$ converges is single-valued. In this sense, a power series *defines* or *represents* a function f; for a specified z within the circle of convergence, the number L to which the power series converges is defined to be the value of f at z, that is, $f(z) = L$. In this section we present some important facts about the nature of this function f.

In the preceding section we saw that every power series has a radius of convergence R. Throughout the discussion in this section we will assume that a power series $\sum_{k=0}^{\infty} a_k(z - z_0)^k$ has either a *positive* or an *infinite* radius R of convergence.

Differentiation and Integration of Power Series The three theorems that follow indicate a function f that is defined by a power series is continuous, differentiable, and integrable within its circle of convergence.

Theorem 6.2.1   Continuity

A power series $\sum_{k=0}^{\infty} a_k(z - z_0)^k$ represents a continuous function f within its circle of convergence $|z - z_0| = R$.

Theorem 6.2.2   Term-by-Term Differentiation

A power series $\sum_{k=0}^{\infty} a_k(z - z_0)^k$ can be differentiated term by term within its circle of convergence $|z - z_0| = R$.

Differentiating a power series term-by-term gives,

$$\frac{d}{dz} \sum_{k=0}^{\infty} a_k(z - z_0)^k = \sum_{k=0}^{\infty} a_k \frac{d}{dz}(z - z_0)^k = \sum_{k=1}^{\infty} a_k k(z - z_0)^{k-1}.$$

Note that the summation index in the last series starts with $k = 1$ because the term corresponding to $k = 0$ is zero. It is readily proved by the ratio test that the original series and the differentiated series,

$$\sum_{k=0}^{\infty} a_k(z - z_0)^k \qquad \text{and} \qquad \sum_{k=1}^{\infty} a_k k(z - z_0)^{k-1}$$

have the same circle of convergence $|z - z_0| = R$. Since the derivative of a power series is another power series, the first series $\sum_{k=1}^{\infty} a_k(z - z_0)^k$ can be differentiated as many times as we wish. In other words, it follows as a corol-

Important ➡ lary to Theorem 6.2.2 that *a power series defines an infinitely differentiable function* within its circle of convergence and each differentiated series has the same radius of convergence R as the original power series.

Theorem 6.2.3   Term-by-Term Integration

A power series $\sum_{k=0}^{\infty} a_k(z - z_0)^k$ can be integrated term-by-term within its circle of convergence $|z - z_0| = R$, for every contour C lying entirely within the circle of convergence.

The theorem states that

$$\int_C \sum_{k=0}^{\infty} a_k(z - z_0)^k \, dz = \sum_{k=0}^{\infty} a_k \int_C (z - z_0)^k \, dz$$

whenever C lies in the interior of $|z - z_0| = R$. Indefinite integration can also be carried out term by term:

$$\int \sum_{k=0}^{\infty} a_k(z - z_0)^k dz = \sum_{k=0}^{\infty} a_k \int (z - z_0)^k dz$$

$$= \sum_{k=0}^{\infty} \frac{a_k}{k+1}(z - z_0)^{k+1} + \text{constant}.$$

The ratio test given in Theorem 6.1.4 can be used to be prove that both

$$\sum_{k=0}^{\infty} a_k (z - z_0)^k \qquad \text{and} \qquad \sum_{k=0}^{\infty} \frac{a_k}{k+1} (z - z_0)^{k+1}$$

have the same circle of convergence $|z - z_0| = R$.

Taylor Series   Suppose a power series represents a function f within $|z - z_0| = R$, that is,

$$f(z) = \sum_{k=0}^{\infty} a_k (z - z_0)^k = a_0 + a_1(z - z_0) + a_2(z - z_0)^2 + a_3(z - z_0)^3 + \cdots . \qquad (1)$$

It follows from Theorem 6.2.2 that the derivatives of f are the series

$$f'(z) = \sum_{k=1}^{\infty} a_k k (z - z_0)^{k-1} = a_1 + 2a_2(z - z_0) + 3a_3(z - z_0)^2 + \cdots , \qquad (2)$$

$$f''(z) = \sum_{k=2}^{\infty} a_k k(k-1)(z - z_0)^{k-2} = 2 \cdot 1 a_2 + 3 \cdot 2a_3(z - z_0) + \cdots , \qquad (3)$$

$$f'''(z) = \sum_{k=3}^{\infty} a_k k(k-1)(k-2)(z - z_0)^{k-3} = 3 \cdot 2 \cdot 1 a_3 + \cdots , \qquad (4)$$

Important ➡ and so on. Since the power series (1) represents a differentiable function f within its circle of convergence $|z - z_0| = R$, where R is either a positive number or infinity, we conclude that a *power series represents an analytic function* within its circle of convergence.

There is a relationship between the coefficients a_k in (1) and the derivatives of f. Evaluating (1), (2), (3), and (4) at $z = z_0$ gives

$$f(z_0) = a_0, \ f'(z_0) = 1!a_1, \ f''(z_0) = 2!a_2, \quad \text{and} \quad f'''(z_0) = 3!a_3,$$

respectively. In general, $f^{(n)}(z_0) = n! \, a_n$, or

$$a_n = \frac{f^{(n)}(z_0)}{n!}, \ n \geq 0. \qquad (5)$$

When $n = 0$ in (5), we interpret the zero-order derivative as $f(z_0)$ and $0! = 1$ so that the formula gives $a_0 = f(z_0)$. Substituting (5) into (1) yields

$$f(z) = \sum_{k=0}^{\infty} \frac{f^{(k)}(z_0)}{k!} (z - z_0)^k. \qquad (6)$$

This series is called the **Taylor series** for f centered at z_0. A Taylor series with center $z_0 = 0$,

$$f(z) = \sum_{k=0}^{\infty} \frac{f^{(k)}(0)}{k!} z^k, \qquad (7)$$

is referred to as a **Maclaurin series**.

We have just seen that a power series with a nonzero radius R of convergence represents an analytic function. On the other hand we ask:

> ### Question
>
> *If we are given a function f that is analytic in some domain D, can we represent it by a power series of the form (6) or (7)?*

Since a power series converges in a circular domain, and a domain D is generally not circular, the question comes down to: Can we expand f in one or more power series that are *valid*—that is, a power series that converges at z and the number to which the series converges is $f(z)$—in circular domains that are all contained in D? The question is answered in the affirmative by the next theorem.

Theorem 6.2.4 Taylor's Theorem

Let f be analytic within a domain D and let z_0 be a point in D. Then f has the series representation

$$f(z) = \sum_{k=0}^{\infty} \frac{f^{(k)}(z_0)}{k!}(z - z_0)^k \qquad (8)$$

valid for the largest circle C with center at z_0 and radius R that lies entirely within D.

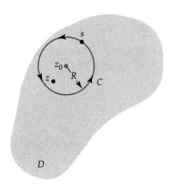

Figure 6.2.1 Contour for the proof of Theorem 6.2.4

Proof Let z be a fixed point within the circle C and let s denote the variable of integration. The circle C is then described by $|s - z_0| = R$. See Figure 6.2.1. To begin, we use the Cauchy integral formula to obtain the value of f at z:

$$f(z) = \frac{1}{2\pi i}\oint_C \frac{f(s)}{s - z}ds = \frac{1}{2\pi i}\oint_C \frac{f(s)}{(s - z_0) - (z - z_0)}ds$$

$$= \frac{1}{2\pi i}\oint_C \frac{f(s)}{(s - z_0)}\left\{ \frac{1}{1 - \dfrac{z - z_0}{s - z_0}} \right\}ds. \qquad (9)$$

By replacing z by $(z - z_0)/(s - z_0)$ in (8) of Section 6.1, we have

$$\frac{1}{1 - \dfrac{z - z_0}{s - z_0}} = 1 + \frac{z - z_0}{s - z_0} + \left(\frac{z - z_0}{s - z_0}\right)^2 + \cdots + \left(\frac{z - z_0}{s - z_0}\right)^{n-1} + \frac{(z - z_0)^n}{(s - z)(s - z_0)^{n-1}},$$

and so (9) becomes

$$f(z) = \frac{1}{2\pi i}\oint_C \frac{f(s)}{s - z_0}ds + \frac{z - z_0}{2\pi i}\oint_C \frac{f(s)}{(s - z_0)^2}ds$$

$$+ \frac{(z - z_0)^2}{2\pi i}\oint_C \frac{f(s)}{(s - z_0)^3}ds + \cdots + \frac{(z - z_0)^{n-1}}{2\pi i}\oint_C \frac{f(s)}{(s - z_0)^n}ds \quad (10)$$

$$+ \frac{(z - z_0)^n}{2\pi i}\oint_C \frac{f(s)}{(s - z)(s - z_0)^n}ds.$$

Utilizing Cauchy's integral formula for derivatives, (6) of Section 5.5, we can rewrite (10) as

$$f(z) = f(z_0) + \frac{f'(z_0)}{1!}(z - z_0) + \frac{f''(z_0)}{2!}(z - z_0)^2 + \cdots + \frac{f^{(n-1)}(z_0)}{(n-1)!}(z - z_0)^{n-1} + R_n(z), \quad (11)$$

where
$$R_n(z) = \frac{(z - z_0)^n}{2\pi i} \oint_C \frac{f(s)}{(s - z)(s - z_0)^n} \, ds.$$

Equation (11) is called Taylor's formula with remainder R_n. We now wish to show that $R_n(z) \to 0$ as $n \to \infty$. This can be accomplished by showing that $|R_n(z)| \to 0$ as $n \to \infty$. Since f is analytic in D, by Theorem 5.5.8 we know that $|f(z)|$ has a maximum value M on the contour C. In addition, since z is inside C, $|z - z_0| < R$ and, consequently,

$$|s - z| = |s - z_0 - (z - z_0)| \geq |s - z_0| - |z - z_0| = R - d,$$

where $d = |z - z_0|$ is the distance from z to z_0. The ML-inequality then gives

$$|R_n(z)| = \left| \frac{(z - z_0)^n}{2\pi i} \oint_C \frac{f(s)}{(s - z)(s - z_0)^n} \, ds \right| \leq \frac{d^n}{2\pi} \cdot \frac{M}{(R - d)R^n} \cdot 2\pi R = \frac{MR}{R - d} \left(\frac{d}{R} \right)^n.$$

Because $d < R$, $(d/R)^n \to 0$ as $n \to \infty$ we conclude that $|R_n(z)| \to 0$ as $n \to \infty$. It follows that the infinite series

$$f(z_0) + \frac{f'(z_0)}{1!}(z - z_0) + \frac{f''(z_0)}{2!}(z - z_0)^2 + \cdots$$

converges to $f(z)$. In other words, the result in (8) is valid for any point z interior to C. ❏

Note ➡   We can find the radius of convergence of a Taylor series in exactly the same manner illustrated in Examples 5–7 of the preceding section. However, we can simplify matters even further by noting that *the radius of convergence R is the distance from the center z_0 of the series to the nearest **isolated singularity** of f*. We shall elaborate more on this concept in the next section, but an isolated singularity is a point at which f fails to be analytic but is, nonetheless, analytic at all other points throughout *some* neighborhood of the point. For example, $z = 5i$ is an isolated singularity of $f(z) = 1/(z - 5i)$. If the function f is entire, then the radius of convergence of a Taylor series centered at any point z_0 is necessarily $R = \infty$. Using (8) and the last fact, we can say that the following Maclaurin series representations are valid for all z, that is, for $|z| < \infty$.

Some Important Maclaurin Series

$$e^z = 1 + \frac{z}{1!} + \frac{z^2}{2!} + \cdots = \sum_{k=0}^{\infty} \frac{z^k}{k!} \qquad (12)$$

$$\sin z = z - \frac{z^3}{3!} + \frac{z^5}{5!} - \cdots = \sum_{k=0}^{\infty} (-1)^k \frac{z^{2k+1}}{(2k + 1)!} \qquad (13)$$

$$\cos z = 1 - \frac{z^2}{2!} + \frac{z^4}{4!} - \cdots = \sum_{k=0}^{\infty} (-1)^k \frac{z^{2k}}{(2k)!} \qquad (14)$$

EXAMPLE 1   Radius of Convergence

Suppose the function $f(z) = \dfrac{3-i}{1-i+z}$ is expanded in a Taylor series with center $z_0 = 4 - 2i$. What is its radius of convergence R?

Solution   Observe that the function is analytic at every point except at $z = -1+i$, which is an isolated singularity of f. The distance from $z = -1+i$ to $z_0 = 4 - 2i$ is

$$|z - z_0| = \sqrt{(-1-4)^2 + (1-(-2))^2} = \sqrt{34}.$$

This last number is the radius of convergence R for the Taylor series centered at $4 - 2i$.   ❑

If two power series with center z_0,

$$\sum_{k=0}^{\infty} a_k(z - z_0)^k \qquad \text{and} \qquad \sum_{k=0}^{\infty} b_k(z - z_0)^k$$

represent the same function f and have the same nonzero radius R of convergence, then

$$a_k = b_k = \frac{f^{(k)}(z_0)}{k!}, \; k = 0, 1, 2, \ldots .$$

Note: Generally, the formula in (8) is used as a last resort. ➡

Stated in another way, the power series expansion of a function, with center z_0, is *unique*. On a practical level this means that a power series expansion of an analytic function f centered at z_0, irrespective of the method used to obtain it, is *the* Taylor series expansion of the function. For example, we can obtain (14) by simply differentiating (13) term by term. The Maclaurin series for e^{z^2} can be obtained by replacing the symbol z in (12) by z^2.

EXAMPLE 2   Maclaurin Series

Find the Maclaurin expansion of $f(z) = \dfrac{1}{(1-z)^2}$.

Solution   We could, of course, begin by computing the coefficients using (8). However, recall from (6) of Section 6.1 that for $|z| < 1$,

$$\frac{1}{1-z} = 1 + z + z^2 + z^3 + \cdots . \tag{15}$$

If we differentiate both sides of the last result with respect to z, then

$$\frac{d}{dz}\frac{1}{1-z} = \frac{d}{dz}1 + \frac{d}{dz}z + \frac{d}{dz}z^2 + \frac{d}{dz}z^3 + \cdots$$

or

$$\frac{1}{(1-z)^2} = 0 + 1 + 2z + 3z^2 + \cdots = \sum_{k=1}^{\infty} k\, z^{k-1}. \tag{16}$$

Since we are using Theorem 6.2.2, the radius of convergence of the last power series is the same as the original series, $R = 1$.   ❑

We can often build on results such as (16). For example, if we want the Maclaurin expansion of $f(z) = \dfrac{z^3}{(1-z)^2}$, we simply multiply (16) by z^3:

$$\frac{z^3}{(1-z)^2} = z^3 + 2z^4 + 3z^5 + \cdots = \sum_{k=1}^{\infty} k\, z^{k+2}.$$

The radius of convergence of the last series is still $R = 1$.

EXAMPLE 3   Taylor Series

Expand $f(z) = \dfrac{1}{1-z}$ in a Taylor series with center $z_0 = 2i$.

Solution In this solution we again use the geometric series (15). By adding and subtracting $2i$ in the denominator of $1/(1-z)$, we can write

$$\frac{1}{1-z} = \frac{1}{1-z+2i-2i} = \frac{1}{1-2i-(z-2i)} = \frac{1}{1-2i}\frac{1}{1-\dfrac{z-2i}{1-2i}}.$$

We now write $\dfrac{1}{1-\dfrac{z-2i}{1-2i}}$ as a power series by using (15) with the symbol z replaced by the expression $\dfrac{z-2i}{1-2i}$:

$$\frac{1}{1-z} = \frac{1}{1-2i}\left[1 + \frac{z-2i}{1-2i} + \left(\frac{z-2i}{1-2i}\right)^2 + \left(\frac{z-2i}{1-2i}\right)^3 + \cdots\right]$$

or

$$\frac{1}{1-z} = \frac{1}{1-2i} + \frac{1}{(1-2i)^2}(z-2i) + \frac{1}{(1-2i)^3}(z-2i)^2 + \frac{1}{(1-2i)^4}(z-2i)^3 + \cdots. \quad (17)$$

Because the distance from the center $z_0 = 2i$ to the nearest singularity $z = 1$ is $\sqrt{5}$, we conclude that the circle of convergence for (17) is $|z-2i| = \sqrt{5}$. This can be verified by the ratio test of the preceding section. ❏

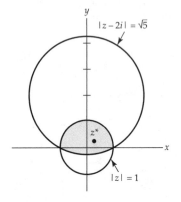

Figure 6.2.2 Series (15) and (17) both converge in the shaded region.

In (15) and (17) we represented the same function $f(z) = 1/(1-z)$ by two different power series. The first series (15) has center $z_0 = 0$ and radius of convergence $R = 1$. The second series (17) has center $z_0 = 2i$ and radius of convergence $R = \sqrt{5}$. The two different circles of convergence are illustrated in Figure 6.2.2. The interior of the intersection of the two circles, shown in color, is the region where *both* series converge; in other words, at a specified point z^* in this region, both series converge to same value $f(z^*) = 1/(1-z^*)$. Outside the colored region at least one of the two series must diverge.

> **Remarks** *Comparison with Real Analysis*
>
> (*i*) As a consequence of Theorem 5.5.3, we know that an analytic function f is infinitely differentiable. As a consequence of Theorem 6.2.4, we know that an analytic function f can always be expanded in a power series with a nonzero radius R of convergence. In real analysis, a function f can be infinitely differentiable, but it may be impossible to represent it by a power series. See Problem 51 in Exercises 6.2.
>
> (*ii*) If you haven't already noticed, the results in (6), (7), (12), (13), and (14) are identical in form with their analogues in elementary calculus.

EXERCISES 6.2 *Answers to selected odd-numbered problems begin on page ANS-16.*

In Problems 1–12, use known results to expand the given function in a Maclaurin series. Give the radius of convergence R of each series.

1. $f(z) = \dfrac{z}{1+z}$

2. $f(z) = \dfrac{1}{4-2z}$

3. $f(z) = \dfrac{1}{(1+2z)^2}$

4. $f(z) = \dfrac{z}{(1-z)^3}$

5. $f(z) = e^{-2z}$

6. $f(z) = ze^{-z^2}$

7. $f(z) = \sinh z$

8. $f(z) = \cosh z$

9. $f(z) = \cos \dfrac{z}{2}$

10. $f(z) = \sin 3z$

11. $f(z) = \sin z^2$

12. $f(z) = \cos^2 z$ [*Hint:* Use a trigonometric identity.]

In Problems 13 and 14, use the Maclaurin series for e^z to expand the given function in a Taylor series centered at the indicated point z_0. [*Hint:* $z = z - z_0 + z_0$.]

13. $f(z) = e^z$, $z_0 = 3i$

14. $f(z) = (z-1)e^{-3z}$, $z_0 = 1$

In Problems 15–22, expand the given function in a Taylor series centered at the indicated point z_0. Give the radius of convergence R of each series.

15. $f(z) = \dfrac{1}{z}$, $z_0 = 1$

16. $f(z) = \dfrac{1}{z}$, $z_0 = 1+i$,

17. $f(z) = \dfrac{1}{3-z}$, $z_0 = 2i$

18. $f(z) = \dfrac{1}{1+z}$, $z_0 = -i$,

19. $f(z) = \dfrac{z-1}{3-z}$, $z_0 = 1$

20. $f(z) = \dfrac{1+z}{1-z}$, $z_0 = i$

21. $f(z) = \cos z$, $z_0 = \pi/4$

22. $f(z) = \sin z$, $z_0 = \pi/2$

In Problems 23 and 24, use (7) to find the first three nonzero terms of the Maclaurin series of the given function.

23. $f(z) = \tan z$

24. $f(z) = e^{1/(1+z)}$

In Problems 25 and 26, use partial fractions as an aid in obtaining the Maclaurin series for the given function. Give the radius of convergence R of the series.

25. $f(z) = \dfrac{i}{(z-i)(z-2i)}$

26. $f(z) = \dfrac{z-7}{z^2-2z-3}$

In Problems 27 and 28, without actually expanding, determine the radius of convergence R of the Taylor series of the given function centered at the indicated point.

27. $f(z) = \dfrac{4 + 5z}{1 + z^2}$, $z_0 = 2 + 5i$ **28.** $f(z) = \cot z$, $z_0 = \pi i$

29. What is the radius of convergence R of the Maclaurin series in Problem 23?

30. What is the radius of convergence R of the Maclaurin series in Problem 24?

In Problems 31 and 32, expand the given function in Taylor series centered at each of the indicated points. Give the radius of convergence R of each series. Sketch the region within which both series converge.

31. $f(z) = \dfrac{1}{2 + z}$, $z_0 = -1$, $z_0 = i$ **32.** $f(z) = \dfrac{1}{z}$, $z_0 = 1 + i$, $z_0 = 3$

In Problems 33 and 34, use results obtained in this section to find the sum of the given power series.

33. $\displaystyle\sum_{k=0}^{\infty} 3^k z^k$ **34.** $\displaystyle\sum_{k=0}^{\infty} \dfrac{z^2}{k!}$

35. Find the Maclaurin series (14) by differentiating the Maclaurin series (13).

36. The **error function** erf(z) is defined by the integral $\text{erf}(z) = \dfrac{2}{\sqrt{\pi}} \displaystyle\int_0^z e^{-t^2}\, dt$.

Find a Maclaurin series for erf(z) by integrating the Maclaurin series for e^{-t^2}.

In Problems 37 and 38, approximate the value of the given expression using the indicated number of terms of a Maclaurin series.

37. $e^{(1+i)/10}$, three terms   **38.** $\sin\left(\dfrac{1+i}{10}\right)$, two terms

Focus on Concepts

39. Every function f has a domain of definition. Describe in words the domain of the function f defined by a power series center at z_0.

40. If $f(z) = \sum_{k=0}^{\infty} a_k z^k$ and $g(z) = \sum_{k=0}^{\infty} b_k z^k$ then the **Cauchy product** of f and g is given by

$$f(z)g(z) = \sum_{k=0}^{\infty} c_k z^k \quad \text{where} \quad c_k = \sum_{n=0}^{k} a_n b_{k-n}.$$

Write out the first five terms of the power series of $f(z)g(z)$.

41. Use Problem 40, (12) of this section, and (6) from Section 6.1 to find the first four nonzero terms of the Maclaurin series of $e^z/(1 - z)$. What is the radius of convergence R of the series?

42. Use Problem 40, and (13) and (14) of this section to find the first four nonzero terms of the Maclaurin series of $\sin z \cos z$. Can you think of another way to obtain this series?

43. The function $f(z) = \sec z$ is analytic at $z = 0$ and hence possesses a Maclaurin series representation. We could, of course, use (7), but there are several alternative ways to obtain the coefficients of the series

$$\sec z = a_0 + a_1 z + a_2 z^2 + a_3 z^3 + \cdots.$$

One way is to equate coefficients on both sides of the identity $1 = (\sec z)\cos z$ or

$$1 = \left(a_0 + a_1 z + a_2 z^2 + a_3 z^3 + \cdots\right)\left(1 - \frac{z^2}{2!} + \frac{z^4}{4!} - \frac{z^6}{6!} + \cdots\right).$$

Find the first three nonzero terms of the Maclaurin series of f. What is the radius of convergence R of the series?

44. (a) Use the definition $f(z) = \sec z = 1/\cos z$ and long division to obtain the first three nonzero terms of the Maclaurin series in Problem 43.

(b) Use $f(z) = \csc z = 1/\sin z$ and long division to obtain the first three nonzero terms of an infinite series. Is this series a Maclaurin series?

45. Suppose that a complex function f is analytic in a domain D that contains $z_0 = 0$ and f satisfies $f'(z) = 4z + f^2(z)$. Suppose further that $f(0) = 1$.

(a) Compute $f'(0)$, $f''(0)$, $f'''(0)$, $f^{(4)}(0)$, and $f^{(5)}(0)$.

(b) Find the first six terms of the Maclaurin expansion of f.

46. Find an alternative way of finding the first three nonzero terms of the Maclaurin series for $f(z) = \tan z$ (see Problem 23):

(a) based on the identity $\tan z = \sin z \sec z$ and Problems 42 and 43

(b) based on Problem 44(a)

(c) based on Problem 45. [*Hint:* $f'(z) = \sec^2 z = 1 + \tan^2 z$.]

47. We saw in Problem 34 in Exercises 1.3 that de Moivre's formula can be used to obtain trigonometric identities for $\cos 3\theta$ and $\sin 3\theta$. Discuss how these identities can be used to obtain Maclaurin series for $\sin^3 z$ and $\cos^3 z$. [*Hint:* You might want to simplify your answers to Problem 34. For example, $\cos^2 \theta \sin \theta = (1 - \sin^2 \theta) \sin \theta$.]

48. (a) Suppose that the principal value of the logarithm $\operatorname{Ln} z = \log_e |z| + i \operatorname{Arg}(z)$ is expanded in a Taylor series with center $z_0 = -1 + i$. Explain why $R = 1$ is the radius of the largest circle centered at $z_0 = -1 + i$ within which f is analytic.

(b) Show that within the circle $|z - (-1 + i)| = 1$ the Taylor series for f is

$$\operatorname{Ln} z = \frac{1}{2} \log_e 2 + \frac{3\pi}{4} i - \sum_{k=1}^{\infty} \frac{1}{k} \left(\frac{1+i}{2} \right)^k (z + 1 - i)^k.$$

(c) Show that the radius of convergence for the power series in part (b) is $R = \sqrt{2}$. Explain why this does not contradict the result in part (a).

49. (a) Consider the function $\operatorname{Ln}(1 + z)$. What is the radius of the largest circle centered at the origin within which f is analytic.

(b) Expand f in a Maclaurin series. What is the radius of convergence of this series?

(c) Use the result in part (b) to find a Maclaurin series for $\operatorname{Ln}(1 - z)$.

(d) Find a Maclaurin series for $\operatorname{Ln}\left(\dfrac{1+z}{1-z} \right)$.

50. In Theorem 3.1.3 we saw that L'Hôpital's rule carries over to complex analysis.

In Problem 33 in Exercises 3.1 you were guided through a proof of the following proposition by using the definition of the derivative:

If functions f and g are analytic at a point z_0 and $f(z_0) = 0$, $g(z_0) = 0$, but $g'(z_0) \neq 0$, then $\lim_{z \to z_0} \dfrac{f(z)}{g(z)} = \dfrac{f'(z_0)}{g'(z_0)}$.

This time, prove the proposition by replacing $f(z)$ and $g(z)$ by their Taylor series centered at z_0.

51. (a) You will find the following real function in most older calculus texts:

$$f(x) = \begin{cases} e^{-1/x^2}, & x \neq 0 \\ 0, & x = 0. \end{cases}$$

Do some reading in these calculus texts as an aid in showing that f is infinitely differentiable at every value of x. Show that f is not represented by its Maclaurin expansion at any value of $x \neq 0$.

(b) Investigate whether the complex analogue of the real function in part (a),

$$f(z) = \begin{cases} e^{-1/z^2}, & z \neq 0 \\ 0, & z = 0. \end{cases}$$

is infinitely differentiable at $z = 0$.

6.3   Laurent Series

If a complex function f fails to be analytic at a point $z = z_0$, then this point is said to be a **singularity** or **singular point** of the function. For example, the complex numbers $z = 2i$ and $z = -2i$ are singularities of the function $f(z) = z/(z^2 + 4)$ because f is discontinuous at each of these points. Recall from Section 4.1 that the principal value of the logarithm, Ln z, is analytic at all points except those points on the branch cut consisting of the nonpositive x-axis; that is, the branch point $z = 0$ as well as all negative real numbers are singular points of Ln z.

In this section we will be concerned with a new kind of "power series" expansion of f about an **isolated singularity** z_0. This new series will involve negative as well as nonnegative integer powers of $z - z_0$.

Isolated Singularities   Suppose that $z = z_0$ is a singularity of a complex function f. The point $z = z_0$ is said to be an **isolated singularity** of the function f if there exists *some* deleted neighborhood, or punctured open disk, $0 < |z - z_0| < R$ of z_0 throughout which f is analytic. For example, we have just seen that $z = 2i$ and $z = -2i$ are singularities of $f(z) = z/(z^2 + 4)$. Both $2i$ and $-2i$ are isolated singularities since f is analytic at every point in the neighborhood defined by $|z - 2i| < 1$, except at $z = 2i$, and at every point in the neighborhood defined by $|z - (-2i)| < 1$, except at $z = -2i$. In other words, f is analytic in the deleted neighborhoods $0 < |z - 2i| < 1$ and $0 < |z + 2i| < 1$. On the other hand, the branch point $z = 0$ is *not* an isolated singularity of Ln z since every neighborhood of $z = 0$ must contain points on the negative x-axis. We say that a singular point $z = z_0$ of a function f is **nonisolated** if *every* neighborhood of z_0 contains at least one singularity of f other than z_0. For example, the branch point $z = 0$ is a nonisolated singularity of Ln z.

A New Kind of Series   If $z = z_0$ is a singularity of a function f, then certainly f cannot be expanded in a power series with z_0 as its center. However, about an isolated singularity $z = z_0$, it is possible to represent f

by a series involving both negative and nonnegative integer powers of $z - z_0$; that is,

$$f(z) = \cdots + \frac{a_{-2}}{(z - z_0)^2} + \frac{a_{-1}}{z - z_0} + a_0 + a_1(z - z_0) + a_2(z - z_0)^2 + \cdots. \quad (1)$$

As a *very* simple example of (1) let us consider the function $f(z) = 1/(z - 1)$. As can be seen, the point $z = 1$ is an isolated singularity of f and consequently the function cannot be expanded in a Taylor series centered at that point. Nevertheless, f can be expanded in a series of the form given in (1) that is valid for all z *near* 1:

$$f(z) = \cdots + \frac{0}{(z - 1)^2} + \frac{1}{z - 1} + 0 + 0 \cdot (z - 1) + 0 \cdot (z - 1)^2 + \cdots. \quad (2)$$

The series representation in (2) is valid for $0 < |z - 1| < \infty$.

Using summation notation, we can write (1) as the sum of two series

$$f(z) = \sum_{k=1}^{\infty} a_{-k}(z - z_0)^{-k} + \sum_{k=0}^{\infty} a_k(z - z_0)^k. \quad (3)$$

The two series on the right-hand side in (3) are given special names. The part with negative powers of $z - z_0$, that is,

$$\sum_{k=1}^{\infty} a_{-k}(z - z_0)^{-k} = \sum_{k=1}^{\infty} \frac{a_{-k}}{(z - z_0)^k} \quad (4)$$

is called the **principal part** of the series (1) and will converge for $|1/(z - z_0)| < r^*$ or equivalently for $|z - z_0| > 1/r^* = r$. The part consisting of the nonnegative powers of $z - z_0$,

$$\sum_{k=0}^{\infty} a_k(z - z_0)^k, \quad (5)$$

is called the **analytic part** of the series (1) and will converge for $|z - z_0| < R$. Hence, the sum of (4) and (5) converges when z satisfies both $|z - z_0| > r$ *and* $|z - z_0| < R$, that is, when z is a point in an annular domain defined by $r < |z - z_0| < R$.

By summing over negative and nonnegative integers, (1) can be written compactly as

$$f(z) = \sum_{k=-\infty}^{\infty} a_k(z - z_0)^k.$$

The principal part of the series (2) consists of exactly one nonzero term, whereas its analytic part consists of all zero terms. Our next example illustrates a series of the form (1) in which the principal part of the series also consists of a finite number of nonzero terms, but this time the analytic part consists of an infinite number of nonzero terms.

EXAMPLE 1   Series of the Form Given in (1)

The function $f(z) = \dfrac{\sin z}{z^4}$ is not analytic at the isolated singularity $z = 0$ and hence cannot be expanded in a Maclaurin series. However, $\sin z$ is an entire function, and from (13) of Section 6.2 we know that its Maclaurin series,

$$\sin z = z - \frac{z^3}{3!} + \frac{z^5}{5!} - \frac{z^7}{7!} + \frac{z^9}{9!} - \cdots,$$

converges for $|z| < \infty$. By dividing this power series by z^4 we obtain a series for f with negative and positive integer powers of z:

$$f(z) = \frac{\sin z}{z^4} = \overbrace{\frac{1}{z^3} - \frac{1}{3!\,z}}^{\substack{\text{principal}\\\text{part}}} + \overbrace{\frac{z}{5!} - \frac{z^3}{7!} + \frac{z^5}{9!} - \cdots}^{\substack{\text{analytic}\\\text{part}}}. \tag{6}$$

The analytic part of the series in (6) converges for $|z| < \infty$. (Verify.) The principal part is valid for $|z| > 0$. Thus (6) converges for all z except at $z = 0$; that is, the series representation is valid for $0 < |z| < \infty$. ❏

A series representation of a function f that has the form given in (1), and (2) and (6) are such examples, is called a **Laurent series** or a **Laurent expansion** of f about z_0 on the annulus $r < |z - z_0| < R$.

Theorem 6.3.1   Laurent's Theorem

Let f be analytic within the annular domain D defined by $r < |z - z_0| < R$. Then f has the series representation

$$f(z) = \sum_{k=-\infty}^{\infty} a_k(z - z_0)^k \tag{7}$$

valid for $r < |z - z_0| < R$. The coefficients a_k are given by

$$a_k = \frac{1}{2\pi i} \oint_C \frac{f(s)}{(s - z_0)^{k+1}}\, ds, \quad k = 0, \pm 1, \pm 2, \ldots, \tag{8}$$

where C is a simple closed curve that lies entirely within D and has z_0 in its interior. See Figure 6.3.1.

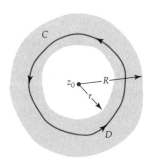

Figure 6.3.1 Contour for Theorem 6.3.1

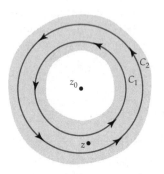

Figure 6.3.2 C_1 and C_2 are concentric circles.

Proof   Let C_1 and C_2 be concentric circles with center z_0 and radii r_1 and R_2, where $r < r_1 < R_2 < R$. Let z be a fixed point in D that also satisfies the inequality $r_1 < |z - z_0| < R_2$. See Figure 6.3.2. By introducing a crosscut between C_2 and C_1 it follows from Cauchy's integral formula that

$$f(z) = \frac{1}{2\pi i} \oint_{C_2} \frac{f(s)}{s - z}\, ds - \frac{1}{2\pi i} \oint_{C_1} \frac{f(s)}{s - z}\, ds. \tag{9}$$

As in the proof of Theorem 6.2.4, we can write

$$\frac{1}{2\pi i} \oint_{C_2} \frac{f(s)}{s - z}\, ds = \sum_{k=0}^{\infty} a_k(z - z_0)^k, \tag{10}$$

where

$$a_k = \frac{1}{2\pi i} \oint_{C_2} \frac{f(s)}{(s - z_0)^{k+1}}\, ds, \quad k = 0, 1, 2, \ldots. \tag{11}$$

We then proceed in a manner similar to (9) of Section 6.2:

$$-\frac{1}{2\pi i}\oint_{C_1}\frac{f(s)}{s-z}\,ds = \frac{1}{2\pi i}\oint_{C_1}\frac{f(s)}{(z-z_0)-(s-z_0)}\,ds$$

$$=\frac{1}{2\pi i}\oint_{C_1}\frac{f(s)}{z-z_0}\left\{\frac{1}{1-\dfrac{s-z_0}{z-z_0}}\right\}ds$$

$$=\frac{1}{2\pi i}\oint_{C_1}\frac{f(s)}{z-z_0}\left\{1+\frac{s-z_0}{z-z_0}+\left(\frac{s-z_0}{z-z_0}\right)^2+\cdots+\left(\frac{s-z_0}{z-z_0}\right)^{n-1}+\frac{(s-z_0)^n}{(z-s)(z-z_0)^{n-1}}\right\}ds \quad (12)$$

$$=\sum_{k=1}^{n}\frac{a_{-k}}{(z-z_0)^k}+R_n(z),$$

where $\qquad a_{-k}=\dfrac{1}{2\pi i}\oint_{C_1}\dfrac{f(s)}{(s-z_0)^{-k+1}}\,ds,\quad k=1,\,2,\,3,\,\ldots, \qquad (13)$

and $\qquad R_n(z)=\dfrac{1}{2\pi i(z-z_0)^n}\oint_{C_1}\dfrac{f(s)(s-z_0)^n}{z-s}\,ds.$

Now let d denote the distance from z to z_0, that is, $|z-z_0|=d$, and let M denote the maximum value of $|f(z)|$ on the contour C_1. Using $|s-z_0|=r_1$ and the inequality (10) of Section 1.2,

$$|z-s|=|z-z_0-(s-z_0)|\geq|z-z_0|-|s-z_0|=d-r_1.$$

The *ML*-inequality then gives

$$|R_n(z)|=\left|\frac{1}{2\pi i(z-z_0)^n}\oint_C\frac{f(s)(s-z_0)^n}{z-s}\,ds\right|\leq\frac{1}{2\pi d^n}\cdot\frac{Mr_1^n}{d-r_1}\cdot2\pi r_1$$

$$=\frac{Mr_1}{d-r_1}\left(\frac{r_1}{d}\right)^n.$$

Because $r_1<d$, $(r_1/d)^n\to0$ as $n\to\infty$, and so $|R_n(z)|\to0$ as $n\to\infty$. Thus we have shown that

$$-\frac{1}{2\pi i}\oint_{C_1}\frac{f(s)}{s-z}\,ds=\sum_{k=1}^{\infty}\frac{a_{-k}}{(z-z_0)^k}, \qquad (14)$$

where the coefficients a_{-k} are given in (13). Combining (14) and (10), we see that (9) yields

$$f(z)=\sum_{k=1}^{\infty}\frac{a_{-k}}{(z-z_0)^k}+\sum_{k=0}^{\infty}a_k(z-z_0)^k. \qquad (15)$$

Finally, by summing over nonnegative and negative integers, (15) can be written as $f(z)=\sum_{k=-\infty}^{\infty}a_k(z-z_0)^k$. Moreover, (11) and (13) can be written as a single integral:

$$a_k=\oint_C\frac{f(z)}{(z-z_0)^{k+1}}\,dz,\quad k=0,\pm1,\pm2,\ldots,$$

where, in view of (5) of Section 5.3, we have replaced the contours C_1 and C_2 by any simple closed contour C in D with z_0 in its interior. ❏

In the case when $a_{-k}=0$ for $k=1,2,3,\ldots$, the principal part (4) is zero and the Laurent series (7) reduces to a Taylor series. Thus, a Laurent expansion can be considered as a generalization of a Taylor series.

The annular domain in Theorem 6.3.1 defined by $r < |z - z_0| < R$ need not have the "ring" shape illustrated in Figure 6.3.2. Here are some other possible annular domains:

$$(i) \ r = 0, \ R \text{ finite}, \quad (ii) \ r \neq 0, \ R = \infty, \text{ and } (iii) \ r = 0, \ R = \infty.$$

In the first case, the series converges in annular domain defined by an $0 < |z - z_0| < R$. This is the interior of the circle $|z - z_0| = R$ except the point z_0; in other words, the domain is a punctured open disk. In the second case, the annular domain is defined by $r < |z - z_0|$ and consists of all points exterior to the circle $|z - z_0| = r$. In the third case, the domain is defined by $0 < |z - z_0|$. This represents the entire complex plane except the point z_0. The Laurent series in (2) and (6) are valid on this last type of domain.

The coefficients defined by (8) are seldom used. See (ii) in the Remarks at the end of this section. ⇒ The integral formula in (8) for the coefficients of a Laurent series are rarely used in actual practice. As a consequence, finding the Laurent series of a function in a specified annular domain is generally not an easy task. But this is not as disheartening as it might seem. In many instances we can obtain a desired Laurent series either by employing a known power series expansion of a function (as we did in Example 1) or by creative manipulation of geometric series (as we did in Example 2 of Section 6.2). The next example once again illustrates the use of geometric series.

EXAMPLE 2   Four Laurent Expansions

Expand $f(z) = \dfrac{1}{z(z-1)}$ in a Laurent series valid for the following annular domains.

(a) $0 < |z| < 1$ **(b)** $1 < |z|$ **(c)** $0 < |z - 1| < 1$ **(d)** $1 < |z - 1|$

Solution The four specified annular domains are shown in Figure 6.3.3. The black dots in each figure represent the two isolated singularities, $z = 0$ and $z = 1$, of f. In parts (a) and (b) we want to represent f in a series involving only negative and nonnegative integer powers of z, whereas in parts (c) and (d) we want to represent f in a series involving negative and nonnegative integer powers of $z - 1$.

(a) By writing

$$f(z) = -\frac{1}{z} \frac{1}{1 - z},$$

we can use (6) of Section 6.1 to write $1/(1 - z)$ as a series:

$$f(z) = -\frac{1}{z}\big[1 + z + z^2 + z^3 + \cdots\big].$$

The infinite series in the brackets converges for $|z| < 1$, but after we multiply this expression by $1/z$, the resulting series

$$f(z) = -\frac{1}{z} - 1 - z - z^2 - z^3 - \cdots$$

converges for $0 < |z| < 1$.

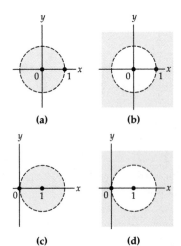

Figure 6.3.3 Annular domains for Example 2

(b) To obtain a series that converges for $1 < |z|$, we start by constructing a series that converges for $|1/z| < 1$. To this end we write the given function f as

$$f(z) = \frac{1}{z^2} \frac{1}{1 - \dfrac{1}{z}}$$

and again use (6) of Section 6.1 with z replaced by $1/z$:

$$f(z) = \frac{1}{z^2}\left[1 + \frac{1}{z} + \frac{1}{z^2} + \frac{1}{z^3} + \cdots\right].$$

The series in the brackets converges for $|1/z| < 1$ or equivalently for $1 < |z|$. Thus the required Laurent series is

$$f(z) = \frac{1}{z^2} + \frac{1}{z^3} + \frac{1}{z^4} + \frac{1}{z^5} + \cdots.$$

(c) This is basically the same problem as in part (a), except that we want all powers of $z - 1$. To that end, we add and subtract 1 in the denominator and use (7) of Section 6.1 with z replaced by $z - 1$:

$$f(z) = \frac{1}{(1 - 1 + z)(z - 1)}$$

$$= \frac{1}{z - 1} \frac{1}{1 + (z - 1)}$$

$$= \frac{1}{z - 1}\left[1 - (z - 1) + (z - 1)^2 - (z - 1)^3 + \cdots\right]$$

$$= \frac{1}{z - 1} - 1 + (z - 1) - (z - 1)^2 + \cdots.$$

The requirement that $z \neq 1$ is equivalent to $0 < |z - 1|$, and the geometric series in brackets converges for $|z - 1| < 1$. Thus the last series converges for z satisfying $0 < |z - 1|$ *and* $|z - 1| < 1$, that is, for $0 < |z - 1| < 1$.

(d) Proceeding as in part (b), we write

$$f(z) = \frac{1}{z - 1} \frac{1}{1 + (z - 1)} = \frac{1}{(z - 1)^2} \frac{1}{1 + \dfrac{1}{z - 1}}$$

$$= \frac{1}{(z - 1)^2}\left[1 - \frac{1}{z - 1} + \frac{1}{(z - 1)^2} - \frac{1}{(z - 1)^3} + \cdots\right]$$

$$= \frac{1}{(z - 1)^2} - \frac{1}{(z - 1)^3} + \frac{1}{(z - 1)^4} - \frac{1}{(z - 1)^5} + \cdots.$$

Because the series within the brackets converges for $|1/(z - 1)| < 1$, the final series converges for $1 < |z - 1|$. ❏

EXAMPLE 3 Laurent Expansions

Expand $f(z) = \dfrac{1}{(z - 1)^2(z - 3)}$ in a Laurent series valid for (a) $0 < |z - 1| < 2$ and (b) $0 < |z - 3| < 2$.

Solution

(a) As in parts (c) and (d) of Example 2, we want only powers of $z-1$ and so we need to express $z-3$ in terms of $z-1$. This can be done by writing

$$f(z) = \frac{1}{(z-1)^2(z-3)} = \frac{1}{(z-1)^2}\,\frac{1}{-2+(z-1)} = \frac{-1}{2(z-1)^2}\,\frac{1}{1-\dfrac{z-1}{2}}$$

and then using (6) of Section 6.1 with the symbol z replaced by $(z-1)/2$,

$$f(z) = \frac{-1}{2(z-1)^2}\left[1 + \frac{z-1}{2} + \frac{(z-1)^2}{2^2} + \frac{(z-1)^3}{2^3} + \cdots\right]$$

$$= -\frac{1}{2(z-1)^2} - \frac{1}{4(z-1)} - \frac{1}{8} - \frac{1}{16}(z-1) - \cdots. \qquad (16)$$

(b) To obtain powers of $z-3$, we write $z-1 = 2 + (z-3)$ and

$$\overbrace{}$$
We now factor 2
from this expression

$$f(z) = \frac{1}{(z-1)^2(z-3)} = \frac{1}{z-3}\,\overbrace{[2+(z-3)]^{-2}}$$

$$= \frac{1}{4(z-3)}\left[1 + \frac{z-3}{2}\right]^{-2}.$$

At this point we can obtain a power series for $\left[1 + \dfrac{z-3}{2}\right]^{-2}$ by using the binomial expansion,*

$$f(z) = \frac{1}{4(z-3)}\left[1 + \frac{(-2)}{1!}\left(\frac{z-3}{2}\right) + \frac{(-2)(-3)}{2!}\left(\frac{z-3}{2}\right)^2 + \frac{(-2)(-3)(-4)}{3!}\left(\frac{z-3}{2}\right)^3 + \cdots\right].$$

The binomial series in the brackets is valid for $|(z-3)/2| < 1$ or $|z-3| < 2$. Multiplying this series by $\dfrac{1}{4(z-3)}$ gives a Laurent series that is valid for $0 < |z-3| < 2$:

$$f(z) = \frac{1}{4(z-3)} - \frac{1}{4} + \frac{3}{16}(z-3) - \frac{1}{8}(z-3)^2 + \cdots.$$

❑

EXAMPLE 4   A Laurent Expansion

Expand $f(z) = \dfrac{8z+1}{z(1-z)}$ in a Laurent series valid for $0 < |z| < 1$.

---

*For α real, the binomial series $(1+z)^\alpha = 1 + \alpha z + \dfrac{\alpha(\alpha-1)}{2!}z^2 + \dfrac{\alpha(\alpha-1)(\alpha-2)}{3!}z^3 + \cdots$ is valid for $|z| < 1$.

Solution By partial fractions we can rewrite f as

$$f(z) = \frac{8z+1}{z(1-z)} = \frac{1}{z} + \frac{9}{1-z}.$$

Then by (6) of Section 6.1,

$$\frac{9}{1-z} = 9 + 9z + 9z^2 + \cdots.$$

The foregoing geometric series converges for $|z| < 1$, but after we add the term $1/z$ to it, the resulting Laurent series

$$f(z) = \frac{1}{z} + 9 + 9z + 9z^2 + \cdots$$

is valid for $0 < |z| < 1$. ❏

In the preceding examples the point at the center of the annular domain of validity for each Laurent series was an isolated singularity of the function f. A re-examination of Theorem 6.3.1 shows that this need not be the case.

EXAMPLE 5 A Laurent Expansion

Expand $f(z) = \dfrac{1}{z(z-1)}$ in a Laurent series valid for $1 < |z-2| < 2$.

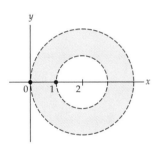

Figure 6.3.4 Annular domain for Example 5

Solution The specified annular domain is shown in Figure 6.3.4. The center of this domain, $z = 2$, is the point of analyticity of the function f. Our goal now is to find two series involving integer powers of $z - 2$, one converging for $1 < |z - 2|$ and the other converging for $|z - 2| < 2$. To accomplish this, we proceed as in the last example by decomposing f into partial fractions:

$$f(z) = -\frac{1}{z} + \frac{1}{z-1} = f_1(z) + f_2(z). \tag{17}$$

Now,

$$f_1(z) = -\frac{1}{z} = -\frac{1}{2+z-2}$$

$$= -\frac{1}{2}\, \frac{1}{1 + \dfrac{z-2}{2}}$$

$$= -\frac{1}{2}\left[1 - \frac{z-2}{2} + \frac{(z-2)^2}{2^2} - \frac{(z-2)^3}{2^3} + \cdots\right]$$

$$= -\frac{1}{2} + \frac{z-2}{2^2} - \frac{(z-2)^2}{2^3} + \frac{(z-2)^3}{2^4} - \cdots.$$

This series converges for $|(z-2)/2| < 1$ or $|z-2| < 2$. Furthermore,

$$f_2(z) = \frac{1}{z-1} = \frac{1}{1+z-2} = \frac{1}{z-2}\, \frac{1}{1 + \dfrac{1}{z-2}}$$

$$= \frac{1}{z-2}\left[1 - \frac{1}{z-2} + \frac{1}{(z-2)^2} - \frac{1}{(z-2)^3} + \cdots\right]$$

$$= \frac{1}{z-2} - \frac{1}{(z-2)^2} + \frac{1}{(z-2)^3} - \frac{1}{(z-2)^4} + \cdots$$

converges for $|1/(z-2)| < 1$ or $1 < |z-2|$. Substituting these two results in (17) then gives

$$f(z) = \cdots - \frac{1}{(z-2)^4} + \frac{1}{(z-2)^3} - \frac{1}{(z-2)^2} + \frac{1}{z-2} - \frac{1}{2} + \frac{z-2}{2^2} - \frac{(z-2)^2}{2^3} + \frac{(z-2)^3}{2^4} - \cdots .$$

This representation is valid for z satisfying $|z-2| < 2$ and $1 < |z-2|$; in other words, for $1 < |z-2| < 2$.   ❏

EXAMPLE 6   A Laurent Expansion

Expand $f(z) = e^{3/z}$ in a Laurent series valid for $0 < |z| < \infty$.

Solution   From (12) of Section 6.2 we know that for all finite z, that is, $|z| < \infty$,

$$e^z = 1 + z + \frac{z^2}{2!} + \frac{z^3}{3!} + \cdots . \tag{18}$$

We obtain the Laurent series for f by simply replacing z in (18) by $3/z$, $z \neq 0$,

$$e^{3/z} = 1 + \frac{3}{z} + \frac{3^2}{2!z^2} + \frac{3^3}{3!z^3} + \cdots . \tag{19}$$

This series (19) is valid for $z \neq 0$, that is, for $0 < |z| < \infty$.   ❏

Remarks

(i) In conclusion, we point out a result that will be of special interest to us in Sections 6.5 and 6.6. Replacing the complex variable s with the usual symbol z, we see that when $k = -1$, formula (8) for the Laurent series coefficients yields $a_{-1} = \dfrac{1}{2\pi i} \displaystyle\oint_C f(z)\,dz$, or more importantly,

$$\oint_C f(z)\,dz = 2\pi i\, a_{-1}. \tag{20}$$

(ii) Regardless how a Laurent expansion of a function f is obtained in a specified annular domain it is *the* Laurent series; that is, the series we obtain is unique.

EXERCISES 6.3   *Answers to selected odd-numbered problems begin on page ANS-17.*

In Problems 1–6, expand the given function in a Laurent series valid for the given annular domain.

1. $f(z) = \dfrac{\cos z}{z},\ 0 < |z|$ **2.** $f(z) = \dfrac{z - \sin z}{z^5},\ 0 < |z|$

3. $f(z) = e^{-1/z^2},\ 0 < |z|$ **4.** $f(z) = \dfrac{1 - e^z}{z^2},\ 0 < |z|$

5. $f(z) = \dfrac{e^z}{z-1},\ 0 < |z-1|$ **6.** $f(z) = z\cos\dfrac{1}{z},\ 0 < |z|$

In Problems 7–12, expand $f(z) = \dfrac{1}{z(z-3)}$ in a Laurent series valid for the indicated annular domain.

7. $0 < |z| < 3$ **8.** $|z| > 3$

9. $0 < |z-3| < 3$ **10.** $|z-3| > 3$

11. $1 < |z-4| < 4$ **12.** $1 < |z+1| < 4$

In Problems 13–16, expand $f(z) = \dfrac{1}{(z-1)(z-2)}$ in a Laurent series valid for the given annular domain.

13. $1 < |z| < 2$ **14.** $|z| > 2$

15. $0 < |z-1| < 1$ **16.** $0 < |z-2| < 1$

In Problems 17–20, expand $f(z) = \dfrac{z}{(z+1)(z-2)}$ in a Laurent series valid for the given annular domain.

17. $0 < |z+1| < 3$ **18.** $|z+1| > 3$

19. $1 < |z| < 2$ **20.** $0 < |z-2| < 3$

In Problems 21 and 22, expand $f(z) = \dfrac{1}{z(1-z)^2}$ in a Laurent series valid for the given annular domain.

21. $0 < |z| < 1$ **22.** $|z| > 1$

In Problems 23 and 24, expand $f(z) = \dfrac{1}{(z-2)(z-1)^3}$ in a Laurent series valid for the given annular domain.

23. $0 < |z-2| < 1$ **24.** $0 < |z-1| < 1$

In Problems 25 and 26, expand $f(z) = \dfrac{7z-3}{z(z-1)}$ in a Laurent series valid for the given annular domain.

25. $0 < |z| < 1$ **26.** $0 < |z-1| < 1$ [*Hint:* $\dfrac{7z-3}{z} = \dfrac{7(z-1)+4}{1+(z-1)}$.]

In Problems 27 and 28, expand $f(z) = \dfrac{z^2 - 2z + 2}{z-2}$ in a Laurent series valid for the given annular domain.

27. $1 < |z-1|$ **28.** $0 < |z-2|$

In Problems 29 and 30, use $\cos z = 1 - \dfrac{z^2}{2!} + \dfrac{z^4}{4!} - \cdots$, $\sin z = z - \dfrac{z^3}{3!} + \dfrac{z^5}{5!} - \cdots$, and long division to find the first three nonzero terms of a Laurent series of the given function f valid for $0 < |z| < \pi$.

29. $f(z) = \csc z$ **30.** $f(z) = \cot z$

Focus on Concepts

31. The function $f(z) = \dfrac{1}{(z+2)(z-4i)}$ possesses a Laurent series $f(z) = \displaystyle\sum_{k=-\infty}^{\infty} a_k(z+2)^k$ valid in the annulus $r < |z+2| < R$. What are r and R?

32. Consider the function $f(z) = \dfrac{e^{-2z}}{(z+1)^2}$. Use (7) to find the principal part of the Laurent series expansion of f about $z_0 = -1$ that is valid on the annulus $0 < |z+1| < \infty$.

33. Consider the function $f(z) = \dfrac{1}{(z-5)^3}$. What is the Laurent series expansion of f about $z_0 = 5$ that is valid on the annulus $0 < |z - 5| < \infty$?

34. Consider the function $f(z) = e^{(\alpha/2)(z-1/z)}$ where α is a real constant.

   (a) Use (8) to show that the Laurent series expansion of f about $z_0 = 0$ is:

$$e^{(\alpha/2)(z-1/z)} = \sum_{k=-\infty}^{\infty} J_k(\alpha) z^k$$

where

$$J_k(\alpha) = \frac{1}{2\pi i} \oint_C \frac{e^{(\alpha/2)(s-1/s)}}{s^{k+1}} ds$$

and C is the unit circle.

   (b) Use the parametrization $s(\theta) = e^{i\theta}$, $-\pi \le \theta \le \pi$ of C to show that

$$J_k(\alpha) = \frac{1}{2\pi} \int_{-\pi}^{\pi} \cos(\alpha \sin(\theta) - k\theta)\, d\theta + \frac{i}{2\pi} \int_{-\pi}^{\pi} \sin(\alpha \sin(\theta) - k\theta)\, d\theta.$$

   (c) Using integral properties of even and odd functions simplify the result in part (b) to obtain:

$$J_k(\alpha) = \frac{1}{\pi} \int_0^{\pi} \cos(k\theta - \alpha \sin\theta)\, d\theta.$$

The functions $J_k(\alpha)$ are called **Bessel functions** of the first kind of order k and are solutions to certain differential equations that occur frequently in applied mathematics.

6.4   Zeros and Poles

Suppose $z = z_0$ is an isolated singularity of a complex function f, and that

$$f(z) = \sum_{k=-\infty}^{\infty} a_k(z-z_0)^k = \sum_{k=1}^{\infty} a_{-k}(z-z_0)^{-k} + \sum_{k=0}^{\infty} a_k(z-z_0)^k \qquad (1)$$

is the Laurent series representation of f valid for the punctured open disk $0 < |z - z_0| < R$. We saw in the preceding section that a Laurent series (1) consists of two parts. That part of the series in (1) with negative powers of $z - z_0$, namely,

$$\sum_{k=1}^{\infty} a_{-k}(z-z_0)^{-k} = \sum_{k=1}^{\infty} \frac{a_{-k}}{(z-z_0)^k} \qquad (2)$$

is the principal part of the series. In the discussion that follows we will assign different names to the isolated singularity $z = z_0$ according to the number of terms in the principal part.

Classification of Isolated Singular Points   An isolated singular point $z = z_0$ of a complex function f is given a classification depending on whether the principal part (2) of its Laurent expansion (1) contains zero, a finite number, or an infinite number of terms.

   (i) If the principal part is zero, that is, *all* the coefficients a_{-k} in (2) are zero, then $z = z_0$ is called a **removable singularity**.

   (ii) If the principal part contains a finite number of nonzero terms, then $z = z_0$ is called a **pole**. If, in this case, the last nonzero coefficient in

(2) is a_{-n}, $n \geq 1$, then we say that $z = z_0$ is a **pole of order** n. If $z = z_0$ is pole of order 1, then the principal part (2) contains exactly one term with coefficient a_{-1}. A pole of order 1 is commonly called a **simple pole**.

(*iii*) If the principal part (2) contains an infinitely many nonzero terms, then $z = z_0$ is called an **essential singularity**.

Table 6.4.1 summarizes the form of a Laurent series for a function f when $z = z_0$ is one of the above types of isolated singularities. Of course, R in the table could be ∞.

$z = z_0$	Laurent Series for $0 < |z - z_0| < R$
Removable singularity	$a_0 + a_1(z - z_0) + a_2(z - z_0)^2 + \cdots$
Pole of order n	$\dfrac{a_{-n}}{(z - z_0)^n} + \dfrac{a_{-(n-1)}}{(z - z_0)^{n-1}} + \cdots + \dfrac{a_{-1}}{z - z_0} + a_0 + a_1(z - z_0) + \cdots$
Simple pole	$\dfrac{a_{-1}}{z - z_0} + a_0 + a_1(z - z_0) + a_2(z - z_0)^2 + \cdots$
Essential singularity	$\cdots + \dfrac{a_{-2}}{(z - z_0)^2} + \dfrac{a_{-1}}{z - z_0} + a_0 + a_1(z - z_0) + a_2(z - z_0)^2 + \cdots$

Table 6.4.1 Forms of Laurent series

EXAMPLE 1   Removable Singularity

Proceeding as we did in Example 1 of Section 6.3 by dividing the Maclaurin series for $\sin z$ by z, we see from

$$\frac{\sin z}{z} = 1 - \frac{z^2}{3!} + \frac{z^4}{5!} - \cdots \tag{3}$$

that all the coefficients in the principal part of the Laurent series are zero. Hence $z = 0$ is a removable singularity of the function $f(z) = (\sin z)/z$. ❑

Important paragraph. Reread it several times. ➡

If a function f has a removable singularity at the point $z = z_0$, then we can always supply an appropriate definition for the value of $f(z_0)$ so that f becomes analytic at $z = z_0$. For instance, since the right-hand side of (3) is 1 when we set $z = 0$, it makes sense to *define* $f(0) = 1$. Hence the function $f(z) = (\sin z)/z$, as given in (3), is now defined and continuous at every complex number z. Indeed, f is also analytic at $z = 0$ because it is represented by the Taylor series $1 - z^2/3! + z^4/5! - \cdots$ centered at 0 (a Maclaurin series).

EXAMPLE 2   Poles and Essential Singularity

(**a**) Dividing the terms of $\sin z = z - \dfrac{z^3}{3!} + \dfrac{z^5}{5!} - \cdots$ by z^2 shows that

$$\frac{\sin z}{z^2} = \overbrace{\frac{1}{z}}^{\substack{\text{principal} \\ \text{part}}} - \frac{z}{3!} + \frac{z^3}{5!} - \cdots$$

for $0 < |z| < \infty$. From this series we see that $a_{-1} \neq 0$ and so $z = 0$ is a simple pole of the function $f(z) = (\sin z)/z^2$. In like manner, we see that $z = 0$ is a pole of order 3 of the function $f(z) = (\sin z)/z^4$ considered in Example 1 of Section 6.3.

(b) In Example 3 of Section 6.3 we showed that the Laurent expansion of $f(z) = 1/(z-1)^2(z-3)$ valid for $0 < |z-1| < 2$ was

$$f(z) = \overbrace{-\frac{1}{2(z-1)^2} - \frac{1}{4(z-1)}}^{\text{principal part}} - \frac{1}{8} - \frac{z-1}{16} - \cdots.$$

Since $a_{-2} = -\frac{1}{2} \neq 0$, we conclude that $z = 1$ is a pole of order 2.

(c) In Example 6 of Section 6.3 we see from (19) that the principal part of the Laurent expansion of the function $f(z) = e^{3/z}$ valid for $0 < |z| < \infty$ contains an infinite number of nonzero terms. This shows that $z = 0$ is an essential singularity of f. ❏

Zeros Recall, a number z_0 is **zero** of a function f if $f(z_0) = 0$. We say that an analytic function f has a **zero of order n** at $z = z_0$ if

$$\overbrace{f(z_0) = 0, \ f'(z_0) = 0, \ f''(z_0) = 0, \ \ldots, \ f^{(n-1)}(z_0) = 0,}^{z_0 \text{ is a zero of } f \text{ and of its first } n-1 \text{ derivatives}} \text{ but } f^{(n)}(z_0) \neq 0. \quad (4)$$

A zero of order n is also referred to as a **zero of multiplicity n**. For example, for $f(z) = (z-5)^3$ we see that $f(5) = 0$, $f'(5) = 0$, $f''(5) = 0$, but $f'''(5) = 6 \neq 0$. Thus f has a zero of order (or multiplicity) 3 at $z_0 = 5$. A zero of order 1 is called a **simple zero**.

The next theorem is a consequence of (4).

Theorem 6.4.1 Zero of Order n

A function f that is analytic in some disk $|z - z_0| < R$ has a zero of order n at $z = z_0$ if and only if f can be written

$$f(z) = (z - z_0)^n \phi(z), \quad (5)$$

where ϕ is analytic at $z = z_0$ and $\phi(z_0) \neq 0$.

Partial Proof We will establish the "only if" part of the theorem. Given that f is analytic at z_0, it can be expanded in a Taylor series that is centered at z_0 and is convergent for $|z - z_0| < R$. Since the coefficients in a Taylor series $f(z) = \sum_{k=0}^{\infty} a_k(z - z_0)^k$ are $a_k = f^{(k)}(z_0)/k!$, $k = 0, 1, 2, \ldots$, it follows from (4) that the first n terms series are zero, and so the expansion must have the form

$$f(z) = a_n(z - z_0)^n + a_{n+1}(z - z_0)^{n+1} + a_{n+2}(z - z_0)^{n+2} + \cdots$$
$$= (z - z_0)^n \left[a_n + a_{n+1}(z - z_0) + a_{n+2}(z - z_0)^2 + \cdots \right].$$

With the power-series identification

$$\phi(z) = a_n + a_{n+1}(z - z_0) + a_{n+2}(z - z_0)^2 + \cdots$$

we conclude that ϕ is an analytic function and that $\phi(z_0) = a_n \neq 0$ because $a_n = f^{(n)}(z_0)/n! \neq 0$ from (4). ❏

EXAMPLE 3   Order of a Zero

The analytic function $f(z) = z \sin z^2$ has a zero at $z = 0$. If we replace z by z^2 in (13) of Section 6.2, we obtain the Maclaurin expansion

$$\sin z^2 = z^2 - \frac{z^6}{3!} + \frac{z^{10}}{5!} - \cdots .$$

Then by factoring z^2 out of the foregoing series we can rewrite f as

$$f(z) = z \sin z^2 = z^3 \phi(z) \qquad \text{where} \qquad \phi(z) = 1 - \frac{z^4}{3!} + \frac{z^8}{5!} - \cdots \qquad (6)$$

and $\phi(0) = 1$. When compared to (5), the result in (6) shows that $z = 0$ is a zero of order 3 of f. ❏

Poles   We can characterize a pole of order n in a manner analogous to (5).

Theorem 6.4.2   Pole of Order n

A function f analytic in a punctured disk $0 < |z - z_0| < R$ has a pole of order n at $z = z_0$ if and only if f can be written

$$f(z) = \frac{\phi(z)}{(z - z_0)^n}, \qquad (7)$$

where ϕ is analytic at $z = z_0$ and $\phi(z_0) \neq 0$.

Partial Proof   As in the proof of (5), we will establish the "only if" part of the preceding sentence. Since f is assumed to have a pole of order n at z_0 it can be expanded in a Laurent series

$$f(z) = \frac{a_{-n}}{(z - z_0)^n} + \cdots + \frac{a_{-2}}{(z - z_0)^2} + \frac{a_{-1}}{z - z_0} + a_0 + a_1(z - z_0) + \cdots, \qquad (8)$$

valid in the disk $0 < |z - z_0| < R$. By factoring out $1/(z - z_0)^n$, (8) confirms that f can be written in the form $\phi(z)/(z - z_0)^n$. Here we identify

$$\phi(z) = a_{-n} + \cdots + a_{-2}(z - z_0)^{n-2} + a_{-1}(z - z_0)^{n-1} + a_0(z - z_0)^n + a_1(z - z_0)^{n+1} + \cdots, \qquad (9)$$

as a power series valid for the open disk $|z - z_0| < R$. By assumption, $z = z_0$ is a pole of order n of f, and so we must have $a_{-n} \neq 0$. If we define $\phi(z_0) = a_{-n}$, then it follows from (9) that ϕ is analytic throughout the disk $|z - z_0| < R$. ❏

Zeros Again   A zero $z = z_0$ of an analytic function f is *isolated* in the sense that there exists some neighborhood of z_0 for which $f(z) \neq 0$ at every point z in that neighborhood except at $z = z_0$. As a consequence, if z_0 is a zero of a nontrivial analytic function f, then the function $1/f(z)$ has an isolated singularity at the point $z = z_0$.

The following result enables us, in some circumstances, to determine the poles of a function by inspection.

Theorem 6.4.3   Pole of Order n

If the functions g and h are analytic at $z = z_0$ and h has a zero of order n at $z = z_0$ and $g(z_0) \neq 0$, then the function $f(z) = g(z)/h(z)$ has a pole of order n at $z = z_0$.

Proof Because the function h has zero of order n, (5) gives $h(z) = (z - z_0)^n \phi(z)$, where ϕ is analytic at $z = z_0$ and $\phi(z_0) \neq 0$. Thus f can be written

$$f(z) = \frac{g(z)/\phi(z)}{(z - z_0)^n}. \tag{10}$$

Since g and ϕ are analytic at $z = z_0$ and $\phi(z_0) \neq 0$, it follows that the function g/ϕ is analytic at z_0. Moreover, $g(z_0) \neq 0$ implies $g(z_0)/\phi(z_0) \neq 0$. We conclude from Theorem 6.4.2 that the function f has a pole of order n at z_0. ❏

When $n = 1$ in (10), we see that a zero of order 1, or a simple zero, in the denominator h of $f(z) = g(z)/h(z)$ corresponds to a simple pole of f.

EXAMPLE 4   Order of Poles

(a) Inspection of the rational function

$$f(z) = \frac{2z + 5}{(z - 1)(z + 5)(z - 2)^4}$$

shows that the denominator has zeros of order 1 at $z = 1$ and $z = -5$, and a zero of order 4 at $z = 2$. Since the numerator is not zero at any of these points, it follows from Theorem 6.4.3 and (10) that f has simple poles at $z = 1$ and $z = -5$, and a pole of order 4 at $z = 2$.

(b) In Example 3 we saw that $z = 0$ is a zero of order 3 of $z \sin z^2$. From Theorem 6.4.3 and (10) we conclude that the reciprocal function $f(z) = 1/(z \sin z^2)$ has a pole of order 3 at $z = 0$. ❏

Remarks

(i) From the preceding discussion, it should be intuitively clear that if a function f has a pole at $z = z_0$, then $|f(z)| \to \infty$ as $z \to z_0$ from any direction. From (i) of the Remarks following Section 2.6 we can write $\lim_{z \to z_0} f(z) = \infty$.

> (*ii*) If you peruse other texts on complex variables, and you are encouraged to do this, you may encounter the term *meromorphic*. A function f is **meromorphic** if it is analytic throughout a domain D, except possibly for poles in D. It can be proved that a meromorphic function can have at most a finite number of poles in D. For example, the rational function $f(z) = 1/(z^2 + 1)$ is meromorphic in the complex plane.

EXERCISES 6.4   *Answers to selected odd-numbered problems begin on page ANS-17.*

In Problems 1–4, show that $z = 0$ is a removable singularity of the given function. Supply a definition of $f(0)$ so that f is analytic at $z = 0$.

1. $f(z) = \dfrac{e^{2z} - 1}{z}$

2. $f(z) = \dfrac{z^3 - 4z^2}{1 - e^{z^2/2}}$

3. $f(z) = \dfrac{\sin 4z - 4z}{z^2}$

4. $f(z) = \dfrac{1 - \frac{1}{2}z^{10} - \cos z^5}{\sin z^2}$

In Problems 5–10, determine the zeros and their order for the given function.

5. $f(z) = (z + 2 - i)^2$

6. $f(z) = z^4 - 16$

7. $f(z) = z^4 + z^2$

8. $f(z) = \sin^2 z$

9. $f(z) = e^{2z} - e^z$

10. $f(z) = ze^z - z$

In Problems 11–14, the indicated number is a zero of the given function. Use a Maclaurin or Taylor series to determine the order of the zero.

11. $f(z) = z(1 - \cos^2 z); \; z = 0$

12. $f(z) = z - \sin z; \; z = 0$

13. $f(z) = 1 - e^{z-1}; \; z = 1$

14. $f(z) = 1 - \pi i + z + e^z; \; z = \pi i$

In Problems 15–26, determine the order of the poles for the given function.

15. $f(z) = \dfrac{3z - 1}{z^2 + 2z + 5}$

16. $f(z) = 5 - \dfrac{6}{z^2}$

17. $f(z) = \dfrac{1 + 4i}{(z + 2)(z + i)^4}$

18. $f(z) = \dfrac{z - 1}{(z + 1)(z^3 + 1)}$

19. $f(z) = \tan z$

20. $f(z) = \dfrac{\cot \pi z}{z^2}$

21. $f(z) = \dfrac{1 - \cosh z}{z^4}$

22. $f(z) = \dfrac{e^z}{z^2}$

23. $f(z) = \dfrac{1}{1 + e^z}$

24. $f(z) = \dfrac{e^z - 1}{z^2}$

25. $f(z) = \dfrac{\sin z}{z^2 - z}$

26. $f(z) = \dfrac{\cos z - \cos 2z}{z^6}$

In Problems 27 and 28, show that the indicated number is an essential singularity of the given function.

27. $f(z) = z^3 \sin\left(\dfrac{1}{z}\right); \; z = 0$

28. $f(z) = (z - 1)\cos\left(\dfrac{1}{z + 2}\right); \; z = -2$

29. Determine whether $z = 0$ is an essential singularity of $f(z) = e^{z+1/z}$.

30. Determine whether $z = 0$ is an isolated or nonisolated singularity of $f(z) = \tan(1/z)$.

Focus on Concepts

31. In part (b) of Example 2 in Section 6.3, we showed that the Laurent series representation of $f(z) = \dfrac{1}{z(z-1)}$ valid for $|z| > 1$ is

$$f(z) = \frac{1}{z^2} + \frac{1}{z^3} + \frac{1}{z^4} + \frac{1}{z^5} + \cdots .$$

The point $z = 0$ is an isolated singularity of f, and the Laurent series contains an infinite number of terms involving negative integer powers of z. Discuss: Does this mean that $z = 0$ is an essential singularity of f? Defend your answer with sound mathematics.

32. Suppose f and g are analytic functions and f has a zero of order m and g has zero of order n at $z = z_0$. Discuss: What is the order of the zero of fg at z_0? of $f + g$ at z_0?

33. An interesting theorem, known as **Picard's theorem,** states that in any arbitrarily small neighborhood of an isolated essential singularity z_0, an analytic function f assumes *every* finite complex value, with *one* exception, an infinite number of times. Since $z = 0$ is an isolated essential singularity of $f(z) = e^{1/z}$, find an infinite number of z in any neighborhood of $z = 0$ for which $f(z) = i$. What is the one exception? That is, what is the one value that $f(z) = e^{1/z}$ does not take on?

34. Suppose $|f(z)|$ is bounded in a deleted neighborhood of an isolated singularity z_0. Classify z_0 as one of the three kinds of isolated singularities listed on page 303. Justify your answer with sound mathematics.

35. Suppose the analytic function $f(z)$ has a zero of order n at $z = z_0$. Prove that the function $[f(z)]^m$, m a positive integer, has a zero of order mn at $z = z_0$.

36. In this problem you are guided through the start of the proof of the proposition:

    The only isolated singularities of a rational function f are poles or removable singularities.

    Proof We begin with the hypothesis that f is a rational function, that is, $f(z) = p(z)/q(z)$, where p and q are polynomials. We know that f is analytic for all z except at the zeros of q. Suppose z_0 is a zero of q but not of p. Then Theorem 6.4.1 tell us that there exists a positive integer n such that $q(z) = (z - z_0)^n Q(z)$, where Q is a polynomial and $Q(z_0) \neq 0$. Now invoke Theorem 6.4.2. Consider one more case to finish the proof.

6.5   Residues and Residue Theorem

We saw in the last section that if a complex function f has an *isolated singularity* at a point z_0, then f has a Laurent series representation

$$f(z) = \sum_{k=-\infty}^{\infty} a_k(z - z_0)^k = \cdots + \frac{a_{-2}}{(z - z_0)^2} + \frac{a_{-1}}{z - z_0} + a_0 + a_1(z - z_0) + \cdots ,$$

which converges for all z *near* z_0. More precisely, the representation is valid in some deleted neighborhood of z_0 or punctured open disk $0 < |z - z_0| < R$. In this section our entire focus will be on the coefficient a_{-1} and its importance in the evaluation of contour integrals.

Residue The coefficient a_{-1} of $1/(z - z_0)$ in the preceding Laurent series is called the **residue** of the function f at the isolated singularity z_0. We shall use the notation

$$a_{-1} = \text{Res}(f(z),\ z_0)$$

to denote the residue of f at z_0. Recall, if the principal part of the Laurent series valid for $0 < |z - z_0| < R$ contains a finite number of terms with a_{-n} the last nonzero coefficient, then z_0 is a pole of order n; if the principal part of the series contains an infinite number of terms with nonzero coefficients, then z_0 is an essential singularity.

EXAMPLE 1   Residues

(a) In part (b) of Example 2 in Section 6.4 we saw that $z = 1$ is a pole of order two of the function $f(z) = \dfrac{1}{(z - 1)^2(z - 3)}$. From the Laurent series obtained in that example valid for the deleted neighborhood of $z = 1$ defined by $0 < |z - 1| < 2$,

$$f(z) = \frac{-1/2}{(z - 1)^2} + \overbrace{\frac{-1/4}{z - 1}}^{a_{-1}} - \frac{1}{8} - \frac{z - 1}{16} - \cdots$$

we see that the coefficient of $1/(z - 1)$ is $a_{-1} = \text{Res}(f(z),\ 1) = -\frac{1}{4}$.

(b) In Example 6 of Section 6.3 we saw that $z = 0$ is an essential singularity of $f(z) = e^{3/z}$. Inspection of the Laurent series obtained in that example,

$$e^{3/z} = 1 + \overbrace{\frac{3}{z}}^{a_{-1}} + \frac{3^2}{2!z^2} + \frac{3^3}{3!z^3} + \cdots,$$

$0 < |z| < \infty$, shows that the coefficient of $1/z$ is $a_{-1} = \text{Res}(f(z),\ 0) = 3$. ❏

We will see why the coefficient a_{-1} is so important later on in this section. In the meantime we are going to examine ways of obtaining this complex number when z_0 is a *pole* of a function f without the necessity of expanding f in a Laurent series at z_0. We begin with the residue at a simple pole.

Theorem 6.5.1   Residue at a Simple Pole

If f has a simple pole at $z = z_0$, then

$$\text{Res}(f(z), z_0) = \lim_{z \to z_0}(z - z_0)f(z). \qquad (1)$$

Proof Since f has a simple pole at $z = z_0$, its Laurent expansion convergent on a punctured disk $0 < |z - z_0| < R$ has the form

$$f(z) = \frac{a_{-1}}{z - z_0} + a_0 + a_1(z - z_0) + a_2(z - z_0) + \cdots,$$

where $a_{-1} \neq 0$. By multiplying both sides of this series by $z - z_0$ and then taking the limit as $z \to z_0$ we obtain

$$\lim_{z \to z_0} (z - z_0)f(z) = \lim_{z \to z_0} [a_{-1} + a_0(z - z_0) + a_1(z - z_0)^2 + \cdots]$$

$$= a_{-1} = \mathrm{Res}(f(z), z_0). \qquad \square$$

Theorem 6.5.2   Residue at a Pole of Order n

If f has a pole of order n at $z = z_0$, then

$$\mathrm{Res}(f(z), z_0) = \frac{1}{(n-1)!} \lim_{z \to z_0} \frac{d^{n-1}}{dz^{n-1}} \Big[(z - z_0)^n f(z) \Big]. \qquad (2)$$

Proof   Because f is assumed to have pole of order n at $z = z_0$, its Laurent expansion convergent on a punctured disk $0 < |z - z_0| < R$ must have the form

$$f(z) = \frac{a_{-n}}{(z - z_0)^n} + \cdots + \frac{a_{-2}}{(z - z_0)^2} + \frac{a_{-1}}{z - z_0} + a_0 + a_1(z - z_0) + \cdots,$$

where $a_{-n} \neq 0$. We multiply the last expression by $(z - z_0)^n$,

$$(z - z_0)^n f(z) = a_{-n} + \cdots + a_{-2}(z - z_0)^{n-2} + a_{-1}(z - z_0)^{n-1} + a_0(z - z_0)^n + a_1(z - z_0)^{n+1} + \cdots$$

and then differentiate both sides of the equality $n - 1$ times:

$$\frac{d^{n-1}}{dz^{n-1}} \Big[(z - z_0)^n f(z) \Big] = (n-1)!a_{-1} + n!a_0(z - z_0) + \cdots. \qquad (3)$$

Since all the terms on the right-hand side after the first involve positive integer powers of $z - z_0$, the limit of (3) as $z \to z_0$ is

$$\lim_{z \to z_0} \frac{d^{n-1}}{dz^{n-1}} \Big[(z - z_0)^n f(z) \Big] = (n-1)!a_{-1}.$$

Solving the last equation for a_{-1} gives (2). $\qquad \square$

Notice that (2) reduces to (1) when $n = 1$.

EXAMPLE 2   Residue at a Pole

The function $f(z) = \dfrac{1}{(z-1)^2(z-3)}$ has a simple pole at $z = 3$ and a pole of order 2 at $z = 1$. Use Theorems 6.5.1 and 6.5.2 to find the residues.

Solution   Since $z = 3$ is a simple pole, we use (1):

$$\mathrm{Res}(f(z), 3) = \lim_{z \to 3} (z - 3)f(z) = \lim_{z \to 3} \frac{1}{(z-1)^2} = \frac{1}{4}.$$

Now at the pole of order 2, the result in (2) gives

$$\text{Res}(f(z), 1) = \frac{1}{1!} \lim_{z \to 1} \frac{d}{dz} \left[(z-1)^2 f(z) \right]$$

$$= \lim_{z \to 1} \frac{d}{dz} \frac{1}{z-3}$$

$$= \lim_{z \to 1} \frac{-1}{(z-3)^2} = -\frac{1}{4}.$$

When f is not a rational function, calculating residues by means of (1) or (2) can sometimes be tedious. It is possible to devise alternative residue formulas. In particular, suppose a function f can be written as a quotient $f(z) = g(z)/h(z)$, where g and h are analytic at $z = z_0$. If $g(z_0) \neq 0$ and if the function h has a zero of order 1 at z_0, then f has a simple pole at $z = z_0$ and

An alternative method for computing a residue at a simple pole ➡

$$\text{Res}(f(z), z_0) = \frac{g(z_0)}{h'(z_0)}. \qquad (4)$$

To derive this result we shall use the definition of a zero of order 1, the definition of a derivative, and then (1). First, since the function h has a zero of order 1 at z_0, we must have $h(z_0) = 0$ and $h'(z_0) \neq 0$. Second, by definition of the derivative given in (12) of Section 3.1,

$$h'(z_0) = \lim_{z \to z_0} \frac{h(z) - \overbrace{h(z_0)}^{0}}{z - z_0} = \lim_{z \to z_0} \frac{h(z)}{z - z_0}.$$

We then combine the preceding two facts in the following manner in (1):

$$\text{Res}(f(z), z_0) = \lim_{z \to z_0} (z - z_0) \frac{g(z)}{h(z)} = \lim_{z \to z_0} \frac{g(z)}{\dfrac{h(z)}{z - z_0}} = \frac{g(z_0)}{h'(z_0)}.$$

There are several alternative ways of arriving at formula (4). For instance, it can be obtained by a single application of L'Hôpital's rule (page 133), but you are asked in Problem 40 in Exercises 6.5 to derive (4) using (5) of Section 6.4. Residue formulas for poles of order greater than 1 are far more complicated than (4) and will not be presented here. Practicality aside, a derivation of one of these higher-order formulas provides an opportunity to review and use important concepts. See Problem 41 in Exercises 6.5.

EXAMPLE 3   Using (4) to Compute Residues

The polynomial $z^4 + 1$ can be factored as $(z - z_1)(z - z_2)(z - z_3)(z - z_4)$, where z_1, z_2, z_3, and z_4 are the four distinct roots of the equation $z^4 + 1 = 0$ (or equivalently, the four fourth roots of -1). It follows from Theorem 6.4.3 that the function

$$f(z) = \frac{1}{z^4 + 1}$$

has four simple poles. Now from (4) of Section 1.4 we have $z_1 = e^{\pi i/4}$, $z_2 = e^{3\pi i/4}$, $z_3 = e^{5\pi i/4}$, and $z_4 = e^{7\pi i/4}$. To compute the residues, we use (4) of this section along with Euler's formula (6) of Section 1.6:

$$\text{Res}(f(z), z_1) = \frac{1}{4z_1^3} = \frac{1}{4}e^{-3\pi i/4} = -\frac{1}{4\sqrt{2}} - \frac{1}{4\sqrt{2}}i$$

$$\text{Res}(f(z), z_2) = \frac{1}{4z_2^3} = \frac{1}{4}e^{-9\pi i/4} = \frac{1}{4\sqrt{2}} - \frac{1}{4\sqrt{2}}i$$

$$\text{Res}(f(z), z_3) = \frac{1}{4z_3^3} = \frac{1}{4}e^{-15\pi i/4} = \frac{1}{4\sqrt{2}} + \frac{1}{4\sqrt{2}}i$$

$$\text{Res}(f(z), z_4) = \frac{1}{4z_4^3} = \frac{1}{4}e^{-21\pi i/4} = -\frac{1}{4\sqrt{2}} + \frac{1}{4\sqrt{2}}i.$$

Of course, we could have calculated each of the residues in Example 3 using formula (1). But the procedure in this case would have entailed substantially more algebra. For example, we first use the factorization of $z^4 + 1$ to write f as:

$$f(z) = \frac{1}{(z - z_1)(z - z_2)(z - z_3)(z - z_4)}.$$

By (1) the residue at, say, the pole z_1 is

$$\text{Res}(f(z), z_1) = \lim_{z \to z_1}(z - z_1)\frac{1}{(z - z_1)(z - z_2)(z - z_3)(z - z_4)}$$

$$= \frac{1}{(z_1 - z_2)(z_1 - z_3)(z_1 - z_4)}$$

$$= \frac{1}{(e^{\pi i/4} - e^{3\pi i/4})(e^{\pi i/4} - e^{5\pi i/4})(e^{\pi i/4} - e^{7\pi i/4})}.$$

Then we face the daunting task of simplifying the denominator of the last expression. Finally, we must do this process three more times.

Residue Theorem   We come now to the reason why the residue concept is important. The next theorem states that under some circumstances we can evaluate complex integrals $\oint_C f(z)\,dz$ by summing the residues at the isolated singularities of f within the closed contour C.

Theorem 6.5.3   Cauchy's Residue Theorem

Let D be a simply connected domain and C a simple closed contour lying entirely within D. If a function f is analytic on and within C, except at a finite number of isolated singularities $z_1, z_2, \ldots, z_n$ within C, then

$$\oint_C f(z)\,dz = 2\pi i \sum_{k=1}^{n} \text{Res}(f(z), z_k). \tag{5}$$

Proof   Suppose $C_1, C_2, \ldots, C_n$ are circles centered at $z_1, z_2, \ldots, z_n$, respectively. Suppose further that each circle C_k has a radius r_k small enough

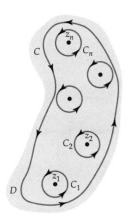

Figure 6.5.1 n singular points within contour C

so that $C_1, C_2, \ldots, C_n$ are mutually disjoint and are interior to the simple closed curve C. See Figure 6.5.1. Now in (20) of Section 6.3 we saw that $\oint_{C_k} f(z)\,dz = 2\pi i \operatorname{Res}(f(z), z_k)$, and so by Theorem 5.3.2 we have

$$\oint_C f(z)\,dz = \sum_{k=1}^{n} \oint_{C_k} f(z)\,dz = 2\pi i \sum_{k=1}^{n} \operatorname{Res}\left(f(z), z_k\right).$$ ❑

EXAMPLE 4   Evaluation by the Residue Theorem

Evaluate $\oint_C \dfrac{1}{(z-1)^2(z-3)}\,dz$, where

(a) the contour C is the rectangle defined by $x = 0$, $x = 4$, $y = -1$, $y = 1$,

(b) and the contour C is the circle $|z| = 2$.

Solution

(a) Since both $z = 1$ and $z = 3$ are poles within the rectangle we have from (5) that

$$\oint_C \frac{1}{(z-1)^2(z-3)}\,dz = 2\pi i \left[\operatorname{Res}(f(z),1) + \operatorname{Res}(f(z),3)\right]$$

We found these residues in Example 2. Therefore,

$$\oint_C \frac{1}{(z-1)^2(z-3)}\,dz = 2\pi i\left[\left(-\frac{1}{4}\right) + \frac{1}{4}\right] = 0.$$

(b) Since only the pole $z = 1$ lies within the circle $|z| = 2$, we have from (5)

$$\oint_C \frac{1}{(z-1)^2(z-3)}\,dz = 2\pi i \operatorname{Res}(f(z),1) = 2\pi i\left(-\frac{1}{4}\right) = -\frac{\pi}{2}i.$$ ❑

EXAMPLE 5   Evaluation by the Residue Theorem

Evaluate $\oint_C \dfrac{2z+6}{z^2+4}\,dz$, where the contour C is the circle $|z - i| = 2$.

Solution By factoring the denominator as $z^2 + 4 = (z - 2i)(z + 2i)$ we see that the integrand has simple poles at $-2i$ and $2i$. Because only $2i$ lies within the contour C, it follows from (5) that

$$\oint_C \frac{2z+6}{z^2+4}\,dz = 2\pi i \operatorname{Res}(f(z), 2i).$$

But

$$\operatorname{Res}(f(z), 2i) = \lim_{z \to 2i}(z - 2i)\frac{2z+6}{(z-2i)(z+2i)}$$

$$= \frac{6+4i}{4i} = \frac{3+2i}{2i}.$$

Hence,

$$\oint_C \frac{2z+6}{z^2+4}\,dz = 2\pi i\left(\frac{3+2i}{2i}\right) = \pi(3+2i).$$ ❑

EXAMPLE 6   Evaluation by the Residue Theorem

Evaluate $\oint_C \dfrac{e^z}{z^4 + 5z^3}\, dz$, where the contour C is the circle $|z| = 2$.

Solution Writing the denominator as $z^4 + 5z^3 = z^3(z + 5)$ reveals that the integrand $f(z)$ has a pole of order 3 at $z = 0$ and a simple pole at $z = -5$. But only the pole $z = 0$ lies within the given contour and so from (5) and (2) we have,

$$\oint_C \frac{e^z}{z^4 + 5z^3}\, dz = 2\pi i\, \mathrm{Res}(f(z), 0) = 2\pi i\, \frac{1}{2!} \lim_{z \to 0} \frac{d^2}{dz^2}\left[z^3 \cdot \frac{e^z}{z^3(z + 5)}\right]$$

$$= \pi i \lim_{z \to 0} \frac{(z^2 + 8z + 17)e^z}{(z + 5)^3} = \frac{17\pi}{125}i.$$

❏

EXAMPLE 7   Evaluation by the Residue Theorem

Evaluate $\oint_C \tan z\, dz$, where the contour C is the circle $|z| = 2$.

Solution The integrand $f(z) = \tan z = \sin z / \cos z$ has simple poles at the points where $\cos z = 0$. We saw in (21) of Section 4.3 that the only zeros of $\cos z$ are the real numbers $z = (2n + 1)\pi/2$, $n = 0, \pm 1, \pm 2, \ldots$. Since only $-\pi/2$ and $\pi/2$ are within the circle $|z| = 2$, we have

$$\oint_C \tan z\, dz = 2\pi i\left[\mathrm{Res}\left(f(z), -\frac{\pi}{2}\right) + \mathrm{Res}\left(f(z), \frac{\pi}{2}\right)\right].$$

With the identifications $g(z) = \sin z$, $h(z) = \cos z$, and $h'(z) = -\sin z$, we see from (4) that

$$\mathrm{Res}\left(f(z), -\frac{\pi}{2}\right) = \frac{\sin(-\pi/2)}{-\sin(-\pi/2)} = -1$$

and

$$\mathrm{Res}\left(f(z), \frac{\pi}{2}\right) = \frac{\sin(\pi/2)}{-\sin(\pi/2)} = -1.$$

Therefore,

$$\oint_C \tan z\, dz = 2\pi i[-1 - 1] = -4\pi i.$$

❏

☞ Theorem 6.5.3 is applicable at an essential singularity.

Although there is no nice formula analogous to (1), (2), or (4), for computing the residue at an essential singularity z_0, the next example shows that Theorem 6.5.3 is still applicable at z_0.

EXAMPLE 8   Evaluation by the Residue Theorem

Evaluate $\oint_C e^{3/z}\, dz$, where the contour C is the circle $|z| = 1$.

Solution As we have seen, $z = 0$ is an essential singularity of the integrand $f(z) = e^{3/z}$ and so neither formulas (1) and (2) are applicable to find the residue of f at that point. Nevertheless, we saw in Example 1 that the Laurent series of f at $z = 0$ gives $\operatorname{Res}(f(z),\, 0) = 3$. Hence from (5) we have

$$\oint_C e^{3/z}\, dz = 2\pi i \operatorname{Res}(f(z), 0) = 2\pi i\,(3) = 6\pi i.$$

❏

EXERCISES 6.5   *Answers to selected odd-numbered problems begin on page ANS-18.*

In Problems 1–6, use an appropriate Laurent series to find the indicated residue.

1. $f(z) = \dfrac{2}{(z-1)(z+4)}$; $\operatorname{Res}(f(z), 1)$

2. $f(z) = \dfrac{1}{z^3(1-z)^3}$; $\operatorname{Res}(f(z), 0)$

3. $f(z) = \dfrac{4z-6}{z(2-z)}$; $\operatorname{Res}(f(z), 0)$

4. $f(z) = (z+3)^2 \sin\left(\dfrac{2}{z+3}\right)$; $\operatorname{Res}(f(z), -3)$

5. $f(z) = e^{-2/z^2}$; $\operatorname{Res}(f(z), 0)$

6. $f(z) = \dfrac{e^{-z}}{(z-2)^2}$; $\operatorname{Res}(f(z), 2)$

In Problems 7–16, use (1), (2), or (4) to find the residue at each pole of the given function.

7. $f(z) = \dfrac{z}{z^2 + 16}$

8. $f(z) = \dfrac{4z+8}{2z-1}$

9. $f(z) = \dfrac{1}{z^4 + z^3 - 2z^2}$

10. $f(z) = \dfrac{1}{(z^2 - 2z + 2)^2}$

11. $f(z) = \dfrac{5z^2 - 4z + 3}{(z+1)(z+2)(z+3)}$

12. $f(z) = \dfrac{2z-1}{(z-1)^4(z+3)}$

13. $f(z) = \dfrac{\cos z}{z^2(z-\pi)^3}$

14. $f(z) = \dfrac{e^z}{e^z - 1}$

15. $f(z) = \sec z$

16. $f(z) = \dfrac{1}{z \sin z}$

In Problems 17–20, use Cauchy's residue theorem, where appropriate, to evaluate the given integral along the indicated contours.

17. $\displaystyle\oint_C \dfrac{1}{(z-1)(z+2)^2}\, dz$ (a) $|z| = \frac{1}{2}$ (b) $|z| = \frac{3}{2}$ (c) $|z| = 3$

18. $\displaystyle\oint_C \dfrac{z+1}{z^2(z-2i)}\, dz$ (a) $|z| = 1$ (b) $|z - 2i| = 1$ (c) $|z - 2i| = 4$

19. $\displaystyle\oint_C z^3 e^{-1/z^2}\, dz$ (a) $|z| = 5$ (b) $|z + i| = 2$ (c) $|z - 3| = 1$

20. $\displaystyle\oint_C \dfrac{1}{z \sin z}\, dz$ (a) $|z - 2i| = 1$ (b) $|z - 2i| = 3$ (c) $|z| = 5$

In Problems 21–34, use Cauchy's residue theorem to evaluate the given integral along the indicated contour.

21. $\displaystyle\oint_C \dfrac{1}{z^2 + 4z + 13}\, dz$, C: $|z - 3i| = 3$

22. $\displaystyle\oint_C \dfrac{1}{z^3(z-1)^4}\, dz$, C: $|z - 2| = \frac{3}{2}$

23. $\oint_C \dfrac{z}{z^4-1}\, dz$, C: $|z|=2$

24. $\oint_C \dfrac{z}{(z+1)(z^2+1)}\, dz$, C: $16x^2+y^2=4$

25. $\oint_C \dfrac{ze^z}{z^2-1}\, dz$, C: $|z|=2$

26. $\oint_C \dfrac{e^z}{z^3+2z^2}\, dz$, C: $|z|=3$

27. $\oint_C \dfrac{\tan z}{z}\, dz$, C: $|z-1|=2$

28. $\oint_C \dfrac{\cot \pi z}{z^2}\, dz$, C: $|z|=\frac{1}{2}$

29. $\oint_C \cot \pi z\, dz$, C is the rectangle defined by $x=\frac{1}{2}$, $x=\pi$, $y=-1$, $y=1$

30. $\oint_C \dfrac{2z-1}{z^2(z^3+1)}\, dz$, C is the rectangle defined by $x=-2$, $x=1$, $y=-\frac{1}{2}$, $y=1$

31. $\oint_C \left(z^2 e^{1/\pi z}+\dfrac{ze^z}{z^4-\pi^4}\right) dz$, C: $4x^2+y^2=16$

32. $\oint_C \dfrac{\cos z}{(z-1)^2(z^2+9)}\, dz$, C: $|z-1|=1$

33. $\oint_C \dfrac{1}{z^6+1}\, dz$, C is the semicircle defined by $y=0$, $y=\sqrt{4-x^2}$

34. $\oint_C e^{4/(z-2)}\, dz$, C: $|z-1|=3$

Focus on Concepts

35. (a) Use series to show that $z=0$ is a zero of order 2 of $1-\cos z$.

(b) In view of part (a), $z=0$ is a pole of order two of the function $f(z)=e^z/(1-\cos z)$ and hence has a Laurent series

$$f(z)=\frac{e^z}{1-\cos z}=\frac{a_{-2}}{z^2}+\frac{a_{-1}}{z}+a_0+a_1 z+a_2 z^2+\cdots$$

valid for $0<|z|<2\pi$. Use series for e^z and $1-\cos z$ and equate coefficients in the product

$$e^z=(1-\cos z)\left(\frac{a_{-2}}{z^2}+\frac{a_{-1}}{z}+a_0+\cdots\right)$$

to determine a_{-2}, a_{-1}, and a_0.

(c) Evaluate $\oint_C \dfrac{e^z}{1-\cos z}\, dz$, where C is $|z|=1$.

36. Discuss how to evaluate $\oint_C e^{1/z}\sin\left(\dfrac{1}{z}\right) dz$, where C is $|z|=1$. Carry out your ideas.

37. Consider the function $f(z)=z^4\big/(1-z^{1/2})$, where $z^{1/2}$ denotes the principal branch of the square root function. Discuss and justify your answer: Does f have a pole at $z=1$? If so, find $\text{Res}(f(z),\,1)$.

38. Residues can be used to find coefficients in partial fraction decompositions of rational functions. Suppose that $p(z)$ is a polynomial of degree ≤ 2 and that z_1, z_2, and z_3 are distinct complex numbers that are not zeros of $p(z)$. Then

$$f(z) = \frac{p(z)}{(z - z_1)(z - z_2)(z - z_3)} = \frac{A}{z - z_1} + \frac{B}{z - z_2} + \frac{C}{z - z_3}.$$

(a) Use (1) to show

$$A = \text{Res}(f(z), z_1) = \frac{p(z_1)}{(z_1 - z_2)(z_1 - z_3)}$$

$$B = \text{Res}(f(z), z_2) = \frac{p(z_2)}{(z_2 - z_1)(z_2 - z_3)}$$

$$C = \text{Res}(f(z), z_3) = \frac{p(z_3)}{(z_3 - z_1)(z_3 - z_2)}.$$

(b) Now use a Laurent series to prove that $A = \text{Res}(f(z), z_1)$. [*Hint:* The second and third terms, $B/(z - z_2)$ and $C/(z - z_3)$, respectively, are analytic at z_1. Also see (3) and (4) in Section 6.4.]

39. Use Problem 38 to find the partial fraction decomposition of $f(z) = \dfrac{5z^2 - z + 2}{z(z + 1)(z - i)}.$

40. If $h(z)$ has a zero of order 1, it follows from (5) of Section 6.4 that $h(z) = (z - z_0)\,\phi(z)$, where ϕ is analytic at $z = z_0$ and $\phi(z_0) \neq 0$. With $f(z) = g(z)/h(z)$, use (1) to derive (4):

$$\text{Res}\Big(f(z), z_0\Big) = \frac{g(z_0)}{\phi(z_0)} = \frac{g(z_0)}{h'(z_0)}.$$

41. If $h(z)$ has a zero of order 2, it follows from (5) of Section 6.4 that $h(z) = (z - z_0)^2\,\phi(z)$, where ϕ is analytic at $z = z_0$ and $\phi(z_0) \neq 0$. With $f(z) = g(z)/h(z)$, use (2) with $n = 2$ to derive a formula analogous to (4) for $\text{Res}(f(z),\ z_0)$ at a pole of order 2 at $z = z_0$.

42. Let $f(z) = 1/\left[p(z)\right]^2$ where $p(z)$ has a zero of order 1 at $z = z_0$.

(a) Use Problem 41 to derive the formula $\text{Res}(f(z), z_0) = -\dfrac{p''(z_0)}{\left[p'(z_0)\right]^3}.$

(b) Use part (b) to determine the residue of $f(z) = \dfrac{1}{(z^2 - z)^2}$ at $z = 0$.

6.6 Some Consequences of the Residue Theorem

In this section we shall see how residue theory can be used to evaluate *real* integrals of the forms

$$\int_0^{2\pi} F(\cos\theta, \sin\theta)\, d\theta, \tag{1}$$

$$\int_{-\infty}^{\infty} f(x)\, dx, \tag{2}$$

$$\int_{-\infty}^{\infty} f(x) \cos \alpha x\, dx \quad \text{and} \quad \int_{-\infty}^{\infty} f(x) \sin \alpha x\, dx, \tag{3}$$

where F in (1) and f in (2) and (3) are rational functions. For the rational function $f(x) = p(x)/q(x)$ in (2) and (3), we will assume that the polynomials p and q have no common factors.

In addition to evaluating the three integrals just given, we shall demonstrate how to use residues to evaluate real improper integrals that require integration along a branch cut.

The discussion ends with the relationship between the residue theory and the zeros of an analytic function and a consideration of how residues can, in certain cases, be used to find the sum of an infinite series.

6.6.1 Evaluation of Real Trigonometric Integrals

Integrals of the Form $\int_0^{2\pi} F(\cos\theta, \sin\theta)d\theta$ The basic idea here is to convert a real trigonometric integral of form (1) into a complex integral, where the contour C is the unit circle $|z| = 1$ centered at the origin. To do this we begin with (10) of Section 2.2 to parametrize this contour by $z = e^{i\theta}$, $0 \le \theta \le 2\pi$. We can then write

$$dz = ie^{i\theta}d\theta, \quad \cos\theta = \frac{e^{i\theta} + e^{-i\theta}}{2}, \quad \sin\theta = \frac{e^{i\theta} - e^{-i\theta}}{2i}.$$

The last two expressions follow from (2) and (3) of Section 4.3. Since $dz = ie^{i\theta}d\theta = iz\,d\theta$ and $z^{-1} = 1/z = e^{-i\theta}$, these three quantities are equivalent to

$$d\theta = \frac{dz}{iz}, \quad \cos\theta = \frac{1}{2}(z + z^{-1}), \quad \sin\theta = \frac{1}{2i}(z - z^{-1}). \tag{4}$$

The conversion of the integral in (1) into a contour integral is accomplished by replacing, in turn, $d\theta$, $\cos\theta$, and $\sin\theta$ by the expressions in (4):

$$\oint_C F\left(\frac{1}{2}(z + z^{-1}), \frac{1}{2i}(z - z^{-1})\right) \frac{dz}{iz},$$

where C is the unit circle $|z| = 1$.

EXAMPLE 1   A Real Trigonometric Integral

Evaluate $\displaystyle\int_0^{2\pi} \frac{1}{(2 + \cos\theta)^2}\,d\theta$.

Solution When we use the substitutions given in (4), the given trigonometric integral becomes the contour integral

$$\oint_C \frac{1}{\left(2 + \frac{1}{2}(z + z^{-1})\right)^2} \frac{dz}{iz} = \oint_C \frac{1}{\left(2 + \dfrac{z^2 + 1}{2z}\right)^2} \frac{dz}{iz}.$$

Carrying out the algebraic simplification of the integrand then yields

$$\frac{4}{i}\oint_C \frac{z}{(z^2 + 4z + 1)^2}\,dz.$$

From the quadratic formula we can factor the polynomial $z^2 + 4z + 1$ as $z^2 + 4z + 1 = (z - z_1)(z - z_2)$, where $z_1 = -2 - \sqrt{3}$ and $z_2 = -2 + \sqrt{3}$. Thus, the integrand can be written

$$\frac{z}{(z^2 + 4z + 1)^2} = \frac{z}{(z - z_1)^2(z - z_2)^2}.$$

Because only z_2 is inside the unit circle C, we have

$$\oint_C \frac{z}{(z^2 + 4z + 1)^2}\, dz = 2\pi i\, \mathrm{Res}(f(z), z_2).$$

To calculate the residue, we first note that z_2 is a pole of order 2 and so we use (2) of Section 6.5:

$$\mathrm{Res}(f(z), z_2) = \lim_{z \to z_2} \frac{d}{dz}\left[(z - z_2)^2 f(z)\right] = \lim_{z \to z_2} \frac{d}{dz} \frac{z}{(z - z_1)^2}$$

$$= \lim_{z \to z_2} \frac{-z - z_1}{(z - z_1)^3} = \frac{1}{6\sqrt{3}}.$$

Hence, $\qquad \dfrac{4}{i}\oint_C \dfrac{z}{(z^2 + 4z + 1)}\, dz = \dfrac{4}{i}\cdot 2\pi i\, \mathrm{Res}(f(z), z_1) = \dfrac{4}{i}\cdot 2\pi i \cdot \dfrac{1}{6\sqrt{3}}$

and, finally, $\qquad\qquad \displaystyle\int_0^{2\pi} \frac{1}{(2 + \cos\theta)^2}\, d\theta = \frac{4\pi}{3\sqrt{3}}.$ ❑

6.6.2 Evaluation of Real Improper Integrals

Integrals of the Form $\int_{-\infty}^{\infty} f(x)\, dx$ Suppose $y = f(x)$ is a real function that is defined and continuous on the interval $[0, \infty)$. In elementary calculus the improper integral $I_1 = \int_0^{\infty} f(x)\, dx$ is defined as the limit

$$I_1 = \int_0^{\infty} f(x)\, dx = \lim_{R \to \infty} \int_0^{R} f(x)\, dx. \qquad (5)$$

If the limit exists, the integral I_1 is said to be **convergent**; otherwise, it is **divergent**. The improper integral $I_2 = \int_{-\infty}^{0} f(x)\, dx$ is defined similarly:

$$I_2 = \int_{-\infty}^{0} f(x)\, dx = \lim_{R \to \infty} \int_{-R}^{0} f(x)\, dx. \qquad (6)$$

Finally, if f is continuous on $(-\infty, \infty)$, then $\int_{-\infty}^{\infty} f(x)\, dx$ is defined to be

$$\int_{-\infty}^{\infty} f(x)\, dx = \int_{-\infty}^{0} f(x)\, dx + \int_0^{\infty} f(x)\, dx = I_1 + I_2, \qquad (7)$$

provided *both* integrals I_1 and I_2 are convergent. If either one, I_1 or I_2, is divergent, then $\int_{-\infty}^{\infty} f(x)\, dx$ is divergent. It is important to remember that the right-hand side of (7) is *not* the same as

$$\lim_{R \to \infty}\left[\int_{-R}^{0} f(x)\, dx + \int_0^{R} f(x)\, dx\right] = \lim_{R \to \infty} \int_{-R}^{R} f(x)\, dx. \qquad (8)$$

For the integral $\int_{-\infty}^{\infty} f(x)\, dx$ to be convergent, the limits (5) and (6) must exist independently of one another. *But*, in the event that we know (a priori) that an improper integral $\int_{-\infty}^{\infty} f(x)\, dx$ converges, we can then evaluate it by means of the single limiting process given in (8):

$$\int_{-\infty}^{\infty} f(x)\, dx = \lim_{R \to \infty} \int_{-R}^{R} f(x)\, dx. \qquad (9)$$

On the other hand, the symmetric limit in (9) may exist even though the improper integral $\int_{-\infty}^{\infty} f(x)\,dx$ is divergent. For example, the integral $\int_{-\infty}^{\infty} x\,dx$ is divergent since $\lim_{R\to\infty}\int_0^R x\,dx = \lim_{R\to\infty}\frac{1}{2}R^2 = \infty$. However, (9) gives

$$\lim_{R\to\infty}\int_{-R}^{R} x\,dx = \lim_{R\to\infty}\frac{1}{2}[R^2-(-R)^2] = 0. \tag{10}$$

The limit in (9), if it exists, is called the **Cauchy principal value** (P.V.) of the integral and is written

$$\text{P.V.}\int_{-\infty}^{\infty} f(x)\,dx = \lim_{R\to\infty}\int_{-R}^{R} f(x)\,dx. \tag{11}$$

In (10) we have shown that P.V.$\int_{-\infty}^{\infty} x\,dx = 0$. To summarize:

Cauchy Principal Value

When an integral of form (2) converges, its Cauchy principal value is the same as the value of the integral. If the integral diverges, it may still possess a Cauchy principal value (11).

Important observation about even functions ➡

One final point about the Cauchy principal value: Suppose $f(x)$ is continuous on $(-\infty,\infty)$ and is an *even* function, that is, $f(-x)=f(x)$. Then its graph is symmetric with respect to the y-axis and as a consequence

$$\int_{-R}^{0} f(x)\,dx = \int_{0}^{R} f(x)\,dx \tag{12}$$

and

$$\int_{-R}^{R} f(x)\,dx = \int_{-R}^{0} f(x)\,dx + \int_{0}^{R} f(x)\,dx = 2\int_{0}^{R} f(x)\,dx. \tag{13}$$

From (12) and (13) we conclude that *if* the Cauchy principal value (11) exists, then both $\int_0^\infty f(x)\,dx$ and $\int_{-\infty}^\infty f(x)\,dx$ converge. The values of the integrals are

$$\int_0^\infty f(x)\,dx = \frac{1}{2}\,\text{P.V.}\int_{-\infty}^\infty f(x)\,dx \quad\text{and}\quad \int_{-\infty}^\infty f(x)\,dx = \text{P.V.}\int_{-\infty}^\infty f(x)\,dx.$$

To evaluate an integral $\int_{-\infty}^\infty f(x)\,dx$, where the rational function $f(x)=p(x)/q(x)$ is continuous on $(-\infty,\infty)$, by residue theory we replace x by the complex variable z and integrate the complex function f over a closed contour C that consists of the interval $[-R,R]$ on the real axis and a semicircle C_R of radius large enough to enclose all the poles of $f(z)=p(z)/q(z)$ in the upper half-plane $\text{Im}(z)>0$. See Figure 6.6.1. By Theorem 6.5.3 of Section 6.5 we have

$$\oint_C f(z)\,dz = \int_{C_R} f(z)\,dz + \int_{-R}^{R} f(x)\,dx = 2\pi i\sum_{k=1}^{n}\text{Res}(f(z),z_k),$$

where z_k, $k=1,2,\ldots,n$ denotes poles in the upper half-plane. If we can show that the integral $\int_{C_R} f(z)\,dz \to 0$ as $R\to\infty$, then we have

$$\text{P.V.}\int_{-\infty}^\infty f(x)\,dx = \lim_{R\to\infty}\int_{-R}^{R} f(x)\,dx = 2\pi i\sum_{k=1}^{n}\text{Res}(f(z),z_k). \tag{14}$$

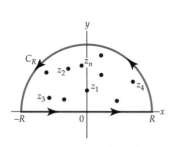

Figure 6.6.1 Semicircular contour

EXAMPLE 2   Cauchy P.V. of an Improper Integral

Evaluate the Cauchy principal value of $\displaystyle\int_{-\infty}^{\infty} \frac{1}{(x^2+1)(x^2+9)}\, dx$.

Solution Let $f(z) = 1/(z^2+1)(z^2+9)$. Since

$$(z^2+1)(z^2+9) = (z-i)(z+i)(z-3i)(z+3i),$$

we take C be the closed contour consisting of the interval $[-R,\ R]$ on the x-axis and the semicircle C_R of radius $R > 3$. As seen from Figure 6.6.2,

$$\oint_C \frac{1}{(z^2+1)(z^2+9)}\, dz = \int_{-R}^{R} \frac{1}{(x^2+1)(x^2+9)}\, dx + \int_{C_R} \frac{1}{(z^2+1)(z^2+9)}\, dz$$

$$= I_1 + I_2.$$

However, by the Residue Theorem we have:

$$I_1 + I_2 = 2\pi i\left[\text{Res}(f(z), i) + \text{Res}(f(z), 3i)\right].$$

At the simple poles $z = i$ and $z = 3i$ we find, respectively,

$$\text{Res}(f(z),\ i) = \frac{1}{16i} \quad \text{and} \quad \text{Res}(f(z),\ 3i) = -\frac{1}{48i},$$

so that

$$I_1 + I_2 = 2\pi i\left[\frac{1}{16i} + \left(-\frac{1}{48i}\right)\right] = \frac{\pi}{12}. \tag{15}$$

We now want to let $R \to \infty$ in (15). Before doing this, we use the inequality (8) of Section 1.2 to note that on the contour C_R,

$$\left|(z^2+1)(z^2+9)\right| = \left|z^2+1\right|\cdot\left|z^2+9\right| \geq \left|\left|z^2\right|-1\right|\cdot\left|\left|z^2\right|-9\right| = (R^2-1)(R^2-9).$$

Since the length L of the semicircle is πR, it follows from the *ML*-inequality, Theorem 5.2.3 of Section 5.2, that

$$|I_2| = \left|\int_{C_R} \frac{1}{(z^2+1)(z^2+9)}\, dz\right| \leq \frac{\pi R}{(R^2-1)(R^2-9)}.$$

This last result shows that $|I_2| \to 0$ as $R \to \infty$, and so we conclude that $\lim_{R\to\infty} I_2 = 0$. It follows from (15) that $\lim_{R\to\infty} I_1 = \pi/12$; in other words,

$$\lim_{R\to\infty}\int_{-R}^{R} \frac{1}{(x^2+1)(x^2+9)}\, dx = \frac{\pi}{12} \quad \text{or} \quad \text{P.V.}\int_{-\infty}^{\infty} \frac{1}{(x^2+1)(x^2+9)}\, dx = \frac{\pi}{12}.$$

Because the integrand in Example 2 is an even function, the existence of the Cauchy principal value implies that the original integral converges to $\pi/12$.

It is often tedious to have to show that the contour integral along C_R approaches zero as $R \to \infty$. Sufficient conditions under which this behavior is always true are summarized in the next theorem.

Theorem 6.6.1   Behavior of Integral as $R \to \infty$

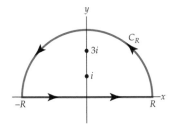

Figure 6.6.2 Contour for Example 2

Suppose $f(z) = \dfrac{p(z)}{q(z)}$ is a rational function, where the degree of $p(z)$ is n and the degree of $q(z)$ is $m \geq n + 2$. If C_R is a semicircular contour $z = Re^{i\theta}$, $0 \leq \theta \leq \pi$, then $\int_{C_R} f(z)\,dz \to 0$ as $R \to \infty$.

In other words, the integral along C_R approaches zero as $R \to \infty$ when the denominator of f is of a power at least 2 more than its numerator. The proof of this fact follows in the same manner as in Example 2. Notice in that example, the conditions stipulated in Theorem 6.6.1 are satisfied, since the degree of $p(z) = 1$ is 0 and the degree of $q(z) = (z^2 + 1)(z^2 + 9)$ is 4.

EXAMPLE 3   Cauchy P.V. of an Improper Integral

Evaluate the Cauchy principal value of $\displaystyle\int_{-\infty}^{\infty} \dfrac{1}{x^4 + 1}\,dx$.

Solution By inspection of the integrand we see that the conditions given in Theorem 6.6.1 are satisfied. Moreover, we know from Example 3 of Section 6.5 that $f(z) = 1/(z^4 + 1)$ has simple poles in the upper half-plane at $z_1 = e^{\pi i/4}$ and $z_2 = e^{3\pi i/4}$. We also saw in that example that the residues at these poles are

$$\text{Res}(f(z), z_1) = -\frac{1}{4\sqrt{2}} - \frac{1}{4\sqrt{2}}i \quad \text{and} \quad \text{Res}(f(z), z_2) = \frac{1}{4\sqrt{2}} - \frac{1}{4\sqrt{2}}i.$$

Thus, by (14),

$$\text{P.V.} \int_{-\infty}^{\infty} \frac{1}{x^4 + 1}\,dx = 2\pi i\left[\text{Res}(f(z), z_1) + \text{Res}(f(z), z_2)\right] = \frac{\pi}{\sqrt{2}}.$$

Since the integrand is an even function, the original integral converges to $\pi/\sqrt{2}$. ❑

Integrals of the Form $\int_{-\infty}^{\infty} f(x) \cos \alpha x\, dx$ and $\int_{-\infty}^{\infty} f(x) \sin \alpha x\, dx$

Because improper integrals of the form $\int_{-\infty}^{\infty} f(x) \sin\alpha x\, dx$ are encountered in applications of Fourier analysis, they often are referred to as **Fourier integrals.** Fourier integrals appear as the real and imaginary parts in the improper integral $\int_{-\infty}^{\infty} f(x)e^{i\alpha x}\, dx$.* In view of Euler's formula $e^{i\alpha x} = \cos \alpha x + i \sin \alpha x$, where α is a positive real number, we can write

$$\int_{-\infty}^{\infty} f(x)\, e^{i\alpha x} dx = \int_{-\infty}^{\infty} f(x) \cos \alpha x\, dx + i \int_{-\infty}^{\infty} f(x) \sin \alpha x\, dx \qquad (16)$$

whenever both integrals on the right-hand side converge. Suppose $f(x) = p(x)/q(x)$ is a rational function that is continuous on $(-\infty, \infty)$. Then both Fourier integrals in (10) can be evaluated at the same time by considering the complex integral $\int_C f(z)e^{i\alpha z} dz$, where $\alpha > 0$, and the contour C again

---

*See Section 6.7.

consists of the interval $[-R, R]$ on the real axis and a semicircular contour C_R with radius large enough to enclose the poles of $f(z)$ in the upper-half plane.

Before proceeding, we give, without proof, sufficient conditions under which the contour integral along C_R approaches zero as $R \to \infty$.

Theorem 6.6.2   Behavior of Integral as $R \to \infty$

Suppose $f(z) = \dfrac{p(z)}{q(z)}$ is a rational function, where the degree of $p(z)$ is n and the degree of $q(z)$ is $m \geq n + 1$. If C_R is a semicircular contour $z = Re^{i\theta}$, $0 \leq \theta \leq \pi$, and $\alpha > 0$, then $\int_{C_R} f(z) e^{i\alpha z} dz \to 0$ as $R \to \infty$.

EXAMPLE 4   Using Symmetry

Evaluate the Cauchy principal value of $\displaystyle\int_0^\infty \frac{x \sin x}{x^2 + 9} dx$.

Solution First note that the limits of integration in the given integral are not from $-\infty$ to ∞ as required by the method just described. This can be remedied by observing that since the integrand is an even function of x (verify), we can write

$$\int_0^\infty \frac{x \sin x}{x^2 + 9} dx = \frac{1}{2} \int_{-\infty}^\infty \frac{x \sin x}{x^2 + 9} dx. \tag{17}$$

With $\alpha = 1$ we now form the contour integral

$$\oint_C \frac{z}{z^2 + 9} e^{iz} dz,$$

where C is the same contour shown in Figure 6.6.2. By Theorem 6.5.3,

$$\int_{C_R} \frac{z}{z^2 + 9} e^{iz} dz + \int_{-R}^R \frac{x}{x^2 + 9} e^{ix} dx = 2\pi i \operatorname{Res}(f(z)e^{iz}, 3i),$$

where $f(z) = z/(z^2 + 9)$, and

$$\operatorname{Res}\left(f(z)e^{iz}, \ 3i\right) = \left.\frac{ze^{iz}}{2z}\right|_{z=3i} = \frac{e^{-3}}{2}$$

from (4) of Section 6.5.   Then, from Theorem 6.6.2 we conclude $\int_{C_R} f(z)e^{iz} dz \to 0$ as $R \to \infty$, and so

$$\text{P.V.} \int_{-\infty}^\infty \frac{x}{x^2 + 9} e^{ix} dx = 2\pi i \left(\frac{e^{-3}}{2}\right) = \frac{\pi}{e^3} i.$$

But by (16),

$$\int_{-\infty}^\infty \frac{x}{x^2 + 9} e^{ix} dx = \int_{-\infty}^\infty \frac{x \cos x}{x^2 + 9} dx + i \int_{-\infty}^\infty \frac{x \sin x}{x^2 + 9} dx = \frac{\pi}{e^3} i.$$

Equating real and imaginary parts in the last line gives the bonus result

$$\text{P.V.} \int_{-\infty}^\infty \frac{x \cos x}{x^2 + 9} dx = 0 \qquad \text{along with} \qquad \text{P.V.} \int_{-\infty}^\infty \frac{x \sin x}{x^2 + 9} dx = \frac{\pi}{e^3}. \tag{18}$$

Finally, in view of the fact that the integrand is an even function, we obtain the value of the prescribed integral:

$$\int_0^\infty \frac{x \sin x}{x^2 + 9}\, dx = \frac{1}{2} \int_{-\infty}^\infty \frac{x \sin x}{x^2 + 9}\, dx = \frac{\pi}{2e^3}.$$

❏

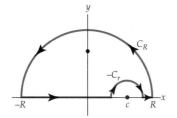

Figure 6.6.3 Indented contour

Indented Contours   The improper integrals of forms (2) and (3) that we have considered up to this point were continuous on the interval $(-\infty, \infty)$. In other words, the complex function $f(z) = p(z)/q(z)$ did not have poles on the real axis. In the situation where f has poles on the real axis, we must modify the procedure illustrated in Examples 2–4. For example, to evaluate $\int_{-\infty}^\infty f(x)\, dx$ by residues when $f(z)$ has a pole at $z = c$, where c is a real number, we use an **indented contour** as illustrated in Figure 6.6.3. The symbol C_r denotes a semicircular contour centered at $z = c$ and oriented in the *positive* direction. The next theorem is important to this discussion.

Theorem 6.6.3   Behavior of Integral as $r \to 0$

Suppose f has a simple pole $z = c$ on the real axis. If C_r is the contour defined by $z = c + re^{i\theta}$, $0 \le \theta \le \pi$, then

$$\lim_{r \to 0} \int_{C_r} f(z)\, dz = \pi i \operatorname{Res}(f(z), c).$$

Proof   Since f has a simple pole at $z = c$, its Laurent series is

$$f(z) = \frac{a_{-1}}{z - c} + g(z),$$

where $a_{-1} = \operatorname{Res}(f(z),\ c)$ and g is analytic at the point c. Using the Laurent series and the parametrization of C_r we have

$$\int_{C_r} f(z)\, dz = a_{-1} \int_0^\pi \frac{ire^{i\theta}}{re^{i\theta}}\, d\theta + ir \int_0^\pi g(c + re^{i\theta})\, e^{i\theta} d\theta = I_1 + I_2. \qquad (19)$$

First, we see that

$$I_1 = a_{-1} \int_0^\pi i\, d\theta = \pi i a_{-1} = \pi i \operatorname{Res}(f(z), c).$$

Next, g is analytic at c, and so it is continuous at this point and bounded in a neighborhood of the point; that is, there exists an $M > 0$ for which $\left| g(c + re^{i\theta}) \right| \le M$. Hence,

$$|I_2| = \left| ir \int_0^\pi g(c + re^{i\theta})\, d\theta \right| \le r \int_0^\pi M\, d\theta = \pi r M.$$

It follows from this last inequality that $\lim_{r \to 0} |I_2| = 0$ and consequently $\lim_{r \to 0} I_2 = 0$. By taking the limit of (19) as $r \to 0$, the theorem is proved.   ❏

EXAMPLE 5   Using an Indented Contour

Evaluate the Cauchy principal value of $\displaystyle\int_{-\infty}^{\infty} \frac{\sin x}{x(x^2 - 2x + 2)}\, dx$.

Solution Since the integral is of the type given in (3), we consider the contour integral

$$\oint_C \frac{e^{iz}}{z(z^2 - 2z + 2)}\, dz.$$

The function $f(z) = 1/z(z^2 - 2z + 2)$ has a pole at $z = 0$ and at $z = 1 + i$ in the upper half-plane. The contour C, shown in Figure 6.6.4, is indented at the origin. Adopting an obvious condensed notation, we have

$$\oint_C = \int_{C_R} + \int_{-R}^{-r} + \int_{-C_r} + \int_r^R = 2\pi i\, \text{Res}(f(z)e^{iz}, 1 + i), \qquad (20)$$

where $\int_{-C_r} = -\int_{C_r}$. If we take the limits of (20) as $R \to \infty$ and as $r \to 0$, it follows from Theorems 6.6.2 and 6.6.3 that

$$\text{P.V.}\int_{-\infty}^{\infty} \frac{e^{ix}}{x(x^2 - 2x + 2)}\, dx - \pi i\, \text{Res}(f(z)e^{iz}, 0) = 2\pi i\, \text{Res}(f(z)e^{iz}, 1 + i).$$

Now,

$$\text{Res}(f(z)e^{iz}, 0) = \frac{1}{2} \quad \text{and} \quad \text{Res}(f(z)e^{iz}, 1 + i) = -\frac{e^{-1+i}}{4}(1 + i).$$

Therefore,

$$\text{P.V.}\int_{-\infty}^{\infty} \frac{e^{ix}}{x(x^2 - 2x + 2)}\, dx = \pi i\left(\frac{1}{2}\right) + 2\pi i\left(-\frac{e^{-1+i}}{4}(1 + i)\right).$$

Using $e^{-1+i} = e^{-1}(\cos 1 + i\sin 1)$, simplifying, and then equating real and imaginary parts, we get from the last equality

$$\text{P.V.}\int_{-\infty}^{\infty} \frac{\cos x}{x(x^2 - 2x + 2)}\, dx = \frac{\pi}{2}\,e^{-1}(\sin 1 + \cos 1)$$

and

$$\text{P.V.}\int_{-\infty}^{\infty} \frac{\sin x}{x(x^2 - 2x + 2)}\, dx = \frac{\pi}{2}[1 + e^{-1}(\sin 1 - \cos 1)]. \qquad \square$$

6.6.3 Integration along a Branch Cut

Branch Point at $z = 0$ In the next discussion we examine integrals of the form $\int_0^\infty f(x)\, dx$, where the integrand $f(x)$ is algebraic. But similar to Example 5, these integrals require a special type of contour because when $f(x)$ is converted to a complex function, the resulting integrand $f(z)$ has, in addition to poles, a nonisolated singularity at $z = 0$. Before proceeding, the reader is encouraged to review the discussion on branch cuts in Sections 2.6 and 4.1.

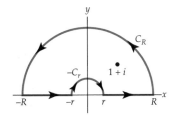

Figure 6.6.4 Indented contour for Example 5

In the example that follows we consider a special case of the real integral

$$\int_0^\infty \frac{x^{\alpha-1}}{x+1}\,dx, \tag{21}$$

where α is a real constant restricted to the interval $0 < \alpha < 1$. Observe that when $\alpha = \frac{1}{2}$ and x is replaced by z, the integrand of (12) becomes the multiple-valued function

$$\frac{1}{z^{1/2}(z+1)}. \tag{22}$$

The origin is a branch point of (22) since $z^{1/2}$ has two values for any $z \neq 0$. If you envision traveling in a complete circle around the origin $z = 0$, starting from a point $z = re^{i\theta}$, $r > 0$, you return to the same starting point z, but θ has increased by 2π. Correspondingly, the value of $z^{1/2}$ changes from $z^{1/2} = \sqrt{r}\,e^{i\theta/2}$ to a different value or different branch:

$$z^{1/2} = \sqrt{r}\,e^{i(\theta+2\pi)/2} = \sqrt{r}\,e^{i\theta/2}e^{i\pi} = -\sqrt{r}\,e^{i\theta/2}.$$

Recall, we can force $z^{1/2}$ to be single valued by restricting θ to some interval of length 2π. For (22), if we choose the positive x-axis as a branch cut, in other words by restricting θ to $0 < \theta < 2\pi$, we then guarantee that $z^{1/2} = \sqrt{r}\,e^{i\theta/2}$ is single valued. See page 114.

EXAMPLE 6   **Integration along a Branch Cut**

Evaluate $\displaystyle\int_0^\infty \frac{1}{\sqrt{x}(x+1)}\,dx$.

Solution First observe that the real integral is improper for two reasons. Notice an infinite discontinuity at $x = 0$ and the infinite limit of integration. Moreover, it can be argued from the facts that the integrand behaves like $x^{-1/2}$ near the origin and like $x^{-3/2}$ as $x \to \infty$, that the integral converges.

We form the integral $\displaystyle\oint_C \frac{1}{z^{1/2}(z+1)}\,dz$, where C is the closed contour shown in Figure 6.6.5 consisting of four components: C_r and C_R are portions of circles, and AB and ED are parallel horizontal line segments running along opposite sides of the branch cut. The integrand $f(z)$ of the contour integral is single valued and analytic on and within C, except for the simple pole at $z = -1 = e^{\pi i}$. Hence we can write

$$\oint_C \frac{1}{z^{1/2}(z+1)}\,dz = 2\pi i\,\mathrm{Res}(f(z),-1)$$

or

$$\int_{C_R} + \int_{ED} + \int_{C_r} + \int_{AB} = 2\pi i\,\mathrm{Res}(f(z),-1). \tag{23}$$

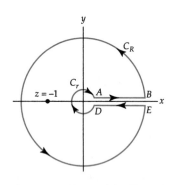

Figure 6.6.5 Contour for Example 6

Despite what is shown in Figure 6.6.5, it is permissible to think that the line segments AB and ED actually rest on the positive real axis, more precisely, AB coincides with the *upper side* of the positive real axis for which $\theta = 0$ and ED coincides with the *lower side* of the positive real axis for which $\theta = 2\pi$. On AB, $z = xe^{0i}$, and on ED, $z = xe^{(0+2\pi)i} = xe^{2\pi i}$, so that

$$\int_{ED} = \int_R^r \frac{(xe^{2\pi i})^{-1/2}}{xe^{2\pi i}+1}\,(e^{2\pi i}dx) = -\int_R^r \frac{x^{-1/2}}{x+1}\,dx = \int_r^R \frac{x^{-1/2}}{x+1}\,dx \tag{24}$$

and
$$\int_{AB} = \int_r^R \frac{(xe^{0i})^{-1/2}}{xe^{0i}+1}(e^{0i}dx) = \int_r^R \frac{x^{-1/2}}{x+1}\,dx. \tag{25}$$

Now with $z = re^{i\theta}$ and $z = Re^{i\theta}$ on C_r and C_R, respectively, it can be shown, by analysis similar to that given in Example 2 and in the proof of Theorem 6.6.1, that $\int_{C_r} \to 0$ as $r \to 0$ and $\int_{C_R} \to 0$ as $R \to \infty$. Thus from (23), (24), and (25) we see that

$$\lim_{\substack{r\to 0\\ R\to\infty}}\left[\int_{C_R} + \int_{ED} + \int_{C_r} + \int_{AB} = 2\pi i\,\mathrm{Res}(f(z),-1)\right]$$

is the same as

$$2\int_0^\infty \frac{1}{\sqrt{x}(x+1)}\,dx = 2\pi i\,\mathrm{Res}(f(z),-1). \tag{26}$$

Finally, from (4) of Section 6.5,

$$\mathrm{Res}(f(z),-1) = z^{-1/2}\Big|_{z=e^{\pi i}} = e^{-\pi i/2} = -i$$

and so (26) yields the result

$$\int_0^\infty \frac{1}{\sqrt{x}(x+1)}\,dx = \pi.$$

6.6.4 The Argument Principle and Rouché's Theorem

Argument Principle Unlike the foregoing discussion in which the focus was on the evaluation of real integrals, we next apply residue theory to the location of zeros of an analytic function. To get to that topic, we must first consider two theorems that are important in their own right.

In the first theorem we need to count the number of zeros and poles of a function f that are located within a simple closed contour C; in this counting we include the order or multiplicity of each zero and pole. For example, if

$$f(z) = \frac{(z-1)(z-9)^4(z+i)^2}{(z^2-2z+2)^2(z-i)^6(z+6i)^7} \tag{27}$$

and C is taken to be the circle $|z| = 2$, then inspection of the numerator of f reveals that the zeros inside C are $z = 1$ (a simple zero) and $z = -i$ (a zero of order or multiplicity 2). Therefore, the number N_0 of zeros inside C is taken to be $N_0 = 1 + 2 = 3$. Similarly, inspection of the denominator of f shows, after factoring $z^2 - 2z + 2$, that the poles inside C are $z = 1 - i$ (pole of order 2), $z = 1 + i$ (pole of order 2), and $z = i$ (pole of order 6). The number N_p of poles inside C is taken to be $N_p = 2 + 2 + 6 = 10$.

Theorem 6.6.4 Argument Principle

Let C be a simple closed contour lying entirely within a domain D. Suppose f is analytic in D except at a finite number of poles inside C, and that $f(z) \neq 0$ on C. Then

$$\frac{1}{2\pi i}\oint_C \frac{f'(z)}{f(z)}\,dz = N_0 - N_p, \tag{28}$$

where N_0 is the total number of zeros of f inside C and N_p is the total number of poles of f inside C. In determining N_0 and N_p, zeros and poles are counted according to their order or multiplicities.

Proof   We start with a reminder that when we use the symbol $\oint_C$ for a contour, this signifies that we are integrating in the positive direction around the closed curve C.

The integrand $f'(z)/f(z)$ in (28) is analytic in and on the contour C except at the points in the interior of C where f has a zero or a pole. If z_0 is a zero of order n of f inside C, then by (5) of Section 6.4 we can write $f(z) = (z-z_0)^n \phi(z)$, where ϕ is analytic at z_0 and $\phi(z_0) \neq 0$. We differentiate f by the product rule,

$$f'(z) = (z - z_0)^n \phi'(z) + n(z - z_0)^{n-1} \phi(z),$$

and divide this expression by f. In some punctured disk centered at z_0, we have

$$\frac{f'(z)}{f(z)} = \frac{(z - z_0)^n \phi'(z) + n(z - z_0)^{n-1} \phi(z)}{(z - z_0)^n \phi(z)} = \frac{\phi'(z)}{\phi(z)} + \frac{n}{z - z_0}. \tag{29}$$

The result in (29) shows that the integrand $f'(z)/f(z)$ has a simple pole at z_0 and the residue at that point is

$$\text{Res}\left(\frac{f'(z)}{f(z)}, z_0\right) = \lim_{z \to z_0} (z - z_0)\left[\frac{\phi'(z)}{\phi(z)} + \frac{n}{z - z_0}\right]$$

$$= \lim_{z \to z_0}\left[(z - z_0)\frac{\phi'(z)}{\phi(z)} + n\right] = 0 + n = n, \tag{30}$$

which is the order of the zero z_0.

Now if z_p is a pole of order m of f within C, then by (7) of Section 6.4 we can write $f(z) = g(z)/(z - z_p)^m$, where g is analytic at z_p and $g(z_p) \neq 0$. By differentiating, in this case $f(z) = (z - z_p)^{-m} g(z)$, we have

$$f'(z) = (z - z_p)^{-m} g'(z) - m(z - z_p)^{-m-1} g(z).$$

Therefore, in some punctured disk centered at z_p,

$$\frac{f'(z)}{f(z)} = \frac{(z - z_p)^{-m} g'(z) - m(z - z_p)^{-m-1} g(z)}{(z - z_p)^{-m} g(z)} = \frac{g'(z)}{g(z)} + \frac{-m}{z - z_p}. \tag{31}$$

We see from (31) that the integrand $f'(z)/f(z)$ has a simple pole at z_p. Proceeding as in (30), we also see that the residue at z_p is equal to $-m$, which is the negative of the order of the pole of f.

Finally, suppose that $z_{0_1}, z_{0_2}, \ldots, z_{0_r}$ and $z_{p_1}, z_{p_2}, \ldots, z_{p_s}$ are the zeros and poles of f within C and suppose further that the order of the zeros are $n_1, n_2, \ldots, n_r$ and that order of the poles are $m_1, m_2, \ldots, m_s$. Then each of these points is a simple pole of the integrand $f'(z)/f(z)$ with corresponding residues $n_1, n_2, \ldots, n_r$ and $-m_1, -m_2, \ldots, -m_s$. It follows from the residue theorem (Theorem 6.5.3) that $\oint_c [f'(z)/f(z)]\, dz$ is equal to $2\pi i$ times the sum of the residues at the poles:

$$\oint_C \frac{f'(z)}{f(z)}\, dz = 2\pi i\left[\sum_{k=1}^{r} \text{Res}\left(\frac{f'(z)}{f(z)}, z_{0_k}\right) + \sum_{k=1}^{s} \text{Res}\left(\frac{f'(z)}{f(z)}, z_{p_k}\right)\right]$$

$$= 2\pi i\left[\sum_{k=1}^{r} n_k + \sum_{k=1}^{s} (-m_k)\right] = 2\pi i[N_0 - N_p].$$

Dividing by $2\pi i$ establishes (28).   ❑

To illustrate Theorem 6.6.4, suppose the simple closed contour is $|z| = 2$ and the function f is the one given in (27). The result in (28) indicates that in the evaluation of $\oint_C [f'(z)/f(z)]\,dz$, each zero of f within C contributes $2\pi i$ times the order of multiplicity of the zero and each pole contributes $2\pi i$ times the negative of the order of the pole:

$$\oint_C \frac{f'(z)}{f(z)}\,dz = \overbrace{[2\pi i(1) + 2\pi i(2)]}^{\substack{\text{contribution}\\\text{of zeros of } f}} + \overbrace{[2\pi i(-2) + 2\pi i(-2) + 2\pi i(-6)]}^{\substack{\text{contribution}\\\text{of poles of } f}} = -14\pi i.$$

Why the Name? Why is Theorem 6.6.4 called *the argument principle*? This question may have occurred to you since no reference is made in the proof of the theorem to any arguments of complex quantities. But in point of fact there is a relation between the number $N_0 - N_p$ in Theorem 6.6.4 and $\arg(f(z))$. More precisely,

$$N_0 - N_p = \frac{1}{2\pi}\,[\text{change in arg } (f(z)) \text{ as } z \text{ traverses } C \text{ once in the positive direction}].$$

This principle can be easily verified using the simple function $f(z) = z^2$ and the unit circle $|z| = 1$ as the simple closed contour C in the z-plane. Because the function f has a zero of multiplicity 2 within C and no poles, we have $N_0 - N_p = 2$. Now, if C is parametrized by $z = e^{i\theta}$, $0 \le \theta \le 2\pi$, then its image C' in the w-plane under the mapping $w = z^2$ is $w = e^{i2\theta}$, $0 \le \theta \le 2\pi$, which is the unit circle $|w| = 1$. As z traverses C once starting at $z = 1$ ($\theta = 0$) and finishing at $z = 1$ ($\theta = 2\pi$), we see $\arg(f(z)) = \arg(w) = 2\theta$ increases from 0 to 4π. Put another way, w traverses or *winds* around the circle $|w| = 1$ twice. Thus,

$$\frac{1}{2\pi}[\text{change in arg}(f(z)) \text{ as } z \text{ traverses } C \text{ once in the positive direction}] = \frac{1}{2\pi}[4\pi - 0] = 2.$$

Rouché's Theorem The next result follows as a consequence of the argument principle. The theorem is helpful in determining the number of zeros of an analytic function.

Theorem 6.6.5 Rouché's Theorem

Let C be a simple closed contour lying entirely within a domain D. Suppose f and g are analytic in D. If the strict inequality $|f(z) - g(z)| < |f(z)|$ holds for all z on C, then f and g have the same number of zeros (counted according to their order or multiplicities) inside C.

Proof We start with the observation that the hypothesis "the strict inequality $|f(z) - g(z)| < |f(z)|$ holds for all z on C" indicates that both f and g have no zeros on the contour C. From $|f(z) - g(z)| = |g(z) - f(z)|$, we see that by dividing the inequality by $|f(z)|$ we have, for all z on C,

$$|F(z) - 1| < 1, \tag{32}$$

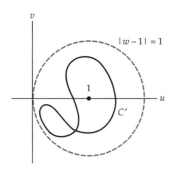

Figure 6.6.6 Image of C lies within the disk $|w - 1| < 1$.

where $F(z) = g(z)/f(z)$. The inequality in (32) shows that the image C' in the w-plane of the curve C under the mapping $w = F(z)$ is a closed path and must lie within the unit open disk $|w - 1| < 1$ centered at $w = 1$. See Figure 6.6.6. As a consequence, the curve C' does not enclose $w = 0$, and therefore $1/w$ is analytic in and on C'. By the Cauchy-Goursat theorem,

$$\int_{C'} \frac{1}{w}\,dw = 0 \qquad \text{or} \qquad \oint_C \frac{F'(z)}{F(z)}\,dz = 0, \tag{33}$$

since $w = F(z)$ and $dw = F'(z)\,dz$. From the quotient rule,

$$F'(z) = \frac{f(z)g'(z) - g(z)f'(z)}{[f(z)]^2},$$

we get

$$\frac{F'(z)}{F(z)} = \frac{g'(z)}{g(z)} - \frac{f'(z)}{f(z)}.$$

Using the last expression in the second integral in (33) then gives

$$\oint_C \left[\frac{g'(z)}{g(z)} - \frac{f'(z)}{f(z)}\right]dz = 0 \qquad \text{or} \qquad \oint_C \frac{g'(z)}{g(z)}\,dz = \oint_C \frac{f'(z)}{f(z)}\,dz.$$

It follows from (28) of Theorem 6.6.4, with $N_p = 0$, that the number of zeros of g inside C is the same as the number of zeros of f inside C. ❏

EXAMPLE 7   Location of Zeros

Locate the zeros of the polynomial function $g(z) = z^9 - 8z^2 + 5$.

Solution We begin by choosing $f(z) = z^9$ because it has the same number of zeros as g. Since f has a zero of order 9 at the origin $z = 0$, we begin our search for the zeros of g by examining circles centered at $z = 0$. In other words, *if* we can establish that $|f(z) - g(z)| < |f(z)|$ for all z on *some* circle $|z| = R$, then Theorem 6.6.5 states that f and g have the same number of zeros within the disk $|z| < R$. Now by the triangle inequality (9) of Section 1.2,

$$|f(z) - g(z)| = \left|z^9 - (z^9 - 8z^2 + 5)\right| = \left|8z^2 - 5\right| \leq 8|z|^2 + 5.$$

Also, $|f(z)| = |z|^9$. Observe that $|f(z) - g(z)| < |f(z)|$ or $8|z|^2 + 5 < |z|^9$ is *not* true for all points on the circle $|z| = 1$, so we can draw no conclusion. However, by expanding the search to the larger circle $|z| = \frac{3}{2}$ we see

$$|f(z) - g(z)| \leq 8|z|^2 + 5 = 8\left(\tfrac{3}{2}\right)^2 + 5 = 23 < \left(\tfrac{3}{2}\right)^9 = |f(z)| \tag{34}$$

since $\left(\tfrac{3}{2}\right)^9 \approx 38.44$. We conclude from (34) that because f has a zero of order 9 within the disk $|z| < \frac{3}{2}$, all nine zeros of g lie within the same disk. ❏

By slightly subtler reasoning, we can demonstrate that the function g in Example 7 has *some* zeros inside the unit disk $|z| < 1$. To see this suppose we choose $f(z) = -8z^2 + 5$. Then for all z on $|z| = 1$,

$$|f(z) - g(z)| = \left|(-8z^2 + 5) - (z^9 - 8z^2 + 5)\right| = \left|-z^9\right| = |z|^9 = (1)^9 = 1. \tag{35}$$

But from (10) of Section 1.2 we have, for all z on $|z| = 1$,

$$|f(z)| = |-f(z)| = |8z^2 - 5| \geq \left|8|z|^2 - |5|\right| = |8 - 5| = 3. \qquad (36)$$

The values in (35) and (36) show, for all z on $|z| = 1$, that $|f(z) - g(z)| < |f(z)|$. Because f has two zeros within $|z| < 1$ (namely, $\pm\sqrt{\frac{5}{8}} \approx \pm 0.79$), we can conclude from Theorem 6.6.5 that two zeros of g also lie within this disk.

We can continue the reasoning of the previous paragraph. Suppose now we choose $f(z) = 5$ and $|z| = \frac{1}{2}$. Then for all z on $|z| = \frac{1}{2}$,

$$|f(z) - g(z)| = |5 - (z^9 - 8z^2 + 5)| = |-z^9 + 8z^2| \leq |z|^9 + 8|z|^2 = \left(\tfrac{1}{2}\right)^9 + 2 \approx 2.002.$$

We now have $|f(z) - g(z)| < |f(z)| = 5$ for all z on $|z| = \frac{1}{2}$. Since f has no zeros within the disk $|z| < \frac{1}{2}$, neither does g. At this point we are able to conclude that all nine zeros of $g(z) = z^9 - 8z^2 + 5$ lie within the annular region $\frac{1}{2} < |z| < \frac{3}{2}$; two of these zeros lie within $\frac{1}{2} < |z| < 1$.

6.6.5 Summing Infinite Series

Using $\cot \pi z$ In some specialized circumstances, the residues at the simple poles of the trigonometric function $\cot \pi z$ enable us to find the sum of an infinite series.

In Section 4.3 we saw that the zeros of $\sin z$ were the real numbers $z = k\pi$, $k = 0, \pm 1, \pm 2, \ldots$. Thus the function $\cot \pi z$ has simple poles at the zeros of $\sin \pi z$, which are $\pi z = k\pi$ or $z = k$, $k = 0, \pm 1, \pm 2, \ldots$. If a polynomial function $p(z)$ has (*i*) real coefficients, (*ii*) degree $n \geq 2$, and (*iii*) no *integer* zeros, then the function

$$f(z) = \frac{\pi \cot \pi z}{p(z)} \qquad (37)$$

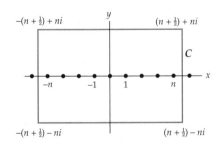

Figure 6.6.7 Rectangular contour C enclosing poles of (37)

has an infinite number of simple poles $z = 0, \pm 1, \pm 2, \ldots$ from $\cot \pi z$ and a finite number of poles $z_{p_1}, z_{p_2}, \ldots, z_{p_r}$ from the zeros of $p(z)$. The closed rectangular contour C shown in Figure 6.6.7 has vertices $\left(n + \frac{1}{2}\right) + ni$, $-\left(n + \frac{1}{2}\right) + ni$, $-\left(n + \frac{1}{2}\right) - ni$, and $\left(n + \frac{1}{2}\right) - ni$, where n is taken large enough so that C encloses the simple poles $z = 0, \pm 1, \pm 2, \ldots, \pm n$ and *all* of the poles $z_{p_1}, z_{p_2}, \ldots, z_{p_r}$. By the residue theorem,

$$\oint_C \frac{\pi \cot \pi z}{p(z)}\, dz = 2\pi i \left[\sum_{k=-n}^{n} \mathrm{Res}\left(\frac{\pi \cot \pi z}{p(z)}, k\right) + \sum_{j=1}^{r} \mathrm{Res}\left(\frac{\pi \cot \pi z}{p(z)}, z_{p_j}\right) \right]. \qquad (38)$$

In a manner similar to that used several times in the discussion in Subsection 6.6.2, it can be shown that $\oint_C \pi \cot \pi z\, dz / p(z) \to 0$ as $n \to \infty$ and so (38) becomes $0 = \sum_k \text{residues} + \sum_j \text{residues}$. That is,

$$\sum_{k=-\infty}^{\infty} \mathrm{Res}\left(\frac{\pi \cot \pi z}{p(z)}, k\right) = -\sum_{j=1}^{r} \mathrm{Res}\left(\frac{\pi \cot \pi z}{p(z)}, z_{p_j}\right). \qquad (39)$$

Now from (4) of Section 6.5 (with the identifications $g(z) = \pi \cos \pi z/p(z)$, $h(z) = \sin \pi z$, $h'(z) = \pi \cos \pi z$), it is a straightforward task to compute the residues at the simple poles $0, \pm 1, \pm 2, \ldots$:

$$\operatorname{Res}\left(\frac{\pi \cot \pi z}{p(z)}, \, k\right) = \frac{\pi \cos k\pi/p(k)}{\pi \cos k\pi} = \frac{1}{p(k)}. \tag{40}$$

By combining (40) and (39) we arrive at our desired result

$$\sum_{k=-\infty}^{\infty} \frac{1}{p(k)} = -\sum_{j=1}^{r} \operatorname{Res}\left(\frac{\pi \cot \pi z}{p(z)}, \, z_{p_j}\right). \tag{41}$$

Using csc πz   There exist several more summation formulas similar to (41). If $p(z)$ is a polynomial function satisfying the same assumptions (*i*)–(*iii*) given above, then the function

$$f(z) = \frac{\pi \csc \pi z}{p(z)} \tag{42}$$

has an infinite number of simple poles $z = 0, \pm 1, \pm 2, \ldots$ from $\csc \pi z$ and a finite number of poles $z_{p_1}, z_{p_2}, \ldots, z_{p_r}$ from the zeros of $p(z)$. In this case it can be shown that

$$\sum_{k=-\infty}^{\infty} \frac{(-1)^k}{p(k)} = -\sum_{j=1}^{r} \operatorname{Res}\left(\frac{\pi \csc \pi z}{p(z)}, \, z_{p_j}\right). \tag{43}$$

In our last example we show how to use the result in (41) to find the sum of an infinite series.

EXAMPLE 8   Summing an Infinite Series

Find the sum of the series $\sum\limits_{k=0}^{\infty} \dfrac{1}{k^2 + 4}$.

Solution   Observe that if we identify $p(z) = z^2 + 4$, then the three assumptions (*i*)–(*iii*) preceding (37) hold true. The zeros of $p(z)$ are $\pm 2i$ and correspond to simple poles of $f(z) = \pi \cot \pi z/(z^2 + 4)$. According to the formula in (41),

$$\sum_{k=-\infty}^{\infty} \frac{1}{k^2 + 4} = -\left[\operatorname{Res}\left(\frac{\pi \cot \pi z}{z^2 + 4}, -2i\right) + \operatorname{Res}\left(\frac{\pi \cot \pi z}{z^2 + 4}, 2i\right)\right]. \tag{44}$$

Now again by (4) of Section 6.5 we have

$$\operatorname{Res}\left(\frac{\pi \cot \pi z}{z^2 + 4}, -2i\right) = \frac{\pi \cot 2\pi i}{4i} \quad \text{and} \quad \operatorname{Res}\left(\frac{\pi \cot \pi z}{z^2 + 4}, 2i\right) = \frac{\pi \cot 2\pi i}{4i}.$$

The sum of the residues is $(\pi/2i) \cot 2\pi i$. This sum is a real quantity because from (27) of Section 4.3:

$$\frac{\pi}{2i} \cot 2\pi i = \frac{\pi}{2i} \frac{\cosh(-2\pi)}{(-i \sinh(-2\pi))} = -\frac{\pi}{2} \coth 2\pi.$$

Hence (44) becomes

$$\sum_{k=-\infty}^{\infty} \frac{1}{k^2 + 4} = \frac{\pi}{2} \coth 2\pi. \qquad (45)$$

This is not quite the desired result. To that end we must manipulate the summation $\sum_{k=-\infty}^{\infty}$ in order to put it in the form $\sum_{k=0}^{\infty}$. Observe

$$\sum_{k=-\infty}^{\infty} \frac{1}{k^2 + 4} = \sum_{k=-\infty}^{-1} \frac{1}{k^2 + 4} + \overbrace{\frac{1}{4}}^{\substack{k=0 \\ \text{term}}} + \sum_{k=1}^{\infty} \frac{1}{k^2 + 4}$$

$$= \sum_{k=1}^{\infty} \frac{1}{(-k)^2 + 4} + \frac{1}{4} + \sum_{k=1}^{\infty} \frac{1}{k^2 + 4}$$

$$= 2\sum_{k=1}^{\infty} \frac{1}{k^2 + 4} + \frac{1}{4} = 2\sum_{k=0}^{\infty} \frac{1}{k^2 + 4} - \frac{1}{4}. \qquad (46)$$

Finally, we obtain the sum of the original series by combining (45) with (46),

$$\sum_{k=-\infty}^{\infty} \frac{1}{k^2 + 4} = 2\sum_{k=0}^{\infty} \frac{1}{k^2 + 4} - \frac{1}{4} = \frac{\pi}{2} \coth 2\pi,$$

and solving for $\sum_{k=0}^{\infty}$:

$$\sum_{k=0}^{\infty} \frac{1}{k^2 + 4} = \frac{1}{8} + \frac{\pi}{4} \coth 2\pi. \qquad (47)$$

With the help of a calculator, we find that the right side of (47) is approximately 0.9104. ❑

EXERCISES 6.6   *Answers to selected odd-numbered problems begin on page ANS-18.*

6.6.1 Evaluation of Real Trigonometric Integrals

In Problems 1–12, evaluate the given trigonometric integral.

1. $\displaystyle\int_0^{2\pi} \frac{1}{1 + 0.5 \sin\theta} \, d\theta$

2. $\displaystyle\int_0^{2\pi} \frac{1}{10 - 6 \cos\theta} \, d\theta$

3. $\displaystyle\int_0^{2\pi} \frac{\cos\theta}{3 + \sin\theta} \, d\theta$

4. $\displaystyle\int_0^{2\pi} \frac{1}{1 + 3 \cos^2\theta} \, d\theta$

5. $\displaystyle\int_0^{\pi} \frac{1}{2 - \cos\theta} \, d\theta$ [*Hint:* Let $t = 2\pi - \theta$.]

6. $\displaystyle\int_0^{\pi} \frac{1}{1 + \sin^2\theta} \, d\theta$

7. $\displaystyle\int_0^{2\pi} \frac{\sin^2\theta}{5 + 4 \cos\theta} \, d\theta$

8. $\displaystyle\int_0^{2\pi} \frac{\cos^2\theta}{3 - \sin\theta} \, d\theta$

9. $\displaystyle\int_0^{2\pi} \frac{\cos 2\theta}{5 - 4 \cos\theta} \, d\theta$

10. $\displaystyle\int_0^{2\pi} \frac{1}{\cos\theta + 2 \sin\theta + 3} \, d\theta$

11. $\displaystyle\int_0^{2\pi} \frac{\cos^2\theta}{2 + \sin\theta} \, d\theta$

12. $\displaystyle\int_0^{2\pi} \frac{\cos 3\theta}{5 - 4 \cos\theta} \, d\theta$

In Problems 13 and 14, establish the given general result. Use Problem 13 to verify the answer in Example 1. Use Problem 14 to verify the answer to Problem 7.

13. $\displaystyle\int_0^\pi \frac{d\theta}{(a+\cos\theta)^2}\,d\theta = \frac{a\pi}{(\sqrt{a^2-1})^3},\ a>1$

14. $\displaystyle\int_0^{2\pi} \frac{\sin^2\theta}{a+b\cos\theta}\,d\theta = \frac{2\pi}{b^2}\left(a-\sqrt{a^2-b^2}\right),\ a>b>0$

6.6.2 Evaluation of Real Improper Integrals

In Problems 15–26, evaluate the Cauchy principal value of the given improper integral.

15. $\displaystyle\int_{-\infty}^\infty \frac{1}{x^2-2x+2}\,dx$

16. $\displaystyle\int_{-\infty}^\infty \frac{1}{x^2-6x+25}\,dx$

17. $\displaystyle\int_{-\infty}^\infty \frac{1}{(x^2+4)^2}\,dx$

18. $\displaystyle\int_{-\infty}^\infty \frac{x^2}{(x^2+1)^2}\,dx$

19. $\displaystyle\int_{-\infty}^\infty \frac{1}{(x^2+1)^3}\,dx$

20. $\displaystyle\int_{-\infty}^\infty \frac{x}{(x^2+4)^3}\,dx$

21. $\displaystyle\int_{-\infty}^\infty \frac{2x^2-1}{x^4+5x^2+4}\,dx$

22. $\displaystyle\int_{-\infty}^\infty \frac{1}{(x^2+1)^2(x^2+9)}\,dx$

23. $\displaystyle\int_0^\infty \frac{x^2+1}{x^4+1}\,dx$

24. $\displaystyle\int_0^\infty \frac{1}{x^6+1}\,dx$

25. $\displaystyle\int_0^\infty \frac{x^2}{x^6+1}\,dx$

26. $\displaystyle\int_{-\infty}^\infty \frac{x^2}{(x^2+2x+2)(x^2+1)^2}\,dx$

In Problems 27–38, evaluate the Cauchy principal value of the given improper integral.

27. $\displaystyle\int_{-\infty}^\infty \frac{\cos x}{x^2+1}\,dx$

28. $\displaystyle\int_{-\infty}^\infty \frac{\cos 2x}{x^2+1}\,dx$

29. $\displaystyle\int_{-\infty}^\infty \frac{x\sin x}{x^2+1}\,dx$

30. $\displaystyle\int_0^\infty \frac{\cos x}{(x^2+4)^2}\,dx$

31. $\displaystyle\int_0^\infty \frac{\cos 3x}{(x^2+1)^2}\,dx$

32. $\displaystyle\int_{-\infty}^\infty \frac{\sin x}{x^2+4x+5}\,dx$

33. $\displaystyle\int_0^\infty \frac{\cos 2x}{x^4+1}\,dx$

34. $\displaystyle\int_0^\infty \frac{x\sin x}{x^4+1}\,dx$

35. $\displaystyle\int_{-\infty}^\infty \frac{\cos x}{(x^2+1)(x^2+9)}\,dx$

36. $\displaystyle\int_0^\infty \frac{x\sin x}{(x^2+1)(x^2+4)}\,dx$

37. $\displaystyle\int_{-\infty}^\infty \frac{\sin x}{x+i}\,dx$ [*Hint:* First substitute $\sin x = (e^{ix}-e^{-ix})/2i$, x real.]

38. $\displaystyle\int_{-\infty}^\infty \frac{\cos x + x\sin x}{x^2+1}\,dx$ [*Hint:* Consider $e^{iz}/(z-i)$.]

In Problems 39–42, use an indented contour and residues to establish the Cauchy principal value of the given improper integral.

39. $\displaystyle\int_{-\infty}^\infty \frac{\sin x}{x}\,dx = \pi$

40. $\displaystyle\int_{-\infty}^\infty \frac{\sin x}{x(x^2+1)}\,dx = \pi(1-e^{-1})$

41. $\displaystyle\int_0^\infty \frac{1-\cos x}{x^2}\,dx = \frac{\pi}{2}$

42. $\displaystyle\int_{-\infty}^\infty \frac{x\cos x}{x^2-3x+2}\,dx = \pi[\sin 1 - 2\sin 2]$

6.6.3 Integration along a Branch Cut

In Problems 43–46, proceed as in Example 6 to establish the Cauchy principal value for the given improper integral.

43. $\displaystyle\int_0^\infty \frac{1}{\sqrt{x}(x^2+1)}\,dx = \frac{\pi}{\sqrt{2}}$

44. $\displaystyle\int_0^\infty \frac{1}{\sqrt{x}(x+1)(x+4)}\,dx = \frac{\pi}{3}$

45. $\displaystyle\int_0^\infty \frac{\sqrt{x}}{(x^2+1)^2}\,dx = \frac{\pi}{4\sqrt{2}}$

46. $\displaystyle\int_0^\infty \frac{x^{1/3}}{(x+1)^2}\,dx = \frac{2\pi}{3\sqrt{3}}$

In Problems 47 and 48, establish the Cauchy principal value for the given improper integral. Use Problem 47 to verify the answer in Example 6. Use Problem 48 to verify the answer to Problem 45.

47. $\displaystyle\int_0^\infty \frac{x^{\alpha-1}}{x+1}\,dx = \frac{\pi}{\sin \alpha\pi},\ 0 < \alpha < 1,$

48. $\displaystyle\int_0^\infty \frac{x^\alpha}{(x^2+1)^2}\,dx = \frac{\pi(1-\alpha)}{4\cos(\alpha\pi/2)},\ -1 < \alpha < 3,\ \alpha \neq 1$

Miscellaneous Real Integrals

49. Use the contour C shown in Figure 6.6.8 to show that

$$\text{P.V.}\int_{-\infty}^\infty \frac{e^{\alpha x}}{1+e^x}\,dx = \frac{\pi}{\sin \alpha\pi},\ 0 < \alpha < 1.$$

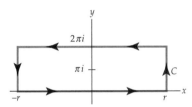

Figure 6.6.8 Figure for Problem 49

50. The integral result $\int_{-\infty}^\infty e^{-x^2}\,dx = \sqrt{\pi}$ can be established using elementary calculus and polar coordinates. Use this result, the contour integral $\oint_C e^{-z^2} e^{i\alpha z}\,dz$, and the contour C shown in Figure 6.6.9, to show that

$$\text{P.V.}\int_0^\infty e^{-x^2}\cos \alpha x\,dx = \frac{\sqrt{\pi}}{2} e^{-\alpha^2/4}.$$

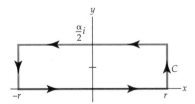

Figure 6.6.9 Figure for Problem 50

51. Discuss how to evaluate the Cauchy principal value of

$$\int_0^\infty \frac{x^{\alpha-1}}{x-1}\,dx,\ 0 < \alpha < 1.$$

Carry out your ideas.

52. **(a)** Use a graphics calculator or computer graphing program to plot on the same coordinates axes the graphs of $\sin\theta$ and $2\theta/\pi$ on the interval $0 \le \theta \le \pi/2$. Explain in graphical terms the validity of the inequality $\sin\theta \ge 2\theta/\pi$ on the interval $0 \le \theta \le \pi/2$. Use this inequality to prove that for $R > 0$,

$$\int_0^{\pi/2} e^{-R\sin\theta}\,d\theta < \frac{\pi}{2R}.$$

(b) Explain how the result in part (a) leads us to conclude that for $R > 0$,

$$\int_0^{\pi} e^{-R\sin\theta}\,d\theta < \frac{\pi}{R}. \tag{48}$$

The result in (48) is known as **Jordan's inequality,** which is often useful when evaluating integrals of the form $\int_{-\infty}^{\infty} f(x)\cos\alpha x\,dx$ and $\int_{-\infty}^{\infty} f(x)\sin\alpha x\,dx$.

53. Reconsider the integral in Problem 39 along with the indented contour in Figure 6.6.4. Use Jordan's inequality in Problem 52 to show that $\int_{C_R} \to 0$ as $R \to \infty$.

54. Investigate the integral $\int_0^{2\pi} \frac{1}{a - \sin\theta}\,d\theta$, $|a| \le 1$, in light of the evaluation procedure outlined in Subsection 6.6.1.

55. Use Euler's formula as a starting point in the evaluation of the integral

$$\int_0^{2\pi} e^{\cos\theta}[\cos(\sin\theta - n\theta) + i\sin(\sin\theta - n\theta)]\,d\theta, n = 0, 1, 2, \dots\,.$$

56. From your work in Problem 55, discern the values of the real integrals

$$\int_0^{2\pi} e^{\cos\theta}\cos(\sin\theta - n\theta)\,d\theta \quad \text{and} \quad \int_0^{2\pi} e^{\cos\theta}\sin(\sin\theta - n\theta)\,d\theta.$$

57. Suppose a real function f is continuous on the interval $[a, b]$ except at a point c within the interval. Then the **principal value** of the integral is defined by

$$\text{P.V.} \int_a^b f(x)\,dx = \lim_{\varepsilon \to 0}\left[\int_a^{c-\varepsilon} f(x)\,dx + \int_{c+\varepsilon}^b f(x)\,dx\right], \ \varepsilon > 0.$$

Compute the principal value of $\int_0^3 \frac{1}{x-1}\,dx$.

58. Determine whether the integral in Problem 57 converges.

6.6.4 The Argument Principle and Rouché's Theorem

In Problems 59 and 60, use the argument principle in (28) of Theorem 6.6.4 to evaluate the integral $\oint_C \frac{f'(z)}{f(z)}\,dz$ for the given function f and closed contour C.

59. $f(z) = z^6 - 2iz^4 + (5 - i)z^2 + 10$, C encloses all the zeros of f

60. $f(z) = \dfrac{(z - 3iz - 2)^2}{z(z^2 - 2z + 2)^5}$, C is $|z| = \frac{3}{2}$

In Problems 61–64, use the argument principle in (28) of Theorem 6.6.4 to evaluate the given integral on the indicated closed contour C. You will have to identify $f(z)$ and $f'(z)$.

61. $\oint_C \dfrac{2z+1}{z^2+z}\, dz$, C is $|z| = 2$ **62.** $\oint_C \dfrac{z}{z^2+4}\, dz$, C is $|z| = 3$

63. $\oint_C \cot z\, dz$, C is the rectangular contour with vertices $10+i$, $-4+i$, $-4-i$, and $10-i$.

64. $\oint_C \tan \pi z\, dz$, C is $|z-1| = 2$

65. Use Rouché's theorem (Theorem 6.6.5) to show that all seven of the zeros of $g(z) = z^7 + 10z^3 + 14$ lie within the annular region $1 < |z| < 2$.

66. **(a)** Use Rouché's theorem (Theorem 6.6.5) to show that all four of the zeros of $g(z) = 4z^4 + 2(1-i)z + 1$ lie within the disk $|z| < 1$.

   (b) Show that three of the zeros of the function g in part (a) lie within the annular region $\frac{1}{2} < |z| < 1$.

67. In the proof of Theorem 6.6.5, explain how the hypothesis that the strict inequality $|f(z) - g(z)| < |f(z)|$ holds for all z on C implies that f and g cannot have zeros on C.

6.6.5 Summing Infinite Series

68. **(a)** Use the procedure illustrated in Example 8 to obtain the general result

$$\sum_{k=0}^{\infty} \frac{1}{k^2 + a^2} = \frac{1}{2a^2} + \frac{\pi}{2a} \coth a\pi.$$

   (b) Use part (a) to verify (47) when $a = 2$.

   (c) Find the sum of the series $\displaystyle\sum_{k=0}^{\infty} \frac{1}{k^2 + 1}$.

In Problems 69 and 70, use (41) find the sum of the given series.

69. $\displaystyle\sum_{k=1}^{\infty} \frac{1}{(2k-1)^2}$ **70.** $\displaystyle\sum_{k=0}^{\infty} \frac{1}{16k^2 + 16k + 3}$

In Problems 71 and 72, use (43) find the sum of the given series.

71. $\displaystyle\sum_{k=-\infty}^{\infty} \frac{(-1)^k}{(4k+1)^2}$ **72.** $\displaystyle\sum_{k=0}^{\infty} \frac{(-1)^k}{(2k+1)^3}$

73. **(a)** Use (41) to obtain the general result

$$\sum_{k=-\infty}^{\infty} \frac{1}{(k-a)^2} = \frac{\pi^2}{\sin^2 \pi a}$$

   where $a \neq 0, \pm 1, \pm 2, \ldots$.

   (b) Use part (a) to verify your answer to Problem 69.

74. **(a)** Use (43) to obtain the general result

$$\sum_{k=-\infty}^{\infty} \frac{(-1)^k}{(k+a)^2} = \frac{\pi^2 \cos \pi a}{\sin^2 \pi a},$$

   where $a \neq 0, \pm 1, \pm 2, \ldots$.

   (b) Use part (a) to verify your answer to Problem 71

6.7   Applications

In other courses in mathematics or engineering you may have used the **Laplace transform** of a real function f defined for $t \geq 0$,

$$\mathcal{L}\{f(t)\} = \int_0^\infty e^{-st} f(t) \, dt. \tag{1}$$

When the integral in (1) converges, the result is a function of s. It is common practice to emphasize the relationship between a function and its transform by using a lowercase letter to denote the function and the corresponding uppercase letter to denote its Laplace transform, for example $\mathcal{L}\{f(t)\} = F(s)$, $\mathcal{L}\{y(t)\} = Y(s)$, and so on.

In the application of (1) we face two problems:

(*i*) *The direct problem*: Given a function $f(t)$ satisfying certain conditions, find its Laplace transform.

(*ii*) *The inverse problem*: Find the function $f(t)$ that has a given transform $F(s)$.

The function $F(s)$ is called the **inverse Laplace transform** and is denoted by $\mathcal{L}^{-1}\{F(s)\}$.

The Laplace transform is an invaluable aid in solving certain kinds of applied problems involving differential equations. In these problems we deal with the transform $Y(s)$ of an unknown function $y(t)$. The determination of $y(t)$ requires the computation of $\mathcal{L}^{-1}\{Y(s)\}$. In the case when $Y(s)$ is a rational function of s, you may recall employing partial fractions, operational properties, or tables to compute this inverse.

We will see in this section that the inverse Laplace transform is not merely a symbol but actually another integral transform. The reason why you did not use this inverse integral transform in previous courses is that it is a special type of complex contour integral.

We begin with a review of the notion of integral transform pairs. The section concludes with a brief introduction to the **Fourier transform**.

Integral Transforms   Suppose $f(x, y)$ is a real-valued function of two real variables. Then a definite integral of f with respect to one of the variables leads to a function of the other variable. For example, if we hold y constant, integration with respect to the real variable x gives $\int_1^2 4xy^2 dx = 6y^2$. Thus a definite integral such as $F(\alpha) = \int_a^b f(x) K(\alpha, x) \, dx$ transforms a function f of the variable x into a function F of the variable α. We say that

$$F(\alpha) = \int_a^b f(x) K(\alpha, x) \, dx \tag{2}$$

is an **integral transform** of the function f. Integral transforms appear in **transform pairs**. This means that the original function f can be recovered by another integral transform

$$f(x) = \int_c^d F(\alpha) H(\alpha, x) \, d\alpha, \tag{3}$$

called the **inverse transform**. The function $K(\alpha, x)$ in (2) and the function $H(\alpha, x)$ in (3) are called the **kernels** of their respective transforms. We note that if α represents a complex variable, then the definite integral (3) is replaced by a contour integral.

The Laplace Transform   Suppose now in (2) that the symbol α is replaced by the symbol s, and that f represents a represents a real function* that is defined on the unbounded interval $[0, \infty)$. Then (2) is an improper integral and is defined as the limit

$$\int_0^\infty K(s,t)f(t)\, dt = \lim_{b\to\infty} \int_0^b K(s,t)f(t)\, dt. \qquad (4)$$

If the limit in (4) exists, we say that the integral exists or is convergent; if the limit does not exist the integral does not exist and is said to be divergent. The choice $K(s,t) = e^{-st}$, where s is a complex variable, for the kernel in (4) gives the Laplace transform $\mathscr{L}\{f(t)\}$ defined previously in (1). The integral that defines the Laplace transform may not converge for certain kinds of functions f. For example, neither $\mathscr{L}\left\{e^{t^2}\right\}$ nor $\mathscr{L}\{1/t\}$ exist. Also, the limit in (4) will exist for only certain values of the variable s.

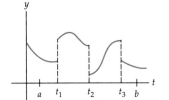

Figure 6.7.1 Piecewise continuity on $[0, \infty)$

EXAMPLE 1   **Existence of a Laplace Transform**

The Laplace transform of $f(t) = 1$, $t \geq 0$ is

$$\mathscr{L}\{1\} = \int_0^\infty e^{-st}(1)\, dt = \lim_{b\to\infty} \int_0^b e^{-st}\, dt$$

$$= \lim_{b\to\infty} \frac{-e^{-st}}{s}\bigg|_0^b = \lim_{b\to\infty} \frac{1 - e^{-sb}}{s}. \qquad (5)$$

If s is a complex variable, $s = x + iy$, then recall

$$e^{-sb} = e^{-bx}(\cos by + i\sin by). \qquad (6)$$

From (6) we see in (5) that $e^{-sb} \to 0$ as $b \to \infty$ if $x > 0$. In other words, (5) gives $\mathscr{L}\{1\} = \dfrac{1}{s}$, provided $\text{Re}(s) > 0$. ❑

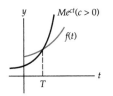

Figure 6.7.2 Exponential order

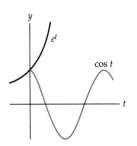

Figure 6.7.3 $f(t) = \cos t$ is of exponential order $c = 0$.

Existence of $\mathscr{L}\{f(t)\}$ Conditions that are *sufficient* to guarantee the existence of $\mathscr{L}\{f(t)\}$ are that f be piecewise continuous on $[0, \infty)$ and that f be of exponential order. Recall from elementary calculus, **piecewise continuity** on $[0, \infty)$ means that on any interval there are at most a finite number of points t_k, $k = 1, 2, \ldots, n$, $t_{k-1} < t_k$, at which f has finite discontinuities and is continuous on each open interval $t_{k-1} < t < t_k$. See Figure 6.7.1. A function f is said to be **exponential order c** if there exist constants c, $M > 0$, and $T > 0$ so that $|f(t)| \leq Me^{ct}$, for $t > T$. The condition $|f(t)| \leq Me^{ct}$ for $t > T$ states that the graph of f on the interval (T, ∞) does not grow faster than the graph of the exponential function Me^{ct}. See Figure 6.7.2. Alternatively, $e^{-ct}|f(t)|$ is bounded; that is, $e^{-ct}|f(t)| \leq M$ for $t > T$. As can be seen in Figure 6.7.3, the function $f(t) = \cos t$, $t \geq 0$ is of exponential order $c = 0$ for $t > 0$. Indeed, it follows that all bounded functions are necessarily of exponential order $c = 0$.

---

*On occasion $f(t)$ could be a complex-valued function of a real variable t.

> **Theorem 6.7.1   Sufficient Conditions for Existence**
>
> Suppose f is piecewise continuous on $[0, \infty)$ and of exponential order c for $t > T$. Then $\mathscr{L}\{f(t)\}$ exists for $\mathrm{Re}(s) > c$.

Proof   By the additive interval property of definite integrals,

$$\mathscr{L}\{f(t)\} = \int_0^T e^{-st}f(t)\,dt + \int_T^\infty e^{-st}f(t)\,dt = I_1 + I_2.$$

The integral I_1 exists since it can be written as a sum of integrals over intervals on which $e^{-st}f(t)$ is continuous. To prove the existence of I_2, we let s be a complex variable $s = x + iy$. Then using $|e^{-st}| = |e^{-xt}(\cos yt - i\sin yt)| = e^{-xt}$ and the definition of exponential order that $|f(t)| \le Me^{ct}$, $t > T$, we get

$$|I_2| \le \int_T^\infty \left|e^{st}f(t)\right|\,dt \le M\int_T^\infty e^{-xt}e^{ct}\,dt$$

$$= M\int_T^\infty e^{-(x-c)t}\,dt = -M\frac{e^{-(x-c)t}}{x-c}\bigg|_T^\infty = M\frac{e^{-(x-c)T}}{x-c}$$

for $x = \mathrm{Re}(s) > c$. Since $\int_T^\infty Me^{-(x-c)t}\,dt$ converges, the integral $\int_T^\infty |e^{-st}f(t)|\,dt$ converges by the comparison test for improper integrals. This, in turn, implies that I_2 exists for $\mathrm{Re}(s) > c$. The existence of I_1 and I_2 implies that $\mathscr{L}\{f(t)\} = \int_0^\infty e^{-st}f(t)\,dt$ exists for $\mathrm{Re}(s) > c$.   ❏

With the foregoing concepts in mind we state the next theorem without proof.

> **Theorem 6.7.2   Analyticity of the Laplace Transform**
>
> Suppose f is piecewise continuous on $[0, \infty)$ and of exponential order c for $t \ge 0$. Then the Laplace transform of f,
>
> $$F(s) = \int_0^\infty e^{-st}f(t)\,dt$$
>
> is an analytic function in the right half-plane defined by $\mathrm{Re}(s) > c$.

The Inverse Laplace Transform   Although Theorem 6.7.2 indicates that the complex function $F(s)$ is analytic to the right of the line $x = c$ in the complex plane, $F(s)$ will, in general, have singularities to the left of that line. We are now in a position to give the integral form of the inverse Laplace transform.

Theorem 6.7.3 Inverse Laplace Transform

If f and f' are piecewise continuous on $[0, \infty)$ and f is of exponential order c for $t \geq 0$, and $F(s)$ is a Laplace transform, then the **inverse Laplace transform** $\mathscr{L}^{-1}\{F(s)\}$ is

$$f(t) = \mathscr{L}^{-1}\{F(s)\} = \frac{1}{2\pi i} \lim_{R \to \infty} \int_{\gamma - iR}^{\gamma + iR} e^{st} F(s)\, ds, \qquad (7)$$

where $\gamma > c$.

The limit in (7), which defines a principal value of the integral, is usually written as

$$f(t) = \mathscr{L}^{-1}\{F(s)\} = \frac{1}{2\pi i} \int_{\gamma - i\infty}^{\gamma + i\infty} e^{st} F(s)\, ds, \qquad (8)$$

where the limits of integration indicate that the integration is along the infinitely long vertical-line contour $\mathrm{Re}(s) = x = \gamma$. Here γ is a positive real constant greater than c and greater than all the real parts of the singularities in the left half-plane. The integral in (8) is called a **Bromwich contour integral**. Relating (8) back to (3), we see that the kernel of the inverse transform is $H(s, t) = e^{st}/2\pi i$.

The fact that $F(s)$ has singularities $s_1, s_2, \ldots, s_n$ to the left of the line $x = \gamma$ makes it possible for us to evaluate (7) by using an appropriate closed contour encircling the singularities. A closed contour C that is commonly used consists of a semicircle C_R of radius R centered at $(\gamma, 0)$ and a vertical line segment L_R parallel to the y-axis passing through the point $(\gamma, 0)$ and extending from $y = \gamma - iR$ to $y = \gamma + iR$. See Figure 6.7.4. We take the radius R of the semicircle to be larger than the largest number in set of moduli of the singularities $\{\,|s_1|, |s_2|, \ldots, |s_n|\,\}$, that is, large enough so that all the singularities lie within the semicircular region. With the contour C chosen in this manner, (7) can often be evaluated using Cauchy's residue theorem. If we allow the radius R of the semicircle to approach ∞, the vertical part of the contour approaches the infinite vertical line that is the contour in (8).

We use the contour just described in the proof of the following theorem.

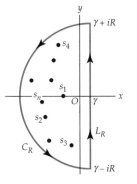

Figure 6.7.4 Possible contour that could be used to evaluate (7)

Theorem 6.7.4 Inverse Laplace Transform

Suppose $F(s)$ is a Laplace transform that has a finite number of poles $s_1, s_2, \ldots, s_n$ to the left of the vertical line $\mathrm{Re}(s) = \gamma$ and that C is the contour illustrated in Figure 6.7.4. If $sF(s)$ is bounded on C_R as $R \to \infty$, then

$$\mathscr{L}^{-1}\{F(s)\} = \sum_{k=1}^{n} \mathrm{Res}\left(e^{st} F(s), s_k\right). \qquad (9)$$

Proof From Figure 6.7.4 and Cauchy's residue theorem, we have

$$\int_{C_R} e^{st}F(s)\,ds + \int_{L_R} e^{st}F(s)\,ds = 2\pi i\sum_{k=1}^{n}\operatorname{Res}\left(e^{st}F(s),\ s_k\right)$$

or $\dfrac{1}{2\pi i}\displaystyle\int_{\gamma-iR}^{\gamma+iR} e^{st}F(s)\,ds = \sum_{k=1}^{n}\operatorname{Res}\left(e^{st}F(s),\ s_k\right) - \dfrac{1}{2\pi i}\int_{C_R} e^{st}F(s)\,ds.$ (10)

The theorem is justified by letting $R \to \infty$ in (10) and showing that $\lim_{R\to\infty}\int_{C_R} e^{st}F(s)\,ds = 0$. Now if the semicircle C_R is parametrized by $s = \gamma + Re^{i\theta}$, $\pi/2 \le \theta \le 3\pi/2$, then $ds = Rie^{i\theta}d\theta = (s-\gamma)id\theta$, and so,

$$\frac{1}{2\pi i}\int_{C_R} e^{st}F(s)\,ds = \frac{1}{2\pi i}\int_{\pi/2}^{3\pi/2} e^{\gamma t + Rte^{i\theta}}F(\gamma + Re^{i\theta})Rie^{i\theta}\,d\theta.$$

Thus,

$$\frac{1}{2\pi}\left|\int_{C_R} e^{st}F(s)\,ds\right| \le \frac{1}{2\pi}\int_{\pi/2}^{3\pi/2}\left|e^{\gamma t + Rte^{i\theta}}\right|\left|F(\gamma + Re^{i\theta})\right|\left|Rie^{i\theta}\right|d\theta.$$ (11)

To find an upper bound for the expression in (11) we examine the three moduli of the integrand of the right-hand side. First,

$$\left|e^{\gamma t + Rte^{i\theta}}\right| = \left|e^{\gamma t}e^{Rt(\cos\theta + i\sin\theta)}\right| \overset{\downarrow \text{ since }|e^{iRt\sin\theta}|=1}{=} e^{\gamma t}e^{Rt\cos\theta}.$$

Next, for R sufficiently large, we can write

$$\left|Rie^{i\theta}\right| = |s-\gamma||i| \le |s| + |\gamma| < |s| + |s| = 2|s| \text{ and } |sF(s)| < M.$$

The first of these two inequalities follows from the triangle inequality, and the second from the hypothesis that $sF(s)$ is bounded C_R as $R \to \infty$. Thus the inequality in (11) continues as

$$\frac{1}{2\pi}\left|\int_{C_R} e^{st}F(s)\,ds\right| \le \frac{M}{\pi}e^{\gamma t}\int_{\pi/2}^{3\pi/2} e^{Rt\cos\theta}\,d\theta.$$ (12)

If we let $\theta = \phi + \pi/2$, then the integral on the right-hand side of (12) becomes $\int_0^\pi e^{-Rt\sin\phi}\,d\phi$. Because the integrand is symmetric about the line $\theta = \pi/2$, we have

$$\int_0^\pi e^{-Rt\sin\phi}\,d\phi = 2\int_0^{\pi/2} e^{-Rt\sin\phi}\,d\phi.$$ (13)

Now since $\sin\phi \ge 2\phi/\pi$,[*] it follows that

$$2\int_0^{\pi/2} e^{-Rt\sin\phi}\,d\phi \le 2\int_0^{\pi/2} e^{-2Rt\phi/\pi}\,d\phi = -\frac{\pi}{Rt}e^{-2Rt\phi/\pi}\Big|_0^{\pi/2} = \frac{\pi}{Rt}\left[1 - e^{-Rt}\right].$$ (14)

Thus (11), (12), (13), and (14) together give

$$\frac{1}{2\pi}\left|\int_{C_R} e^{st}F(s)\,ds\right| \le \frac{Me^{\gamma t}}{Rt}\left[1 - e^{-Rt}\right].$$ (15)

---

*See Problem 52 in Exercises 6.6.

Since the right-hand side of (15) approaches zero as $R \to \infty$ for $t > 0$ we conclude that $\lim\limits_{R \to \infty} \int_{C_R} e^{st} F(s)\, ds = 0$. Finally, as $R \to \infty$ we see from (10) that

$$\mathscr{L}^{-1}\{F(s)\} = \frac{1}{2\pi i} \int_{\gamma - i\infty}^{\gamma + i\infty} e^{st} F(s)\, ds = \sum_{k=1}^{n} \mathrm{Res}\left(e^{st} F(s),\ s_k\right).$$

❏

In the following examples we assume that the hypotheses of Theorem 6.7.4 are satisfied.

EXAMPLE 2   Inverse Laplace Transform

Evaluate $\mathscr{L}^{-1}\left\{\dfrac{1}{s^3}\right\}$, $\mathrm{Re}(s) > 0$.

Solution   Considered as a function of a complex variable s, the function $F(s) = 1/s^3$ has a pole of order 3 at $s = 0$. Thus by (9) and (2) of Section 6.5:

$$f(t) = \mathscr{L}^{-1}\left\{\frac{1}{s^3}\right\} = \mathrm{Res}\left(e^{st}\frac{1}{s^3}, 0\right) = \frac{1}{2}\lim_{s \to 0}\frac{d^2}{ds^2}\left[(s-0)^3\frac{e^{st}}{s^3}\right]$$

$$= \frac{1}{2}\lim_{s \to 0}\frac{d^2}{ds^2}e^{st}$$

$$= \frac{1}{2}\lim_{s \to 0} t^2 e^{st}$$

$$= \frac{1}{2}t^2.$$

❏

Those readers familiar with the Laplace transform recognize that the answer in Example 1 is consistent (for $n = 2$) with the result $\mathscr{L}\{t^n\} = n!/s^{n+1}$ found in all tables of Laplace transforms.

The Laplace transform (1) utilizes only the values of a function $f(t)$ for $t > 0$, and so f is often taken to be 0 for $t < 0$. This is no major handicap because the functions we deal with in applications are for the most part defined only for $t > 0$. Although we shall not delve into details, the inversion integral (7) can be derived from a result known as the Fourier integral formula. In that analysis it is shown that

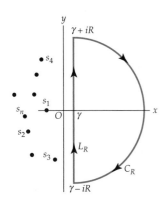

Figure 6.7.5 Contour for inversion integral (7) for $t < 0$

$$\frac{1}{2\pi i} \int_{\gamma - i\infty}^{\gamma + i\infty} e^{st} F(s)\, ds = \begin{cases} f(t), & t > 0 \\ 0, & t < 0. \end{cases} \tag{16}$$

This result is hinted at in the proof of Theorem 6.7.4 Notice from (15) that the conclusion $\lim\limits_{R \to \infty} \int_{C_R} e^{st} F(s)\, ds = 0$ is *not valid* for $t < 0$. However, if we close the contour to the right for $t < 0$, as shown in Figure 6.7.5, then $\dfrac{1}{2\pi i} \int_{\gamma - i\infty}^{\gamma + i\infty} e^{st} F(s)\, ds = 0$, which is consistent with (16). We use these results in the next example.

EXAMPLE 3   Inverse Laplace Transform

Evaluate $\mathscr{L}^{-1}\left\{\dfrac{e^{-2s}}{(s-1)(s-3)}\right\}$, $\mathrm{Re}(s) > 3$.

Solution Before we calculate the residues at the simple poles at $s = 1$ and $s = 3$, we note, after combining the two exponential functions and replacing the symbol t by $t - 2$, that (16) gives

$$\frac{1}{2\pi i}\int_{\gamma - i\infty}^{\gamma + i\infty}\frac{e^{s(t-2)}}{(s-1)(s-3)}ds = \begin{cases} f(t), & t - 2 > 0 \\ 0, & t - 2 < 0. \end{cases} \tag{17}$$

Thus from (17), (9), and (1) of Section 6.5,

$$f(t) = \mathscr{L}^{-1}\left\{\frac{e^{-2s}}{(s-1)(s-3)}\right\} = \mathrm{Res}\left(e^{st}\frac{e^{-2s}}{(s-1)(s-3)}, 1\right) + \mathrm{Res}\left(e^{st}\frac{e^{-2s}}{(s-1)(s-3)}, 3\right)$$

$$= \lim_{s\to 1}(s-1)\frac{e^{s(t-2)}}{(s-1)(s-3)} + \lim_{s\to 3}(s-3)\frac{e^{s(t-2)}}{(s-1)(s-3)}$$

$$= -\frac{1}{2}e^{t-2} + \frac{1}{2}e^{3(t-2)}.$$

In other words,

$$f(t) = \begin{cases} -\frac{1}{2}e^{t-2} + \frac{1}{2}e^{3(t-2)}, & t > 2 \\ 0, & t < 2. \end{cases} \tag{18}$$

❏

In the study of the Laplace transform the **unit step function**,

$$\mathscr{U}(t-a) = \begin{cases} 1, & t \geq a \\ 0, & t < a \end{cases}$$

proves to be extremely useful when working with piecewise continuous functions. The discontinuous function in (18) can be written as

$$f(t) = -\tfrac{1}{2}e^{t-2}\mathscr{U}(t-2) + \tfrac{1}{2}e^{3(t-2)}\mathscr{U}(t-2).$$

Fourier Transform Suppose now that $f(x)$ is a real function defined on the interval $(-\infty, \infty)$. Another important transform pair is the **Fourier transform**

$$\mathscr{F}\{f(x)\} = \int_{-\infty}^{\infty} f(x)e^{i\alpha x}dx = F(\alpha) \tag{19}$$

and the **inverse Fourier transform**

$$\mathscr{F}^{-1}\{F(\alpha)\} = \frac{1}{2\pi}\int_{-\infty}^{\infty} F(\alpha)e^{-i\alpha x}d\alpha = f(x). \tag{20}$$

Matching (19) and (20) with (2) and (3), we see that the kernel of the Fourier transform is $K(\alpha, x) = e^{i\alpha x}$, whereas the kernel of the inverse transform is $H(\alpha, x) = e^{-i\alpha x}/2\pi$. In (19) and (20) we assume that α is a real variable.

Also, observe that in contrast to (7), the inverse transform (20) is not a contour integral.

EXAMPLE 4   Fourier Transform

Find the Fourier transform of $f(x) = e^{-|x|}$.

Solution   The graph of f,

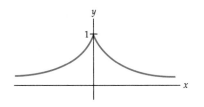

Figure 6.7.6 Graph of f in Example 4

$$f(x) = \begin{cases} e^x, & x < 0 \\ e^{-x}, & x \geq 0 \end{cases},\tag{21}$$

is given in Figure 6.7.6. From the expanded definition of f in (21), it follows from (19) that the Fourier transform of f is

$$\mathscr{F}\{f(x)\} = \int_{-\infty}^{0} e^x e^{i\alpha x}dx + \int_{0}^{\infty} e^{-x}e^{i\alpha x}dx = I_1 + I_2.\tag{22}$$

We shall begin by evaluating the improper integral I_2. One of several ways of proceeding is to write:

$$I_2 = \lim_{b\to\infty}\int_{0}^{b} e^{-x(1-\alpha i)}dx = \lim_{b\to\infty}\frac{e^{-x(1-\alpha i)}}{\alpha i - 1}\Big|_{0}^{b} = \lim_{b\to\infty}\frac{e^{-b(1-\alpha i)}-1}{\alpha i - 1}$$

$$= \frac{1}{\alpha i - 1}\lim_{b\to\infty}\left[e^{-b}\cos b\alpha + ie^{-b}\sin b\alpha - 1\right] = \frac{1}{1-\alpha i}.$$

Here we have used $\lim_{b\to\infty} e^{-b}\cos b\alpha = 0$ and $\lim_{b\to\infty} e^{-b}\sin b\alpha = 0$ for $b > 0$.

The integral I_1 can be evaluate in the same manner to obtain

$$I_1 = \frac{1}{1+\alpha i}.$$

Adding I_1 and I_2 gives the value of the Fourier transform (22):

$$\mathscr{F}\{f(x)\} = \frac{1}{1-\alpha i} + \frac{1}{1+\alpha i} \quad\text{or}\quad F(\alpha) = \frac{2}{1+\alpha^2}.$$

❑

EXAMPLE 5   Inverse Fourier Transform

Find the inverse Fourier transform of $F(\alpha) = \dfrac{2}{1+\alpha^2}$.

Solution   The idea here is to recover the function f in Example 4 from the inverse transform (20),

$$\mathscr{F}^{-1}\{F(\alpha)\} = \frac{1}{2\pi}\int_{-\infty}^{\infty}\frac{2}{1+\alpha^2}e^{-i\alpha x}\,d\alpha = f(x).\tag{23}$$

To evaluate (23), we let z be a complex variable and introduce the contour integral $\displaystyle\oint_{C}\frac{1}{\pi(1+z^2)}e^{-izx}dz$. Note that the integrand has simple poles at $z = \pm i$. From here on, the procedure used is basically the same as that used to evaluate trigonometric integrals in the preceding section by the theory of

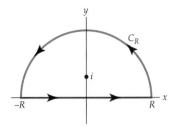

Figure 6.7.7 First contour used to evaluate (23)

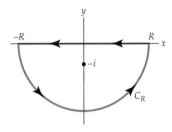

Figure 6.7.8 Second contour used to evaluate (23)

residues. The contour C shown in Figure 6.7.7 encloses the simple pole $z = i$ in the upper plane and consists of the interval $[-R, \ R]$ on the real axis and a semicircular contour C_R, where $R > 1$. Formally, we have

$$\oint_C \frac{1}{\pi(1+z^2)} e^{-izx} dz = 2\pi i \operatorname{Res}\left(\frac{1}{\pi(1+z^2)} e^{-izx}, \ i\right) = e^x. \qquad (24)$$

Obviously the result in (24) is not the function f that we started with in Example 4. A more detailed analysis in this case would reveal that the contour integral along C_R approaches zero as $R \to \infty$ only if we assume that $x < 0$. In other words, the answer in (24) is actually e^x, $x < 0$.

If we consider $\oint_C \frac{1}{\pi(1+z^2)} e^{-izx} dz$, where C is the contour in Figure 6.7.8, it can be shown that the integral along C_R now approaches zero as $R \to \infty$ when x is assumed to be positive. Hence,

$$\oint_C \frac{1}{\pi(1+z^2)} e^{-izx} dz = -2\pi i \operatorname{Res}\left(\frac{1}{\pi(1+z^2)} e^{-izx}, \ -i\right) = e^{-x}, \ x > 0. \qquad (25)$$

Note the extra minus sign appearing in front of the factor $2\pi i$ on the right side of (25). This sign comes from the fact that on C in Figure 6.7.8, $\oint_C = \int_{C_R} + \int_R^{-R} = \int_{C_R} - \int_{-R}^R = 2\pi i \operatorname{Res}(z = -i)$. As $R \to \infty$, $\int_{C_R} \to 0$ for $x > 0$, we then have $-\lim_{R\to\infty} \int_{-R}^R = 2\pi i \operatorname{Res}(z = -i)$ or $\lim_{R\to\infty} \int_{-R}^R = -2\pi i \operatorname{Res}(z = -i)$. By combining (23), (24), and (25), we arrive at

$$\mathscr{F}^{-1}\{F(\alpha)\} = \frac{1}{2\pi} \int_{-\infty}^{\infty} \frac{2}{1+\alpha^2} e^{-i\alpha x}\, d\alpha = \begin{cases} e^x, & x < 0 \\ e^{-x}, & x > 0 \end{cases}$$

which agrees with (21). Note that when $x = 0$ in (23) conventional integration gives the value 1, which is $f(0)$ in (21). ❏

Remarks

(i) The two conditions of piecewise continuity and exponential order are *sufficient* but not *necessary* for the existence of $F(s) = \mathscr{L}\{f(t)\}$. For example, the function $f(t) = t^{-1/2}$ is not piecewise continuous on $[0, \infty)$ (Why not?); nevertheless $\mathscr{L}\{t^{-1/2}\}$ exists.

(ii) We have assumed that $F(s)$ has a finite number of poles in the complex plane. This is usually the case when $F(s)$ arises from the solution of an *ordinary* differential equation. In the solution of applied problems involving a *partial* differential equation it is not uncommon to obtain a function $F(s)$ with an infinite number of poles. Although the proof of Theorem 6.7.4 is not valid when $F(s)$ has an infinite number of poles in the left half-plane $\operatorname{Re}(s) < c$, the result stated in the theorem is valid. In this case the value of the integral is an infinite series obtained from the infinite sum of the residues.

(*iii*) Although we have illustrated the use of (1) when the singularities of $F(s)$ are poles, its principal use is to compute inverse transforms of more complicated functions such as $F(s) = (s^2 + a^2)^{-1/2}$.

(*iv*) We did not mention conditions under which the Fourier transform (19) of a function $f(x)$ exists. These conditions are considerably more demanding than those stated for the existence of the Laplace transform. For example, $\mathscr{L}\{1\} = 1/s$ but $\mathscr{F}\{1\}$ does not exist. For more information on the theory and applications of the Fourier integral you are urged to consult texts on Fourier analysis or advanced engineering mathematics.*

EXERCISES 6.7   *Answers to selected odd-numbered problems begin on page ANS-19.*

In Problems 1–4, find the Laplace transform of the given function. Determine a condition on s that is sufficient to guarantee the existence of $F(s) = \mathscr{L}\{f(t)\}$.

1. $f(t) = e^{5t}$

2. $f(t) = e^{(-2+3i)t}$

3. $f(t) = \sin 3t$

4. $f(t) = e^t \cos t$

5. Generalize the result in Problem 1 and state of condition on s that is sufficient to guarantee the existence of $\mathscr{L}\{e^{kt}\}$ when k is a real constant.

6. Generalize the result in Problem 2 and state of condition on s that is sufficient to guarantee the existence of $\mathscr{L}\{e^{kt}\}$ when k is a complex constant.

7. The Laplace transform is a linear transformation; that is, for constants α and β,

$$\mathscr{L}\{\alpha f(t) + \beta g(t)\} = \alpha\mathscr{L}\{f(t)\} + \beta\mathscr{L}\{g(t)\}$$

whenever both transforms exist. Use the linearity defined above along with the definitions

$$\sinh kt = \frac{e^{kt} - e^{-kt}}{2}, \quad \cosh kt = \frac{e^{kt} - e^{-kt}}{2},$$

k a real constant, to find $\mathscr{L}\{\sinh kt\}$ and $\mathscr{L}\{\cosh kt\}$.

8. State a condition on s that is sufficient to guarantee the existence of the Laplace transforms in Problem 7.

In Problems 9–18, use the theory of residues to compute the inverse Laplace transform $\mathscr{L}^{-1}\{F(s)\}$ for the given function $F(s)$.

9. $\dfrac{1}{s^6}$

10. $\dfrac{1}{(s-5)^3}$

11. $\dfrac{1}{s^2 + 4}$

12. $\dfrac{s}{(s^2 + 1)^2}$

13. $\dfrac{1}{s^2 - 3}$

14. $\dfrac{1}{(s-a)^2 + b^2}$

15. $\dfrac{e^{-as}}{s^2 - 5s + 6}$, $a > 0$

16. $\dfrac{e^{-as}}{(s-a)^2}$, $a > 0$

17. $\dfrac{1}{s^4 - 1}$

18. $\dfrac{s + 4}{s^2 + 6s + 11}$

---

*See *Advanced Engineering Mathematics*, 3rd Edition, by Dennis G. Zill and Michael R. Cullen, Jones and Bartlett Publishers.

In Problems 19 and 20, find the Fourier transform (19) of the given function.

19. $f(x) = \begin{cases} 0, & x \leq 0 \\ e^{-x}, & x > 0 \end{cases}$ **20.** $f(x) = \begin{cases} \sin x, & |x| \leq \pi \\ 0, & |x| > \pi \end{cases}$

21. Use the inverse Fourier transform (20) and the theory of residues to recover the function f in Problem 19.

22. The Fourier transform of a function f is $F(\alpha) = \dfrac{1}{(1 - i\alpha)^2}$. Use the inverse Fourier transform (20) and the theory of residues to find the function f.

Focus on Concepts

23. For the result obtained in Problem 8, find values of γ that can be used in the inverse transform (7).

24. (a) If $F(\alpha)$ is the Fourier transform of $f(x)$, then the function $|F(\alpha)|$ is called the **amplitude spectrum** of f. Find the amplitude spectrum of

$$f(x) = \begin{cases} 1, & |x| \leq 1 \\ 0, & |x| > 1 \end{cases}.$$

Graph $|F(\alpha)|$.

(b) Do some additional reading and find an application of the concept of the amplitude spectrum of a function.

25. Find the Fourier transform of $f(x) = \begin{cases} x, & 0 < x < 1 \\ 0, & x < 0 \text{ or } x > 1 \end{cases}$. Discuss how to find the inverse Fourier transform (20).

Projects

26. In the application of the Laplace transform to problems involving partial differential equations, one often encounters an inverse such as

$$f(x,t) = \mathscr{L}^{-1}\left\{\frac{\sinh xs}{(s^2 + 1)\sinh s}\right\}.$$

Investigate how (8) and (9) can be used to determine $f(x,t)$.

CHAPTER 6 REVIEW QUIZ

Answers to selected odd-numbered problems begin on page ANS-19.

In Problems 1–20, answer true or false. If the statement is false, justify your answer by either explaining why it is false or giving a counterexample; if the statement is true, justify your answer by either proving the statement or citing an appropriate result in this chapter.

1. For the sequence $\{z_n\}$, where $z_n = i^n = x_n + iy_n$, $\text{Re}(z_n) = x_n = \cos(n\pi/2)$ and $\text{Im}(z_n) = y_n = \sin(n\pi/2)$.

2. The sequence $\{i^n\}$ converges.

3. $\lim_{n\to\infty} \left(\dfrac{1+i}{\sqrt{\pi}}\right)^n = 0$.

4. $\lim_{n\to\infty} z_n = 0$ if and only if $\lim_{n\to\infty} |z_n| = 0$.

5. The power series $\sum_{k=1}^{\infty} \dfrac{z^k}{k^2}$ converges absolutely at every point on its circle of convergence.

6. There exists a power series centered at $z_0 = 1 + i$ that converges at $z = 25 - 4i$ and diverges at $z = 15 + 21i$.

7. A function f is analytic at a point z_0 if f can be expanded in a convergent power series centered at z_0.

8. Suppose a function f has a Taylor series representation with circle of convergence $|z - z_0| = R$, $R > 0$. Then f is analytic everywhere on the circle of convergence.

9. Suppose a function f has a Taylor series representation centered at z_0.

Then f is analytic everywhere inside the circle of convergence $|z - z_0| = R$, $R > 0$, and is not analytic everywhere outside $|z - z_0| = R$.

10. If the function f is entire, then the radius of convergence of a Taylor series expansion of f centered at $z_0 = 1 - i$ is necessarily $R = \infty$.

11. Both power series

$$\frac{1}{1+z} = 1 - z + z^2 - z^3 + \cdots$$

and
$$\frac{1}{1+z} = \frac{1}{2} - \frac{z-1}{2^2} + \frac{(z-1)^2}{2^3} - \frac{(z-1)^3}{2^4} + \cdots$$

converge at $z = 0.86 - 0.52i$.

12. If the power series $\sum_{k=0}^{\infty} a_k z^k$ has radius of convergence R, then the power series $\sum_{k=0}^{\infty} a_k z^{2k}$ has radius of convergence $\sqrt{R}$.

13. The power series $\sum_{k=0}^{\infty} a_k z^k$ and $\sum_{k=1}^{\infty} k a_k z^{k-1}$ have the same radius of convergence R.

14. The principal branch $f_1(z)$ of the complex logarithm does not possess a Maclaurin expansion.

15. If f is analytic throughout some deleted neighborhood of z_0 and z_0 is a pole of order n, then $\lim_{z\to z_0} (z - z_0)^n f(z) \neq 0$.

16. A singularity of a rational function is either removable or is a pole.

17. The function $f(z) = \dfrac{1}{z^2 + 2iaz - 1}$, $a > 1$, has two simple poles within the unit circle $|z| = 1$.

18. $z = 0$ is a simple pole of $f(z) = -\dfrac{1}{z} + \cot z$.

19. If z_0 is a simple pole of a function f, then it is possible that $\text{Res}(f(z), z_0) = 0$.

20. The principal part of the Laurent series of $f(z) = \dfrac{1}{1 - \cos z}$ valid for $0 < |z| < 2\pi$ contains precisely two nonzero terms.

In Problems 21–40, try to fill in the blanks without referring back to the text.

21. The sequence $\left\{ \dfrac{2in}{n+i} - \dfrac{(9 - 12i)n + 2}{3n + 1 + 7i} \right\}$ converges to _____.

22. The series $i + 2i + 3i + 4i + \cdots$ diverges because _____.

23. $5 - i - \dfrac{1}{5} + \dfrac{i}{25} + \dfrac{1}{125} - \cdots =$ _____.

24. The equality $\sum_{k=0}^{\infty} \left(\dfrac{z-1}{z+1}\right)^k = \frac{1}{2}(z+1)$ comes from _____ and is valid in the region of the complex plane defined by _____ .

25. The power series $\sum_{k=0}^{\infty} (5+12i)^k (z-2-i)^k$ converges absolutely within the circle _____ .

26. The power series $\sum_{k=0}^{\infty} \dfrac{4^k}{2k+5}(z-2+3i)^{2k}$ diverges for $|z-2+3i| >$ _____ .

27. If the power series $\sum_{k=0}^{\infty} a_k z^k$, $a_k \neq 0$, has radius of convergence $R > 0$, then the power series $\sum_{k=0}^{\infty} \dfrac{z^k}{a_k}$ has radius of convergence _____ .

28. Without finding the actual expansion, the Taylor series of $f(z) = \csc z$ centered at $z_0 = 3 + 2i$ has radius of convergence $R =$ _____ .

29. Use the first series in Problem 11 to obtain the first three terms of a Taylor series of $f(z) = \dfrac{z+1}{6+z}$ centered at $z_0 = -1$: _____ . The radius of convergence of the series is $R =$ _____ .

30. A power series centered at $-5i$ for $f(z) = e^z$ is given by $e^z = \sum_{k=0}^{\infty}$ ___ $(z+5i)^k$.

31. $z = -1$ is an isolated singularity of $f(z) = \dfrac{(z+1)^3 - 2(z+1)^2 + 4(z+1) + 7}{(z+1)^2}$. The Laurent series valid for $0 < |z+1| < \infty$ is _____ .

32. The analytic function $f(z) = \frac{1}{6}z^9 - z^3 + \sin z^3$ has a zero of order _____ at $z = 0$.

33. The zeros of the function $f(z) = \sin \pi \left(\dfrac{1}{z} - 1\right)$ are _____ and are of order _____ .

34. If $f(z)$ has a zero of order 5 at z_0, then the derivative of lowest order that is not zero is $f^{(\,\text{---}\,)}(z_0)$.

35. The function $f(z) = (z - \sin z)/z^3$ has a removable singularity at $z = 0$. The value $f(0)$ is defined to be _____ .

36. If $f(z) = z^3 e^{-1/z^2}$, then $\text{Res}\,(f(z),\, 0) =$ _____ .

37. Suppose $z = \pi$ is a simple pole of $f(z) = \cot z$. From an appropriate residue formula, $\text{Res}(f(z), \pi) =$ _____ and so the principal part of the Laurent series about $z = \pi$ is _____ and the Laurent series is valid for $0 < |z - \pi| <$ _____ .

38. On $|z| = 1$, the contour integral $\oint_C \dfrac{\cos z}{z^2 - (2+\pi)z + 2\pi}\, dz$ equals _____ , on $|z| = 3$ the integral equals _____ , and on $|z| = 4$ the integral equals _____ .

39. On $|z| = 1$,

(a) $\oint_C \dfrac{z^2 + 2iz + 1 - i}{e^{2z} - 1}\, dz =$ _____ ,

(b) $\oint_C \dfrac{\sin z}{z^n}\, dz =$ _____ , $n = 0, 1, 2, \ldots$.

40. $\displaystyle\int_0^{2\pi} \dfrac{1}{4\cos^2\theta + \sin^2\theta}\, d\theta =$ _____ .

Conformal
Mappings

Chapter Outline

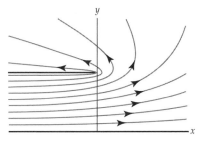

The planar flow of an ideal fluid.
See page 397.

Introduction In Section 4.5 we saw that analytic mappings can be used to solve certain types of boundary-value problems. In this chapter we introduce the fundamental notion of a conformal mapping, and we show that conformal mappings can be used to solve a larger class of boundary-value problems. The methods we introduce are applied to problems in heat flow, electrostatics, and fluid flow.

7.1   Conformal Mapping

In Section 2.3 we saw that a nonconstant linear mapping acts by rotating, magnifying, and translating points in the complex plane. As a result, the angle between any two intersecting arcs in the z-plane is equal to the angle between the images of the arcs in the w-plane under a linear mapping. Complex mappings that have this angle-preserving property are called **conformal mappings**. In this section we will formally define and discuss conformal mappings. We show that any analytic complex function is conformal at points where the derivative is nonzero. Consequently, all of the elementary functions studied in Chapter 4 are conformal in some domain D. Later in this chapter we will see that conformal mappings have important applications to boundary-value problems involving Laplace's equation.

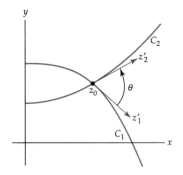

Figure 7.1.1 The angle θ between C_1 and C_2

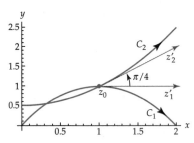

(a) Curves C_1 and C_2 in the z-plane

$$w = \bar{z}$$

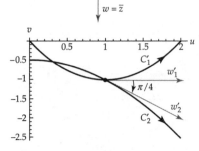

(b) Images of the curves in (a) under $w = \bar{z}$

Figure 7.1.2 Figure for Example 1

Conformal Mapping Suppose that $w = f(z)$ is a complex mapping defined in a domain D. The mapping is said to be conformal at a point z_0 in D if it "preserves the angle" between any two curves intersecting at z_0. To make this concept precise, assume that C_1 and C_2 are smooth curves in D that intersect at z_0 and have a fixed orientation as described in Section 5.1. Let $z_1(t)$ and $z_2(t)$ be parametrizations of C_1 and C_2 such that $z_1(t_0) = z_2(t_0) = z_0$, and such that the orientations on C_1 and C_2 correspond to the increasing values of the parameter t. Because C_1 and C_2 are smooth, the tangent vectors $z_1' = z_1'(t_0)$ and $z_2' = z_2'(t_0)$ are both nonzero. We define the **angle** between C_1 and C_2 to be the angle θ in the interval $[0, \pi]$ between the tangent vectors z_1' and z_2'. See Figure 7.1.1. Now suppose that under the complex mapping $w = f(z)$ the curves C_1 and C_2 in the z-plane are mapped onto the curves C_1' and C_2' in the w-plane, respectively. Because C_1 and C_2 intersect at z_0, we must have that C_1' and C_2' intersect at $f(z_0)$. If C_1' and C_2' are smooth, then the angle between C_1' and C_2' at $f(z_0)$ is similarly defined to be the angle ϕ in the interval $[0, \pi]$ between similarly defined tangent vectors w_1' and w_2'. We say that the angles θ and ϕ are **equal in magnitude** if $\theta = \phi$.

In the z-plane, the vector z_1', whose initial point is z_0, can be rotated through the angle θ onto the vector z_2'. This rotation is either in the counterclockwise or the clockwise direction. Similarly, in the w-plane, the vector w_1', whose initial point is $f(z_0)$, can be rotated in either the counterclockwise or clockwise direction through an angle of ϕ onto the vector w_2'. If the rotation in the z-plane is in the same direction as the rotation in the w-plane, we say that the angles θ and ϕ are **equal in sense**. The following example illustrates these concepts.

EXAMPLE 1   Magnitude and Sense of Angles

The smooth curves C_1 and C_2 shown in Figure 7.1.2(a) are given by $z_1(t) = t + \left(2t - t^2\right) i$ and $z_2(t) = t + \frac{1}{2}\left(t^2 + 1\right) i$, $0 \leq t \leq 2$, respectively. These curves intersect at the point $z_0 = z_1(1) = z_2(1) = 1 + i$. The tangent vectors at z_0 are $z_1' = z_1'(1) = 1$ and $z_2' = z_2'(1) = 1 + i$. Furthermore, from Figure 7.1.2(a) we see that the angle between C_1 and C_2 at z_0 is $\theta = \pi/4$. Under the complex mapping $w = \bar{z}$, the images of C_1 and C_2 are the curves C_1' and C_2', respectively, shown in Figure 7.1.2(b). The image curves are parametrized by $w_1(t) = t - \left(2t - t^2\right) i$ and $w_2(t) = t - \frac{1}{2}\left(t^2 + 1\right) i$, $0 \leq t \leq 2$, and intersect at the point $w_0 = f(z_0) = 1 - i$. In addition, at w_0 we have the tangent vectors $w_1' = w_1'(1) = 1$ and $w_2' = w_2'(1) = 1 - i$ to C_1' and C_2', respectively. Inspection of Figure 7.1.2(b) indicates that the angle between C_1' and C_2' at w_0 is $\phi = \pi/4$. Therefore, the angles θ and ϕ are equal in magnitude.

However, because the rotation through $\pi/4$ of the vector z_1' onto z_2' must be *counterclockwise*, whereas the rotation through $\pi/4$ of w_1' onto w_2' must be *clockwise*, we conclude that θ and ϕ are not equal in sense. ❑

With the terminology regarding the magnitude and sense of an angle established, we are now in a position to give the following precise definition of a conformal mapping.

Definition 7.1.1   Conformal Mapping

Let $w = f(z)$ be a complex mapping defined in a domain D and let z_0 be a point in D. Then we say that $w = f(z)$ is **conformal** at z_0 if for every pair of smooth oriented curves C_1 and C_2 in D intersecting at z_0 the angle between C_1 and C_2 at z_0 is equal to the angle between the image curves C_1' and C_2' at $f(z_0)$ in both magnitude and sense.

We will also use the term **conformal mapping** to refer to a complex mapping $w = f(z)$ that is conformal at z_0. In addition, if $w = f(z)$ maps a domain D onto a domain D' and if $w = f(z)$ is conformal at every point in D, then we call $w = f(z)$ a conformal mapping of D onto D'. From Section 2.3 it should be intuitively clear that if $f(z) = az + b$ is a linear function with $a \neq 0$, then $w = f(z)$ is conformal at every point in the complex plane. In Example 1 we have shown that $w = \bar{z}$ is not a conformal mapping at the point $z_0 = 1 + i$ because the angles θ and ϕ are equal in magnitude but not in sense.

Angles between Curves Definition 7.1.1 is seldom used directly to show that a complex mapping is conformal. Rather, we will prove in Theorem 7.1.1 that an analytic function f is a conformal mapping at z whenever $f'(z) \neq 0$. In order to prove this result we need a procedure to determine the angle (in both magnitude and sense) between two smooth curves in the complex plane. For our purposes, the most efficient way to do this is to use the argument of a complex number.

Let us again adopt the notation of Figure 7.1.1, where C_1 and C_2 are smooth curves parametrized by $z_1(t)$ and $z_2(t)$, respectively, which intersect at $z_1(t_0) = z_2(t_0) = z_0$. The requirement that C_1 is smooth ensures that the tangent vector to C_1 at z_0, given by $z_1' = z_1'(t_0)$, is nonzero, and so $\arg(z_1')$ is defined and represents an angle between the position vector z_1' and the positive x-axis. Similarly, the tangent vector to C_2 at z_0, given by $z_2' = z_2'(t_0)$, is nonzero, and $\arg(z_2')$ represents an angle between the position vector z_2' and the positive x-axis. Inspection of Figure 7.1.3 shows that the angle θ between C_1 and C_2 at z_0 is the value of

$$\arg(z_2') - \arg(z_1') \tag{1}$$

in the interval $[0, \pi]$, provided that we can rotate z_1' counterclockwise about 0 through the angle θ onto z_2'. In the case that a clockwise rotation is needed, then $-\theta$ is the value of (1) in the interval $(-\pi, 0)$. In either case, we see that (1) gives both the magnitude and sense of the angle between C_1 and C_2 at z_0. As an example, consider the curves C_1, C_2, and their images under the complex mapping $w = \bar{z}$ in Example 1. Notice that the unique value of

$$\arg(z_2') - \arg(z_1') = \arg(1 + i) - \arg(1) = \frac{\pi}{4} + 2n\pi,$$

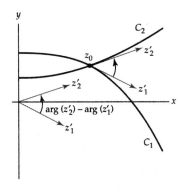

Figure 7.1.3 The angle between C_1 and C_2

$n = 0, \pm 1, \pm 2, \dots$, that lies in the interval $[0, \pi]$ is $\pi/4$. Therefore, the angle between C_1 and C_2 is $\theta = \pi/4$, and the rotation of z_1' onto z_2' is counterclockwise. On the other hand,

$$\arg(w_2') - \arg(w_2') = \arg(1-i) - \arg(1) = -\frac{\pi}{4} + 2n\pi,$$

$n = 0, \pm 1, \pm 2, \dots$, has no value in $[0, \pi]$, but has the unique value $-\pi/4$ in the interval $(-\pi, 0)$. Thus, the angle between C_1' and C_2' is $\phi = \pi/4$, and the rotation of w_1' onto w_2' is clockwise.

Analytic Functions We now use (1) to prove the following theorem.

> **Theorem 7.1.1 Conformal Mapping**
>
> If f is an analytic function in a domain D containing z_0, and if $f'(z_0) \neq 0$, then $w = f(z)$ is a conformal mapping at z_0.

Proof Suppose that f is analytic in a domain D containing z_0, and that $f'(z_0) \neq 0$. Let C_1 and C_2 be two smooth curves in D parametrized by $z_1(t)$ and $z_2(t)$, respectively, with $z_1(t_0) = z_2(t_0) = z_0$. In addition, assume that $w = f(z)$ maps the curves C_1 and C_2 onto the curves C_1' and C_2'. We wish to show that the angle θ between C_1 and C_2 at z_0 is equal to the angle ϕ between C_1' and C_2' at $f(z_0)$ in both magnitude and sense. We may assume, by renumbering C_1 and C_2 if necessary, that $z_1' = z_1'(t_0)$ can be rotated counterclockwise about 0 through the angle θ onto $z_2' = z_2'(t_0)$. Thus, by (1), the angle θ is the unique value of $\arg(z_2') - \arg(z_1')$ in the interval $[0, \pi]$. From (11) of Section 2.2, C_1' and C_2' are parametrized by $w_1(t) = f(z_1(t))$ and $w_2(t) = f(z_2(t))$. In order to compute the tangent vectors w_1' and w_2' to C_1' and C_2' at $f(z_0) = f(z_1(t_0)) = f(z_2(t_0))$ we use the chain rule

$$w_1' = w_1'(t_0) = f'(z_1(t_0)) \cdot z_1'(t_0) = f'(z_0) \cdot z_1',$$

and $$w_2' = w_2'(t_0) = f'(z_2(t_0)) \cdot z_2'(t_0) = f'(z_0) \cdot z_2'.$$

Since C_1 and C_2 are smooth, both z_1' and z_2' are nonzero. Furthermore, by our hypothesis, we have $f'(z_0) \neq 0$. Therefore, both w_1' and w_2' are nonzero, and the angle ϕ between C_1' and C_2' at $f(z_0)$ is a value of

$$\arg(w_2') - \arg(w_1') = \arg(f'(z_0) \cdot z_2') - \arg(f'(z_0) \cdot z_1').$$

Now by two applications of (8) from Section 1.3 we obtain:

$$\arg(f'(z_0) \cdot z_2') - \arg(f'(z_0) \cdot z_1') = \arg(f'(z_0)) + \arg(z_2') - [\arg(f'(z_0)) + \arg(z_1')]$$
$$= \arg(z_2') - \arg(z_1').$$

This expression has a unique value in $[0, \pi]$, namely θ. Therefore, $\theta = \phi$ in both magnitude and sense, and consequently the $w = f(z)$ is a conformal mapping at z_0. ❏

In light of Theorem 7.1.1 it is relatively easy to determine where an analytic function is a conformal mapping.

EXAMPLE 2   Conformal Mappings

(a) By Theorem 7.1.1 the entire function $f(z) = e^z$ is conformal at every point in the complex plane since $f'(z) = e^z \neq 0$ for all z in $\mathbf{C}$.

(b) By Theorem 7.1.1 the entire function $g(z) = z^2$ is conformal if $z \neq 0$ since $g'(z) = 2z \neq 0$ when $z \neq 0$.   ❑

Critical Points   In general, if a complex function f is analytic at a point z_0 and if $f'(z_0) = 0$, then z_0 is called a **critical point** of f. Although it does not follow from Theorem 7.1.1, it is true that analytic functions are not conformal at critical points. More specifically, we can show that the following magnification of angles occurs at a critical point.

Theorem 7.1.2   Angle Magnification at a Critical Point

Let f be analytic at the critical point z_0. If $n > 1$ is an integer such that $f'(z_0) = f''(z_0) = \dots = f^{(n-1)}(z_0) = 0$ and $f^{(n)}(z_0) \neq 0$, then the angle between any two smooth curves intersecting at z_0 is increased by a factor of n by the complex mapping $w = f(z)$. In particular, $w = f(z)$ is not a conformal mapping at z_0.

A proof of Theorem 7.1.2 is sketched in Problem 22 of Exercises 7.1.

EXAMPLE 3   Conformal Mappings

Find all points where the mapping $f(z) = \sin z$ is conformal.

Solution   The function $f(z) = \sin z$ is entire, and from Section 4.3 we have that $f'(z) = \cos z$. In (21) of Section 4.3 we found that $\cos z = 0$ if and only if $z = (2n+1)\pi/2$, $n = 0, \pm1, \pm2, \dots$, and so each of these points is a critical point of f. By Theorem 7.1.1, $w = \sin z$ is a conformal mapping for all $z \neq (2n+1)\pi/2$, $n = 0, \pm1, \pm2, \dots$. Furthermore, $w = \sin z$ is not a conformal mapping at $z = (2n+1)\pi/2$, $n = 0, \pm1, \pm2, \dots$. Because $f''(z) = -\sin z = \pm1$ at the critical points of f, Theorem 7.1.2 indicates that angles at these points are increased by a factor of 2.   ❑

The angle magnification at a critical point of the complex mapping $w = \sin z$ in Example 3 can be seen directly. For example, consider the critical point $z = \pi/2$. Under $w = \sin z$, the vertical ray C_1 in the z-plane emanating from $z = \pi/2$ and given by $z = \pi/2 + iy$, $y \geq 0$, is mapped onto the set in the w-plane given by $w = \sin(\pi/2)\cosh y + i\cos(\pi/2)\sinh y$, $y \geq 0$. Because $\sin(\pi/2) = 1$ and $\cos(\pi/2) = 0$, the image can be rewritten as $w = \cosh y$, $y \geq 0$. In words, the image C_1' is a ray in the w-plane emanating from $w = 1$ and containing the point $w = 2$. A similar analysis reveals that the image C_2' of the vertical ray C_2 given by $z = \pi/2 + iy$, $y \leq 0$, is also the ray emanating from $w = 1$ and containing the point $w = 2$. That is, $C_1' = C_2'$. The angle between the rays C_1 and C_2 in the z-plane is π, and so Theorem 7.1.2 implies that the angle between their images in the w-plane is increased to 2π, or, equivalently, 0. This agrees with the observation that $C_1' = C_2'$. See Figure 7.1.4.

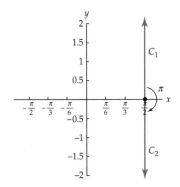

(a) The angle between the vertical rays in the z-plane is π

$$\downarrow w = \sin z$$

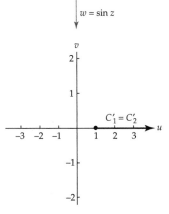

(b) The angle between the images of the rays in (a) is 2π or 0

Figure 7.1.4 The mapping $w = \sin z$

Conformal Mappings Using Tables   In Section 4.5 we introduced a method of solving a particular type of boundary-value problem using complex mappings. Specifically, we saw that a Dirichlet problem in a complicated domain D can be solved by finding an *analytic* mapping of D onto a simpler domain D' in which the associated Dirichlet problem has already been solved. At the end of this chapter we will see a similar application of *conformal* mappings to a generalized type of Dirichlet problem. In these applications our method for producing a solution in a domain D will first require that we find a *conformal* mapping of D onto a simpler domain D' in which the associated boundary-value problem has a solution. An important aid in this task is the table of conformal mappings given in Appendix III.

The mappings in Appendix III have been categorized as elementary mappings (E-1 to E-9), mappings onto half-planes (H-1 to H-6), mappings onto circular regions (C-1 to C-5), and miscellaneous mappings (M-1 to M-10). Many properties of the mappings appearing in this table have been derived in Chapter 2 and Chapter 4, whereas other properties will be derived in the coming sections. When using the table, bear in mind that in some cases the desired mapping will appear as a single entry in the table, whereas in other cases one or more successive mappings from the table may be required. You should also note that the mappings in Appendix III are, in general, only con-

Note ➡   formal mappings of the interiors of the regions shown. For example, it is clear that the complex mapping shown in Entry E-4 is not conformal at $B = 0$. As a general rule, when we refer to a conformal mapping of a region R onto a region R' we are requiring only that the mapping be conformal at the points in the interior of R.

EXAMPLE 4   Using a Table of Conformal Mappings

Use Appendix III to find a conformal mapping from the infinite horizontal strip $0 \le y \le 2$, $-\infty < x < \infty$, onto the upper half-plane $v \ge 0$. Under this mapping, what is the image of the negative x-axis?

Solution   Entry H-2 in Appendix III gives a mapping from an infinite horizontal strip onto the upper half-plane. Setting $a = 2$, we obtain the desired mapping $w = e^{\pi z/2}$. From H-2 we also see that the points labeled D and $E = 0$ on the negative x-axis in the z-plane are mapped onto the points D' and $E' = 1$ on the positive u-axis in the w-plane. Noting the relative positions of these points, we conclude that the negative x-axis is mapped onto the interval $(0, 1]$ in the u-axis by $w = e^{\pi z/2}$. See Figure 7.1.5. This observation can also be verified using parametrizations.   ❑

EXAMPLE 5   Using a Table of Conformal Mappings

Use Appendix III to find a conformal mapping from the infinite horizontal strip $0 \le y \le 2$, $-\infty < x < \infty$, onto the unit disk $|w| \le 1$. Under this mapping, what is the image of the negative x-axis?

Solution   Appendix III does not have an entry that maps an infinite horizontal strip onto the unit disk. Therefore, we construct the a conformal mapping that does this by composing two mappings in the table. In Example 4 we found that the infinite horizontal strip $0 \le y \le 2$, $-\infty < x < \infty$, is mapped onto the upper half-plane by $f(z) = e^{\pi z/2}$. In addition, from entry C-4 we

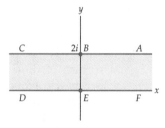

(a) The horizontal strip $0 \le y \le 2$

$$w = e^{\pi z/2}$$

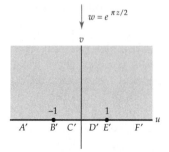

(b) Image of the strip in (a)

Figure 7.1.5 Figure for Example 4

see that the upper half-plane is mapped onto the unit disk by $g(z) = \dfrac{i-z}{i+z}$. The composition of these two functions

$$w = g(f(z)) = \frac{i - e^{\pi z/2}}{i + e^{\pi z/2}}$$

therefore maps the strip $0 \le y \le 2$, $-\infty < x < \infty$, onto the unit disk $|w| \le 1$. Under the first of these successive mappings, the negative real axis is mapped onto the interval $(0, 1]$ in the real axis as was noted in Example 4. Inspection of entry C-4 (or Figure 7.1.6) reveals that the interval from 0 to $C = 1$ is mapped onto the circular arc from 1 to $C' = i$ on the unit circle $|w| = 1$. Therefore, we conclude that the negative real axis is mapped onto the circular arc from 1 to i on the unit circle under $w = \dfrac{i - e^{\pi z/2}}{i + e^{\pi z/2}}$. ❑

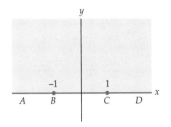

(a) Image of the strip $0 \le y \le 2$ under $w = e^{\pi z/2}$

$w = \dfrac{i-z}{i+z}$

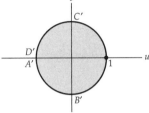

(b) Image of the half-plane in (a) under $w = \dfrac{i-z}{i+z}$

Figure 7.1.6 Figure for Example 5

Remarks

In the foregoing discussion regarding conformal mappings using tables we alluded to the fact that in many applications one needs to find a conformal mapping of a domain D onto a simpler domain D'. A natural question to ask is whether such a mapping always exists. That is, given domains D and D', does there exist a conformal mapping of D onto D'? An answer to this question was given by the mathematician Bernhard Riemann (1826–1866). Although there was a gap in Riemann's original proof (which was subsequently filled), this amazing theorem still bears his name.

The Riemann Mapping Theorem Let D be a simply connected domain in the z-plane such that D is not all of $\mathbf{C}$. Then there exists a one-to-one conformal mapping $w = f(z)$ from D onto the open unit disk $|w| < 1$ in the w-plane.

It is not immediately clear that this theorem answers our question of the existence of a mapping from D onto D'. To see that it does, we first use the theorem to find a conformal mapping f from D onto the open unit disk $|w| < 1$. We then apply the theorem a second time to obtain a mapping g from D' onto the open unit disk $|w| < 1$. Since the theorem ensures that g is one-to-one, it has a well defined inverse function g^{-1} that maps the open unit disk onto D'. The desired mapping from D onto D' is then given by the composition $w = g^{-1} \circ f(z)$.

Riemann's theorem is of critical theoretical importance, but its proof is not *constructive*. This means that the theorem establishes the *existence* of the mapping f but offers no method of actually finding a formula for f. A proof of the Riemann mapping theorem is well beyond the scope of this text. The interested reader is encouraged to refer to the text *Complex Analysis* by Lars V. Ahlfors, McGraw-Hill, 1979.

EXERCISES 7.1 *Answers to selected odd-numbered problems begin on page ANS-19.*

In Problems 1–6, determine where the complex mapping $w = f(z)$ is conformal.

1. $f(z) = z^3 - 3z + 1$ **2.** $f(z) = z^2 + 2iz - 3$

3. $f(z) = z - e^{-z} + 1 - i$ **4.** $f(z) = ze^{z^2 - 2}$

5. $f(z) = \tan z$ **6.** $f(z) = z - \text{Ln}(z + i)$

In Problems 7–10, proceed as in Example 1 to show that the given function f is not conformal at the indicated point.

7. $f(z) = (z - i)^3$; $z_0 = i$ **8.** $f(z) = (iz - 3)^2$; $z_0 = -3i$

9. $f(z) = e^{z^2}$; $z_0 = 0$ **10.** the principle square root function $f(z) = z^{1/2}$; $z_0 = 0$

In Problems 11–16, use Appendix III to find a conformal mapping of the region R shown in color onto the region R' shown in gray. Then find the image of the curve from A to B.

11.

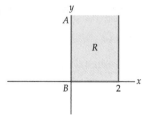 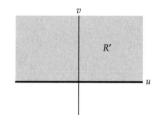

Figure 7.1.7 Figure for Problem 11

12.

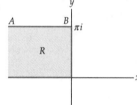 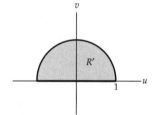

Figure 7.1.8 Figure for Problem 12

13.

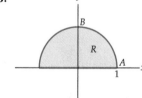 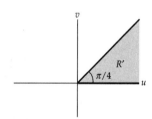

Figure 7.1.9 Figure for Problem 13

14.

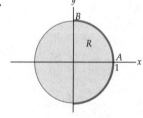 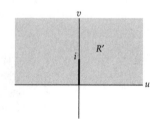

Figure 7.1.10 Figure for Problem 14

15.

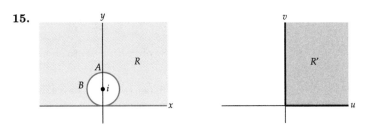

Figure 7.1.11 Figure for Problem 15

16.

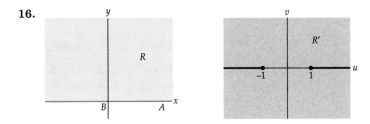

Figure 7.1.12 Figure for Problem 16

Focus on Concepts

17. Where is the mapping $w = \bar{z}$ conformal? Justify your answer.

18. Suppose $w = f(z)$ is a conformal mapping at every point in the complex plane. Where is the mapping $w = \overline{f(\bar{z})}$ conformal? Justify your answer.

19. Suppose that $w = f(z)$ is a conformal mapping at every point in the complex plane. Where is the mapping $w = e^{f(z)}$ conformal?

20. This problem concerns determining the angle between two curves C_1 and C_2 at a point where one (or both) of the curves has a zero tangent vector.

   (a) Assume that two curves C_1 and C_2 are parametrized by $z_1(t)$ and $z_2(t)$, respectively, and that the curves intersect at $z_1(t_0) = z_2(t_0) = z_0$. Assume further that both z_1 and z_2 are differentiable functions of t, and let $z_1' = z_1'(t_0)$ and $z_2' = z_2'(t_0)$. Explain why $\arg(z_2') - \arg(z_1')$ *does not* represent the angle between C_1 and C_2 if either z_1' or z_2' is zero.

   (b) Explain why $\lim_{t \to t_0} [\arg(z_2(t) - z_0)] - \lim_{t \to t_0} [\arg(z_1(t) - z_0)]$ does represent the angle between C_1 and C_2 regardless of whether z_1' or z_2' is zero.

   (c) Use part (b) to determine the angle between the curves parametrized by $z_1(t) = t + it^2$ and $z_2(t) = t^2 + it^2$, $-1 \le t \le 1$, at $z_0 = 0$. Does this computation match your intuition?

21. On page 355 we showed that the function $f(z) = z^2$ was not conformal at $z_0 = 0$ because the angle between the positive x- and y-axes was doubled. In this problem you will show that *every* pair of smooth curves intersecting at $z_0 = 0$ has the angle between them doubled by $f(z) = z^2$. This is a very specific case of Theorem 7.1.2.

   (a) Suppose that the smooth curves C_1 and C_2 are parametrized by $z_1(t)$ and $z_2(t)$ with $z_1(t_0) = z_2(t_0) = 0$. If $z_1' = z_1'(t_0)$ and $z_2' = z_2'(t_0)$ are both nonzero, then the angle θ between C_1 and C_2 is given by (1). Explain why $\phi = \arg(f'(0) \cdot z_2') - \arg(f'(0) \cdot z_1')$ *does not* represent the angle between the images C_1' and C_2' of C_1 and C_2 under the mapping $w = f(z) = z^2$, respectively.

   (b) Use Problem 20 to write down an expression involving arguments that does represent the angle ϕ between C_1' and C_2'. [*Hint:* C_1' and C_2' are parametrized by $w_1(t) = f(z_1(t)) = [z_1(t)]^2$ and $w_2(t) = f(z_2(t)) = [z_2(t)]^2$.]

(c) Use (8) of Section 1.3 to show that your expression for ϕ from (b) is equal to 2θ.

22. In this problem you will prove Theorem 7.1.2. Let f be an analytic function at the point z_0 such that $f'(z_0) = f''(z_0) = \cdots = f^{(n-1)}(z_0) = 0$ and $f^{(n)}(z_0) \neq 0$ for some $n > 1$.

(a) Explain why f can be written as

$$f(z) = f(z_0) + \frac{f^{(n)}(z_0)}{n!}(z - z_0)^n (1 + g(z)),$$

where g is an analytic function at z_0 and $g(z_0) = 0$.

(b) Use (a) and Problem 20 to show that the angle between two smooth curves intersecting at z_0 is increased by a factor of n by the mapping $w = f(z)$.

7.2   Linear Fractional Transformations

In many applications that involve boundary-value problems associated with Laplace's equation, it is necessary to find a conformal mapping that maps a disk onto the half-plane $v \geq 0$. Such a mapping would have to map the circular boundary of the disk to the boundary line of the half-plane. An important class of elementary conformal mappings that map circles to lines (and vice versa) are the linear fractional transformations. In this section we will define and study this special class of mappings.

Linear Fractional Transformations   In Section 2.3 we examined complex linear mappings $w = az + b$ where a and b are complex constants and $a \neq 0$. Recall that such mappings act by rotating, magnifying, and translating points in the complex plane. We then discussed the complex reciprocal mapping $w = 1/z$ in Section 2.5. An important property of the reciprocal mapping, when defined on the extended complex plane, is that it maps certain lines to circles and certain circles to lines. A more general type of mapping that has similar properties is a linear fractional transformation defined next.

Definition 7.2.1   Linear Fractional Transformation

If a, b, c, and d are complex constants with $ad - bc \neq 0$, then the complex function defined by:

$$T(z) = \frac{az + b}{cz + d} \tag{1}$$

is called a **linear fractional transformation**.

Linear fractional transformations are also called **Möbius transformations** or **bilinear transformations**. If $c = 0$, then the transformation T given by (1) is a linear mapping, and so a linear mapping is a special case of a linear fractional transformation. If $c \neq 0$, then we can write

$$T(z) = \frac{az + b}{cz + d} = \frac{bc - ad}{c} \frac{1}{cz + d} + \frac{a}{c}. \tag{2}$$

Setting $A = \dfrac{bc - ad}{c}$ and $B = \dfrac{a}{c}$, we see that the linear transformation T in (2) can be written as the composition $T(z) = f \circ g \circ h(z)$, where $f(z) = Az + B$ and $h(z) = cz + d$ are linear functions and $g(z) = 1/z$ is the reciprocal function.

The domain of a linear fractional transformation T given by (1) is the set of all complex z such that $z \neq -d/c$. Furthermore, since

$$T'(z) = \frac{ad - bc}{(cz + d)^2}$$

it follows from Theorem 7.1.1 and the requirement that $ad - bc \neq 0$ that linear fractional transformations are conformal on their domains. The requirement that $ad - bc \neq 0$ also ensures that the T is a one-to-one function on its domain. See Problem 27 in Exercises 7.2.

Observe that if $c \neq 0$, then (1) can be written as

$$T(z) = \frac{az + b}{cz + d} = \frac{(a/c)\,(z + b/a)}{z + d/c} = \frac{\phi(z)}{z - (-d/c)},$$

where $\phi(z) = (a/c)\,(z + b/a)$. Because $ad - bc \neq 0$, we have that $\phi\,(-d/c) \neq 0$, and so from Theorem 6.4.2 of Section 6.4 it follows that the point $z = -d/c$ is a simple pole of T.

When $c \neq 0$, that is, when T is not a linear function, it is often helpful to view T as a mapping of the extended complex plane. Since T is defined for all points in the extended plane except the pole $z = -d/c$ and the ideal point ∞, we need only extend the definition of T to include these points. We make this definition by considering the limit of T as z tends to the pole and as z tends to the ideal point. Because

$$\lim_{z \to -d/c} \frac{cz + d}{az + b} = \frac{0}{a\,(-d/c) + b} = \frac{0}{-ad + bc} = 0,$$

it follows from (25) of Section 2.6 that

$$\lim_{z \to -d/c} \frac{az + b}{cz + d} = \infty.$$

Moreover, from (24) of Section 2.6 we have that

$$\lim_{z \to \infty} \frac{az + b}{cz + d} = \lim_{z \to 0} \frac{a/z + b}{c/z + d} = \lim_{z \to 0} \frac{a + zb}{c + zd} = \frac{a}{c}.$$

The values of these two limits indicate how to extend the definition of T. In particular, if $c \neq 0$, then we regard T as a one-to-one mapping of the extended complex plane defined by:

$$T(z) = \begin{cases} \dfrac{az + b}{cz + d}, & z \neq -\dfrac{d}{c}, z \neq \infty \\[2ex] \infty, & z = -\dfrac{d}{c} \\[2ex] \dfrac{a}{c}, & z = \infty. \end{cases} \tag{3}$$

A special case of (3) corresponding to $a = d = 0$ and $b = c \neq 0$ is the reciprocal function defined on the extended complex plane. Refer to Definition 2.5.1.

EXAMPLE 1   A Linear Fractional Transformation

Find the images of the points 0, $1 + i$, i, and ∞ under the linear fractional transformation $T(z) = (2z + 1)/(z - i)$.

Solution For $z = 0$ and $z = 1 + i$ we have:

$$T(0) = \frac{2(0) + 1}{0 - i} = \frac{1}{-i} = i \quad \text{and} \quad T(1 + i) = \frac{2(1 + i) + 1}{(1 + i) - i} = \frac{3 + 2i}{1} = 3 + 2i.$$

Identifying $a = 2$, $b = 1$, $c = 1$, and $d = -i$ in (3), we also have:

$$T(i) = T\left(-\frac{d}{c}\right) = \infty \quad \text{and} \quad T(\infty) = \frac{a}{c} = 2.$$

❑

Circle-Preserving Property In the discussion preceding Example 1 we indicated that the reciprocal function $1/z$ is a special case of a linear fractional transformation. We saw two interesting properties of the reciprocal mapping in Section 2.7. First, the image of a circle centered at the pole $z = 0$ of $1/z$ is a circle, and second, the image of a circle with center on the x- or y-axis and containing the pole $z = 0$ is a vertical or horizontal line. Linear fractional transformations have a similar mapping property. This is the content of the following theorem.

Theorem 7.2.1   Circle-Preserving Property

If C is a circle in the z-plane and if T is a linear fractional transformation given by (3), then the image of C under T is either a circle or a line in the extended w-plane. The image is a line if and only if $c \neq 0$ and the pole $z = -d/c$ is on the circle C.

Proof When $c = 0$, T is a linear function, and we saw in Section 2.3 that linear functions map circles onto circles. It remains to be seen that the theorem still holds for $c \neq 0$. Assume then that $c \neq 0$. From (2) we have that $T(z) = f \circ g \circ h(z)$, where $f(z) = Az + B$ and $h(z) = cz + d$ are linear functions and $g(z) = 1/z$ is the reciprocal function. Observe that since h is a linear mapping, the image C' of the circle C under h is a circle. We now examine two cases:

Case 1 Assume that the origin $w = 0$ is on the circle C'. This occurs if and only if the pole $z = -d/c$ is on the circle C. From the Remarks in Section 2.5, if $w = 0$ is on C', then the image of C' under $g(z) = 1/z$ is either a horizontal or vertical line L. Furthermore, because f is a linear function, the image of the line L under f is also a line. Thus, we have shown that if the pole $z = -d/c$ is on the circle C, then the image of C under T is a line.

Case 2 Assume that the point $w = 0$ is not on C'. That is, the pole $z = -d/c$ is not on the circle C. Let C' be the circle given by $|w - w_0| = \rho$. If we set $\xi = f(w) = 1/w$ and $\xi_0 = f(w_0) = 1/w_0$, then for any point w on C' we have

$$|\xi - \xi_0| = \left|\frac{1}{w} - \frac{1}{w_0}\right| = \frac{|w - w_0|}{|w|\,|w_0|} = \rho|\xi_0||\xi|. \tag{4}$$

It can be shown that the set of points satisfying the equation

$$|\xi - a| = \lambda|\xi - b| \qquad (5)$$

is a line if $\lambda = 1$ and is a circle if $\lambda > 0$ and $\lambda \neq 1$. See Problem 28 in Exercises 7.2. Thus, with the identifications $a = \xi_0$, $b = 0$, and $\lambda = \rho|\xi_0|$ we see that (4) can be put into the form (5). Since $w = 0$ is not on C', we have $|w_0| \neq \rho$, or, equivalently, $\lambda = \rho|\xi_0| \neq 1$. This implies that the set of points given by (4) is a circle. Finally, since f is a linear function, the image of this circle under f is again a circle, and so we conclude that the image of C under T is a circle. ❏

The key observation in the foregoing proof was that a linear fractional transformation (1) can be written as a composition of the reciprocal function and two linear functions as shown in (2). In Problem 27 of Exercises 2.5 you were asked to show that the image of any line L under the reciprocal mapping $w = 1/z$ is a line or a circle. Therefore, using similar reasoning, we can also show:

> ### *Mapping Lines to Circles with $T(z)$*
>
> If T is a linear fractional transformation given by (3), then the image of a line L under T is either a line or a circle. The image is a circle if and only if $c \neq 0$ and the pole $z = -d/c$ is not on the line L.

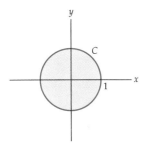

(a) The unit circle $|z| = 1$

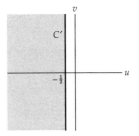

(b) The image of the circle in (a)

Figure 7.2.1 The linear fractional transformation $T(z) = (z + 2)/(z - 1)$

EXAMPLE 2   Image of a Circle

Find the image of the unit circle $|z| = 1$ under the linear fractional transformation $T(z) = (z + 2)/(z - 1)$. What is the image of the interior $|z| < 1$ of this circle?

Solution The pole of T is $z = 1$ and this point is on the unit circle $|z| = 1$. Thus, from Theorem 7.2.1 we conclude that the image of the unit circle is a line. Since the image is a line, it is determined by any two points. Because $T(-1) = -\frac{1}{2}$ and $T(i) = -\frac{1}{2} - \frac{3}{2}i$, we see that the image is the line $u = -\frac{1}{2}$. To answer the second question we first note that a linear fractional transformation is a rational function, and so it is continuous on its domain. As a consequence, the image of the interior $|z| < 1$ of the unit circle is either the half-plane $u < -\frac{1}{2}$ or the half-plane $u > -\frac{1}{2}$. Using $z = 0$ as a **test point**, we find that $T(0) = -2$, which is to the left of the line $u = -\frac{1}{2}$, and so the image is the half-plane $u < -\frac{1}{2}$. This mapping is illustrated in Figure 7.2.1. The circle $|z| = 1$ is shown in color in Figure 7.2.1(a) and its image $u = -\frac{1}{2}$ is shown in black in Figure 7.2.1(b). ❏

EXAMPLE 3   Image of a Circle

Find the image of the unit circle $|z| = 2$ under the linear fractional transformation $T(z) = (z + 2)/(z - 1)$. What is the image of the disk $|z| \leq 2$ under T?

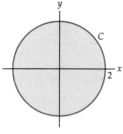

(a) The circle $|z| = 2$

$$w = \frac{z+2}{z-1}$$

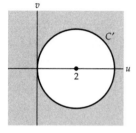

(b) The image of the circle in (a)

Figure 7.2.2 The linear fractional transformation $T(z) = (z+2)/(z-1)$

Solution In this example the pole $z = 1$ does not lie on the circle $|z| = 2$, and so Theorem 7.2.1 indicates that the image of $|z| = 2$ is a circle C'. To find an algebraic description of C', we first note that the circle $|z| = 2$ is symmetric with respect to the x-axis. That is, if z is on the circle $|z| = 2$, then so is $\bar{z}$. Furthermore, we observe that for all z,

$$T(\bar{z}) = \frac{\bar{z}+2}{\bar{z}-1} = \frac{\overline{z+2}}{\overline{z-1}} = \overline{\left(\frac{z+2}{z-1}\right)} = \overline{T(z)}.$$

Hence, if z and $\bar{z}$ are on the circle $|z| = 2$, then we must have that both $w = T(z)$ and $\bar{w} = \overline{T(z)} = T(\bar{z})$ are on the circle C'. It follows that C' is symmetric with respect to the u-axis. Since $z = 2$ and -2 are on the circle $|z| = 2$, the two points $T(2) = 4$ and $T(-2) = 0$ are on C'. The symmetry of C' implies that 0 and 4 are endpoints of a diameter, and so C' is the circle $|w - 2| = 2$. Using $z = 0$ as a test point, we find that $w = T(0) = -2$, which is outside the circle $|w - 2| = 2$. Therefore, the image of the interior of the circle $|z| = 2$ is the exterior of the circle $|w - 2| = 2$. In summary, the disk $|z| \leq 2$ shown in color in Figure 7.2.2(a) is mapped onto the region $|w - 2| \geq 2$ shown in gray in Figure 7.2.2(b) by the linear fractional transformation $T(z) = (z + 2)/(z - 1)$. ❑

Linear Fractional Transformations as Matrices Matrices can be used to simplify many of the computations associated with linear fractional transformations. In order to do so, we associate the matrix

$$\mathbf{A} = \begin{pmatrix} a & b \\ c & d \end{pmatrix} \tag{6}$$

with the linear fractional transformation

$$T(z) = \frac{az+b}{cz+d}. \tag{7}$$

The assignment in (6) is not unique because if e is a nonzero complex number, then the linear fractional transformation $T(z) = (az + b)/(cz + d)$ is also given by $T(z) = (eaz + eb)/(ecz + ed)$. However, if $e \neq 1$, then the two matrices

$$\mathbf{A} = \begin{pmatrix} a & b \\ c & d \end{pmatrix} \quad \text{and} \quad \mathbf{B} = \begin{pmatrix} ea & eb \\ ec & ed \end{pmatrix} = e\mathbf{A} \tag{8}$$

are not equal even though they represent the same linear fractional transformation.

It is easy to verify that the composition $T_2 \circ T_1$ of two linear fractional transformations

$$T_1(z) = (a_1 z + b_1)/(c_1 z + d_1) \quad \text{and} \quad T_2(z) = (a_2 z + b_2)/(c_2 z + d_2)$$

is represented by the product of matrices

$$\begin{pmatrix} a_2 & b_2 \\ c_2 & d_2 \end{pmatrix} \begin{pmatrix} a_1 & b_1 \\ c_1 & d_1 \end{pmatrix} = \begin{pmatrix} a_2 a_1 + b_2 c_1 & a_2 b_1 + b_2 d_1 \\ c_2 a_1 + d_2 c_1 & c_2 b_1 + d_2 d_1 \end{pmatrix}. \tag{9}$$

In Problem 27 of Exercises 7.2 you are asked to find the formula for $T^{-1}(z)$ by solving the equation $w = T(z)$ for z. The formula for the inverse function $T^{-1}(z)$ of a linear fractional transformation T of (7) is represented by the inverse of the matrix $\mathbf{A}$ in (6)

$$\mathbf{A}^{-1} = \begin{pmatrix} a & b \\ c & d \end{pmatrix}^{-1} = \frac{1}{ad-bc} \begin{pmatrix} d & -b \\ -c & a \end{pmatrix}.$$

By identifying $e = \dfrac{1}{ad-bc}$ in (8) we can also represent $T^{-1}(z)$ by the matrix

$$\begin{pmatrix} d & -b \\ -c & a \end{pmatrix}.^* \tag{10}$$

EXAMPLE 4   Using Matrices

Suppose $S(z) = (z-i)/(iz-1)$ and $T(z) = (2z-1)/(z+2)$. Use matrices to find $S^{-1}(T(z))$.

Solution We represent the linear fractional transformations S and T by the matrices

$$\begin{pmatrix} 1 & -i \\ i & -1 \end{pmatrix} \quad \text{and} \quad \begin{pmatrix} 2 & -1 \\ 1 & 2 \end{pmatrix},$$

respectively. By (10), the transformation S^{-1} is given by

$$\begin{pmatrix} -1 & i \\ -i & 1 \end{pmatrix},$$

and so, from (9), the composition $S^{-1} \circ T$ is given by

$$\begin{pmatrix} -1 & i \\ -i & 1 \end{pmatrix}\begin{pmatrix} 2 & -1 \\ 1 & 2 \end{pmatrix} = \begin{pmatrix} -2+i & 1+2i \\ 1-2i & 2+i \end{pmatrix}.$$

Therefore,

$$S^{-1}(T(z)) = \frac{(-2+i)z + 1 + 2i}{(1-2i)z + 2 + i}.$$

Cross-Ratio   In applications we often need to find a conformal mapping from a domain D that is bounded by circles onto a domain D' that is bounded by lines. Linear fractional transformations are particularly well-suited for such applications. However, in order to use them, we must determine a general

---

*You may recall that this matrix is called the **adjoint matrix** of $\mathbf{A}$.

Recall, a circle is uniquely determined ⮕
by three noncolinear points.

method to construct a linear fractional transformation $w = T(z)$, which maps three given distinct points z_1, z_2, and z_3 on the boundary of D to three given distinct points w_1, w_2, and w_3 on the boundary of D'. This is accomplished using the **cross-ratio**, which is defined as follows.

Definition 7.2.2   Cross-Ratio

The **cross-ratio** of the complex numbers z, z_1, z_2, and z_3 is the complex number

$$\frac{z - z_1}{z - z_3} \frac{z_2 - z_3}{z_2 - z_1}. \tag{11}$$

When computing a cross-ratio, we must be careful with the order of the complex numbers. For example, you should verify that the cross-ratio of 0, 1, i, and 2 is $\frac{3}{4} + \frac{1}{4}i$, whereas the cross-ratio of 0, i, 1, and 2 is $\frac{1}{4} - \frac{1}{4}i$.

We extend the concept of the cross-ratio to include points in the extended complex plane by using the limit formula (24) from the Remarks in Section 2.6. For example, the cross-ratio of, say, ∞, z_1, z_2, and z_3 is given by the limit

$$\lim_{z \to \infty} \frac{z - z_1}{z - z_3} \frac{z_2 - z_3}{z_2 - z_1}.$$

The following theorem illustrates the importance of cross-ratios in the study of linear fractional transformations. In particular, we prove that the cross-ratio is invariant under a linear fractional transformation.

**Theorem 7.2.2   Cross-Ratios and Linear
Fractional Transformations**

If $w = T(z)$ is a linear fractional transformation that maps the distinct points z_1, z_2, and z_3 onto the distinct points w_1, w_2, and w_3, respectively, then

$$\frac{z - z_1}{z - z_3} \frac{z_2 - z_3}{z_2 - z_1} = \frac{w - w_1}{w - w_3} \frac{w_2 - w_3}{w_2 - w_1} \tag{12}$$

for all z.

Proof   Let R be the linear fractional transformation

$$R(z) = \frac{z - z_1}{z - z_3} \frac{z_2 - z_3}{z_2 - z_1}, \tag{13}$$

and note that $R(z_1) = 0$, $R(z_2) = 1$, and $R(z_3) = \infty$. Consider also the linear fractional transformation

$$S(z) = \frac{z - w_1}{z - w_3} \frac{w_2 - w_3}{w_2 - w_1}. \tag{14}$$

For the transformation S, we have $S(w_1) = 0$, $S(w_2) = 1$, and $S(w_3) = \infty$. Therefore, the points z_1, z_2, and z_3 are mapped onto the points w_1, w_2, and w_3, respectively, by the linear fractional transformation $S^{-1}(R(z))$. From this it follows that 0, 1, and ∞ are mapped onto 0, 1, and ∞, respectively, by the composition $T^{-1}(S^{-1}(R(z)))$. Now it is a straightforward exercise to verify that the only linear fractional transformation that maps 0, 1, and ∞ onto 0, 1, and ∞ is the identity mapping. See Problem 30 in Exercises 7.2. From this we conclude that $T^{-1}(S^{-1}(R(z))) = z$, or, equivalently, that $R(z) = S(T(z))$. Identifying $w = T(z)$, we have shown that $R(z) = S(w)$. Therefore, from (13) and (14) we have

$$\frac{z - z_1}{z - z_3} \frac{z_2 - z_3}{z_2 - z_1} = \frac{w - w_1}{w - w_3} \frac{w_2 - w_3}{w_2 - w_1}.$$ ❏

EXAMPLE 5   Constructing a Linear Fractional Transformation

Construct a linear fractional transformation that maps the points 1, i, and -1 on the unit circle $|z| = 1$ onto the points -1, 0, 1 on the real axis. Determine the image of the interior $|z| < 1$ under this transformation.

Solution Identifying $z_1 = 1$, $z_2 = i$, $z_3 = -1$, $w_1 = -1$, $w_2 = 0$, and $w_3 = 1$, in (12) we see from Theorem 7.2.2 that the desired mapping $w = T(z)$ must satisfy

$$\frac{z - 1}{z - (-1)} \frac{i - (-1)}{i - 1} = \frac{w - (-1)}{w - 1} \frac{0 - 1}{0 - (-1)}.$$

After solving for w and simplifying we obtain

$$w = T(z) = \frac{z - i}{iz - 1}.$$

Note: A linear fractional transformation can have many equivalent forms. ➡

Using the test point $z = 0$, we obtain $T(0) = i$. Therefore, the image of the interior $|z| < 1$ is the upper half-plane $v > 0$. ❏

EXAMPLE 6   Constructing a Linear Fractional Transformation

Construct a linear fractional transformation that maps the points $-i$, 1, and ∞ on the line $y = x - 1$ onto the points 1, i, and -1 on the unit circle $|w| = 1$.

Solution We proceed as in Example 5. Using (24) of Section 2.6, we find that the cross-ratio of z, $z_1 = -i$, $z_2 = 1$, and $z_3 = \infty$ is

$$\lim_{z_3 \to \infty} \frac{z + i}{z - z_3} \frac{1 - z_3}{1 + i} = \lim_{z_3 \to 0} \frac{z + i}{z - 1/z_3} \frac{1 - 1/z_3}{1 + i} = \lim_{z_3 \to 0} \frac{z + i}{z z_3 - 1} \frac{z_3 - 1}{1 + i} = \frac{z + i}{1 + i}.$$

Now from (12) with identifications $w_1 = 1$, $w_2 = i$, and $w_3 = -1$, the desired mapping $w = T(z)$ must satisfy

$$\frac{z + i}{1 + i} = \frac{w - 1}{w + 1} \frac{i + 1}{i - 1}.$$

After solving for w and simplifying we obtain

$$w = T(z) = \frac{z + 1}{-z + 1 - 2i}.$$

❏

EXERCISES 7.2   *Answers to selected odd-numbered problems begin on page ANS-20.*

In Problems 1–4, find the images of the points 0, 1, i, and ∞ under the given linear fractional transformation T.

1. $T(z) = \dfrac{i}{z}$

2. $T(z) = \dfrac{2}{z-i}$

3. $T(z) = \dfrac{z+i}{z-i}$

4. $T(z) = \dfrac{z-1}{z}$

In Problems 5–8, find the image of the disks $|z| \leq 1$ and $|z-i| \leq 1$ under the given linear fractional transformation T.

5. T is the mapping in Problem 1

6. T is the mapping in Problem 2

7. T is the mapping in Problem 3

8. T is the mapping in Problem 4

In Problems 9–12, find the image of the half-planes $x \geq 0$ and $y \leq 1$ under the given linear fractional transformation T.

9. T is the mapping in Problem 1

10. T is the mapping in Problem 2

11. T is the mapping in Problem 3

12. T is the mapping in Problem 4

In Problems 13–16, find the image of the region shown in color under the given linear fractional transformation.

13. $T(z) = \dfrac{z}{z-2}$

14. $T(z) = \dfrac{z-i}{z+1}$

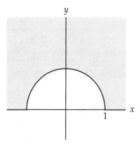

Figure 7.2.3 Figure for Problem 13

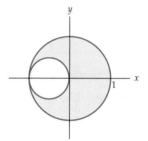

Figure 7.2.4 Figure for Problem 14

15. $T(z) = \dfrac{z+1}{z-2}$

16. $T(z) = \dfrac{-z-1+i}{z-1+i}$

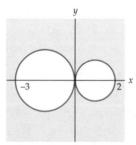

Figure 7.2.5 Figure for Problem 15

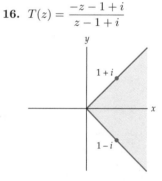

Figure 7.2.6 Figure for Problem 16

In Problems 17–20, use matrices to find **(a)** $S^{-1}(z)$ and **(b)** $S^{-1}(T(z))$.

17. $T(z) = \dfrac{z}{iz-1}$, $S(z) = \dfrac{iz+1}{z-1}$

18. $T(z) = \dfrac{iz}{z-2i}$, $S(z) = \dfrac{2z+1}{z+1}$

19. $T(z) = \dfrac{2z-3}{z-3}$, $S(z) = \dfrac{z-2}{z-1}$

20. $T(z) = \dfrac{z-1+i}{iz-2}$, $S(z) = \dfrac{(2-i)z}{z-1-i}$

In Problems 21–26, construct a linear fractional transformation that takes the given points z_1, z_2, and z_3 onto the given points w_1, w_2, and w_3, respectively.

21. $z_1 = -1$, $z_2 = 0$, $z_3 = 2$; $w_1 = 0$, $w_2 = 1$, $w_3 = \infty$

22. $z_1 = i$, $z_2 = 0$, $z_3 = -i$; $w_1 = 0$, $w_2 = 1$, $w_3 = \infty$

23. $z_1 = 0$, $z_2 = i$, $z_3 = \infty$; $w_1 = 0$, $w_2 = 1$, $w_3 = 2$

24. $z_1 = -1$, $z_2 = 0$, $z_3 = 1$; $w_1 = i$, $w_2 = 0$, $w_3 = \infty$

25. $z_1 = 1$, $z_2 = i$, $z_3 = -i$; $w_1 = -1$, $w_2 = 0$, $w_3 = 3$

26. $z_1 = 1$, $z_2 = i$, $z_3 = -i$; $w_1 = -i$, $w_2 = i$, $w_3 = \infty$

Focus on Concepts

27. Let a, b, c, and d be complex numbers such that $ad - bc \neq 0$.

    (a) Solve the equation $w = \dfrac{az + b}{cz + d}$ for z.

    (b) Explain why (a) implies that the linear fractional transformation $T(z) = (az + b)/(cz + d)$ is a one-to-one function.

28. Consider the equation

$$|z - a| = \lambda |z - b| \tag{15}$$

    where λ is a positive real constant.

    (a) Show that the set of points satisfying (15) is a line if $\lambda = 1$.

    (b) Show that the set of points satisfying (15) is a circle if $\lambda \neq 1$.

29. Let $T(z) = (az + b)/(cz + d)$ be a linear fractional transformation.

    (a) If $T(0) = 0$, then what, if anything, can be said about the coefficients a, b, c, and d?

    (b) If $T(1) = 1$, then what, if anything, can be said about the coefficients a, b, c, and d?

    (c) If $T(\infty) = \infty$, then what, if anything, can be said about the coefficients a, b, c, and d?

30. Use Problem 29 to show that if T is a linear fractional transformation and $T(0) = 0$, $T(1) = 1$, and $T(\infty) = \infty$, then T must be the identity function. That is, $T(z) = z$.

31. Use Theorem 7.2.2 to derive the mapping in entry H-1 in Appendix III.

32. Use Theorem 7.2.2 to derive the mapping in entry H-3 in Appendix III.

7.3 Schwarz-Christoffel Transformations

One problem that arises frequently in the study of fluid flow is that of constructing the flow of an ideal fluid that remains inside a polygonal domain D'. We will see in Section 7.5 that this problem can be solved by finding a one-to-one complex mapping of the half-plane $y \geq 0$ onto the polygonal region that is a conformal mapping in the domain $y > 0$. The existence of such a mapping is guaranteed by the Riemann mapping theorem discussed in the Remarks at the end of Section 7.1. However, even though the Riemann mapping theorem does assert the existence of a mapping, it gives no practical means of finding a formula for the mapping. In this section we present the Schwarz-Christoffel formula, which provides an explicit formula for the derivative of a conformal mapping from the upper half-plane onto a polygonal region.

370 Chapter 7 Conformal Mappings

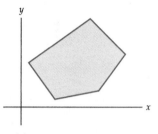

(a) A bounded polygonal region

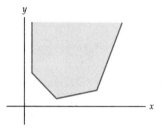

(b) An unbounded polygonal region

Figure 7.3.1 Polygonal regions

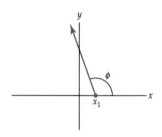

(a) A ray emanating from x_1

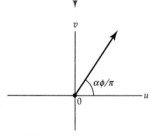

(b) Image of the ray in (a)

Figure 7.3.2 The mapping $w = (z - x_1)^{\alpha/\pi}$

Polygonal Regions A **polygonal region** in the complex plane is a region that is bounded by a simple, connected, piecewise smooth curve consisting of a finite number of line segments. The boundary curve of a polygonal region is called a **polygon** and the endpoints of the line segments in the polygon are called **vertices** of the polygon. If a polygon is a closed curve, then the region enclosed by the polygon is called a **bounded polygonal region**, and a polygonal region that is not bounded is called an **unbounded polygonal region**. See Figure 7.3.1. In the case of an unbounded polygonal region, the ideal point ∞ is also called a vertex of the polygon.

Simple examples of polygonal regions include the region bounded by the triangle with vertices 0, 1, and i, which is an example of a bounded polygonal region, and the region defined by $0 \leq x \leq 1$, $0 \leq y < \infty$, which is an example of an unbounded polygonal region whose vertices are 0, 1, and ∞.

Special Cases In order to motivate a general formula for a conformal mapping of the upper half-plane $y \geq 0$ onto a polygonal region, we first examine the complex mapping

$$w = f(z) = (z - x_1)^{\alpha/\pi}, \tag{1}$$

where x_1 and α are a real numbers and $0 < \alpha < 2\pi$. The mapping in (1) is the composition of a translation $T(z) = z - x_1$ followed by the real power function $F(z) = z^{\alpha/\pi}$. Because x_1 is real, T translates in a direction parallel to the real axis. Under this translation the x-axis is mapped onto the u-axis with the point $z = x_1$ mapping onto the point $w = 0$. In order to understand the power function F as a complex mapping we replace the symbol z with the exponential notation $re^{i\theta}$ to obtain:

$$F(z) = \left(re^{i\theta}\right)^{\alpha/\pi} = r^{\alpha/\pi}e^{i(\alpha\theta/\pi)}. \tag{2}$$

From (2) we see that the complex mapping $w = z^{\alpha/\pi}$ can be visualized as the process of magnifying or contracting the modulus r of z to the modulus $r^{\alpha/\pi}$ of w, and rotating z through α/π radians about the origin to increase or decrease an argument θ of z to an argument $\alpha\theta/\pi$ of w. Thus, under the composition $w = F(T(z)) = (z - x_1)^{\alpha/\pi}$, a ray emanating from x_1 and making an angle of ϕ radians with the real axis is mapped onto a ray emanating from the origin and making an angle of $\alpha\phi/\pi$ radians with the real axis. See Figure 7.3.2.

Now consider the mapping (1) on the half-plane $y \geq 0$. Since this set consists of the point $z = x_1$ together with the family of rays $\arg(z - x_1) = \phi$, $0 \leq \phi \leq \pi$, the image under $w = (z - x_1)^{\alpha/\pi}$ consists of the point $w = 0$ together with the family of rays $\arg(w) = \alpha\phi/\pi$, $0 \leq \alpha\phi/\pi \leq \alpha$. Put another way, the image of the half-plane $y \geq 0$ is the point $w = 0$ together with the wedge $0 \leq \arg(w) \leq \alpha$. See Figure 7.3.3 on page 371.

The function f given by (1), which maps the half-plane $y \geq 0$ onto an unbounded polygonal region with a single vertex, has derivative:

$$f'(z) = \frac{\alpha}{\pi}(z - x_1)^{(\alpha/\pi)-1}. \tag{3}$$

Since $f'(z) \neq 0$ if $z = x + iy$ and $y > 0$, it follows that $w = f(z)$ is a conformal mapping at any point z with $y > 0$. In general, we will use the derivative f', not f, to describe a conformal mapping of the upper half-plane $y \geq 0$ onto an arbitrary polygonal region. With this in mind, we will now present a generalization of the mapping in (1) based on its derivative in (3).

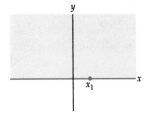

(a) The half-plane $y \geq 0$

$$\downarrow w = (z - x_1)^{\alpha/\pi}$$

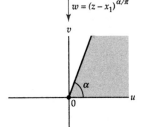

(b) The image of the half-plane

Figure 7.3.3 The mapping $w = (z - x_1)^{\alpha/\pi}$

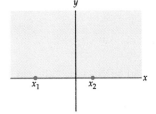

(a) The upper half-plane $y \geq 0$

$$\downarrow w = f(z)$$

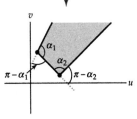

(b) The image of the region in (a)

Figure 7.3.4 The mapping associated with (4)

Consider a new function f, which is analytic in the domain $y > 0$ and whose derivative is:

$$f'(z) = A \left(z - x_1\right)^{(\alpha_1/\pi)-1} \left(z - x_2\right)^{(\alpha_2/\pi)-1}, \qquad (4)$$

where x_1, x_2, α_1, and α_2 are real, $x_1 < x_2$, and A is a complex constant. A useful fact that will help us determine the image of the half-plane $y \geq 0$ under f is that a parametrization $w(t)$, $a < t < b$, gives a line segment if and only if some value of $\arg(w'(t))$ is constant for all t in the interval $a < t < b$. We now use this fact to determine the images of the intervals $(-\infty, x_1)$, (x_1, x_2), and (x_2, ∞) on the real axis under the complex mapping $w = f(z)$. If we parametrize the interval $(-\infty, x_1)$ by $z(t) = t$, $-\infty < t < x_1$, then by (11) of Section 2.2 the image under $w = f(z)$ is parametrized by $w(t) = f(z(t)) = f(t)$, $-\infty < t < x_1$. From (4) with the identification $z = t$, we obtain:

$$w'(t) = f'(t) = A \left(t - x_1\right)^{(\alpha_1/\pi)-1} \left(t - x_2\right)^{(\alpha_2/\pi)-1}.$$

An argument of $w'(t)$ is then given by:

$$\operatorname{Arg}(A) + \left(\frac{\alpha_1}{\pi} - 1\right) \operatorname{Arg}(t - x_1) + \left(\frac{\alpha_2}{\pi} - 1\right) \operatorname{Arg}(t - x_2). \qquad (5)$$

Because $-\infty < t < x_1$, we have that $t - x_1$ is a negative real number, and so $\operatorname{Arg}(t - x_1) = \pi$. In addition, since $x_1 < x_2$, we also have that $t - x_2$ is a negative real number, and thus $\operatorname{Arg}(t - x_2) = \pi$. By substituting these values into (5) we find that $\operatorname{Arg}(A) + \alpha_1 + \alpha_2 - 2\pi$ is a constant value of $\arg(w'(t))$ for all t in the interval $(-\infty, x_1)$. Therefore, we conclude that the interval $(-\infty, x_1)$ is mapped onto a line segment by $w = f(z)$.

By similar reasoning we determine that the intervals (x_1, x_2) and (x_2, ∞) also map onto line segments. A value of the argument of w' for each interval is given in the following table. The change in the value of the argument is also listed.

Interval	Argument of w'	Change in Argument
$(-\infty, x_1)$	$\operatorname{Arg}(A) + \alpha_1 + \alpha_2 - 2\pi$	0
(x_1, x_2)	$\operatorname{Arg}(A) + \alpha_2 - \pi$	$\pi - \alpha_1$
(x_2, ∞)	$\operatorname{Arg}(A)$	$\pi - \alpha_2$

Table 7.3.1 Arguments of w'

Since f is an analytic (and, hence, continuous) mapping, we conclude that the image of the half-plane $y \geq 0$ is an unbounded polygonal region. By Table 7.3.1 we see that the exterior angles between successive sides of the polygonal boundary are given by the change in argument of w' from one interval to the next. Therefore, the interior angles of the polygon are α_1 and α_2. See Figure 7.3.4.

Schwarz-Christoffel Formula The foregoing discussion can be generalized to produce a formula for the derivative f' of a function f that maps the half-plane $y \geq 0$ onto a polygonal region with any number of sides. This formula, given in the following theorem, is called the **Schwarz-Christoffel formula**.

372   Chapter 7   Conformal Mappings

Theorem 7.3.1   Schwarz-Christoffel Formula

Let f be a function that is analytic in the domain $y > 0$ and has the derivative

$$f'(z) = A\,(z - x_1)^{(\alpha_1/\pi)-1}\,(z - x_2)^{(\alpha_2/\pi)-1}\cdots(z - x_n)^{(\alpha_n/\pi)-1}, \quad (6)$$

where $x_1 < x_2 < \cdots < x_n$, $0 < \alpha_i < 2\pi$ for $1 \le i \le n$, and A is a complex constant. Then the upper half-plane $y \ge 0$ is mapped by $w = f(z)$ onto an unbounded polygonal region with interior angles α_1, α_2, ..., α_n.

It follows from Theorem 7.1.1 of Section 7.1 that the function given by the Schwarz-Christoffel formula (6) is a conformal mapping in the domain $y > 0$. For the sake of brevity, we will, henceforth, refer to a mapping obtained from (6) as a *conformal mapping from the upper half-plane* onto a polygonal region. It should be kept in mind that although such a mapping is defined on the upper half-plane $y \ge 0$, it is only conformal in the domain $y > 0$.

Before investigating some examples of the Schwarz-Christoffel formula, we need to point out three things. First, in practice we usually have some freedom in the selection of the points x_k on the x-axis. A judicious choice can simplify the computation of $f(z)$. Second, Theorem 7.3.1 provides a formula only for the derivative of f. A general formula for f is given by an integral

$$f(z) = A \int (z - x_1)^{(\alpha_1/\pi)-1}\,(z - x_2)^{(\alpha_2/\pi)-1}\cdots(z - x_n)^{(\alpha_n/\pi)-1}\,dz + B,$$

where A and B are complex constants. Thus, f is the composition of the function

$$g(z) = \int (z - x_1)^{(\alpha_1/\pi)-1}\,(z - x_2)^{(\alpha_2/\pi)-1}\cdots(z - x_n)^{(\alpha_n/\pi)-1}\,dz$$

and the linear mapping $h(z) = Az + B$. As described in Section 2.3, the linear mapping h allows us to rotate, magnify (or contract), and translate the polygonal region produced by g. Third, although it is not stated in Theorem 7.3.1, the Schwarz-Christoffel formula (6) can also be used to construct a mapping of the upper half-plane $y \ge 0$ onto a bounded polygonal region. To do so, we apply (6) using only $n-1$ of the n interior angles of the bounded polygonal region.* We illustrate these ideas in the following examples.

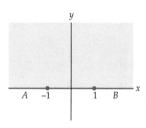

(a) Half-plane $y \ge 0$

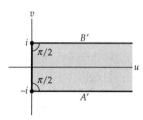

(b) Semi-infinite strip

Figure 7.3.5 Figure for Example 1

EXAMPLE 1   Using the Schwarz-Christoffel Formula

Use the Schwarz-Christoffel formula (6) to construct a conformal mapping from the upper half-plane onto the polygonal region defined by $u \ge 0$, $-1 \le v \le 1$.

Solution Observe that the polygonal region defined by $u \ge 0$, $-1 \le v \le 1$, is the semi-infinite strip shown in gray in Figure 7.3.5(b). The interior angles of this unbounded polygonal region are $\alpha_1 = \alpha_2 = \pi/2$, and the vertices are $w_1 = -i$ and $w_2 = i$. To find the desired mapping, we apply Theorem 7.3.1 with $x_1 = -1$ and $x_2 = 1$. With these identifications, (6) gives

$$f'(z) = A\,(z+1)^{-1/2}\,(z-1)^{-1/2}. \quad (7)$$

---

*For a bounded polygon in the plane, any $n-1$ of its interior angles uniquely determine the remaining one.

A formula for $f(z)$ is found by integrating (7). Since z is in the upper half-plane $y \geq 0$, we first use the principal square root to rewrite (7) as

$$f'(z) = \frac{A}{(z^2 - 1)^{1/2}}.$$

Furthermore, since the principal value of $(-1)^{1/2} = i$, we have

$$f'(z) = \frac{A}{(z^2 - 1)^{1/2}} = \frac{A}{[-1(1 - z^2)]^{1/2}} = \frac{A}{i} \frac{1}{(1 - z^2)^{1/2}} = -Ai \frac{1}{(1 - z^2)^{1/2}}. \tag{8}$$

From (7) of Section 4.4 we recognize that an antiderivative of (8) is given by

$$f(z) = -Ai \sin^{-1} z + B, \tag{9}$$

where $\sin^{-1} z$ is the single-valued function obtained by using the principal square root and principal value of the logarithm and where A and B are complex constants. If we choose $f(-1) = -i$ and $f(1) = i$, then the constants A and B must satisfy the system of equations

$$-Ai \sin^{-1}(-1) + B = Ai\frac{\pi}{2} + B = -i$$

$$-Ai \sin^{-1}(1) + B = -Ai\frac{\pi}{2} + B = i.$$

By adding these two equations we see that $2B = 0$, or, $B = 0$. Now by substituting $B = 0$ into either the first or second equation we obtain $A = -2/\pi$. Therefore, the desired mapping is given by

$$f(z) = i\frac{2}{\pi} \sin^{-1} z.$$

This mapping is shown in Figure 7.3.5. The line segments labeled A and B shown in color in Figure 7.3.5(a) are mapped by $w = i\frac{2}{\pi} \sin^{-1} z$ onto the line segments labeled A' and B' shown in black in Figure 7.3.5(b). ❑

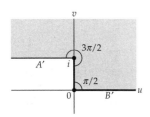

(a) Half-plane $y \geq 0$

(b) Polygonal region for Example 2

Figure 7.3.6 Figure for Example 2

EXAMPLE 2   Using the Schwarz-Christoffel Formula

Use the Schwarz-Christoffel formula (6) to construct a conformal mapping from the upper half-plane onto the polygonal region shown in gray in Figure 7.3.6(b).

Solution We proceed as in Example 1. The region shown in gray in Figure 7.3.6(b) is an unbounded polygonal region with interior angles $\alpha_1 = 3\pi/2$ and $\alpha_1 = \pi/2$ at the vertices $w_1 = i$ and $w_2 = 0$, respectively. If we select $x_1 = -1$ and $x_2 = 1$ to map onto w_1 and w_2, respectively, then (6) gives

$$f'(z) = A(z + 1)^{1/2} (z - 1)^{-1/2}. \tag{10}$$

Since

$$(z + 1)^{1/2} (z - 1)^{-1/2} = \left(\frac{z+1}{z-1}\right)^{1/2} \left(\frac{z+1}{z+1}\right)^{1/2} = \frac{z+1}{(z^2-1)^{1/2}},$$

we can rewrite (10) as

$$f'(z) = A\left[\frac{z}{(z^2-1)^{1/2}} + \frac{1}{(z^2-1)^{1/2}}\right]. \tag{11}$$

An antiderivative of (11) is given by

$$f(z) = A\left[(z^2-1)^{1/2} + \cosh^{-1}z\right] + B,$$

where A and B are complex constants, and where $(z^2-1)^{1/2}$ and $\cosh^{-1}z$ represent branches of the square root and inverse hyperbolic cosine functions defined on the domain $y > 0$. Because $f(-1) = i$ and $f(1) = 0$, the constants A and B must satisfy the system of equations

$$A\left(0 + \cosh^{-1}(-1)\right) + B = A\pi i + B = i$$
$$A\left(0 + \cosh^{-1}1\right) + B = B = 0.$$

Therefore, $A = 1/\pi$, $B = 0$, and the desired mapping is given by

$$f(z) = \frac{1}{\pi}\left[(z^2-1)^{1/2} + \cosh^{-1}z\right].$$

The mapping is illustrated in Figure 7.3.6. The line segments labeled A and B shown in color in Figure 7.3.6(a) are mapped by $w = f(z)$ onto the line segments labeled A' and B' shown in black in Figure 7.3.6(b). ❏

When using the Schwarz-Christoffel formula, it is not always possible to express $f(z)$ in terms of elementary functions. In such cases, however, numerical techniques can be used to approximate f with great accuracy. The following example illustrates that even relatively simple polygonal regions can lead to integrals that cannot be expressed in terms of elementary functions.

EXAMPLE 3   Using the Schwarz-Christoffel Formula

Use the Schwarz-Christoffel formula (6) to construct a conformal mapping from the upper half-plane onto the polygonal region bounded by the equilateral triangle with vertices $w_1 = 0$, $w_2 = 1$, and $w_3 = \frac{1}{2} + \frac{1}{2}\sqrt{3}i$. See Figure 7.3.7.

Solution   The region bounded by the equilateral triangle is a bounded polygonal region with interior angles $\alpha_1 = \alpha_2 = \alpha_3 = \pi/3$. As mentioned on page 372, we can find a desired mapping by using the Schwarz-Christoffel formula (6) with $n - 1 = 2$ of the interior angles. After selecting $x_1 = 0$ and $x_2 = 1$, (6) gives

$$f'(z) = Az^{-2/3}(z-1)^{-2/3}. \tag{12}$$

There is no antiderivative of the function in (12) that can be expressed in terms of elementary functions. However, f' is analytic in the simply connected domain $y > 0$, and so, from Theorem 5.4.3, an antiderivative f does exist in this domain. The antiderivative is given by the integral formula

$$f(z) = A\int_0^z \frac{1}{s^{2/3}(s-1)^{2/3}}ds + B, \tag{13}$$

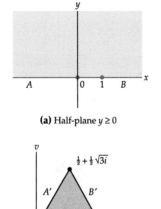

(a) Half-plane $y \geq 0$

(b) Equilateral triangle

Figure 7.3.7 Figure for Example 3

where A and B are complex constants. Requiring that $f(0) = 0$ allows us to solve for the constant B. Since $\int_0^0 = 0$, we have

$$f(0) = A \int_0^0 \frac{1}{s^{2/3} (s-1)^{2/3}} ds + B = 0 + B = B,$$

and so $f(0) = 0$ implies that $B = 0$. If we also require that $f(1) = 1$, then

$$f(1) = A \int_0^1 \frac{1}{s^{2/3} (s-1)^{2/3}} ds = 1.$$

Let Γ denote value of the integral

$$\Gamma = \int_0^1 \frac{1}{s^{2/3} (s-1)^{2/3}} ds.$$

Then $A = 1/\Gamma$ and f can be written as

$$f(z) = \frac{1}{\Gamma} \int_0^z \frac{1}{s^{2/3} (s-1)^{2/3}} ds.$$

Values of f can be approximated using a CAS. For example, using the **NIntegrate** command in *Mathematica* we find that

$$f(i) \approx 0.4244 + 0.3323i \quad \text{and} \quad f(1+i) \approx 0.5756 + 0.3323i.$$

□

The Schwarz-Christoffel formula can also sometimes be used to find mappings onto nonpolygonal regions. This approach requires that the desired nonpolygonal region can be obtained as a "limit" of a sequence of polygonal regions. The following example illustrates this technique.

EXAMPLE 4 **Using the Schwarz-Christoffel Formula**

Use the Schwarz-Christoffel formula (6) to construct a conformal mapping from the upper half-plane onto the nonpolygonal region defined by $v \geq 0$, with the horizontal half-line $v = \pi$, $-\infty < u \leq 0$, deleted. See Figure 7.3.8(c).

Solution Let u_0 be a point on the nonpositive u-axis in the w-plane. We can approximate the non-polygonal region defined by $v \geq 0$, with the half-line $v = \pi$, $-\infty < u \leq 0$, deleted by the polygonal region whose boundary consists of the horizontal half-line $v = \pi$, $-\infty < u \leq 0$, the line segment from πi to u_0, and the horizontal half-line $v = 0$, $u_0 \leq u \leq \infty$. The vertices of this polygonal region are $w_1 = \pi i$ and $w_2 = u_0$, with corresponding interior angles α_1 and α_2. See Figure 7.3.8(b). If we choose the points $z_1 = -1$ and $z_2 = 0$ to map onto the vertices $w_1 = \pi i$ and $w_2 = u_0$, respectively, then (6) gives the derivative

$$A (z+1)^{(\alpha_1/\pi)-1} z^{(\alpha_2/\pi)-1}. \tag{14}$$

Observe in Figure 7.3.8(b) that as u_0 approaches $-\infty$ along the u-axis, the interior angle α_1 approaches 2π and the interior angle α_2 approaches 0. With these limiting values, (14) suggests that our desired mapping f has derivative

$$f'(z) = A (z+1)^1 z^{-1} = A \left(1 + \frac{1}{z}\right). \tag{15}$$

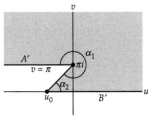

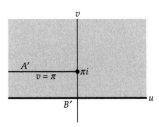

Figure 7.3.8 Figure for Example 4

(a) Half-plane $y \geq 0$

(b) Polygonal region

(c) Limit of polygonal regions

An antiderivative of the function in (15) is given by

$$f(z) = A\left(z + \operatorname{Ln} z\right) + B, \tag{16}$$

where A and B are complex constants.

In order to determine the appropriate values of the constants A and B, we first consider the mapping $g(z) = z + \operatorname{Ln} z$ on the upper half-plane $y \geq 0$. The function g has a point of discontinuity at $z = 0$; thus, we will consider separately the boundary half-lines $y = 0$, $-\infty < x < 0$, and $y = 0$, $0 < x < \infty$, of the half-plane $y \geq 0$. If $z = x + 0i$ is on the half-line $y = 0$, $-\infty < x < 0$, then $\operatorname{Arg}(z) = \pi$, and so $g(z) = x + \log_e |x| + i\pi$.

When $x < 0$, $x + \log_e |x|$ takes on all values from $-\infty$ to -1. Thus, the image of the negative x-axis under g is the horizontal half-line $v = \pi$, $-\infty < u < -1$. On the other hand, if $z = x + 0i$ is on the half-line $y = 0$, $0 < x < \infty$, then $\operatorname{Arg}(z) = 0$, and so $g(z) = x + \log_e |x|$. When $x > 0$, $x + \log_e |x|$ takes on all values from $-\infty$ to ∞. Therefore, the image of the positive x-axis under g is the u-axis. It follows that the image of the half-plane $y \geq 0$ under $g(z) = z + \operatorname{Ln} z$ is the region defined by $v \geq 0$, with the horizontal half-line $v = \pi$, $-\infty < u < -1$ deleted. In order to obtain the region shown in Figure 7.3.8(c), we should compose g with a translation by 1. Therefore, the desired mapping is given by

$$f(z) = z + \operatorname{Ln}(z) + 1.$$

❏

EXERCISES 7.3   *Answers to selected odd-numbered problems begin on page ANS-20.*

In Problems 1–6, use Theorem 7.3.1 to describe the image of the upper half-plane $y \geq 0$ under the conformal mapping $w = f(z)$ that satisfies the given conditions. Do not try to solve for $f(z)$.

1. $f'(z) = (z-1)^{-1/2}$, $f(1) = 0$

2. $f'(z) = (z+1)^{-1/3}$, $f(-1) = 0$

3. $f'(z) = (z+1)^{-1/2}(z-1)^{1/2}$, $f(-1) = 0$, $f(1) = 1$

4. $f'(z) = (z+1)^{-1/2}(z-1)^{-3/4}$, $f(-1) = 0$, $f(0) = 1$

5. $f'(z) = (z+1)^{1/2}z^{-1/2}(z-1)^{-1/4}$, $f(-1) = i$, $f(0) = 0$, $f(1) = 1$

6. $f'(z) = (z-1)^{-1/4}z^{-1/2}(z-1)^{-1/4}$, $f(-1) = -1+i$, $f(0) = 0$, $f(1) = 1+i$

In Problems 7–10, use the Schwarz-Christoffel formula (6) to find $f'(z)$ for a conformal mapping $w = f(z)$ from the upper half-plane onto the given polygonal region shown in gray. Use the values $x_1 = -1$, $x_2 = 0$, $x_3 = 1$, and so on in (6). Do not try to solve for $f(z)$.

7. $f(-1) = 0$, $f(0) = 1$ **8.** $f(-1) = -1$, $f(0) = 0$

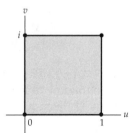

Figure 7.3.9 Figure for Problem 7

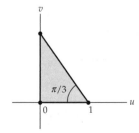

Figure 7.3.10 Figure for Problem 8

9. $f(-1) = -1, \ f(0) = 1$ **10.** $f(-1) = i, \ f(0) = 0$

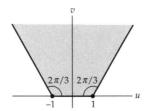

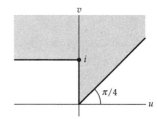

Figure 7.3.11 Figure for Problem 9   Figure 7.3.12 Figure for Problem 10

Focus on Concepts

11. Use the Schwarz-Christoffel formula (6) to construct a conformal mapping from the upper half-plane onto the polygonal region shown in gray in Figure 7.3.13. Require that $f(-1) = \pi i$ and $f(1) = 0$.

12. Use Schwarz-Christoffel formula (6) to construct a conformal mapping from the upper half-plane onto the polygonal region shown in gray in Figure 7.3.14. Require that $f(-1) = -ai$ and $f(1) = ai$.

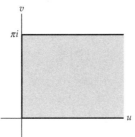

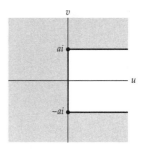

Figure 7.3.13 Figure for Problem 11   Figure 7.3.14 Figure for Problem 12

13. Use the Schwarz-Christoffel formula (6) to verify the conformal mapping in entry M-3 of Appendix III by first constructing the derivative of a mapping of the upper half-plane onto the polygonal region shown in gray in Figure 7.3.15. Require that $f(-1) = -a \ f(0) = v_1 i$, and $f(1) = a$, and then let $v_1 \to -\infty$ along the v-axis.

14. Use the Schwarz-Christoffel formula (6) to verify the conformal mapping in entry M-4 of Appendix III by first constructing the derivative of a mapping of the upper half-plane onto the polygonal region shown in gray in Figure 7.3.16. Require that $f(-1) = -u_1$, $f(0) = ai$, and $f(1) = u_1$, and then let $u_1 \to 0$ along the u-axis.

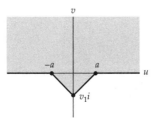

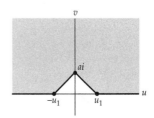

Figure 7.3.15 Figure for Problem 13   Figure 7.3.16 Figure for Problem 14

In Problems 15–18, use a CAS to approximate the images of the points $z_1 = i$ and $z_2 = 1 + i$ under the given function.

15. $w = f(z)$ is the mapping from Problem 3.

16. $w = f(z)$ is the mapping from Problem 6.

17. $w = f(z)$ is the mapping from Problem 8.

18. $w = f(z)$ is the mapping from Problem 9.

7.4   Poisson Integral Formulas

The success of using a conformal mapping to solve a boundary-value problem associated with Laplace's equation often depends on the ability to solve a related boundary-value problem in a simple domain such as the upper half-plane $y > 0$ or the open unit disk $|z| < 1$. In this section we present two important integral formulas for solving a Dirichlet problem in these domains.

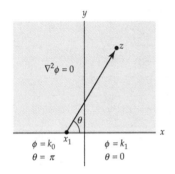

Figure 7.4.1 Dirichlet problem (1)

Formula for the Upper Half-Plane   We begin by investigating the following Dirichlet problem:

Solve:
$$\frac{\partial^2 \phi}{\partial x^2} + \frac{\partial^2 \phi}{\partial y^2} = 0, \quad -\infty < x < \infty, \, y > 0$$

Subject to:
$$\phi(x, 0) = \begin{cases} k_0, & -\infty < x < x_1 \\ k_1, & x_1 < x < \infty, \end{cases}$$

(1)

where k_0 and k_1 are real constants and x_1 is a point on the x-axis. Inspection of Figure 7.4.1 suggests that the function ϕ is constant on each ray emanating from x_1 in the upper half-plane and that the values of ϕ on each ray vary from k_1 to k_0 as the angle θ that the ray makes with the ray emanating from x_1 and containing the point $x_1 + 1$ varies from 0 to π. In other words, we seek a function $\phi(\theta)$ such that $\phi(0) = k_1$ and $\phi(\pi) = k_0$. The simplest such function is the linear function:

$$\phi(\theta) = k_1 + \left(\frac{k_0 - k_1}{\pi} \right) \theta, \quad 0 \le \theta \le \pi.$$

After applying the translation $z - x_1$, we observe that $\theta = \text{Arg}(z - x_1)$ and so ϕ can be rewritten as:

$$\phi(x, y) = k_1 + \frac{1}{\pi}(k_0 - k_1)\text{Arg}(z - x_1), \quad 0 \le \text{Arg}(z - x_1) \le \pi. \quad (2)$$

We will now show that the function $\phi(x, \, y)$ defined in (2) is, in fact, a solution of the Dirichlet problem (1). In the upper half-plane $y > 0$, we have that $\phi(x, \, y)$ is the real part of the function

$$f(z) = k_1 - \frac{i}{\pi}(k_0 - k_1) \, \text{Ln}(z - x_1).$$

Because f is analytic when $y > 0$, it follows from Theorem 3.3.1 that its real part is harmonic. Therefore, $\phi(x, y)$ satisfies Laplace's equation

$$\frac{\partial^2 \phi}{\partial x^2} + \frac{\partial^2 \phi}{\partial y^2} = 0 \tag{3}$$

when $-\infty < x < \infty$ and $y > 0$.

We next verify that the boundary conditions of (1) are satisfied by $\phi(x, y)$. Assume that z is in the interval (x_1, ∞) in the real axis. That is, $z = x + 0i$ with $x_1 < x < \infty$. In this case, $\text{Arg}(z - x_1) = 0$, and so (2) gives

$$\phi(x, 0) = k_1 + \frac{1}{\pi}(k_0 - k_1)\text{Arg}(z - x_1) = k_1 + \frac{1}{\pi}(k_0 - k_1)\,0 = k_1. \tag{4}$$

On the other hand, if $z = x + 0i$ with $-\infty < x < x_1$, then $\text{Arg}(z - x_1) = \pi$, and so

$$\phi(x, 0) = k_1 + \frac{1}{\pi}(k_0 - k_1)\text{Arg}(z - x_1) = k_1 + \frac{1}{\pi}(k_0 - k_1)\pi = k_0. \tag{5}$$

Therefore, from (3), (4), and (5), we conclude that the function $\phi(x, y)$ defined by (2) is a solution of the Dirichlet problem (1).

The foregoing discussion can be generalized. In particular, consider the Dirichlet problem

Solve: $\quad \dfrac{\partial^2 \phi}{\partial x^2} + \dfrac{\partial^2 \phi}{\partial y^2} = 0, \quad -\infty < x < \infty,\, y > 0$

$$\tag{6}$$

Subject to: $\quad \phi(x, 0) = \begin{cases} k_0, & -\infty < x < x_1 \\ k_1, & x_1 < x < x_2 \\ \vdots & \vdots \\ k_n, & x_n < x < \infty, \end{cases}$

where $x_1 < x_2 < \cdots < x_n$ are n distinct points on the x-axis and k_0, k_1, $\ldots$, k_n are $n + 1$ real constants. See Figure 7.4.2. Observe that (1) is simply a special case of (6) corresponding to $n = 1$. With reasoning similar to that used to obtain (2) we construct the function

$$\phi(x, y) = k_n + \frac{1}{\pi}\sum_{j=1}^{n}(k_{j-1} - k_j)\,\text{Arg}(z - x_j). \tag{7}$$

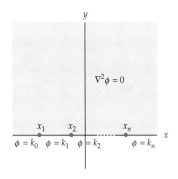

Figure 7.4.2 Dirichlet problem (6)

As with (2), we can verify that this function is harmonic in the domain $y > 0$ by observing that $\phi(x, y)$ is the real part of the analytic function

$$f(z) = k_n - \frac{i}{\pi}\sum_{j=1}^{n}(k_{j-1} - k_j)\,\text{Ln}(z - x_j).$$

Now we show that $\phi(x, y)$ satisfies the boundary conditions in (6). Let N be a fixed value of j. If $z = x + 0i$ is a point with $x_N < x < x_{N+1}$, then $\text{Arg}(z - x_j) = 0$ for $1 \leq j \leq N$, while $\text{Arg}(z - x_j) = \pi$ for $N + 1 \leq j \leq n$.

Therefore, for $z = x + 0i$ with $x_N < x < x_{N+1}$, (7) gives

$$\phi(x, 0) = k_n + \frac{1}{\pi} \sum_{j=1}^{n} (k_{j-1} - k_j) \operatorname{Arg}(z - x_j)$$

$$= k_n + \frac{1}{\pi} \sum_{j=1}^{N} (k_{j-1} - k_j) \operatorname{Arg}(z - x_j) + \frac{1}{\pi} \sum_{j=N+1}^{n} (k_{j-1} - k_j) \operatorname{Arg}(z - x_j)$$

$$= k_n + \frac{1}{\pi} \sum_{j=1}^{N} (k_{j-1} - k_j) \cdot 0 + \frac{1}{\pi} \sum_{j=N+1}^{n} (k_{j-1} - k_j) \cdot \pi$$

$$= k_n + (k_N - k_{N+1}) + (k_{N+1} - k_{N+2}) + \cdots + (k_{n-1} - k_n)$$

$$= k_N.$$

Therefore, the function $\phi(x, y)$ satisfies the boundary conditions of (6). In summary, we have shown that the function $\phi(x, y)$ defined in (7) is a solution of the Dirichlet problem given by (6). This solution will be used to find an integral formula for a solution of a more general type of Dirichlet problem in the upper half-plane $y > 0$.

EXAMPLE 1   **Solving a Dirichlet Problem in the Upper Half-Plane**

Use (7) to solve the Dirichlet problem

Solve: $\dfrac{\partial^2 \phi}{\partial x^2} + \dfrac{\partial^2 \phi}{\partial y^2} = 0, \ -\infty < x < \infty, \ y > 0$

Subject to: $\phi(x, 0) = \begin{cases} 2, & -\infty < x < 0 \\ -1, & 0 < x < 3 \\ 5, & 3 < x < \infty, \end{cases}$

illustrated in Figure 7.4.3.

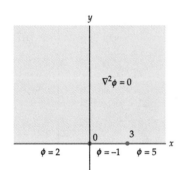

Figure 7.4.3 Figure for Example 1

Solution   Identifying $k_0 = 2$, $k_1 = -1$, $k_2 = 5$, $x_1 = 0$, and $x_2 = 3$ in (7), we obtain the solution

$$\phi(x, y) = 5 + \frac{1}{\pi} (2 + 1) \operatorname{Arg}(z - 0) + \frac{1}{\pi} (-1 - 5) \operatorname{Arg}(z - 3)$$

$$= 5 + \frac{3}{\pi} \operatorname{Arg}(z) - \frac{6}{\pi} \operatorname{Arg}(z - 3).$$

❏

Poisson Integral Formula   A special case of the Dirichlet problem (6) occurs when $k_0 = k_n = 0$.

Solve: $\dfrac{\partial^2 \phi}{\partial x^2} + \dfrac{\partial^2 \phi}{\partial y^2} = 0, \ -\infty < x < \infty, \ y > 0$ (8)

$$\text{Subject to: } \phi(x,0) = \begin{cases} 0, & -\infty < x < x_1 \\ k_1, & x_1 < x < x_2 \\ \vdots & \vdots \\ k_{n-1}, & x_{n-1} < x < x_n \\ 0, & x_n < x < \infty. \end{cases}$$

After setting $z_i = z - x_i$ for $i = 1, 2, \ldots, n$ and identifying $k_0 = k_n = 0$, the solution given by (7) can be written as

$$\phi(x,y) = 0 + \frac{1}{\pi}\left[(0 - k_1)\operatorname{Arg} z_1 + (k_1 - k_2)\operatorname{Arg} z_2 + \cdots + (k_{n-1} - 0)\operatorname{Arg} z_n \right]$$

$$= \frac{1}{\pi}\left[k_1(\operatorname{Arg} z_2 - \operatorname{Arg} z_1) + k_2(\operatorname{Arg} z_3 - \operatorname{Arg} z_2) + \cdots + k_{n-1}(\operatorname{Arg} z_n - \operatorname{Arg} z_{n-1}) \right]$$

$$= \sum_{j=1}^{n-1} \frac{k_j}{\pi} \left(\operatorname{Arg} z_{j+1} - \operatorname{Arg} z_j \right)$$

That is, the function

$$\phi(x,y) = \sum_{j=1}^{n-1} \frac{k_j}{\pi} \left[\operatorname{Arg}(z - x_{j+1}) - \operatorname{Arg}(z - x_j) \right] \tag{9}$$

is a solution of the Dirichlet problem in (8). We can write (9) in terms of a real improper integral. In order to do so, let t be a real variable and observe that if $y > 0$, then

$$\operatorname{Arg}(z - t) = \cot^{-1}\left(\frac{x - t}{y} \right) \quad \text{and} \quad \frac{d}{dt}\operatorname{Arg}(z - t) = \frac{y}{(x - t)^2 + y^2}.$$

Put another way,

$$\operatorname{Arg}(z_{j+1}) - \operatorname{Arg}(z_j) = \int_{x_j}^{x_{j+1}} \frac{d}{dt}\left[\frac{y}{(x - t)^2 + y^2} \right] dt.$$

With these substitutions, (9) becomes

$$\phi(x,y) = \frac{1}{\pi} \sum_{j=1}^{n-1} \int_{x_j}^{x_{j+1}} \frac{k_j y}{(x - t)^2 + y^2}\, dt.$$

Since $\phi(x,0) = 0$ when $x < x_1$ or $x > x_n$, $\phi(x, y)$ can also be written as

$$\phi(x,y) = \frac{y}{\pi} \int_{-\infty}^{\infty} \frac{\phi(t,0)}{(x - t)^2 + y^2}\, dt. \tag{10}$$

The integral formula in (10) is called the **Poisson integral formula** for the upper half-plane $y > 0$, and it gives a solution $\phi(x, y)$ of the Dirichlet problem in (8). The Poisson integral formula can also be used to solve a more general type of Dirichlet problem in which the boundary conditions are specified by any piecewise continuous and bounded function. This is the content of the following theorem.

> **Theorem 7.4.1   Poisson Integral Formula for the Half-Plane**
>
> Let $f(x)$ be a piecewise continuous and bounded function on $-\infty < x < \infty$. Then the function defined by
>
> $$\phi(x,y) = \frac{y}{\pi} \int_{-\infty}^{\infty} \frac{f(t)}{(x-t)^2 + y^2}\, dt \qquad (11)$$
>
> is a solution of the Dirichlet problem in the upper half-plane $y > 0$ with boundary condition $\phi(x,0) = f(x)$ at all points of continuity of f.

Unfortunately, there are few functions f for which the Poisson integral formula (11) can be evaluated. The following example represents an exception to the previous remark.

EXAMPLE 2   Using the Poisson Integral Formula

Use the Poisson integral formula (11) to find a solution of the Dirichlet problem

Solve: $\dfrac{\partial^2 \phi}{\partial x^2} + \dfrac{\partial^2 \phi}{\partial y^2} = 0, \quad -\infty < x < \infty,\ y > 0$

Subject to: $\phi(x,0) = \begin{cases} 0, & -\infty < x < -1 \\ x, & -1 < x < 1 \\ 0, & 1 < x < \infty, \end{cases}$

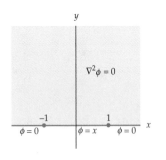

Figure 7.4.4 Figure for Example 2

illustrated in Figure 7.4.4.

Solution   We first define a real function f by $f(x) = \phi(x,0)$. Then $f(x) = x$ on $-1 < x < 1$ and 0 elsewhere. Thus, f is piecewise continuous and bounded on the real line $-\infty < x < \infty$. After replacing the symbol x with the integration variable t, we identify $f(t) = \phi(t,0) = t$ for $-1 < t < 1$ and 0 elsewhere. Then (11) gives

$$\phi(x,y) = \frac{y}{\pi}\int_{-\infty}^{\infty} \frac{\phi(t,0)}{(x-t)^2 + y^2}\, dt = \frac{y}{\pi}\int_{-1}^{1} \frac{t}{(x-t)^2 + y^2}\, dt.$$

With the substitutions $s = x - t$ and $ds = -dt$ this integral becomes

$$\phi(x,y) = -\frac{y}{\pi}\int_{x+1}^{x-1} \frac{x-s}{s^2 + y^2}\, ds = -\frac{y}{\pi}\int_{x+1}^{x-1} \frac{x}{s^2 + y^2}\, ds + \frac{y}{\pi}\int_{x+1}^{x-1} \frac{s}{s^2 + y^2}\, ds.$$

From elementary calculus we have that

$$\int \frac{x}{s^2 + y^2}\, ds = \frac{x}{y}\tan^{-1}\left(\frac{s}{y}\right) + C_1$$

and

$$\int \frac{s}{s^2 + y^2}\, ds = \frac{1}{2}\log_e\left(s^2 + y^2\right) + C_2.$$

Therefore,

$$\phi(x,y) = -\frac{x}{\pi} \tan^{-1}\left(\frac{s}{y}\right)\bigg|_{s=x+1}^{s=x-1} + \frac{y}{2\pi} \log_e\left(s^2 + y^2\right)\bigg|_{s=x+1}^{s=x-1}$$

$$= \frac{x}{\pi}\left[\tan^{-1}\left(\frac{x+1}{y}\right) - \tan^{-1}\left(\frac{x-1}{y}\right)\right] + \frac{y}{2\pi}\log_e\left[\frac{(x-1)^2 + y^2}{(x+1)^2 + y^2}\right]$$

is a solution of the Dirichlet problem. ❏

Formula for the Unit Disk A Poisson integral formula for the unit disk can be derived in a similar manner. This gives an integral formula for a solution of a Dirichlet problem in the open unit disk $|z| < 1$ subject to certain types of boundary conditions. The following theorem gives the precise statement of this result.

Theorem 7.4.2   Poisson Integral Formula for the Unit Disk

Let $f(z)$ be a complex function for which the values $f\left(e^{i\theta}\right)$ on the unit circle $z = e^{i\theta}$ give a piecewise continuous and bounded function for $-\pi \le \theta \le \pi$. Then the function defined by

$$\phi(x,y) = \frac{1}{2\pi}\int_{-\pi}^{\pi} f\left(e^{it}\right)\frac{1 - |z|^2}{\left|e^{it} - z\right|^2}\,dt \qquad (12)$$

is a solution of the Dirichlet problem in the open unit disk $|z| < 1$ with boundary condition $\phi(\cos\theta, \sin\theta) = f\left(e^{i\theta}\right)$ at all points of continuity of f.

As with Theorem 7.4.1, the integral given in (12) can seldom be expressed in terms of elementary functions. When we cannot evaluate the integral, we appeal to numerical methods to approximate values of a solution given by (12).

EXAMPLE 3   Using the Poisson Integral Formula

Use the Poisson integral formula (12) to find a solution of the Dirichlet problem

Solve: $\qquad \dfrac{\partial^2\phi}{\partial x^2} + \dfrac{\partial^2\phi}{\partial y^2} = 0, \quad x^2 + y^2 < 1$

Subject to: $\quad \phi(\cos\theta, \sin\theta) = |\theta|, \quad -\pi < \theta \le \pi,$

illustrated in Figure 7.4.5.

Solution The function $f\left(e^{i\theta}\right) = \phi(\cos\theta, \sin\theta) = |\theta|$ is piecewise continuous and bounded for $-\pi \le \theta \le \pi$. Thus, after identifying $f\left(e^{it}\right) = \phi(\cos t, \sin t) = |t|$ in (12) we obtain the integral formula

$$\phi(x,y) = \frac{1}{2\pi}\int_{-\pi}^{\pi} |t|\frac{1 - |z|^2}{\left|e^{it} - z\right|^2}\,dt.$$

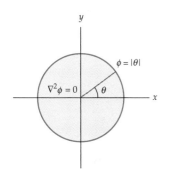

Figure 7.4.5 Figure for Example 3

This integral cannot be evaluated in terms of elementary functions. However, with the use of the **NIntegrate** command in *Mathematica* we can approximate values of the function $\phi(x, y)$. For example, *Mathematica* indicates that $\phi\left(\frac{1}{2}, 0\right) \approx 0.9147$ and $\phi\left(0, \frac{1}{2}\right) \approx 1.5708$.   ❑

EXERCISES 7.4   *Answers to selected odd-numbered problems begin on page ANS-20.*

In Problems 1–4, use (7) to solve the given Dirichlet problem in the upper half-plane $y > 0$.

1.

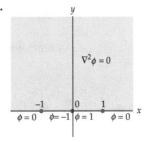

Figure 7.4.6 Figure for Problem 1

2.

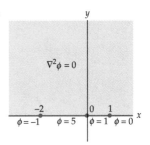

Figure 7.4.7 Figure for Problem 2

3.

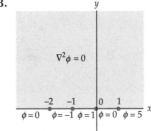

Figure 7.4.8 Figure for Problem 3

4.

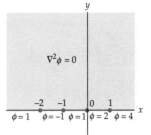

Figure 7.4.9 Figure for Problem 4

In Problems 5–8, use the Poisson integral formula (11) to solve the Dirichlet problem in the upper half-plane $y > 0$ subject to the given boundary conditions.

5. $\phi(x,0) = \begin{cases} 0, & -\infty < x < 0 \\ 2x - 1, & 0 < x < 2 \\ 0, & 2 < x < \infty \end{cases}$

6. $\phi(x,0) = \begin{cases} -1, & -\infty < x < -1 \\ x, & -1 < x < 1 \\ 1, & 1 < x < \infty \end{cases}$

7. $\phi(x,0) = \begin{cases} 0, & -\infty < x < 0 \\ x^2, & 0 < x < 1 \\ 0, & 1 < x < \infty \end{cases}$

8. $\phi(x,0) = \begin{cases} 0, & -\infty < x < 0 \\ x^2, & 0 < x < 1 \\ 1, & 1 < x < \infty \end{cases}$

9. (a) Use the techniques presented in Section 6.6 to establish the integral formulas

$$\int_{-\infty}^{\infty} \frac{\cos s}{s^2 + a^2} ds = \frac{\pi e^{-a}}{a} \quad \text{and} \quad \int_{-\infty}^{\infty} \frac{\sin s}{s^2 + a^2} ds = 0 \text{ for } a > 0.$$

(b) Solve the Dirichlet in the upper half-plane $y > 0$ subject to the boundary condition $\phi(x,0) = \cos x, -\infty < x < \infty$. [*Hint:* Make the substitution $s = t - x$ and use the formulas in part (a).]

10. Solve the Dirichlet in the upper half-plane $y > 0$ subject to the boundary condition $\phi(x,0) = \sin x$, $-\infty < x < \infty$. [*Hint:* Make the substitution $s = t - x$ and use the formulas in part (a) of Problem 9.]

Focus on Concepts

11. Let $f(z)$ be a complex function and suppose that on the unit disk $z = e^{i\theta}$, $-\pi \leq \theta \leq \pi$, we have that $f\left(e^{i\theta}\right)$ is piecewise continuous and bounded. Let $z = re^{i\theta}$, $0 \leq r < 1$, be a point inside the unit disk. Show that the Poisson integral formula (12) can be written as

$$\phi(x,y) = \frac{1}{2\pi} \int_{-\pi}^{\pi} f(e^{it}) \frac{1 - |z|^2}{|e^{it} - z|^2} dt = \frac{1}{2\pi} \int_{-\pi}^{\pi} f(e^{it}) \frac{1 - r^2}{1 + r^2 - 2r \cos(t - \theta)} dt. \tag{13}$$

12. In this problem we determine a solution of the Dirichlet problem on the unit disk subject to a piecewise constant boundary condition. That is, we derive a formula for a solution of a Dirichlet problem in the unit disk that is analogous to the Dirichlet problem (6) in the half-plane.

(**a**) Verify that

$$\frac{1}{2\pi} \int \frac{1 - r^2}{1 + r^2 - 2r \cos(t - \theta)} dt = \frac{1}{\pi} \tan^{-1} \left[\frac{1+r}{1-r} \tan \left(\frac{t - \theta}{2} \right) \right] + C. \tag{14}$$

(**b**) Assume that $\theta_1 < \theta_2 < \cdots < \theta_n$ are n distinct points in the interval $(-\pi, \pi)$. Explain how (13) and (14) can be used to solve the Dirichlet problem

$$\text{Solve:} \qquad \frac{\partial^2 \phi}{\partial x^2} + \frac{\partial^2 \phi}{\partial y^2} = 0, \quad x^2 + y^2 < 1$$

$$\text{Subject to:} \quad \phi(\cos\theta, \sin\theta) = \begin{cases} k_0, & -\pi < \theta < \theta_1 \\ k_1, & \theta_1 < \theta < \theta_2 \\ \vdots & \vdots \\ k_n, & \theta_n < \theta < \pi. \end{cases}$$

13. Use Problems 11 and 12 to solve the Dirichlet problem in the unit disk shown in Figure 7.4.10.

14. Use Problems 11 and 12 to solve the Dirichlet problem in the unit disk shown in Figure 7.4.11.

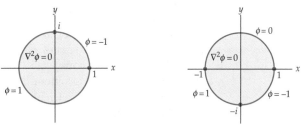

Figure 7.4.10 Figure for Problem 13   **Figure 7.4.11** Figure for Problem 14

In Problems 15 and 16, (**a**) use the Poisson integral formula (12) to find an integral representation of a solution of the given Dirichlet problem in the unit disk, and (**b**) use a CAS to approximate the values of the solution at the points $(0,0)$, $\left(\frac{1}{2}, \frac{1}{2}\right)$, and $\left(0, \frac{1}{3}\right)$.

15.

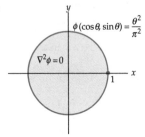

16.

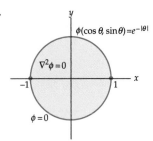

Figure 7.4.12 Figure for Problem 15   Figure 7.4.13 Figure for Problem 16

7.5 Applications

In this section we revisit the method introduced in Section 4.5 for solving Dirichlet problems; here we incorporate the new mappings defined in this chapter or in Appendix III. We also describe a similar process for solving a new type of boundary-value problem that relies on finding a conformal mapping between two domains. This allows us to investigate more complicated boundary-value problems arising in the two-dimensional modeling of electrostatics and heat flow. We conclude this section with an application of conformal mapping to the problem of finding an irrotational flow of an incompressible fluid, that is, the flow of an ideal fluid, in a region of the plane.

7.5.1 Boundary-Value Problems

Dirichlet Problems Revisited   Suppose that D is a domain in the z-plane and that g is a function defined on the boundary C of D. The problem of finding a function $\phi(x, y)$ that satisfies Laplace's equation $\nabla^2 \phi = 0$, or

$$\frac{\partial^2 \phi}{\partial x^2} + \frac{\partial^2 \phi}{\partial y^2} = 0, \tag{1}$$

in D and that equals g on the boundary of D, is called a **Dirichlet problem**.

In Section 4.5, we saw that analytic functions can be used to solve certain Dirichlet problems. We obtained a solution of a Dirichlet problem in a domain D by finding an analytic mapping of D onto a domain D' in which the associated, or transformed, Dirichlet problem can be solved. That is, we found a mapping $w = f(z)$ of D onto D' such that $f(z) = u(x, y) + iv(x, y)$ is analytic in D. By Theorem 4.5.1 of Section 4.5, if $\Phi(u, v)$ is a solution of the transformed Dirichlet problem in D', then $\phi(x, y) = \Phi(u(x, y), v(x, y))$ is a solution of the original Dirichlet problem in D. Thus, our method presented in Section 4.5 for solving Dirichlet problems consisted of the following four steps:

- Find an analytic mapping $w = f(z) = u(x, y) + iv(x, y)$ of the domain D onto a domain D',

- transform the boundary conditions from D to D',

- solve the transformed Dirichlet problem in D', and

- set $\phi(x,\ y) = \Phi(u(x,\ y),\ v(x,\ y))$.

For a more detailed discussion of these steps refer to Section 4.5.

In this chapter we investigate a number of topics that can help complete these four steps. The table of conformal mappings discussed in Section 7.1, the linear fractional transformations studied in Section 7.2, and the Schwarz-Christoffel transformation of Section 7.3 provide a valuable source of mappings to use in Step 1. In addition, if D' is taken to be either the upper half-plane $y > 0$ or the open unit disk $|z| < 1$, then the Poisson integral formulas of Section 7.4 provide a means to determine a solution of the associated Dirichlet problem in D'.

In the following examples, we will apply some ideas from the preceding sections of this chapter to help solve Dirichlet problems arising in the areas of electrostatics, fluid flow, and heat flow. Recall from Section 3.3 that if a function $\phi(x,\ y)$ satisfies Laplace's equation (1) in some domain D, then $\phi(x,\ y)$ is harmonic in D. Moreover, if $\psi(x,\ y)$ is a harmonic conjugate of $\phi(x,\ y)$ in D, then the function

$$\mathbf{\Omega}(z) = \phi(x,y) + i\psi(x,y)$$

is analytic in D and is called a **complex potential function**. The level curves of ϕ and ψ have important physical interpretations in applied mathematics. Their interpretations are summarized in Table 3.4.1.

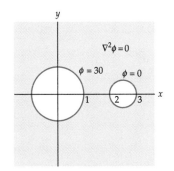

(a) Dirichlet problem

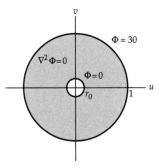

(b) Transformed Dirichlet problem

Figure 7.5.1 Figure for Example 1

EXAMPLE 1   A Heat Flow Application

Determine the steady-state temperature ϕ in the domain D consisting of all points outside of the two circles $|z| = 1$ and $\left|z - \frac{5}{2}\right| = \frac{1}{2}$, shown in color in Figure 7.5.1(a), that satisfies the indicated boundary conditions.

Solution The steady-state temperature ϕ is a solution of Laplace's equation (1) in D that satisfies the boundary conditions

$$\phi(x,y) = 30 \quad \text{if } x^2 + y^2 = 1,$$
$$\phi(x,y) = 0 \quad \text{if } \left(x - \tfrac{5}{2}\right)^2 + y^2 = \tfrac{1}{4}.$$

We solve this problem using the four steps given on pages 386 and 387.

Step 1 Entry C-1 in Appendix III indicates that we can map D onto an annulus. Identifying $b = 2$ and $c = 3$ in Entry C-1, we find that

$$a = \frac{bc + 1 + \sqrt{(b^2 - 1)(c^2 - 1)}}{b + c} = \frac{7 + 2\sqrt{6}}{5},$$

and

$$r_0 = \frac{bc - 1 - \sqrt{(b^2 - 1)(c^2 - 1)}}{c - b} = 5 - 2\sqrt{6}.$$

Thus, the domain D is mapped onto the annulus $5 - 2\sqrt{6} < w < 1$ shown in gray in Figure 7.5.1(b) by the analytic mapping $w = f(z)$, where

$$f(z) = \frac{5z - 7 - 2\sqrt{6}}{\left(7 + 2\sqrt{6}\right)z - 5}. \tag{2}$$

Step 2 Inspection of entry C-1 in Appendix III shows that the boundary circle $\left|z - \frac{5}{2}\right| = \frac{1}{2}$ is mapped onto the boundary circle $|w| = r_0 = 5 - 2\sqrt{6}$. Thus, the boundary condition $\phi = 0$ is transformed to the boundary condition $\Phi = 0$ on the circle $|w| = 5 - 2\sqrt{6}$. Similarly, we see that the boundary condition $\phi = 30$ on the circle $|z| = 1$ is transformed to the boundary condition $\Phi = 30$ on the circle $|w| = 1$. See Figure 7.5.1(b).

Step 3 The shape of the annulus along with the fact that the two boundary conditions are constant in Figure 7.5.1(b) suggests that a solution of the transformed Dirichlet problem is given by a function $\Phi(u, v)$ that is defined in terms of the modulus $r = \sqrt{u^2 + v^2}$ of $w = u + iv$. In Problem 14 in Exercises 3.4 you were asked to show that a solution is given by

$$\Phi(u, v) = A \log_e \sqrt{u^2 + v^2} + B, \tag{3}$$

where

$$A = \frac{k_0 - k_1}{\log_e(a/b)} \quad \text{and} \quad B = \frac{-k_0 \log_e b + k_1 \log_e a}{\log_e(a/b)}.$$

Following the definitions of k_0, k_1, a, and b given in Problem 14, we have $a = 5 - 2\sqrt{6}$, $b = 1$, $k_0 = 0$, and $k_1 = 30$. Thus, we obtain the solution

$$\Phi(u, v) = \frac{-30 \log_e \sqrt{u^2 + v^2}}{\log_e\left(5 - 2\sqrt{6}\right)} + 30 \tag{4}$$

of the transformed Dirichlet problem.

Step 4 The final step is to substitute the real and imaginary parts of the function f given by (2) for the variables u and v in (4). Since

$$u(x, y) + iv(x, y) = \frac{5z - 7 - 2\sqrt{6}}{\left(7 + 2\sqrt{6}\right)z - 5},$$

we have

$$\sqrt{u(x, y)^2 + v(x, y)^2} = \left|\frac{5z - 7 - 2\sqrt{6}}{\left(7 + 2\sqrt{6}\right)z - 5}\right|.$$

Therefore, the steady-state temperature is given by the function

$$\phi(x, y) = \frac{-30}{\log_e\left(5 - 2\sqrt{6}\right)} \log_e \left|\frac{5z - 7 - 2\sqrt{6}}{\left(7 + 2\sqrt{6}\right)z - 5}\right| + 30.$$

❑

A complex potential function $\mathbf{\Omega}(z) = \phi(x, y) + i\psi(x, y)$ for the harmonic function $\phi(x, y)$ found in Example 1 is

$$\mathbf{\Omega}(z) = \frac{-30}{\log_e\left(5 - 2\sqrt{6}\right)} \text{Ln}\left(\frac{5z - 7 - 2\sqrt{6}}{(7 - 2\sqrt{6})z - 5}\right).$$

If we define this function as $\mathbf{\Omega}[\mathbf{z}]$ in *Mathematica*, then the real and imaginary parts $\phi(x, y)$ and $\psi(x, y)$ of $\mathbf{\Omega}(z)$ are given by $\mathbf{Re}[\mathbf{\Omega}[\mathbf{z}]]$ and $\mathbf{Im}[\mathbf{\Omega}[\mathbf{z}]]$, respectively. We can then use the **ContourPlot** command in *Mathematica*

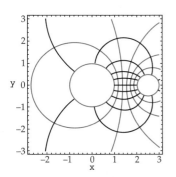

Figure 7.5.2 Isotherms and lines of heat flux for Example 1

to plot the level curves of the real and imaginary parts of Ω. For example, the command

$$\textbf{ContourPlot[Re[}\Omega\textbf{[x + I y]] , \{x, a, b\}, \{y, c, d\}]}$$

produces a plot of the level curves $\phi = c_1$ in the rectangular region $a \leq x \leq b$, $c \leq y \leq d$ of the plane. According to Table 3.4.1, the level curves of ϕ and ψ represent the isotherms and lines of heat flux, respectively. Both sets of level curves are shown in Figure 7.5.2. The isotherms are the curves shown in color and the lines of heat flux are the curves shown in black.

EXAMPLE 2   An Electrostatics Application

Determine the electrostatic potential ϕ in the domain D between the circles $|z| = 1$ and $\left|z - \frac{1}{2}\right| = \frac{1}{2}$, shown in color in Figure 7.5.3(a), that satisfies the indicated boundary conditions.

Solution The electrostatic potential ϕ is a solution of Laplace's equation (1) in D that satisfies the boundary conditions

$$\phi(x,y) = -10 \quad \text{on} \quad x^2 + y^2 = 1,$$

$$\phi(x,y) = 20 \quad \text{on} \quad \left(x - \tfrac{1}{2}\right)^2 + y^2 = \tfrac{1}{4}.$$

We proceed as in Example 1.

Step 1 The given domain D can be mapped onto the infinite horizontal strip $0 < v < 1$, shown in gray in Figure 7.5.3(b), by a linear fractional transformation. One way to do this is to require that the points 1, i, and -1 on the circle $|z| = 1$ map onto the points ∞, 0, and 1, respectively. By Theorem 7.2.2 the desired linear fractional transformation $w = T(z)$ must satisfy

$$\frac{z-1}{z+1}\frac{i+1}{i-1} = \lim_{w_1 \to \infty} \frac{w-w_1}{w-1}\frac{0-1}{0-w_1}, \quad \text{or} \quad \frac{z-1}{z+1}(-i) = \frac{-1}{w-1}.$$

After solving for $w = T(z)$, we obtain

$$T(z) = (1-i)\frac{z-i}{z-1}. \tag{5}$$

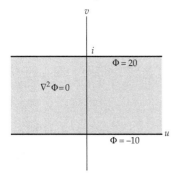

(a) Dirichlet problem

(b) Transformed Dirichlet problem

Figure 7.5.3 Figure for Example 2

By construction, the circle $|z| = 1$ is mapped onto the line $v = 0$ by $w = T(z)$. Furthermore, because the pole $z = 1$ of (5) is on the circle $\left|z - \frac{1}{2}\right| = \frac{1}{2}$, it follows that this circle is mapped onto a line. The image line can be determined by finding the image of two points on the circle $\left|z - \frac{1}{2}\right| = \frac{1}{2}$. For the points $z = 0$ and $z = \frac{1}{2} + \frac{1}{2}i$ on the circle $\left|z - \frac{1}{2}\right| = \frac{1}{2}$, we have $T(0) = 1+i$ and $T\left(\frac{1}{2} + \frac{1}{2}i\right) = -1+i$. Therefore, the image of the circle $\left|z - \frac{1}{2}\right| = \frac{1}{2}$ must be the horizontal line $v = 1$. Using the test point $z = -\frac{1}{2}$, we find that $T\left(-\frac{1}{2}\right) = 1 + \frac{1}{3}i$, and so we conclude that the domain shown in color between the circles in Figure 7.5.3(a) is mapped by $w = T(z)$ onto the domain shown in gray between the horizontal lines in Figure 7.5.3(b).

Step 2 From Step 1 we have $w = T(z)$ maps the circle $|z| = 1$ onto the horizontal line $v = 0$, and it maps the circle $\left|z - \frac{1}{2}\right| = \frac{1}{2}$ onto the horizontal line $v = 1$. Thus, the transformed boundary conditions are $\Phi = -10$ on the line $v = 0$ and $\Phi = 20$ on the line $v = 1$. See Figure 7.5.3(b).

Step 3 Modeled after Example 2 in Section 3.4 and Problem 12 in Exercises 3.4, a solution of the transformed Dirichlet problem is given by

$$\Phi(u, v) = 30v - 10.$$

Step 4 A solution of the original Dirichlet problem is now obtained by substituting the real and imaginary parts of $T(z)$ defined in (5) for the variables u and v in $\Phi(u, v)$. By replacing the symbol z with $x + iy$ in $T(z)$ and simplifying we obtain:

$$T(x + iy) = (1 - i)\frac{x + iy - i}{x + iy - 1} = (1 - i)\frac{x + i(y - 1)}{x - 1 + yi}\frac{x - 1 - iy}{x - 1 - iy}$$

$$= \frac{x^2 + y^2 - 2x - 2y + 1}{(x - 1)^2 + y^2} + \frac{1 - x^2 - y^2}{(x - 1)^2 + y^2}i.$$

Therefore,

$$\phi(x, y) = 30\frac{1 - x^2 - y^2}{(x - 1)^2 + y^2} - 10 \tag{6}$$

is the desired electrostatic potential function.   ❑

A complex potential function for the harmonic function $\phi(x, y)$ given by (6) in Example 2 can be found as follows. If $\mathbf{\Omega}(z)$ is a complex potential for ϕ, then $\mathbf{\Omega}(z) = \phi(x, y) + i\psi(x, y)$ and $\mathbf{\Omega}(z)$ is analytic in D. From Step 4 of Example 2 we have that the complex function $T(z)$ given by (5) has real and imaginary parts $u = \dfrac{x^2 + y^2 - 2x - 2y + 1}{(x - 1)^2 + y^2}$ and $v = \dfrac{1 - x^2 - y^2}{(x - 1)^2 + y^2}$, respectively. That is, $T(z) = u + iv$. We also have from Step 4 that $\phi(x, y) = 30v - 10$. In order to obtain a function with $30v - 10$ as its real part, we multiply $T(z)$ by $-30i$ then subtract 10:

$$-30iT(z) - 10 = -30i(u + iv) - 10 = 30v - 10 - 30ui.$$

Since $T(z) = (1 - i)\dfrac{z - i}{z - 1}$ is analytic in D, it follows that the function $-30iT(z) - 10$ is also analytic in D. Therefore,

$$\mathbf{\Omega}(z) = -30i(1 - i)\frac{z - i}{z - 1} - 10 \tag{7}$$

is a complex potential function for $\phi(x, y)$. Since ϕ represents the electrostatic potential, the level curves of the real and imaginary parts of $\mathbf{\Omega}$ represent the equipotential curves and lines of force, respectively. The *Mathematica*-generated plot in Figure 7.5.4 shows the equipotential curves in color and the lines of force in black.

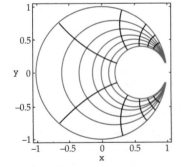

Figure 7.5.4 Equipotential curves and lines of force for Example 2

Neumann Problems   Theorem 7.1.1 states that an analytic mapping is conformal at a point where the derivative is nonzero. This fact did not appear to be of immediate importance in previous examples when we solved Dirichlet problems, but it is extremely important in another class of boundary-value problems associated with Laplace's equation called Neumann problems.

Neumann Problem

Suppose that D is a domain in the plane and that h is a function defined on the boundary C of D. The problem of finding a function $\phi(x, y)$ that satisfies Laplace's equation in D and whose normal derivative $d\phi/dn$ equals h on the boundary C of D is called a **Neumann problem.**

    Certain types of Neumann problems occur naturally in the study of electrostatics, fluid flow, and heat flow. For example, consider the problem of determining the steady-state temperature ϕ in a domain D with boundary C. If the temperatures on the boundary C of D are specified, then we have a Dirichlet problem. However, it may also be the case that all or part of the boundary is **insulated**. This means that there is no heat flow across the boundary, and, it can be shown that this implies that the directional derivative of ϕ in the direction of the normal vector $\mathbf{n}$ to C is 0. We call this derivative the **normal derivative** and denote it by $d\phi/dn$. In summary, an insulated boundary curve in a heat flow problem corresponds to a boundary condition of the form $d\phi/dn = 0$, and, thus, is an example of a Neumann problem. As the following theorem asserts, conformal mappings preserve boundary conditions of the form $d\phi/dn = 0$.

Theorem 7.5.1   Preservation of Boundary Conditions

Suppose that the function $f(z) = u(x, y) + iv(x, y)$ is conformal at every point of a smooth curve C. Let C' be the image of C under $w = f(z)$. If the normal derivative $d\Phi/dN$ of the function $\Phi(u, v)$ satisfies

$$\frac{d\Phi}{dN} = 0$$

at every point on C' in the w-plane, then the normal derivative $d\phi/dn$ of the function $\phi(x, y) = \Phi(u(x, y), v(x, y))$ satisfies

$$\frac{d\phi}{dn} = 0$$

at every point of C in the z-plane.

Proof   Assume that f and h satisfy the hypothesis of the theorem. Let $z_0 = x_0 + iy_0$ be a point on C and let $w_0 = u_0 + iv_0 = f(z_0)$ be its image on C'. Recall from multivariable calculus that if $\mathbf{N}$ is a normal vector to C' at w_0, then the normal derivative at w_0 is given by the dot product

$$\frac{d\Phi}{dN} = \nabla\Phi \cdot \mathbf{N},$$

where $\nabla\Phi$ is the gradient vector $\Phi_u(u_0, v_0)\mathbf{i} + \Phi_v(u_0, v_0)\mathbf{j}$. The condition $d\Phi/dN = 0$ implies that $\nabla\Phi$ and $\mathbf{N}$ are orthogonal, or, equivalently, that $\nabla\Phi$ is a tangent vector to C' at w_0. Let B' be the level curve $\Phi(u, v) = c_0$ containing (u_0, v_0). In multivariable calculus, you learned that the gradient vector $\nabla\Phi$ is orthogonal to the level curve B'. Thus, since the gradient is tangent to C' and orthogonal to B', we conclude that C' is orthogonal to B' at w_0. See Figure 7.5.5.

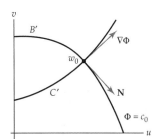

Figure 7.5.5 Figure for the proof of Theorem 7.5.1

Now consider the level curve B in the z-plane given by

$$\phi(x, y) = \Phi(u(x, y), v(x, y)) = c_0.$$

The point (x_0, y_0) is on B and the gradient vector $\nabla\phi$ is orthogonal to B at this point. Moreover, given any point (x, y) on B in the z-plane, we have that the point $(u(x, y), v(x, y))$ is on B' in the w-plane. That is, the image of B under $w = f(z)$ is B'. The curve C intersects B at z_0, and because f is conformal at z_0, it follows that the angle between C and B at z_0 is the same as the angle between C' and B' at w_0. In the preceding paragraph we found that this angle is $\pi/2$, and so C and B are orthogonal at z_0. Since $\nabla\phi$ is orthogonal to B, we must have that $\nabla\phi$ is tangent to C. If $\mathbf{n}$ is a normal vector to C at z_0, then we have shown that $\nabla\phi$ and $\mathbf{n}$ are orthogonal. Therefore,

$$\frac{d\phi}{dn} = \nabla\phi \cdot \mathbf{n} = 0. \qquad \blacksquare$$

Theorem 7.5.1 gives us a procedure for solving Nuemann problems associated with boundary conditions of the form $d\phi/dn = 0$. Namely, we follow the same four steps given on pages 386 and 387 to solve a Dirichlet problem. In Step 1, however, we find a conformal mapping from D onto D'. Since conformal mappings preserve boundary conditions of the form $d\phi/dn = 0$, solving the associated Nuemann problem in D' will give us a solution of the original Nuemann problem. Because analytic mappings are conformal at noncritical points, this approach also works with **mixed** boundary conditions. Roughly, these are boundary conditions where values of ϕ are specified on some boundary curves, whereas the normal derivative is required to satisfy $d\phi/dn = 0$ on other boundary curves.

EXAMPLE 3   **A Heat Flow Application**

Find the steady-state temperature ϕ in the first quadrant, shown in color in Figure 7.5.6(a), which satisfies the indicated mixed boundary conditions.

Solution   The steady-state temperature ϕ is a solution of Laplace's equation in the domain D defined by $0 < x < \infty$, $0 < y < \infty$, which satisfies the boundary conditions

$$\phi(0, y) = 0, \quad y > 1$$
$$\phi(x, 0) = 1, \quad x > 1$$
$$\frac{d\phi}{dn} = 0, \text{ for } 0 < y < 1, x = 0, \text{ and } 0 < x < 1, y = 0.$$

We will find ϕ using the four steps given on pages 386 and 387.

Step 1   As we will see in Step 3, this particular type of boundary-value problem is easy to solve in the half-infinite vertical strip $-a < u < a$, $v > 0$, when the boundary curve $-a < u < a$, $v = 0$, is insulated. Thus, in this step we find a conformal mapping of the first quadrant onto a half-infinite vertical strip. By identifying $\alpha = 2$ in entry E-4 of Appendix III, we see that the first quadrant is mapped onto the upper half-plane $v > 0$ by the mapping $w = z^2$. Next we apply the mapping $w = \sin^{-1} z$ of entry E-6. Under this

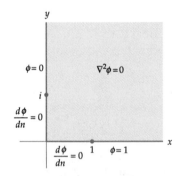

(a) Boundary-value problem

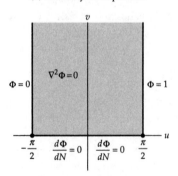

(b) Transformed boundary-value problem

Figure 7.5.6 Figure for Example 3

mapping, the upper half-plane $y > 0$ is mapped onto the half-infinite vertical strip $-\frac{1}{2}\pi < u < \frac{1}{2}\pi$, $v > 0$. Therefore, the composition

$$w = \sin^{-1}\left(z^2\right) \tag{8}$$

maps the first quadrant $x > 0$, $y > 0$, onto the domain D' defined by $-\frac{1}{2}\pi < u < \frac{1}{2}\pi$, $v > 0$.

Step 2 From entries E-4 and E-6 of Appendix III we see that the boundary curves $1 < x < \infty, y = 0$, and $1 < y < \infty, x = 0$, are mapped by $w = \sin^{-1}\left(z^2\right)$ onto the half-lines $u = \frac{1}{2}\pi, v > 0$, and $u = -\frac{1}{2}\pi, v > 0$, respectively. We also see that the segments $0 < x < 1, y = 0$, and $0 < y < 1, x = 0$, are mapped onto the segments $0 < u < \frac{1}{2}\pi, v = 0$, and $-\frac{1}{2}\pi < u < 0, v = 0$, respectively. Thus, the transformed boundary conditions are

$$\Phi\left(-\frac{\pi}{2}, v\right) = 0, \ \Phi\left(\frac{\pi}{2}, v\right) = 1, v > 0$$
$$\frac{d\Phi}{dN} = 0, \ -\frac{\pi}{2} < u < \frac{\pi}{2}, v = 0.$$

Step 3 Inspection of the domain D' and the transformed boundary conditions suggests that a solution Φ is a linear function in the variable u. That is,

$$\Phi(u, v) = Au + B$$

for some real constants A and B. Since the vector $\mathbf{N} = 0\mathbf{i} + 1\mathbf{j}$ is normal to the boundary curve $-\frac{1}{2}\pi < u < \frac{1}{2}\pi$, $v = 0$, we have

$$\frac{d\Phi}{dN} = \nabla\Phi \cdot \mathbf{N} = A(0) + 0(1) = 0,$$

and so, for any values of A and B, Φ satisfies the boundary condition for the normal derivative. By requiring that

$$\Phi\left(-\frac{\pi}{2}, v\right) = -A\frac{\pi}{2} + B = 0 \quad \text{and} \quad \Phi\left(\frac{\pi}{2}, v\right) = A\frac{\pi}{2} + B = 1,$$

we can solve for the constants A and B to obtain the solution

$$\Phi(u, v) = \frac{1}{\pi}u + \frac{1}{2}. \tag{9}$$

Step 4 In order to find a solution of the original boundary-value problem, we substitute the real and imaginary parts of the mapping in (8) for the variables u and v in (9). Since the formula for the real part of the expression $\sin^{-1}\left(z^2\right)$ is complicated, the simplest way of writing the solution ϕ is

$$\phi(x, y) = \frac{1}{\pi}\text{Re}\left[\sin^{-1}\left(z^2\right)\right] + \frac{1}{2}.$$

❏

Since both $1/\pi$ and $\frac{1}{2}$ are real, a complex potential function $\mathbf{\Omega}(z) = \phi(x, y) + i\psi(x, y)$ for the steady-state temperature function ϕ found in Example 3 is

$$\mathbf{\Omega}(z) = \frac{1}{\pi}\sin^{-1}\left(z^2\right) + \frac{1}{2}.$$

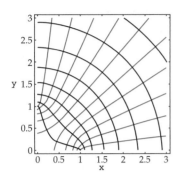

Figure 7.5.7 Isotherms and lines of heat flux for Example 3

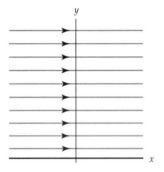

Figure 7.5.8 Uniform flow

The level curves of the real and imaginary parts of $\mathbf{\Omega}$ represent isotherms and lines of heat flux, respectively. In Figure 7.5.7, we have used *Mathematica* to plot these curves. The isotherms are shown in color and the lines of heat flux are shown in black.

7.5.2 Fluid Flow

Streamlining   We now discuss a method of using conformal mappings to model the planar flow of an ideal fluid. Recall from Section 5.6 that an ideal fluid is an incompressible nonviscous fluid whose flow is irrotational. If $\mathbf{\Omega}(z) = \phi(x,\,y) + i\psi(x,\,y)$ is the complex velocity potential of the flow of an ideal fluid in a domain D, then $\mathbf{\Omega}(z)$ is analytic in D and $f(z) = \overline{\mathbf{\Omega}'(z)}$ is a complex representation of the velocity field. Furthermore, the **streamlines** of the flow of an ideal fluid are the level curves $\psi(x,\,y) = c_2$, and, for this reason, ψ is called the **stream function** of the flow.

As a simple example consider the complex analytic function $\mathbf{\Omega}(z) = Az$, where $A > 0$ is a real constant. As presented in part (b) of Example 3 in Section 5.6, this function is the complex velocity potential of the velocity field of the flow of an ideal fluid whose complex representation is $f(z) = \overline{\mathbf{\Omega}'(z)} = A$. Since $\mathbf{\Omega}(z) = Az = Ax + iAy$, the streamlines of this flow are the curves $y = c_2$. All streamlines are therefore horizontal. See Figure 7.5.8. Recall from Section 5.6 that this particular flow is called the **uniform flow**.

The process of constructing a flow of an ideal fluid that remains inside a given domain D is called **streamlining**. If C is a boundary curve of D, then the requirement that the flow remain inside of D means that there is no flow across C, or, equivalently, that the directional derivative of ψ in the direction of the normal vector $\mathbf{n}$ to C is 0. Since the gradient vector $\nabla \psi$ is always normal to the level curve $\psi(x,\,y) = c_2$, this condition is equivalent to ψ being constant on C. Put yet another way, the boundary of D must be a streamline of the flow. The following summarizes this discussion.

> *Streamlining*
>
> Suppose that the complex velocity potential $\mathbf{\Omega}(z) = \phi(x,\,y) + i\psi(x,\,y)$ is analytic in a domain D and that ψ is constant on the boundary of D. Then $f(z) = \overline{\mathbf{\Omega}'(z)}$ is a complex representation of the velocity field of a flow of an ideal fluid in D. Moreover, if a particle is placed in D and allowed to flow with the fluid, then its path $z(t)$ remains in D.

Many streamlining problems can be solved using conformal mappings in a manner similar to that presented for solving Dirichlet and Neumann problems. In order to do so, we consider the complex velocity potential as an conformal mapping of the z-plane to the w-plane. If $z(t) = x(t) + iy(t)$ is a parametrization of a streamline $\psi(x,\,y) = c_2$ in the z-plane, then

$$w(t) = \mathbf{\Omega}(z(t)) = \phi(x(t), y(t)) + i\psi(x(t), y(t)) = \phi(x(t), y(t)) + ic_2.$$

Thus, the image of a streamline under the conformal mapping $w = \mathbf{\Omega}(z)$ is a horizontal line in the w-plane. Since the boundary C is required to be a streamline, the image of C under $w = \mathbf{\Omega}(z)$ must be a horizontal line. That is, we can determine the complex velocity potential by finding a conformal mapping from D onto a domain in the w-plane that maps the boundary C of

D onto a horizontal line. It is often the case, however, that it is easier to find a conformal mapping $z = \Omega^{-1}(w)$ from, say, the upper half-plane $v > 0$ onto D that takes the boundary $v = 0$ onto the boundary C of D. If $z = \Omega^{-1}(z)$ is a one-to-one function, then its inverse $w = \Omega(z)$ is the desired complex velocity potential. In summary, we have the following method for solving streamlining problems.

> ### Solving a Streamlining Problem
> *If $w = \Omega(z) = \phi(x, y) + i\psi(x, y)$ is a one-to-one conformal mapping of the domain D in the z-plane onto a domain D' in the w-plane such that the image of the boundary C of D is a horizontal line in the w-plane, then $f(z) = \overline{\Omega'(z)}$ is a complex representation of a flow of an ideal fluid in D.*

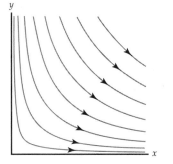

Figure 7.5.9 Flow around a corner

EXAMPLE 4   Flow around a Corner

Construct a flow of an ideal fluid in the first quadrant.

Solution Let D denote the first quadrant $x > 0$, $y > 0$. From Entry E-4 of Appendix III with the identification $\alpha = 2$, we see that $w = \Omega(z) = z^2$ is a one-to-one conformal mapping of the domain D onto the upper half-plane $v > 0$ and that the image of the boundary of D under this mapping is the real axis $v = 0$. Therefore, $f(z) = \overline{\Omega'(z)} = 2\bar{z}$ is a complex representation of the flow of an ideal fluid in the first quadrant. Since $\Omega(z) = z^2 = x^2 - y^2 + 2xyi$, the streamlines of this flow are the curves $2xy = c_2$. Some streamlines have been plotted in Figure 7.5.9. It should be clear from this figure why this flow is referred to as "flow around a corner." ❏

EXAMPLE 5   Flow around a Cylinder

Construct a flow of an ideal fluid in the domain consisting of all points outside the unit circle $|z| = 1$ and in the upper half-plane $y > 0$ shown in Figure 7.5.10.

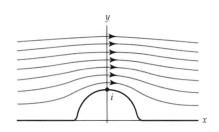

Figure 7.5.10 Flow around a cylinder

Solution Let D be the domain shown in Figure 7.5.10. Identifying $a = 2$ in entry H-3 of Appendix III, we obtain the one-to-one conformal mapping

$$w = \Omega(z) = z + \frac{1}{z}$$

of D onto the upper half-plane $v > 0$. In addition, entry H-3 indicates that the boundary of D is mapped onto the real axis $v = 0$. Therefore,

$$f(z) = \overline{\Omega'(z)} = \overline{1 - \frac{1}{z^2}} = 1 - \frac{1}{\bar{z}^2}$$

is a complex representation of a flow of an ideal fluid in D. Since

$$\Omega(z) = z + \frac{1}{z} = x + \frac{x}{x^2 + y^2} + i\left(y - \frac{y}{x^2 + y^2}\right),$$

the streamlines of this flow are the curves

$$\psi(x, y) = c_2, \quad \text{or} \quad y - \frac{y}{x^2 + y^2} = c_2.$$

Some streamlines for this flow have been plotted in Figure 7.5.10. ❑

It is not always possible to describe the streamlines with a Cartesian equation in the variables x and y. This situation occurs when an appropriate mapping $z = \boldsymbol{\Omega}^{-1}(w)$ of a domain D' in the w-plane onto the domain D in the z-plane can be found, but you cannot solve for the mapping $w = \boldsymbol{\Omega}(z)$. In such cases, it is possible to describe the streamlines parametrically.

EXAMPLE 6   Streamlines Defined Parametrically

Construct a flow of an ideal fluid in the domain D consisting of all points in the upper half-plane $y > 0$ excluding the points on the ray $y = \pi$, $-\infty < x \le 0$, shown in Figure 7.5.11.

Solution In Example 4 of Section 7.3 we used the Schwarz-Christoffel formula to find a conformal mapping of the upper half-plane $y > 0$ onto the domain D. By replacing the symbol z with the symbol w in the solution from Example 4, we obtain the mapping

$$z = \boldsymbol{\Omega}^{-1}(w) = w + \operatorname{Ln}(w) + 1 \tag{10}$$

of the upper half-plane $v > 0$ onto D. The inverse $\boldsymbol{\Omega}$ of the mapping in (10) is a complex velocity potential of a flow of an ideal fluid in D, but we cannot solve for w to obtain an explicit formula for $\boldsymbol{\Omega}$. In order to describe the streamlines, we recall that the streamlines in D are the images of horizontal lines $v = c_2$ in the upper half-plane $v > 0$ under the mapping $z = w + \operatorname{Ln}(w) + 1$. Since a horizontal line can be described by $w(t) = t + ic_2$, $-\infty < t < \infty$, it follows that the streamlines in D are given parametrically by

$$z(t) = \boldsymbol{\Omega}^{-1}(w(t)) = w(t) + \operatorname{Ln}[w(t)] + 1 = t + ic_2 + \operatorname{Ln}[t + ic_2] + 1,$$

or $x(t) = t + \frac{1}{2}\log_e\left(t^2 + c_2^2\right) + 1$, $y(t) = c_2 + \operatorname{Arg}(t + ic_2)$, $-\infty < t < \infty$. Some streamlines for this flow have been plotted using *Mathematica* in Figure 7.5.11. ❑

A stream function $\psi(x, y)$ is harmonic, but unlike a solution of a Dirichlet problem, $\psi(x, y)$ need not be bounded in D nor satisfy a fixed boundary condition. Therefore, there can be many different stream functions for a given domain D. We illustrate this in the following example.

EXAMPLE 7   Streamlines Defined Parametrically

The one-to-one conformal mapping

$$z = \boldsymbol{\Omega}^{-1}(w) = w + e^w + 1$$

also maps the upper half-plane $v > 0$ onto the domain D shown in Figure 7.5.11. The streamlines for this flow are parametrized by

$$z(t) = t + ic_2 + e^{t+ic_2} + 1,$$

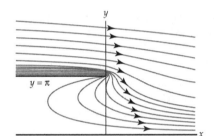

Figure 7.5.11 Flow for Example 6

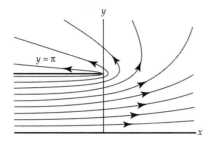

Figure 7.5.12 Flow for Example 7

or $x(t) = t + e^t \cos c_2 + 1$, $y(t) = c_2 + e^t \sin c_2$, $-\infty < t < \infty$. From the *Mathematica*-generated plot of the streamlines in Figure 7.5.12, we observe that this flow is different from the one constructed in Example 6. ❑

Sources and Sinks Recall from Section 5.6 that if **F** is the velocity field of a planar fluid flow, then a **source** is a point z_0 at which fluid is produced and a **sink** is a point z_0 at which fluid disappears. If C is a simple closed contour, then a nonzero value of the net flux across C, that is, a nonzero value of the integral $\oint_C \mathbf{F} \cdot \mathbf{N}\, ds$, indicates the presence of either a source or a sink inside of C. We saw in Section 5.6 that if **F** is the velocity field of the flow of an incompressible fluid in a domain D, then there are no sources or sinks in D. Incompressibility does not, however, rule out the existence of a source or sink on the boundary of D. Sources and sinks on the boundary of D are used to model planar flows in which fluid is entering or leaving D through a small slit in the boundary.

In Problem 23 in Exercises 5.6 we found that a source at a point $z = x_1$ on the boundary $y = 0$ of the upper half-plane $y > 0$ can be described by the complex velocity potential

$$\mathbf{\Omega}(z) = k\mathrm{Ln}\,(z - x_1), \tag{11}$$

where k is a positive constant. Similarly, $z = x_1$ is a sink when k is a negative constant. The strength of the source or sink is proportional to $|k|$. A flow containing both sources and sinks can be described by adding together functions of the form (11). For example,

$$\mathbf{\Omega}(z) = \mathrm{Ln}(z + 1) - \mathrm{Ln}(z - 1) = \mathrm{Ln}\frac{z + 1}{z - 1} \tag{12}$$

is a complex velocity potential for the flow of an ideal fluid in the upper half-plane $y > 0$ that has a source at $x_1 = -1$ and a sink at $x_2 = 1$ of equal strength. See Figure 7.5.13. You should also compare (12) with Problem 24 in Exercises 5.6.

Our method of determining the stream function in a domain with sources or sinks on the boundary is similar to that used in the absence of sources and sinks. Let $\psi(u,\,v)$ be the stream function of a flow of an ideal fluid in a domain D' in the w-plane with sources or sinks on the boundary. If $f(z) = u(x,\,y) + iv(x,\,y)$ is a conformal mapping of a domain D in the z-plane onto the domain D', then $\psi(x,\,y) = \psi(u(x,\,y),\,v(x,\,y))$ is a stream function for a flow of an ideal fluid in D with sources and sinks on the boundary. We illustrate this method in our final example.

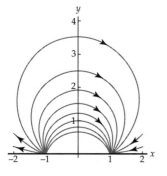

Figure 7.5.13 Source and sink

EXAMPLE 8   **Flow with a Source and Sink of Equal Strength**

Construct a flow of an ideal fluid in the domain D given by $0 < \arg(z) < \pi/4$ with a source at the boundary point $x_1 = 1$ and sink of equal strength at the boundary point $x_2 = 3$.

Solution From entry E-4 in Appendix III with $\alpha = 4$, we have that $f(z) = z^4$ is a one-to-one conformal mapping of D onto the upper half-plane $v > 0$. Under this mapping, the image of $x_1 = 1$ is $u_1 = 1^4 = 1$ and the image of

$x_2 = 3$ is $u_2 = 3^4 = 81$. With the obvious modifications to (12), we obtain the complex velocity potential

$$\text{Ln}(w - 1) - \text{Ln}(w - 81), \tag{13}$$

which describes the flow of an ideal fluid in the upper half-plane $v > 0$ that has a source at $u_1 = 1$ and a sink of equal strength at $u_2 = 81$. Since the domain D is mapped on the domain $v > 0$ by the conformal mapping $w = z^4$, we obtain a complex potential function for a flow in D by replacing the symbol w with z^4 in (13). This yields

$$\Omega(z) = \text{Ln}\left(z^4 - 1\right) - \text{Ln}\left(z^4 - 81\right). \tag{14}$$

Streamlines of this flow are given by $\psi(x, y) = c_2$, or

$$\text{Arg}\left(z^4 - 1\right) - \text{Arg}\left(z^4 - 81\right) = c_2.$$

See Figure 7.5.14.   ❏

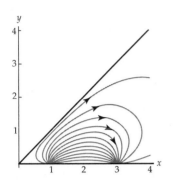

Figure 7.5.14 Figure for Example 8

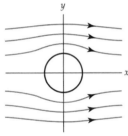

(a) Flow around the unit circle

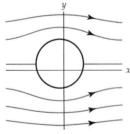

(b) Flow around a circle containing $z_1 = -1$ and passing through $z_2 = 1$

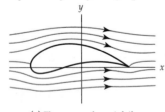

(c) Flow around an airfoil

Figure 7.5.15 Flow around a Joukowski airfoil

Remarks

The complex mapping $w = J(z) = z + k^2/z$ is called the **Joukowski transformation**. Under this mapping, a circle in the z-plane that contains the point $z_1 = -1$ and passes through the point $z_2 = 1$ is mapped onto a curve in the w-plane that resembles the cross-section of an airplane wing. See Figure 7.5.15(c). The image curve is called a **Joukowski airfoil**, and the air flow around this curve can be determined using techniques from this section. We begin with the flow shown in Figure 7.5.15(a) given by $\Omega(z) = z + 1/z$ around the unit circle $|z| = 1$. Using an appropriate linear mapping, we can adjust this flow to be one around a circle containing the point $z_1 = -1$ and passing through the point $z_2 = 1$. See Figure 7.5.15(b). The Joukowski transformation is then used to "transform" this flow to one around the airfoil as shown in Figure 7.5.15(c).

EXERCISES 7.5   *Answers to selected odd-numbered problems begin on page ANS-20.*

7.5.1 Boundary-Value Problems

In Problems 1–6, **(a)** find a conformal mapping of the domain shown in color onto the upper half-plane, and **(b)** use the mapping from (a) and the solution (7) in Section 7.4 to find the steady-state temperature $\phi(x, y)$ in the domain subject to the given boundary conditions.

1.

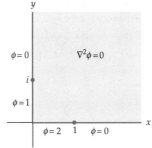

Figure 7.5.16 Figure for Problem 1

2.

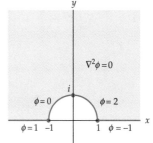

Figure 7.5.17 Figure for Problem 2

3.

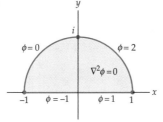

Figure 7.5.18 Figure for Problem 3

4.

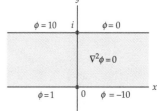

Figure 7.5.19 Figure for Problem 4

5.

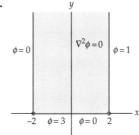

Figure 7.5.20 Figure for Problem 5

6.

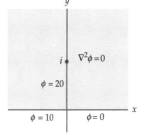

Figure 7.5.21 Figure for Problem 6 [*Hint*: Use the inverse of the mapping in entry M-4 of Appendix III.]

In Problems 7 and 8, (**a**) find a linear fractional transformation of the domain shown in color onto an infinite strip, and (**b**) use the mapping from (a) and the solution from Example 2 of Section 3.4 to find the electrostatic potential $\phi(x, y)$ in the domain subject to the given boundary conditions.

7.

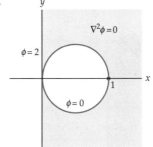

Figure 7.5.22 Figure for Problem 7

8.

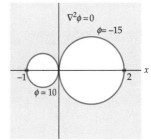

Figure 7.5.23 Figure for Problem 8

In Problems 9 and 10, (**a**) find a linear fractional transformation of the domain shown in color onto an annulus, and (**b**) use the mapping from (a) and a solution similar to that in Example 1 to find the electrostatic potential $\phi(x, y)$ in the domain subject to the given boundary conditions.

9.

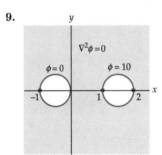

10.

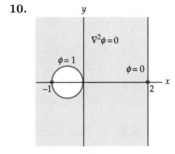

Figure 7.5.24 Figure for Problem 9   Figure 7.5.25 Figure for Problem 10

In Problems 11 and 12, (**a**) find a conformal mapping of the domain shown in color onto the domain used in Example 3, and (**b**) use the mapping from (a) and a solution similar to that in Example 3 to find the steady-state temperature $\phi(x, y)$ in the domain subject to the given boundary conditions.

11.

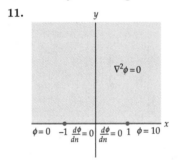

12.

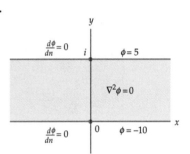

Figure 7.5.26 Figure for Problem 11   Figure 7.5.27 Figure for Problem 12

7.5.2 Fluid Flow

In Problems 13–16, find the complex velocity potential $\Omega(z)$ for the flow of an ideal fluid in the domain shown in color.

13.

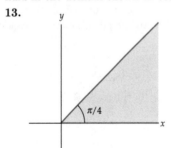

14.

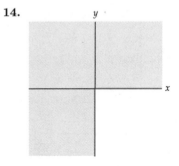

Figure 7.5.28 Figure for Problem 13   Figure 7.5.29 Figure for Problem 14

15.

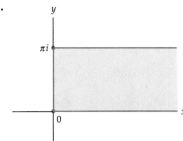

Figure 7.5.30 Figure for Problem 15

16.

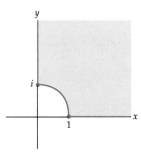

Figure 7.5.31 Figure for Problem 16

In Problems 17–20, the flow of an ideal fluid is shown in a domain in the z-plane. (**a**) Find a conformal mapping of the upper half-plane $w > 0$ onto the domain in the z-plane, and (**b**) find a parametric representation of the streamlines of the flow.

17.

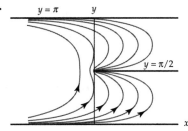

Figure 7.5.32 Figure for Problem 17

18.

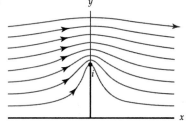

Figure 7.5.33 Figure for Problem 18

19.

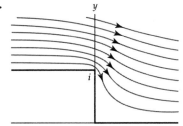

Figure 7.5.34 Figure for Problem 19

20.

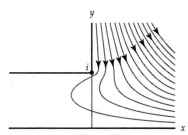

Figure 7.5.35 Figure for Problem 20

In Problems 21 and 22, construct the flow of an ideal fluid in the given domain with sinks or sources on the boundary of the domain.

21. The domain from Problem 13 with a source at $z_1 = 1 + i$ and a sink at $z_2 = 2$

22. The domain from Problem 16 with a source at $z_1 = \frac{1}{2}\sqrt{2} + \frac{1}{2}\sqrt{2}i$ and sinks at $z_2 = 2$ and $z_3 = 3i$

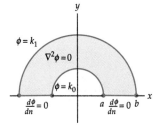

Figure 7.5.36 Figure for Problem 23

Focus on Concepts

23. Show that the function given by (3) with the symbols u and v replaced by the symbols x and y is a solution of the boundary-value problem in the domain shown in color in Figure 7.5.36.

24. Use a conformal mapping and Problem 23 to solve the boundary-value problem in the domain shown in color in Figure 7.5.37.

25. Use a conformal mapping and Problem 23 to solve the boundary-value problem in the domain shown in color in Figure 7.5.38.

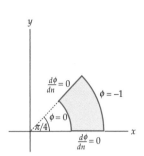

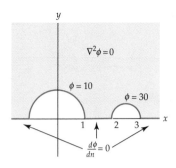

Figure 7.5.37 Figure for Problem 24   **Figure 7.5.38** Figure for Problem 25

26. In Problem 22 in Exercises 5.6 we defined a **stagnation point** of a flow to be a point at which $\mathbf{F}(x,\ y) = \mathbf{0}$. Find the stagnation points for:

   (a) the flow in Example 5

   (b) the flow in Problem 16.

27. In this problem you will construct the flow of an ideal fluid through a slit shown in Figure 7.5.40.

   (a) Determine a velocity potential for the flow of an ideal fluid in the domain $-\pi/2 < x < \pi/2$, $-\infty < y < \infty$, shown in Figure 7.5.39.

   (b) Use the potential from part (a) and a conformal mapping to find the velocity potential for the flow of an ideal fluid in the region shown in Figure 7.5.40.

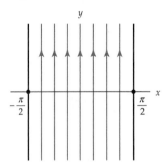

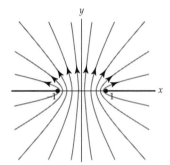

Figure 7.5.39 Figure for Problem 27   **Figure 7.5.40** Figure for Problem 27

28. In this problem you will construct the flow of an ideal fluid around a plate shown Figure 7.5.42.

   (a) Use a linear mapping and the velocity potential from Example 5 to show that the velocity potential of an ideal fluid in the domain shown in Figure 7.5.41 is given by

$$\Omega(z) = \frac{z}{e^{i\alpha}} + \frac{e^{i\alpha}}{z}.$$

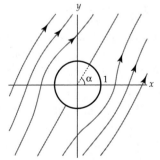

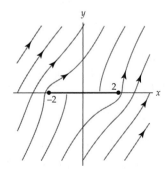

Figure 7.5.41 Figure for Problem 28   **Figure 7.5.42** Figure for Problem 28

(b) The domain outside of the unit circle shown in Figure 7.5.41 is mapped onto the complex plane excluding the line segment $y = 0, -2 \leq x \leq 2$, shown in Figure 7.5.42 by the conformal mapping

$$w = \frac{z + \left(z^2 - 4\right)^{1/2}}{2}.$$

Use the velocity potential from part (a) and this conformal mapping to find the velocity potential for the flow of an ideal fluid in the region shown in Figure 7.5.42.

Computer Lab Assignments

In Problems 29–36, use a CAS to plot the isotherms for the given steady-state temperature $\phi(x, y)$.

29. $\phi(x, y)$ is the steady-state temperature from Problem 1.

30. $\phi(x, y)$ is the steady-state temperature from Problem 2.

31. $\phi(x, y)$ is the steady-state temperature from Problem 3.

32. $\phi(x, y)$ is the steady-state temperature from Problem 4.

33. $\phi(x, y)$ is the steady-state temperature from Problem 5.

34. $\phi(x, y)$ is the steady-state temperature from Problem 6.

35. $\phi(x, y)$ is the steady-state temperature from Problem 11.

36. $\phi(x, y)$ is the steady-state temperature from Problem 12.

In Problems 37–40, use a CAS to plot the equipotential curves for the given electrostatic potential $\phi(x, y)$.

37. $\phi(x, y)$ is the electrostatic potential from Problem 7.

38. $\phi(x, y)$ is the electrostatic potential from Problem 8.

39. $\phi(x, y)$ is the electrostatic potential from Problem 9.

40. $\phi(x, y)$ is the electrostatic potential from Problem 10.

In Problems 41–44, use a CAS to plot the streamlines of the given flow.

41. The flow from Problem 13.

42. The flow from Problem 14.

43. The flow from Problem 15.

44. The flow from Problem 16.

CHAPTER 7 REVIEW QUIZ

Answers to selected odd-numbered problems begin on page ANS-21.

In Problems 1–15, answer true or false. If the statement is false, justify your answer by either explaining why it is false or giving a counterexample; if the statement is true, justify your answer by either proving the statement or citing an appropriate result in this chapter.

1. If $f(z)$ is analytic at a point z_0, then the mapping $w = f(z)$ is conformal at z_0.

2. The mapping $w = z^2 + iz + 1$ is not conformal at $z = -\frac{1}{2}i$.

3. The mapping $w = z^2 + 1$ is not conformal at $z = \pm i$.

4. The mapping $w = \bar{z}$ fails to be conformal at every point in the complex plane.

5. A linear fractional transformation is conformal at every point in its domain.

6. The image of a circle under a linear fractional transformation is a circle.

7. The linear fractional transformation $T(z) = \dfrac{z-i}{z+1}$ maps the points 0, -1, and i onto the points $-i$, ∞, and 0, respectively.

8. Given any three distinct points z_1, z_2, and z_3, there is a linear fractional transformation that maps z_1, z_2, and z_3 onto 0, 1, and ∞.

9. The inverse of the linear fractional transformation $T(z) = (az + b)/(cz + d)$ is $T^{-1}(z) = (cz + d)/(az + b)$.

10. If $f'(z) = A(z+1)^{-1/2}(z-1)^{-3/4}$, then $w = f(z)$ maps the upper half-plane onto an unbounded polygonal region.

11. If $f'(z) = A(z+1)^{-1/2}z^{-1/2}(z-1)^{-1/2}$, then $w = f(z)$ maps the upper half-plane onto a rectangle.

12. Every Dirichlet problem in the upper half-plane can be solved using the Poisson integral formula.

13. If $w = f(z) = u(x,\ y) + iv(x,\ y)$ is a conformal mapping of a domain D onto the upper half-plane $v > 0$ and if $\Phi(u,\ v)$ is a harmonic function for $v > 0$, then $\phi(x,\ y) = \Phi(u(x,\ y),\ v(x,\ y))$ is harmonic on D.

14. If $\psi(x,\ y)$ is a function defined on a domain D and if the boundary of D is a level curve of $\psi(x,\ y)$, then $\psi(x,\ y)$ is the stream function of an ideal fluid in D.

15. Given a domain D, there can be more than one flow of an ideal fluid that remains inside of D.

In Problems 16–30, try to fill in the blanks without referring back to the text.

16. The analytic function $f(z) = \cosh z$ is conformal except at $z = $ _____ .

17. Conformal mappings preserve both the magnitude and the _____ of an angle.

18. The mapping _____ is an example of a mapping that is conformal at every point in the complex plane.

19. If $f'(z_0) = f''(z_0) = 0$ and $f'''(z_0) \neq 0$, then the mapping $w = f(z)$ _____ the magnitude of angles at the point z_0.

20. $T(z) = $ _____ is a linear fractional transformation that maps the points 0, $1 + i$, and i onto the points 1, i, and ∞.

21. The image of the circle $|z - 1| = 2$ under the linear fractional transformation $T(z) = (2z - i)/(iz + 1)$ is a _____ .

22. The image of a line L under the linear fractional transformation $T(z) = (iz - 2)/(3z + 1 - i)$ is a circle if and only if the point $z = $ _____ is on L.

23. The cross-ratio of the points z, z_1, z_2, and z_3 is _____ and _____ .

24. The derivative of a Schwarz-Christoffel mapping from the upper half-plane onto the triangle with vertices at 0, 1, and $1 + i$ is $f'(z) = $ _____ .

25. If $f'(z) = A(z+1)^{-1/2}z^{-1/4}$, then $w = f(z)$ maps the upper half-plane onto a polygonal region with interior angles _____ .

26. The Poisson integral formula gives an integral solution $\phi(x,y)$ to a Dirichlet problem in the upper half-plane $y > 0$ provided the function $f(x) = \phi(x,\ 0)$ is _____ and _____ on $-\infty < x < \infty$.

27. The complex velocity potential $\Omega(z) = z^5$ describes the flow of an ideal fluid in the domain $0 < \arg z < \underline{\hspace{1cm}}$.

28. If $\Omega(z) = e^z + e^{-z}$ is the complex velocity potential for the flow of an ideal fluid in a domain D, then a complex representation of the velocity field is given by $f(z) = \underline{\hspace{1cm}}$.

29. If $z = \left(\dfrac{1+w}{1-w}\right)^2$ is a one-to-one conformal mapping of the upper half-plane onto a domain D, then a streamline of the flow of an ideal fluid in D is parametrized by $z(t) = \underline{\hspace{1cm}}$.

30. The complex velocity potential $\Omega(z) = \text{Ln}(z-2) + \text{Ln}(z-3) - \text{Ln}(z-4)$ describes the flow of an ideal fluid in the upper half-plane $y > 0$ with a $\underline{\hspace{1cm}}$ at $z = 2$ and $z = 3$ and a $\underline{\hspace{1cm}}$ at $z = 4$.

Appendixes

The following theorem was presented in Section 2.6 as a practical method for computing complex limits. In this appendix we give the full epsilon-delta proof of this theorem.

Theorem A.1   **Real and Imaginary Parts of a Limit**

Suppose that $f(z) = u(x, \ y) + iv(x, \ y)$, $z_0 = x_0 + iy_0$, and $L = u_0 + iv_0$. Then $\lim\limits_{z \to z_0} f(z) = L$ if and only if

$$\lim_{(x,y) \to (x_0, y_0)} u(x, y) = u_0 \quad \text{and} \quad \lim_{(x,y) \to (x_0, y_0)} v(x, y) = v_0.$$

Proof Theorem A.1 states that

$$\lim_{z \to z_0} f(z) = L \tag{1}$$

if and only if

$$\lim_{(x,y) \to (x_0, y_0)} u(x, y) = u_0 \qquad \text{and} \qquad \lim_{(x,y) \to (x_0, y_0)} v(x, y) = v_0. \tag{2}$$

Because Theorem A.1 involves an "if and only if" statement, we must prove two things:

(i) that (1) implies (2), and
(ii) that (2) implies (1).

We begin with the former.

(i) If we assume that $\lim_{z \to z_0} f(z) = L$, then by Definition 2.6.1 of Section 2.6 we have:

> For every $\varepsilon > 0$ there exists a $\delta > 0$ such that $|f(z) - L| < \varepsilon$ whenever $0 < |z - z_0| < \delta$. (3)

Using the identifications $f(z) = u(x,\ y) + iv(x,\ y)$ and $L = u_0 + iv_0$ we obtain:

$$|f(z) - L| = \sqrt{\left(u(x,y) - u_0\right)^2 + \left(v(x,y) - v_0\right)^2}.$$

Furthermore, since $0 \le \left(v(x,y) - v_0\right)^2$, it follows that:

$$|u(x,y) - u_0| = \sqrt{\left(u(x,y) - u_0\right)^2} \le \sqrt{\left(u(x,y) - u_0\right)^2 + \left(v(x,y) - v_0\right)^2}.$$

Thus, for all $z = x + iy$ we have:

$$|u(x,y) - u_0| \le |f(z) - L|. \tag{4}$$

In particular, if $|f(z) - L| < \varepsilon$, then $|u(x,\ y) - u_0| < \varepsilon$. Now by making the identifications $z = x + iy$ and $z_0 = x_0 + iy_0$ we also find that

$$|z - z_0| = \sqrt{\left(x - x_0\right)^2 + \left(y - y_0\right)^2}. \tag{5}$$

Therefore, it follows from (3), (4), and (5) that for every $\varepsilon > 0$ there exists a $\delta > 0$ such that $|u(x,\ y) - u_0| < \varepsilon$ whenever $0 < \sqrt{(x - x_0)^2 + (y - y_0)^2} < \delta$. Thus, by (7) in Section 2.6 we have shown that $\lim_{(x,y) \to (x_0,y_0)} u(x,\ y) = u_0$.

Because $0 \le \left(u(x,\ y) - u_0\right)^2$, we can use a similar argument to establish the second limit in (2): $\lim_{(x,y) \to (x_0,y_0)} v(x,\ y) = v_0$. This completes the proof that (1) implies (2).

(ii) In this part we begin by assuming the limits

$$\lim_{(x,y) \to (x_0,y_0)} u(x,y) = u_0 \quad \text{and} \quad \lim_{(x,y) \to (x_0,y_0)} v(x,y) = v_0$$

and then proceed to show that $\lim_{z \to z_0} f(z) = L$. Given any $\varepsilon > 0$, then $\varepsilon/2 > 0$. Therefore, from our assumption that $\lim_{(x,y) \to (x_0,y_0)} u(x,\ y) = u_0$ and (7) of Section 2.6, we have that there is a $\delta_1 > 0$ such that

$$|u(x,y) - u_0| < \varepsilon/2 \quad \text{whenever} \quad 0 < \sqrt{(x - x_0)^2 + (y - y_0)^2} < \delta_1.$$

In a similar manner, since $\lim\limits_{(x,y)\to(x_0,y_0)} v(x,\ y) = v_0$, there also exists a $\delta_2 > 0$ such that

$$|v(x,y) - v_0| < \varepsilon/2 \quad \text{whenever} \quad 0 < \sqrt{(x-x_0)^2 + (y-y_0)^2} < \delta_2.$$

If we set δ to be the minimum of δ_1 and δ_2, that is $\delta \leq \delta_1$ and $\delta \leq \delta_2$, then we are guaranteed that $|u(x,\ y) - u_0| < \varepsilon/2$ and $|v(x,\ y) - v_0| < \varepsilon/2$ whenever $0 < \sqrt{(x-x_0)^2 + (y-y_0)^2} < \delta$. Thus, from (5) we obtain

$$|u(x,y) - u_0| + |v(x,y) - v_0| < \varepsilon/2 + \varepsilon/2 = \varepsilon \quad \text{whenever} \quad 0 < |z - z_0| < \delta. \tag{6}$$

On the other hand, with the identifications $f(z) = u(x,\ y) + iv(x,\ y)$ and $L = u_0 + iv_0$, the triangle inequality gives

$$|f(z) - L| \leq |u(x,y) - u_0| + |v(x,y) - v_0|.$$

Therefore, it follows from (6) that $|f(z) - L| < \varepsilon$ whenever $0 < |z - z_0| < \delta$. Since ε was allowed to be any positive number, we have shown that for every $\varepsilon > 0$ there exists a $\delta > 0$ such that $|f(z) - L| < \varepsilon$ whenever $0 < |z - z_0| < \delta$, and so $\lim\limits_{z\to z_0} f(z) = L$ by Definition 2.6.1. ❑

Appendix II   Proof of the Cauchy-Goursat Theorem

In Section 5.3 we proved Cauchy's theorem using Green's theorem. This simple proof was possible because of the hypothesis of continuity of f' throughout a simply connected domain D. The French mathematician Edouard Goursat (1858–1936) published a proof of Cauchy's theorem in 1900 without this continuity assumption. As a result his name was thereafter linked with Cauchy's in the title of one of the most fundamental of all theorems in complex analysis.

Note ➡ In this appendix we discuss how the full proof of the Cauchy-Goursat theorem is accomplished. To avoid needless repetition throughout the following discussion we will take for granted that we are working in a simply connected domain D and that f represents a complex function analytic in D.

The proof of the Cauchy-Goursat theorem is accomplished in three steps. The first two steps are helping theorems—sometimes called *lemmas*—which are actually special cases of the Cauchy-Goursat theorem. The first of these helping theorems deals with integrals along a triangular contour and the second deals with integrals along a closed polygonal contour. The first theorem is used in the proof of the second, and the second theorem is utilized to establish the Cauchy-Goursat theorem in its full generality.

We will prove the first theorem, but because of its length, we will simply sketch the proofs of the remaining two.

Theorem A.2   Triangular Contour

If Δ is a triangular contour lying entirely within D, then $\int_\Delta f(z)\,dz = 0$.

Proof Let Δ be the triangular contour shown in Figure AII.1; the vertices of Δ are labeled A, B, and C. We form four smaller triangles C_1, C_2, C_3, and C_4 by joining the midpoints E, F, and G of the sides of Δ by straight line segments as shown in Figure AII.2. By Theorem 5.2.2(iii) we can write

$$\oint_\Delta f(z)\,dz = \int_{EBF} + \int_{FCG} + \int_{GAE}$$
$$= \left(\int_{EBF} + \int_{FE}\right) + \left(\int_{FCG} + \int_{GF}\right)$$
$$+ \left(\int_{GAE} + \int_{EG}\right) + \left(\int_{EF} + \int_{FG} + \int_{GE}\right)$$
$$= \int_{EBFE} + \int_{FCGF} + \int_{GAEG} + \int_{EFGE}$$

Integrals $\int_{FE}, \int_{GF}, \int_{EG}$ are canceled by $\int_{EF}, \int_{FG}, \int_{GE}$. ➡

$$\text{or} \quad \oint_\Delta f(z)\,dz = \oint_{C_1} f(z)\,dz + \oint_{C_2} f(z)\,dz + \oint_{C_3} f(z)\,dz + \oint_{C_4} f(z)\,dz. \quad (1)$$

Then by (6) of Section 1.2, the triangle inequality,

$$\left|\oint_\Delta f(z)\,dz\right| \le \left|\oint_{C_1} f(z)\,dz\right| + \left|\oint_{C_2} f(z)\,dz\right| + \left|\oint_{C_3} f(z)\,dz\right| + \left|\oint_{C_4} f(z)\,dz\right|. \quad (2)$$

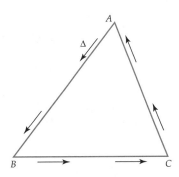
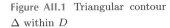

Figure AII.1 Triangular contour Δ within D

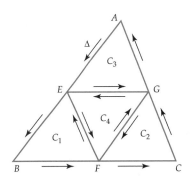

Figure AII.2 Triangular contours C_1, C_2, C_3, and C_4

The four quantities on the right side of (2) are nonnegative real numbers and, as a consequence, one of them must be greater than or equal to the other three. Let us denote the triangular contour of the integral with the largest modulus by the symbol Δ_1. Hence (2) gives

$$\left| \oint_{\Delta} f(z)\,dz \right| \leq 4 \left| \oint_{\Delta_1} f(z)\,dz \right|. \tag{3}$$

We repeat the foregoing process for the triangle Δ_1; that is, we form four triangles within Δ_1 by joining the midpoints of its sides by line segments in the fashion shown in Figure AII.2 and proceed to the equivalent of (1) and the inequality (2), where the left side of the inequality is now $\left| \oint_{\Delta_1} f(z)\,dz \right|$. The integral of f along one of these new triangular contours, let us call it Δ_2, then satisfies

$$\left| \oint_{\Delta_1} f(z)\,dz \right| \leq 4 \left| \oint_{\Delta_2} f(z)\,dz \right|. \tag{4}$$

We combine this last inequality with (3) to obtain

$$\left| \oint_{\Delta} f(z)\,dz \right| \leq 4 \left| \oint_{\Delta_1} f(z)\,dz \right| \leq 4^2 \left| \oint_{\Delta_2} f(z)\,dz \right|.$$

Continuing in this manner, we obtain a sequence of "nested" triangular contours Δ, Δ_1, Δ_2, ..., Δ_n, that is, each triangle in the sequence is contained in the one immediately preceding it. After n steps we arrive at

$$\left| \oint_{\Delta} f(z)\,dz \right| \leq 4^n \left| \oint_{\Delta_n} f(z)\,dz \right|. \tag{5}$$

Consistent with intuition, it can be proved that there exists a point in the domain D that is common to every triangle in the sequence Δ, Δ_1, Δ_2, ..., Δ_n. Let z_0 denote that point. Since the function f is analytic at $z = z_0$, $f'(z_0)$ exists. If we define

$$\eta(z) = \frac{f(z) - f(z_0)}{z - z_0} - f'(z_0), \tag{6}$$

then $|\eta(z)|$ can be made arbitrarily small whenever z is sufficiently close to z_0. This fact—which will be used shortly—follows from the hypothesis that f is analytic in D and so the limit of the difference quotient $\dfrac{f(z) - f(z_0)}{z - z_0}$ as

$z \to z_0$ exists and equals $f'(z_0)$. In the ε-δ symbolism of Definition 2.6.1, for every $\varepsilon > 0$, there exists a $\delta > 0$ such that

$$|\eta(z)| < \varepsilon \quad \text{whenever} \quad |z - z_0| < \delta. \tag{7}$$

At this juncture we wish to solve (6) for $f(z)$ and use it to replace the integrand in the contour integral $\oint_{\Delta_n} f(z)\,dz$ that appears in (5). The two results are:

$$f(z) = f(z_0) + (z - z_0)f'(z_0) + (z - z_0)\eta(z)$$

and

$$\oint_{\Delta_n} f(z)dz = f(z_0) \oint_{\Delta_n} dz + f'(z_0) \oint_{\Delta_n} (z - z_0)dz + \oint_{\Delta_n} (z - z_0)\eta(z)dz. \tag{8}$$

In (8) we were able write $f(z_0)$ and $f'(z_0)$ outside of the integrals since these quantities are constants. Moreover in (8), $\oint_{\Delta_n} dz = 0$ and $\oint_{\Delta_n} (z - z_0)dz = 0$.

The last two results are true for *any* simple closed contour (such as Δ_n) and can be proved either directly from Definition 5.2.1 (see Problem 29 in Exercises 5.2), or from Cauchy's theorem. (That last statement may surprise the reader; no, we are not using what we are trying to prove—remember, the constant function 1 and the polynomial function $z - z_0$ are analytic in D and *do* have continuous derivatives.) Hence the right side of (8) reduces to single term:

$$\oint_{\Delta_n} f(z)\,dz = \oint_{\Delta_n} (z - z_0)\eta(z)\,dz. \tag{9}$$

Now let L and L_1 denote the lengths of the triangular contours Δ and Δ_1, respectively. Then, keeping in mind how the triangle Δ_1 was constructed, it is a straightforward problem in similar triangles to show that L_1 is related to L by $L_1 = \frac{1}{2}L$. Likewise, if L_2 is the length of Δ_2, then $L_2 = \frac{1}{2}L_1 = \frac{1}{2^2}L$. In general, if L_n is the length of Δ_n, then $L_n = \frac{1}{2^n}L$.

We are almost finished. Now for any point z on Δ_n, $|z - z_0| < L_n$, where $L_n = \frac{1}{2^n}L$. If we choose n large enough so that $|z - z_0| < \frac{1}{2^n}L < \delta$, it then follows from (9), (7), and the *ML*-inequality

$$\left| \oint_{\Delta_n} f(z)\,dz \right| = \left| \oint_{\Delta_n} (z - z_0)\eta(z)\,dz \right| \leq \frac{L}{2^n} \cdot \varepsilon \cdot \frac{L}{2^n} = \frac{\varepsilon}{4^n}L^2. \tag{10}$$

Putting (5) together with (10) gives us a bound for the modulus of the integral on Δ:

$$\left| \oint_{\Delta} f(z)\,dz \right| \leq 4^n \frac{\varepsilon}{4^n}L^2 = \varepsilon L^2. \tag{11}$$

The result in (11) completes the proof. Since $\varepsilon > 0$ can be made arbitrarily small, we must have $\left| \oint_{\Delta} f(z)\,dz \right| = 0$ and so $\oint_{\Delta} f(z)\,dz = 0$. ❑

> **Theorem A.3 Closed Polygonal Contour**
>
> If C is a closed polygonal contour lying entirely within D, then $\int_C f(z)\,dz = 0$.

The proof of this theorem depends on Theorem A.2 and on the fact that any closed polygonal contour C, such as the one in Figure AII.3, can be "triangulated." Roughly, this means that the closed polygon C can be decomposed into a finite number of triangles by adding lines as shown in Figure AII.4. We

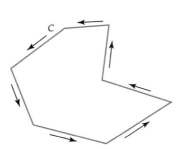

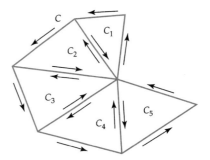

Figure AII.3 Closed polygonal contour C within D

Figure AII.4 Triangulation of the polygonal contour C

can then proceed as in the proof of Theorem A.2 and integrate twice along these added line segments but in opposite directions. If the closed polygon C has n sides, then it can be decomposed into n triangles $C_1, C_2, \ldots, C_n$ and we would eventually arrive at the following analogue of (1):

$$\oint_C f(z)\,dz = \oint_{C_1} f(z)\,dz + \oint_{C_2} f(z)\,dz + \cdots + \oint_{C_n} f(z)\,dz. \qquad (12)$$

By Theorem A.2, each of the integrals of right side of (12) is zero and we are left with the desired conclusion that $\oint_C f(z)\,dz = 0$.

In passing it should be noted that the closed polygonal contour C in Theorem A.3 need not be simple as shown in Figure AII.3; in other words, C can intersect itself.

The proof of the final part of the Cauchy-Goursat theorem demonstrates that any closed contour C can be approximated to any desired degree of accuracy by a closed polygonal path.

> **Theorem A.4 Any Simple Closed Contour**
>
> If C is a simple closed contour lying entirely within D, then $\oint_C f(z)\,dz = 0$.

In Figure AII.5 we have shown a simple closed contour C and n points $z_1, z_2, \ldots, z_n$ on C through which a polygonal curve P has been constructed. Then it can be shown that the difference between the integral along C, $\oint_C f(z)\,dz$, and the integral along the polygonal contour P, $\oint_P f(z)\,dz$, can be made arbitrarily small as $n \to \infty$. As a consequence of Theorem A.3, $\oint_P f(z)\,dz = 0$ for any n, and thus the integral along C must also be zero.

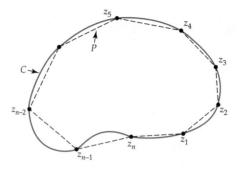

Figure AII.5 Simple closed contour C approximated by a closed polygonal curve P

Appendix III   Table of Conformal Mappings

■ Elementary Mappings

E-1

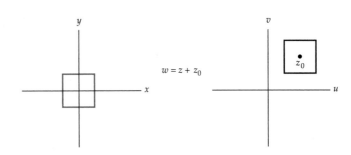

$w = z + z_0$

E-2

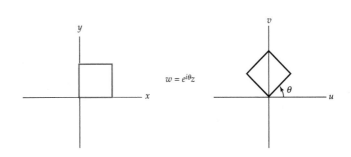

$w = e^{i\theta}z$

E-3

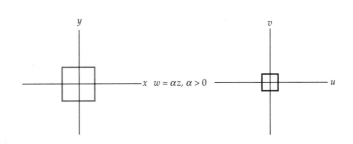

$w = \alpha z,\ \alpha > 0$

E-4

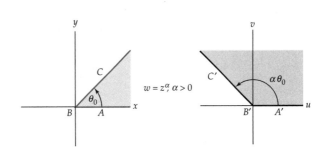

$w = z^{\alpha},\ \alpha > 0$

E-5

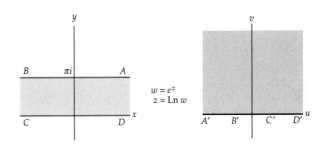

$w = e^z$
$z = \operatorname{Ln} w$

E-6

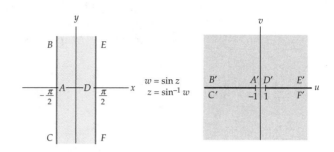

E-7

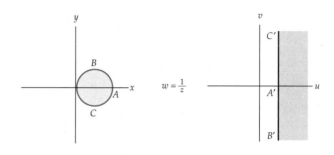

E-8

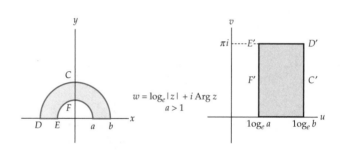

E-9

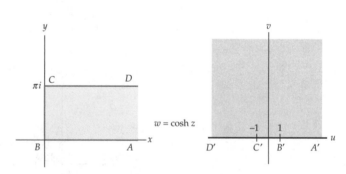

■ Mappings onto Half-Planes

H-1

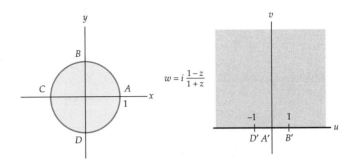

$$w = i\,\frac{1-z}{1+z}$$

H-2

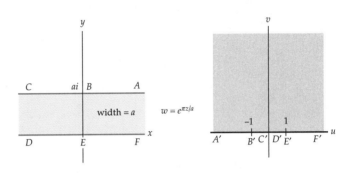

$$w = e^{\pi z/a}$$

H-3

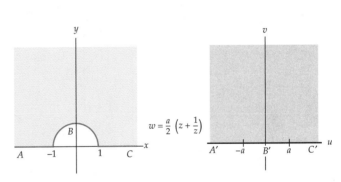

$$w = \frac{a}{2}\left(z + \frac{1}{z}\right)$$

H-4

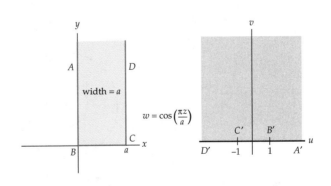

$$w = \cos\left(\frac{\pi z}{a}\right)$$

H-5

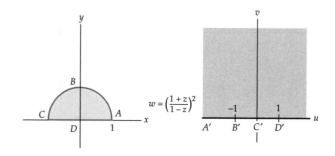

$$w = \left(\frac{1+z}{1-z}\right)^2$$

H-6

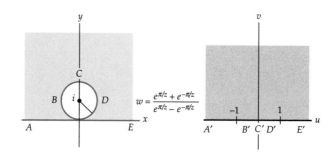

$$w = \frac{e^{\pi/z} + e^{-\pi/z}}{e^{\pi/z} - e^{-\pi/z}}$$

■ Mappings onto Circular Regions

C-1

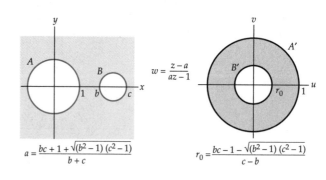

$$w = \frac{z-a}{az-1}$$

$$a = \frac{bc + 1 + \sqrt{(b^2-1)(c^2-1)}}{b+c}$$

$$r_0 = \frac{bc - 1 - \sqrt{(b^2-1)(c^2-1)}}{c-b}$$

C-2

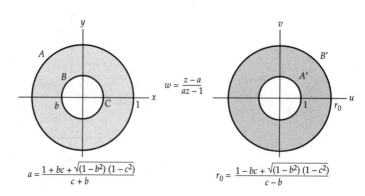

$$w = \frac{z-a}{az-1}$$

$$a = \frac{1 + bc + \sqrt{(1-b^2)(1-c^2)}}{c+b}$$

$$r_0 = \frac{1 - bc + \sqrt{(1-b^2)(1-c^2)}}{c-b}$$

C-3

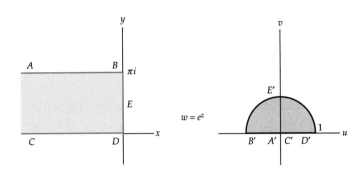

$$w = e^z$$

C-4

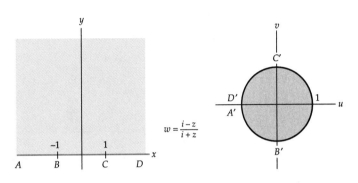

$$w = \frac{i - z}{i + z}$$

C-5

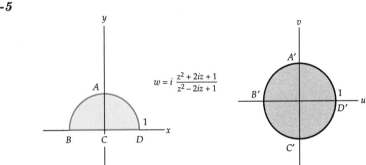

$$w = i\,\frac{z^2 + 2iz + 1}{z^2 - 2iz + 1}$$

■ **Miscellaneous Mappings**

M-1

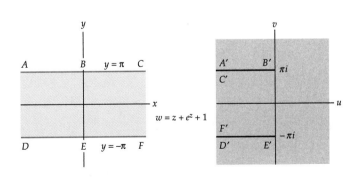

$$w = z + e^z + 1$$

M-2

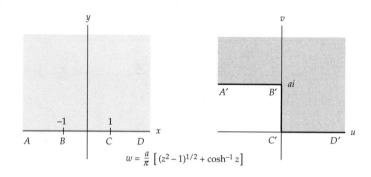

$$w = \frac{a}{\pi}\left[(z^2-1)^{1/2} + \cosh^{-1} z\right]$$

M-3

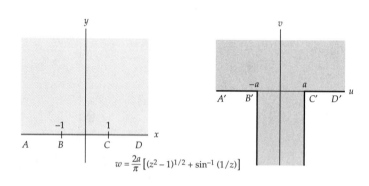

$$w = \frac{2a}{\pi}\left[(z^2-1)^{1/2} + \sin^{-1}(1/z)\right]$$

M-4

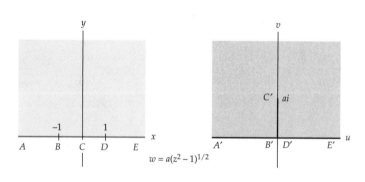

$$w = a(z^2-1)^{1/2}$$

M-5

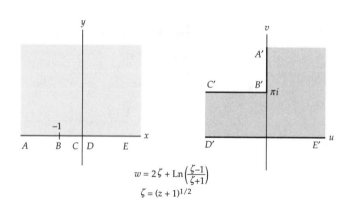

$$w = 2\zeta + \operatorname{Ln}\left(\frac{\zeta-1}{\zeta+1}\right)$$
$$\zeta = (z+1)^{1/2}$$

M-6

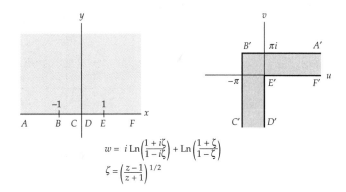

$$w = i\,\mathrm{Ln}\!\left(\frac{1+i\zeta}{1-i\zeta}\right) + \mathrm{Ln}\!\left(\frac{1+\zeta}{1-\zeta}\right)$$

$$\zeta = \left(\frac{z-1}{z+1}\right)^{1/2}$$

M-7

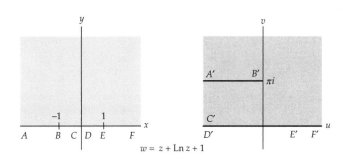

$$w = z + \mathrm{Ln}\,z + 1$$

M-8

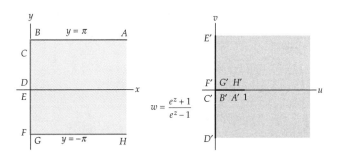

$$w = \frac{e^z + 1}{e^z - 1}$$

M-9

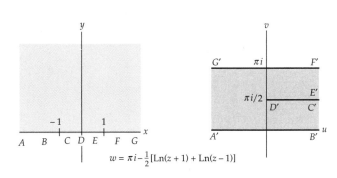

$$w = \pi i - \tfrac{1}{2}[\mathrm{Ln}(z+1) + \mathrm{Ln}(z-1)]$$

M-10

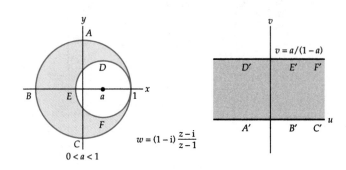

$$w = (1-i)\frac{z-i}{z-1}$$

$0 < a < 1$

Answers to Selected Odd-Numbered Problems

Chapter 1

Exercises 1.1, page 6

1. (a) 1   (b) $-i$ (c) -1 (d) i

3. $7 - 13i$

5. $-7 + 5i$

7. $11 - 10i$

9. $\frac{2}{5} + \frac{16}{5}i$

11. $-\frac{7}{17} - \frac{11}{17}i$

13. $8 - i$

15. $\frac{23}{37} - \frac{64}{37}i$

17. $20i$

19. $\frac{102}{5} + \frac{116}{5}i$

21. $-5 + 12i$

23. $128 - 128i$

25. $\mathrm{Re}(z) = \frac{7}{130}$, $\mathrm{Im}(z) = \frac{9}{130}$

27. $\dfrac{x}{x^2 + y^2}$

29. $-2y - 4$

31. $-\mathrm{Im}(z)$

33. $\mathrm{Re}(z) + \mathrm{Im}(z)$

35. $z_2 = \dfrac{\sqrt{2}}{2} - \dfrac{\sqrt{2}}{2}i$

37. $z = -\frac{9}{2} + i$

39. $z = \dfrac{\sqrt{2}}{2} + \dfrac{\sqrt{2}}{2}i$ or $z = -\dfrac{\sqrt{2}}{2} - \dfrac{\sqrt{2}}{2}i$

41. $z = -\frac{1}{30} + \frac{7}{10}i$

43. $z_1 = 17 + 11i$, $z_2 = 7 + 13i$

Exercises 1.2, page 12

5. $16 - 12i$

7. right triangle

9. 2

11. $\frac{2}{5}$

13. $(x-1)^2 + (y-3)^2$

15. $11 - 6i;\ 10 + 8i$

17. the line $x - y = 1$

19. the line $y = x$

21. the hyperbola $xy = 1$

23. the circle centered at $(1,\ 0)$ of radius 1

25. the parabola $y^2 = 4(x-1)$

29. 6

31. $z = -\frac{3}{4} - i$

Exercises 1.3, page 19

1. $z = 2\left(\cos 2\pi + i\sin 2\pi\right);\ z = 2\left(\cos 0 + i\sin 0\right)$

3. $z = 3\left(\cos\dfrac{3\pi}{2} + i\sin\dfrac{3\pi}{2}\right);\ z = 3\left[\cos\left(-\dfrac{\pi}{2}\right) + i\sin\left(-\dfrac{\pi}{2}\right)\right]$

5. $z = \sqrt{2}\left(\cos\dfrac{9\pi}{4} + i\sin\dfrac{9\pi}{4}\right);\ z = \sqrt{2}\left(\cos\dfrac{\pi}{4} + i\sin\dfrac{\pi}{4}\right)$

7. $z = 2\left[\cos\left(-\dfrac{7\pi}{6}\right) + i\sin\left(-\dfrac{7\pi}{6}\right)\right];\ z = 2\left(\cos\dfrac{5\pi}{6} + i\sin\dfrac{5\pi}{6}\right)$

9. $z = \dfrac{3\sqrt{2}}{2}\left(\cos\dfrac{5\pi}{4} + i\sin\dfrac{5\pi}{4}\right);\ z = \dfrac{3\sqrt{2}}{2}\left[\cos\left(-\dfrac{3\pi}{4}\right) + i\sin\left(-\dfrac{3\pi}{4}\right)\right]$

11. $z = 3\left(\cos 8.34486 + i\sin 8.34486\right);\ z = 3\left(\cos 2.06168 + i\sin 2.06168\right)$

13. $2 + 2\sqrt{3}\,i$

15. $-\dfrac{5\sqrt{3}}{2} - \dfrac{5}{2}i$

17. $5.5433 + 2.2961i$

19. $8i;\ \dfrac{\sqrt{2}}{4} - \dfrac{\sqrt{2}}{4}i$

21. $30\sqrt{2}\left(\cos\dfrac{\pi}{12} + i\sin\dfrac{\pi}{12}\right) \approx 40.9808 + 10.9808i$

23. $\dfrac{\sqrt{2}}{2}\left(\cos\dfrac{5\pi}{4} + i\sin\dfrac{5\pi}{4}\right) = -\dfrac{1}{2} - \dfrac{1}{2}i$

25. -512

27. $\frac{1}{32}i$

29. $-64i$

31. $32\left(\cos\dfrac{13\pi}{6} + i\sin\dfrac{13\pi}{6}\right) = 16\sqrt{3} + 16i$

33. $\cos 2\theta = \cos^2\theta - \sin^2\theta,\quad \sin 2\theta = 2\sin\theta\cos\theta$

35. $n = 6$

Exercises 1.4, page 24

In answers 1–13, the principal nth root is given first.

1. $w_0 = 2,\ w_1 = -1 + \sqrt{3}i\ \ w_2 = -1 - \sqrt{3}i$

3. $w_0 = 3i,\ w_1 = -3i$

5. $w_0 = \dfrac{\sqrt{2}}{2} + \dfrac{\sqrt{2}}{2}i,\ w_1 = -\dfrac{\sqrt{2}}{2} - \dfrac{\sqrt{2}}{2}i$

7. $w_0 = \dfrac{1}{\sqrt[3]{2}} + \dfrac{1}{\sqrt[3]{2}}i,\ w_1 \approx -1.0842 + 0.2905i,\ w_2 \approx 0.2905 - 1.0842i$

9. $w_0 = \dfrac{\sqrt{2}}{2} + \dfrac{\sqrt{6}}{2}i,\ w_1 = -\dfrac{\sqrt{2}}{2} - \dfrac{\sqrt{6}}{2}i$

11. $w_0 = 2 + i$, $w_1 = -2 - i$

13. $w_0 \approx 1.3477 + 0.1327i$, $w_1 \approx 0.8591 + 1.0469i$,

$w_2 \approx -0.1327 + 1.3477i$, $w_3 \approx -1.0469 + 0.8591i$,

$w_4 \approx -1.3477 - 0.1327i$, $w_5 \approx -0.8591 - 1.0469i$,

$w_6 \approx 0.1327 - 1.3477i$, $w_7 \approx 1.0469 - 0.8591i$

15. (b) $4 + 3i$, $-4 - 3i$

17. $\pm\dfrac{\sqrt{2}}{2}(1 + i)\,;\ \pm\dfrac{\sqrt{2}}{2}(1 - i)$

19. (b) $n = 3 : 1,\ -\dfrac{1}{2} + \dfrac{\sqrt{3}}{2}i,\ -\dfrac{1}{2} - \dfrac{\sqrt{3}}{2}i$;

$n = 4 : 1, i, -1, -i$;

$n = 5$:
$1, 0.3090 + 0.9511i, -0.8090 + 0.5878i, -0.8090 - 0.5878i, 0.3090 - 0.9511i$,

25. (c) $-\dfrac{\sqrt{2}}{2} + \dfrac{\sqrt{2}}{2}i,\ \dfrac{\sqrt{2}}{2} - \dfrac{\sqrt{2}}{2}i$

Exercises 1.5, page 31

1.

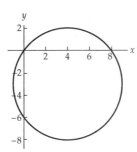

Circle centered at $4 - 3i$ of radius 5

3.

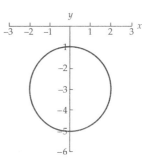

Circle centered at $-3i$ of radius 2

5.

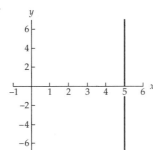

Verticle line $x = 5$

7.

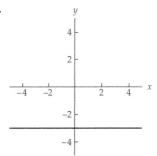

Horizontal line $y = -3$

9.

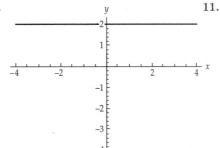

Horizontal lines $y = 2$, $y = -4$

11.

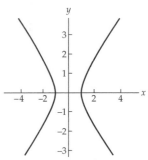

Hyperbola $x^2 - y^2 = 1$

13.

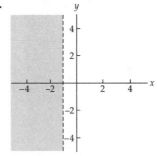

(a) yes   (b) no   (c) yes

(d) no   (e) yes

15.

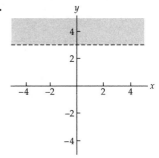

(a) yes   (b) no   (c) yes

(d) no   (e) yes

17.

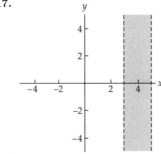

(a) yes   (b) no   (c) yes

(d) no   (e) yes

19.

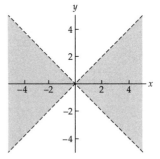

(a) yes   (b) no   (c) no

(d) no   (e) no

21.

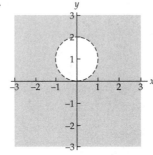

(a) yes   (b) no   (c) yes

(d) no   (e) yes

23.

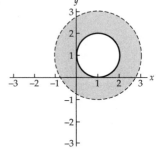

(a) no   (b) no   (c) no

(d) yes   (e) yes

25. For Problem 13: the line $x = -1$;

For Problem 15: the line $y = 3$;

For Problem 17: the lines $x = 3$ and $x = 5$;

For Problem 19: the lines $y = x$ and $y = -x$;

For Problem 21: the circle $|z - i| = 1$;

For Problem 23: the circles $|z - 1 - i| = 1$ and $|z - 1 - i| = 2$

27.

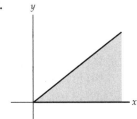

29. $|\arg(z)| \le 2\pi/3$ **31.** $z = 1 + \sqrt{3}i$ or $z = 1 - \sqrt{3}i$

Exercises 1.6, page 38

1. $\dfrac{\sqrt{7}}{2} - \dfrac{1}{2}i,\ -\dfrac{\sqrt{7}}{2} - \dfrac{1}{2}i;\ \left(z - \dfrac{\sqrt{7}}{2} + \dfrac{1}{2}i\right)\left(z + \dfrac{\sqrt{7}}{2} + \dfrac{1}{2}i\right)$

3. $-2 - 3i,\ 3 + 4i;\ (z + 2 + 3i)(z - 3 - 4i)$

5. $-1 - \dfrac{\sqrt{6}}{2} - \dfrac{\sqrt{2}}{2}i,\ -1 + \dfrac{\sqrt{6}}{2} + \dfrac{\sqrt{2}}{2}i;$

$$\left(z + 1 + \dfrac{\sqrt{6}}{2} + \dfrac{\sqrt{2}}{2}i\right)\left(z + 1 - \dfrac{\sqrt{6}}{2} - \dfrac{\sqrt{2}}{2}i\right)$$

7. $10e^{\pi i}$ **9.** $4\sqrt{2}e^{\frac{5}{4}\pi i}$

11. $10e^{i\tan^{-1}\left(-\frac{3}{4}\right)}$ **13.** $y_1 = e^{2x}\cos 3x,\ y_2 = e^{2x}\sin 3x$

15. $y_1 = e^{-\frac{1}{2}x}\cos\dfrac{\sqrt{3}}{2}x,\ y_2 = e^{-\frac{1}{2}x}\sin\dfrac{\sqrt{3}}{2}x$

17. $q_p(t) = \frac{1}{3}\sin 5t,\ i_p(t) = \frac{5}{3}\cos 5t,\ Z_C = 6 + 0j,\ Z = 6$

19. $q_p(t) = 50(\sin t - \cos t),\ i_p(t) = 50(\sin t + \cos t),\ Z_C = 1 - j,\ Z = \sqrt{2}$

Chapter 1 Review Quiz, page 42

1. false   **3.** true

5. false   **7.** true

9. true   **11.** false

13. true   **15.** false

17. true   **19.** false

21. true   **23.** $\frac{9}{13},\ -\frac{7}{13}$

25. a nonnegative real number

27. $-3\pi/4$

29. $\dfrac{5\pi}{4},\ 8,\ -8,\ 16$ **31.** fourth

33. z_2 **35.** $z = -3 - i$

37. 1   **39.** first

41. the set of all points z above the line $y = x$

43. real axis   **45.** $n = 24$

47. $\cos 4\theta = \cos^4\theta - 6\cos^2\theta\sin^2\theta + \sin^4\theta,$

$\sin 4\theta = 4\cos^3\theta\sin\theta - 4\cos\theta\sin^3\theta$

49. The equation cannot have three complex roots since complex roots must appear in conjugate pairs.

Chapter 2

Exercises 2.1, page 51

1. **(a)** $6i$ **(b)** 2 **(c)** $39 - 28i$

3. **(a)** 0 **(b)** $\log_e 4 + \frac{1}{2}\pi i$ **(c)** $\frac{1}{2}\log_e 2 + \frac{1}{4}\pi i$

5. **(a)** $3i$ **(b)** $-12 + 13i$ **(c)** $-24 + 4i$

7. **(a)** $3 + i$ **(b)** 2 **(c)** $\sqrt{5} + \frac{4}{5}i$

9. $u = 6x - 5;\ v = 6y + 9$

11. $u = x^3 - 2x - 3xy^2 + 6;\ v = 3x^2y - 2y - y^3$

13. $u = \dfrac{x^2 + x - y^2}{(x+1)^2 + y^2};\ v = -\dfrac{2xy + y}{(x+1)^2 + y^2}$

15. $u = e^{2x}\cos(2y + 1);\ v = e^{2x}\sin(2y + 1)$

17. $u = r\cos\theta;\ v = -r\sin\theta$

19. $u = r^4\cos 4\theta;\ v = r^4\sin 4\theta$

21. $u = e^{r\cos\theta}\cos(r\sin\theta);\ v = e^{r\cos\theta}\sin(r\sin\theta)$

23. **C**   25. all z such that $z \neq 1$

Exercises 2.2, page 60

1. the vertical line $u = 3$ 3. the half-plane $\text{Im}(w) > 6$

5. the line $v = 4 - u$ 7. the half-plane $\text{Re}(w) \geq 3$

9. the parabola $u = \frac{1}{4}v^2 - 1$ 11. the ray $-\infty < u \leq 0,\ v = 0$

13. the ray $u = 0,\ 0 \leq v < \infty$

15. **(a)**   **(b)** $w(t) = 6(1 - t) + 3it$

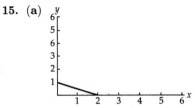

Line segment from 2 to i

(c)

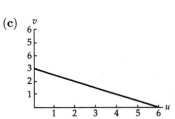

Line segment from 6 to $3i$

17. (a)

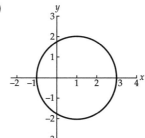

Circle centered at 1 with radius 2

(b) $w(t) = 2 - i + 2e^{it}$

(c)

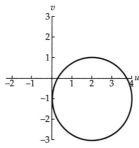

Circle centered at $2 - i$ with radius 2

19. (a)

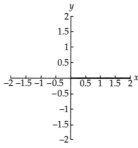

Line segment from 0 to 2

(b) $w(t) = e^{i\pi t}$

(c)

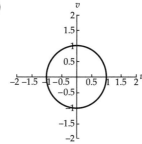

Unit circle

21. the negative imaginary axis

23. the circle $|w| = \frac{1}{2}$

25. the line segment from –2 to 2 on the real axis

Exercises 2.3, page 69

1. $|w - 3i| \leq 1$

3. $|w| \leq 3$

5. $|w + i| \leq 2$

7. triangle with vertices $2i$, $1 + 2i$, and $3i$

9. triangle with vertices 0, $\frac{1}{2}\sqrt{2} + \frac{1}{2}\sqrt{2}i$, and $-\frac{1}{2}\sqrt{2} + \frac{1}{2}\sqrt{2}i$

11. triangle with vertices $i, -3 + i$, and $-2i$

13. $f(z) = T \circ M \circ R(z)$ where $R(z) = e^{\pi i/2}z$, $M(z) = 3z$, and $T(z) = z + 4$

15. $f(z) = T \circ M \circ R(z)$ where $R(z) = e^{\pi i}z$, $M(z) = \frac{1}{2}z$, and $T(z) = z + 1 - \sqrt{3}i$

17. $f(z) = iz + 2i$ 19. $f(z) = e^{-\pi i/4}z + i$

21. $f(z) = z - 1, g(z) = iz$

23. (a) $w(t) = (z_0 + b)(1 - t) + (z_1 + b)t, 0 \le t \le 1$, the line segment from $z_0 + b$ to $z_1 + b$

    (b) $w(t) = az_0(1 - t) + az_1 t, 0 \le t \le 1$, the line segment from az_0 to az_1

    (c) $w(t) = az_0(1 - t) + az_1 t, 0 \le t \le 1$, the line segment from az_0 to az_1

25. (a) $f(z) = 2e^{\pi i/4}z + 1 + i$ (b) $f(z) = 2e^{\pi i/4}z + 1 + i$

    (c) $f(z) = 2e^{\pi i/4}z + 1 + i$

Exercises 2.4, page 88

1. $\arg(w) = \frac{2}{3}\pi$ 3. $u = 9 - \frac{1}{36}v^2, -\infty < v < \infty$

5. $u = 4v^2 - \frac{1}{16}, -\infty < v < \infty$ 7. $v = 0, -\infty < u < 0$

9. $|w| = \frac{1}{4}$

11. Image consists of the arcs: $v = 0$, $0 \le u \le 1$; $u = 0$, $0 \le v \le 2$; $u = 1 - \frac{1}{4}v^2, 0 \le v \le 2$.

13. Image consists of the arcs: $v = 0$, $-1 \le u \le 1$; $u = 1 - \frac{1}{4}v^2$, $0 \le v \le 2$; $u = \frac{1}{4}v^2 - 1, 0 \le v \le 2$.

15. Image is a ray emanating from $1 - i$ and containing $(\sqrt{3} - 1)i$; $1 - i$ is not in the image.

17. $v = 4 - \frac{1}{16}(u + 3)^2, -\infty < u < \infty$ 19. $|w| = 1, \frac{1}{4}\pi \le \arg(w) \le \frac{5}{4}\pi$

21. (a) $\arg(w) = \frac{1}{2}\pi$ (b) $\arg(w) = \frac{2}{3}\pi$ (c) $\arg(w) = \frac{5}{6}\pi$

23. (a) $1 \le |w| \le 4, \frac{1}{2}\pi \le \arg(w) \le \frac{3}{2}\pi$ (b) $1 \le |w| \le 8, \frac{3}{4}\pi \le \arg(w) \le \frac{9}{4}\pi$

    (c) $1 \le |w| \le 16$

25. $\frac{1}{2}\sqrt{2} - \frac{1}{2}\sqrt{2}i$ 27. $\frac{1}{2} + \frac{1}{2}\sqrt{3}i$

29. $\frac{1}{2}\sqrt[4]{18} + \frac{1}{2}\sqrt[4]{2}i$ 31. $\arg(w) = \frac{1}{8}\pi$

33. $\arg(w) = \frac{1}{4}\pi$ 35. $|w| = 3, -\frac{1}{4}\pi \le \arg(w) \le \frac{1}{2}\pi$

37. $u = \frac{3}{2}$

39. the region bounded by the lines $u = 2$ and $v = u$ containing the point $w = 3 + 4i$

Exercises 2.5, page 97

1. $|w| = \frac{1}{5}$ 3. $|w| = \frac{1}{3}, -3\pi/4 \le \arg(w) \le \pi/4$

5. $\frac{1}{2} \le |w| \le 3$ 7. $\arg(w) = -\frac{1}{4}\pi$

9. $\left|w + \frac{1}{8}i\right| = \frac{1}{8}$ 11. $v = \frac{1}{2}$

13. $u = \frac{1}{4}$

15. Image is the region bounded by $\left|w + \frac{1}{4}\right| = \frac{1}{4}$ and $\left|w + \frac{1}{2}\right| = \frac{1}{2}$.

17. Image is the region bounded by $v = 0$, $v = -u$, and $|w| = 2$ and containing the point $-3 + 2i$

19. (a) Invert in the unit circle, reflect across the real axis, rotate through $\pi/2$ counterclockwise about the origin, magnify by 2, then translate by 1.

    (b) $\left|w - 1 - \frac{1}{4}i\right| = \frac{1}{4}$ **(c)** $v = -\frac{1}{2}$

21. (a) if $f(z) = 1/z$ and $g(z) = z^2$, then $h(z) = g(f(z))$

    (b) $u = \frac{1}{4}v^2 - 1$ **(c)** $u = \frac{1}{4} - v^2$

Exercises 2.6, page 116

1. $-4 + 2i$ **3.** $3 - i$

5. -1 **7.** $2i$

9. $1 - 3i$ **11.** $\sqrt{2}$

13. $4i$ **15.** a

17. (a) 1   **(b)** 0   **(c)** does not exist

19. (a) 1   **(b)** 1   **(c)** no

    (d) -1 **(e)** does not exist

21. $\frac{1}{5} - \frac{2}{5}i$ **23.** ∞

25. ∞ **27.** $\displaystyle\lim_{z \to 2-i} f(z) = f(2 - i) = 5 - 8i$

29. $\displaystyle\lim_{z \to i} f(z) = f(i) = \frac{1}{3}i$ **31.** $\displaystyle\lim_{z \to 1} f(z) = f(1) = 3$

33. $\displaystyle\lim_{z \to 3-2i} f(z) = f(3 - 2i) = -6 + 3i$ **35.** $f(-i)$ is not defined

37. $\displaystyle\lim_{z \to -1} f(z)$ does not exist   **39.** $\displaystyle\lim_{z \to i} f(z) \neq f(i)$

41. the entire complex plane $\mathbf{C}$

43. all points in the complex plane except those on the circle $|z| = 2$

Exercises 2.7, page 124

1. (a)   **(b)**

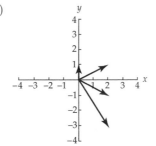
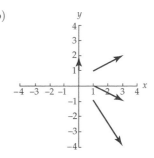

3. (a)   **(b)**

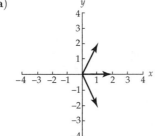
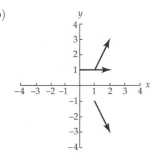

5. (a) 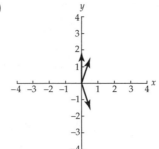   **(b)**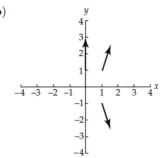

7. (a) 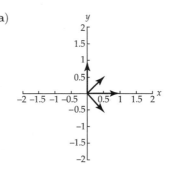   **(b)**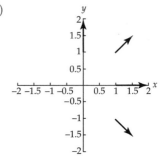

9. (a) $y = -2x + c$ **11. (a)** $x^2 + y^2 = c$

(b) 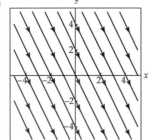   **(b)**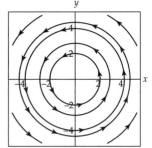

Chapter 2 Review Quiz, page 125

1. false   **3.** false

5. true   **7.** false

9. false   **11.** true

13. false   **15.** true

17. true   **19.** true

21. $x^2 - y^2 + y,\ x + 2xy$ **23.** imaginary

25. $(1 + i)(1 - t) + 2ti, 0 \le t \le 1$ **27.** rotation, magnification, translation

29. doubles   **31.** parabolas

33. $\frac{1}{2}\sqrt{3} + \frac{1}{2}i$ **35.** $z^2 + 4,\ -4 - i$

37. $f(z_0)$ **39.** $0 < x < \infty,\ y \ne 0$

Chapter 3

Exercises 3.1, page 135

1. $f'(z) = 9i$ **3.** $f'(z) = 3iz^2 - 14z$

5. $f'(z) = 1 + \dfrac{1}{z^2}$ **7.** $f'(z) = 10z - 10$

9. $f'(z) = 4z^3 - 2z$ **11.** $f'(z) = (10 - 5i)z^4 + 4iz^3 - 6z$

13. $f'(z) = 8z^7 - 7z^6 + (6 - 30i)z^5 - 2z + 1$

15. $f'(z) = \dfrac{3iz^2 + (2 + 2i)z - 2 + 2i}{(3z + 1 - i)^2}$

17. $f'(z) = 10 \left(z^4 - 2iz^2 + z\right)^9 \left(4z^3 - 4iz + 1\right)$

23. $\frac{1}{2}i$

25. $8i$

27. f is not analytic at $z = -\frac{1}{3} + \frac{1}{3}i$

29. f is analytic for all z

Exercises 3.2, page 141

9. **(b)** $f'(z) = -e^{-x}\cos x + i - e^{-x}\sin y$

11. **(b)** $f'(z) = 2e^{x^2 - y^2}(x\cos 2xy - y\sin 2xy) + i2e^{x^2 - y^2}(x\sin 2xy - y\cos 2xy)$

13. **(b)** $f'(z) = \dfrac{-(x - 1)^2 + y^2}{((x - 1)^2 + y^2)^2} + i\dfrac{2(x - 1)y}{((x - 1)^2 + y^2)^2}$

15. **(b)** $f'(z) = e^{-i\theta}\left(-\dfrac{\cos\theta}{r^2} + i\dfrac{\sin\theta}{r^2}\right)$

17. $a = 1$, $b = 3$

19. **(b)** $f'(z) = 2x$ on the x-axis

21. **(b)** $f'(z) = 3x^2 - 1$ on the x-axis; $f'(z) = 3y^2 - 1$ on the y-axis

Exercises 3.3, page 147

1. **(c)** $f(z) = x + i(y + C)$ **3.** **(c)** $f(z) = x^2 - y^2 + i(2xy + C)$

5. **(c)** $f(z) = \log_e(x^2 + y^2) + i\left(2\tan^{-1}\dfrac{y}{x} + C\right)$

7. **(c)** $f(z) = e^x(x\cos y - y\sin y) + i(e^x(x\sin y + y\cos y) + C)$

9. **(c)** $f(z) = \dfrac{x}{x^2 + y^2} + i\left(\dfrac{-y}{x^2 + y^2} + C\right)$

11. $f(z) = xy + x + 2y - 5 + i\left(\frac{1}{2}y^2 - \frac{1}{2}x^2 + y - 2x + 1\right)$

13. **(b)** $f(z) = \dfrac{y}{x^2 + y^2} + i\dfrac{x}{x^2 + y^2}$ **(c)** $f(z) = \dfrac{i}{z}$

Exercises 3.4, page 153

1. $x = c_1$, $y = c_2$

3. $c_1 x = x^2 + y^2$, $c_2 y = x^2 + y^2$; the level curves $u(x, y) = 0$ and $v(x, y) = 0$ correspond to $x = 0$ and $y = 0$, respectively.

11. **(a)** $\phi(x) = -50x + 50$ **(b)** $\Omega(z) = -50x + 50 - 50yi$

13. **(a)** $\phi(\theta) = \dfrac{120}{\pi}\theta$ **(b)** $\Omega(z) = \dfrac{120}{\pi}\theta - \dfrac{120}{\pi}\log_e r$

Chapter 3 Review Quiz, page 155

1. false   3. true

5. true   7. true

9. true   11. true

13. $-\dfrac{2z + 5i}{\left(z^2 + 5iz - 4\right)^2}$ 15. $2 + i$

17. $f'(z) = \dfrac{(y-1)^2 - (x-1)^2}{\left[(x-1)^2 + (y-1)^2\right]^2} + i\dfrac{2(x-1)(y-1)}{\left[(x-1)^2 + (y-1)^2\right]^2}$

19. constant

21. $v(x,\ y) = e^{-x}(x\cos y + y\sin y)$

Chapter 4

Exercises 4.1, page 172

1. $z^2 e^{z+i} + 2z e^{z+i}$ 3. $ie^{iz} + ie^{-iz}$

5. $e^{x^2 - x - y^2}$ 7. $2x + 2n\pi,\ n = 0,\ \pm 1,\ \pm 2,\ \ldots$

9. $e^y \cos x - ie^y \sin x$ 11. $e^{x^2 - y^2}\cos(2xy) + ie^{x^2 - y^2}\sin(2xy)$

13. f is nowhere differentiable   15. $\arg(w) = -2$

17. $e < |w| \le e^2$ 19. $1 \le |w| \le 2,\ -\pi/4 \le \arg(w) \le \pi/2$

21. $\log_e 5 + (2n+1)\pi i$ 23. $\frac{3}{2}\log_e 2 + \frac{1}{4}(8n+3)\pi i$

25. $\frac{3}{2}\log_e 2 + \frac{1}{3}(6n+1)\pi i$ 27. $\frac{1}{2}\log_e 72 - \frac{1}{4}\pi i$

29. $2.5650 + 2.7468i$ 31. $5\log_e 2 - \frac{1}{3}\pi i$

33. $2\log_e 2 + \frac{1}{2}(4n+1)\pi i$ 35. $4 + \frac{1}{2}(4n-1)\pi i$

37. differentiable on the domain $|z| > 0,\ -\pi < \arg(z) < \pi$,
$f'(z) = 6z - 2ie^{2iz} + \dfrac{i}{z}$

39. differentiable when z is not on the ray emanating from $\frac{1}{2}i$ containing

$-1 + \frac{1}{2}i;\ z \ne -i$, and $z \ne i$, $f'(z) = \dfrac{2\dfrac{z^2 + 1}{2z - i} - 2z\mathrm{Ln}(2z - i)}{\left(z^2 + 1\right)^2}$

41. $v = \frac{1}{6}\pi$ 43. $u = 2\log_e 2,\ -\pi < v \le \pi$

45. $\log_e 3 \le u \le \log_e 5,\ -\pi < v \le \pi$

Exercises 4.2, page 179

1. $e^{-3(2n+1)\pi},\ n = 0,\ \pm 1,\ \pm 2,\ \ldots$

3. $\sqrt{2}e^{(8n+1)\pi/4 + i[(8n+1)\pi/4 - (\log_e 2)/2]},\ n = 0,\ \pm 1,\ \pm 2,\ \ldots$

5. $e^{(-4n+1)\pi/2},\ n = 0,\ \pm 1,\ \pm 2,\ \ldots$ 7. $e^{-3\pi}$

9. $e^{i4\log_e 2}$ 11. $e^{-\pi + i3\log_e 2}$

15. $\frac{3}{2}\sqrt[4]{2}e^{\pi i/8}$ 17. $\sqrt{2}e^{-\pi/3 + i[(\pi/4) + \log_e 2]}$

Exercises 4.3, page 191

1. $i \sinh 4 \approx 27.2899i$

3. $\cos 2 \cosh 4 + i \sin 2 \sinh 4 \approx -11.3642 + 24.8147i$

5. $i \tanh 2 \approx 0.9640i$ **7.** $-i \operatorname{csch} 1 \approx -0.8509i$

9. $z = 2n\pi - i \log_e \left(\sqrt{2} - 1\right)$ or $z = (2n+1)\pi - i \log_e \left(\sqrt{2} + 1\right)$, $n = 0, \pm 1, \pm 2, \ldots$

11. $z = \frac{1}{4}(4n+1)\pi$, $n = 0, \pm 1, \pm 2, \ldots$

17. $2z \cos\left(z^2\right)$ **19.** $\tan\left(\dfrac{1}{z}\right) - \dfrac{1}{z}\sec^2\left(\dfrac{1}{z}\right)$

21. -1 **23.** $\frac{1}{2}\sqrt{3}\cosh 1 + i\frac{1}{2}\sinh 1 \approx 1.3364 + 0.5876i$

25. $z = \log_e\left(1 + \sqrt{2}\right) + \frac{1}{2}(2n+1)\pi i$ or $z = \log_e\left(-1 + \sqrt{2}\right) + \frac{1}{2}(2n-1)\pi i$, $n = 0, \pm 1, \pm 2, \ldots$

27. There are no solutions.

33. $\cos z \sinh z + \sin z \cosh z$ **35.** $i \operatorname{sech}^2\left(iz - 2\right)$

Exercises 4.4, page 199

1. $\frac{1}{2}(4n+1)\pi - i \log_e\left(\sqrt{2} + 1\right)$ and $\frac{1}{2}(4n-1)\pi - i \log_e\left(\sqrt{2} - 1\right)$, $n = 0, \pm 1, \pm 2, \ldots$

3. $\frac{1}{2}(4n+1)\pi - i \log_e\left(\sqrt{2} \pm 1\right)$, $n = 0, \pm 1, \pm 2, \ldots$

5. $-\frac{1}{4}(4n-1)\pi$, $n = 0, \pm 1, \pm 2, \ldots$ **7.** $\frac{1}{2}(4n+1)\pi i$, $n = 0, \pm 1, \pm 2, \ldots$

9. $\frac{1}{4}\log_e 2 + \frac{1}{8}(8n+3)\pi i$, $n = 0, \pm 1, \pm 2, \ldots$

11. (a) $-i\log_e\left(\frac{1}{2}\left(\sqrt{5} - 1\right)\right)$ (b) $\frac{2}{5}\sqrt{5}$

13. (a) $\frac{1}{2}\left(\pi - \arctan 2\right) + i\frac{1}{4}\log_e 5$ (b) $\frac{1}{5} - \frac{2}{5}i$

15. (a) $\log_e\left(\sqrt{2} + 1\right) - \frac{1}{2}\pi i$ (b) $\frac{1}{2}\sqrt{2}i$

Exercises 4.5, page 207

1. $\phi(x, y) = -x + 5$ **3.** $\phi(x, y) = \frac{5}{4}\sqrt{3}x - \frac{5}{4}y + 10$

5. $\phi(x, y) = 12 + \dfrac{33}{\pi}\operatorname{Arg}(\sin(z - \pi) + 1) - \dfrac{25}{\pi}\operatorname{Arg}(\sin(z - \pi) - 1)$

7. $\phi(x, y) = 15 - \dfrac{9}{\pi}\operatorname{Arg}(\sin(iz) + 1) + \dfrac{17}{\pi}\operatorname{Arg}(\sin(iz) - 1)$

Chapter 4 Review Quiz, page 209

1. true   **3.** false

5. false   **7.** true

9. false   **11.** false

13. false   **15.** true

17. true   **19.** true

21. $e^x \cos y$, $e^x \sin y$ **23.** $\log_e 2 + \frac{1}{6}\pi i$

25. $2n\pi - i\log_e 2$, $n = 0, \pm 1, \pm 2, \ldots$ **27.** nonpositive real axis

29. $\frac{1}{2}\log_e 2 + \frac{1}{4}(8n+1)\pi i$, $n = 0, \pm 1, \pm 2, \ldots$

31. $z_2 = e^{2\pi}$ **35.** $\cosh 4$

37. $\sin x \cosh y$, $\cos x \sinh y$ **39.** ± 1

Chapter 5

Exercises 5.1, page 218

1. $\frac{64}{3}$

3. $-\dfrac{1}{\pi}$

5. $\frac{1}{2}\ln 9$

7. $8e^{-1} - 12e^{-2}$

9. $-\frac{1}{2}\ln 3$

11. $-125/3\sqrt{2}$; $-250(\sqrt{2}-4)/12$; $\frac{125}{2}$

13. $3;\ 6;\ 3\sqrt{5}$

15. 21

17. 30

19. 1

21. 1

23. 460

25. $\frac{26}{9}$

27. $-\frac{64}{3}$

29. $-\frac{8}{3}$

31. 0

33. On each curve the line integral has the value $\frac{208}{3}$.

35. With $\rho = kx$, $m = k\pi$.

Exercises 5.2, page 228

1. $-28 + 84i$

3. $-48 + \frac{736}{3}i$

5. $(2+\pi)i$

7. πi

9. $-\frac{7}{12} + \frac{1}{12}i$

11. $-e - 1$

13. $\dfrac{3}{2} - \dfrac{\pi}{4}$

15. 0

17. $\frac{1}{2}i$

19. 0

21. $\frac{4}{3} - \frac{5}{3}i$

23. $\frac{4}{3} - \frac{5}{3}i$

25. $\frac{5}{12}\pi e^5$

27. $6\sqrt{2}$

31. (a) $-11 + 38i$ (b) 0

Exercises 5.3, page 236

9. $2\pi i$

11. $2\pi i$

13. 0

15. (a) $2\pi i$ (b) $4\pi i$ (c) 0

17. (a) $-8\pi i$ (b) $-6\pi i$

19. $-\pi(1+i)$

21. 0

23. $-4\pi i$

25. $-6\pi i$

Exercises 5.4, page 244

1. $-2i$

3. $48 + 24i$

5. $6 + \frac{26}{3}i$

7. 0

9. $-\frac{7}{6} - \frac{22}{3}i$

11. $-\dfrac{1}{\pi} - \dfrac{1}{\pi}i$

13. $2.3504i$

15. 0

17. πi

19. $\frac{1}{2}i$

21. $11.4928 + 0.9667i$

23. $-0.9056 + 1.7699i$

25. $\sqrt{2}\,i$

Exercises 5.5, page 253

1. $8\pi i$

3. $-2\pi i$

5. $-\pi(20 + 8i)$

7. (a) -2π (b) 2π

9. -8π

11. $-2\pi e^{-1}i$

13. $\frac{4}{3}\pi i$

15. (a) $-5\pi i$ (b) $-5\pi i$ (c) $9\pi i$ (d) 0

17. (a) $-\pi(3 + i)$ (b) $\pi(3 + i)$

19. $\pi\left(\frac{8}{3} + 12i\right)$

21. 0

23. $-\pi i$

25. 6

27. (a) $16; 4$ (b) $25; 9$ (c) $7; 3$

Exercises 5.6, page 265

5. $f(z) = \cos\theta_0 + i\sin\theta_0 = e^{i\theta_0}$, $g(z) = \overline{f(z)} = \cos\theta_0 - i\sin\theta_0 = e^{-i\theta_0}$ is constant and so is analytic everywhere.

7. $f(z) = 2\bar{z} + 3i$, $g(z) = \overline{f(z)} = 2z - 3i$ is a polynomial function and so is analytic for all z.

9. $\mathbf{F}(x,\ y) = (x^2 - y^2 - 2xy)\mathbf{i} + (y^2 - x^2 - 2xy)\mathbf{j}$

11. $\mathbf{F}(x,\ y) = (e^x\cos y)\mathbf{i} - (e^x\sin y)\mathbf{j}$

13. $\Omega(z) = e^{-i\theta_0}z$; equipotential lines are the family of straight lines $x\cos\theta_0 + y\sin\theta_0 = c_1$; the streamlines are the family of straight lines $-x\sin\theta_0 + y\cos\theta_0 = c_2$.

15. $\Omega(z) = z^2 - 3iz$; equipotential lines are the family of hyperbolas $x^2 - y^2 + 3y = c_1$; the streamlines are the family of hyperbolas $2xy - 3x = c_2$.

17. $\mathbf{F}(x, y) = -2xy\mathbf{i} + (y^2 - x^2)\mathbf{j}$

21. (a) For a point $(x,\ y)$ far from the origin, the velocity field is given by $\mathbf{F}(x,\ y) \approx A\mathbf{i}$, that is, the flow is a nearly uniform.

23. (a) The streamlines are $\text{Arg}(z - x_1) = c_1$, which are rays with vertex at $z = x_1$.

25. Circulation is 0; net flux is 0.

27. Circulation is 0; net flux is 2π.

29. Circulation is -4π; net flux is 12π.

Chapter 5 Review Quiz, page 267

1. false

3. true

5. true

7. true

9. true

11. true

13. false

15. true

17. true

19. true

21. unit circle centered at the origin

23. $z_1(t)$ and $z_2(t)$ both describe a unit circle centered at the origin but have opposite orientations.

25. 0 **27.** $\frac{8}{3}$

29. $2\cos(2+i) - 2\cos 3i$ **31.** $2\pi i$

33. $6\pi^2 - \pi i$ **35.** $2\pi i/(n-1)!$

37. 0 for $n \neq -1$ and $2\pi i$ for $n = -1$.   **39.** $i - 1$

Chapter 6

Exercises 6.1, page 280

1. $5i,\ -5,\ -5i,\ 5,\ 5i$ **3.** $0, 2, 0, 2, 0$

5. converges   **7.** converges

9. diverges

11. $\lim\limits_{n\to\infty} \mathrm{Re}(z_n) = 2$ and $\lim\limits_{n\to\infty} \mathrm{Im}(z_n) = \frac{3}{2}$ and so $L = 2 + \frac{3}{2}i$.

13. The series converges to $\frac{1}{5} - \frac{2}{5}i$.

15. divergent   **17.** convergent, $-\frac{1}{5} + \frac{2}{5}i$

19. convergent, $\frac{9}{5} - \frac{12}{5}i$ **21.** $|z - 2i| = \sqrt{5},\ R = \sqrt{5}$

23. $|z - 1 - i| = 2,\ R = 2$ **25.** $|z - i| = 1/\sqrt{10},\ R = 1/\sqrt{10}$

27. $|z - 4 - 3i| = 25,\ R = 25$ **29.** $|z - i| = \frac{1}{2},\ R = \frac{1}{2}$

31. $z = -2 + i$ **33.** $\sum_{k=1}^{\infty} z_k$ diverges.

Exercises 6.2, page 289

1. $\sum\limits_{k=1}^{\infty} (-1)^{k+1} z^k,\ R = 1$ **3.** $\sum\limits_{k=1}^{\infty} (-1)^{k-1} k(2z)^{k-1},\ R = \frac{1}{2}$

5. $\sum\limits_{k=0}^{\infty} \frac{(-1)^k}{k!}(2z)^k,\ R = \infty$ **7.** $\sum\limits_{k=0}^{\infty} \frac{1}{(2k+1)!} z^{2k+1},\ R = \infty$

9. $\sum\limits_{k=0}^{\infty} \frac{(-1)^k}{(2k)!} \left(\frac{z}{2}\right)^{2k},\ R = \infty$ **11.** $\sum\limits_{k=0}^{\infty} \frac{(-1)^k}{(2k+1)!} z^{4k+2},\ R = \infty$

13. $e^{3i} \sum\limits_{k=0}^{\infty} \frac{1}{k!}(z - 3i)^k,\ R = \infty$ **15.** $\sum\limits_{k=0}^{\infty} (-1)^k (z-1)^k,\ R = 1$

17. $\sum\limits_{k=0}^{\infty} \frac{1}{(3-2i)^{k+1}}(z-2i)^k,\ R = \sqrt{13}$ **19.** $\sum\limits_{k=1}^{\infty} \frac{1}{2^k}(z-1)^k,\ R = 2$

21. $\frac{\sqrt{2}}{2} - \frac{\sqrt{2}}{2 \cdot 1!}\left(z - \frac{\pi}{4}\right) - \frac{\sqrt{2}}{2 \cdot 2!}\left(z - \frac{\pi}{4}\right)^2 + \frac{\sqrt{2}}{2 \cdot 3!}\left(z - \frac{\pi}{4}\right)^3 + \cdots,\ R = \infty$

23. $z + \frac{1}{3}z^3 + \frac{2}{15}z^5 + \cdots$

25. $\frac{1}{2i} + \frac{3}{(2i)^2}z + \frac{7}{(2i)^3}z^2 + \frac{15}{(2i)^4}z^3 + \cdots,\ R = 1$

27. $R = 2\sqrt{5}$ **29.** $R = \frac{\pi}{2}$

31. $\displaystyle\sum_{k=0}^{\infty}(-1)^k(z+1)^k,\ R=\sqrt{2};\quad \sum_{k=0}^{\infty}\frac{(-1)^k}{(2+i)^{k+1}}(z-i)^k,\ R=\sqrt{5}$

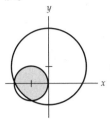

33. $\dfrac{1}{1-3z}$ **37.** $1.1+0.12i$

Exercises 6.3, page 300

1. $\dfrac{1}{z}-\dfrac{z}{2!}+\dfrac{z^3}{4!}-\dfrac{z^5}{6!}+\cdots$ **3.** $1-\dfrac{1}{1!\cdot z^2}+\dfrac{1}{2!\cdot z^4}-\dfrac{1}{3!\cdot z^6}+\cdots$

5. $\dfrac{e}{z-1}+e+\dfrac{e(z-1)}{2!}+\dfrac{e(z-1)^2}{3!}+\cdots$ **7.** $-\dfrac{1}{3z}-\dfrac{1}{3^2}-\dfrac{z}{3^3}-\dfrac{z^2}{3^4}-\cdots$

9. $\dfrac{1}{3(z-3)}-\dfrac{1}{3^2}+\dfrac{z-3}{3^3}-\dfrac{(z-3)^2}{3^4}+\cdots$

11. $\cdots-\dfrac{1}{3(z-4)^2}+\dfrac{1}{3(z-4)}-\dfrac{1}{12}+\dfrac{z-4}{3\cdot 4^2}-\dfrac{(z-4)^2}{3\cdot 4^3}+\cdots$

13. $\cdots-\dfrac{1}{z^2}-\dfrac{1}{z}-\dfrac{1}{2}-\dfrac{z}{2^2}-\dfrac{z^2}{2^3}-\cdots$ **15.** $\dfrac{-1}{z-1}-1-(z-1)-(z-1)^2-\cdots$

17. $\dfrac{1}{3(z+1)}-\dfrac{2}{3^2}-\dfrac{2(z+1)}{3^3}-\dfrac{2(z+1)^2}{3^4}-\cdots$

19. $\cdots-\dfrac{1}{3z^2}+\dfrac{1}{3z}-\dfrac{1}{3}-\dfrac{z}{3\cdot 2}-\dfrac{z^2}{3\cdot 2^2}-\cdots$

21. $\dfrac{1}{z}+2+3z+4z^2+\cdots$ **23.** $\dfrac{1}{z-2}-3+6(z-2)-10(z-2)^2+\cdots$

25. $\dfrac{3}{z}-4-4z-4z^2-\cdots$ **27.** $\cdots+\dfrac{2}{(z-1)^3}+\dfrac{2}{(z-1)^2}+\dfrac{2}{z-1}+1+(z-1)$

29. $\dfrac{1}{z}+\dfrac{z}{6}+\dfrac{7z^3}{360}+\cdots$

Exercises 6.4, page 307

1. Define $f(0)=2$.   **3.** Define $f(0)=0$.

5. $-2+i$ is a zero of order 2.

7. 0 is a zero of order 2; i and $-i$ are simple zeros.

9. $2n\pi i,\ n=0,\pm 1,\ldots,$ are simple zeros.

11. order 5   **13.** order 1

15. $-1+2i$ and $-1-2i$ are simple poles.

17. -2 is a simple pole; $-i$ is a pole of order 4.

19. $(2n+1)\pi/2, n = 0, \pm1, \ldots$, are simple poles.

21. 0 is a pole of order 2.

23. $(2n+1)\pi i, n = 0, \pm1, \ldots$, are simple poles.

25. 1 is a simple pole.

27. essential singularity

Exercises 6.5, page 315

1. $\frac{2}{5}$ **3.** -3

5. 0   **7.** $\operatorname{Res}(f(z), -4i) = \frac{1}{2}$, $\operatorname{Res}(f(z), 4i) = \frac{1}{2}$

9. $\operatorname{Res}(f(z), 1) = \frac{1}{3}$, $\operatorname{Res}(f(z), -2) = -\frac{1}{12}$, $\operatorname{Res}(f(z), 0) = -\frac{1}{4}$

11. $\operatorname{Res}(f(z), -1) = 6$, $\operatorname{Res}(f(z), -2) = -31$, $\operatorname{Res}(f(z), -3) = 30$

13. $\operatorname{Res}(f(z), 0) = -3/\pi^4$, $\operatorname{Res}(f(z), \pi) = (\pi^2 - 6)/2\pi^4$

15. $\operatorname{Res}(f(z), (2n+1)\pi/2) = (-1)^{n+1}$, $n = 0, \pm1, \pm2, \ldots$

17. $0; 2\pi i/9; 0$ **19.** $\pi i; \pi i; 0$

21. $\pi/3$ **23.** 0

25. $2\pi i \cosh 1$ **27.** $-4i$

29. $6i$ **31.** $\left(\dfrac{1}{3\pi^2} + \dfrac{1}{\pi}\right) i$

33. $2\pi/3$

Exercises 6.6, page 333

1. $\dfrac{4\pi}{\sqrt{3}}$ **3.** 0

5. $\dfrac{\pi}{\sqrt{3}}$ **7.** $\dfrac{\pi}{4}$

9. $\dfrac{\pi}{6}$ **11.** $\pi\left(\dfrac{90 - 52\sqrt{3}}{12 - 7\sqrt{3}}\right)$

15. π **17.** $\dfrac{\pi}{16}$

19. $\dfrac{3\pi}{8}$ **21.** $\dfrac{\pi}{2}$

23. $\dfrac{\pi}{\sqrt{2}}$ **25.** $\dfrac{\pi}{6}$

27. πe^{-1} **29.** πe^{-1}

31. πe^{-3} **33.** $\dfrac{\pi e^{-\sqrt{2}}}{2\sqrt{2}}\left(\cos\sqrt{2} + \sin\sqrt{2}\right)$

35. $-\dfrac{\pi}{8}\left(\dfrac{e^{-3}}{3} - e^{-1}\right)$ **37.** πe^{-1}

57. $\log_e 2$ **59.** $12\pi i$

61. $4\pi i$ **63.** $10\pi i$

69. $\dfrac{\pi^2}{8}$ **71.** $\dfrac{\sqrt{2}\pi^2}{16}$

Exercises 6.7, page 347

1. $\dfrac{1}{s-5}$, $s > 5$

3. $\dfrac{3}{s^2+9}$, $s > 0$

5. $s > k$

7. $\dfrac{k}{s^2-k^2}$, $\dfrac{s}{s^2-k^2}$

9. $\frac{1}{120}t^5$

11. $\frac{1}{2}\sin 2t$

13. $\dfrac{1}{\sqrt{3}}\sinh\sqrt{3}t$

15. $e^{3(t-a)}\,\mathcal{U}(t-a) - e^{2(t-a)}\,\mathcal{U}(t-a)$

17. $\frac{1}{2}\sinh t - \frac{1}{2}\sin t$

19. $\dfrac{1}{1-i\alpha}$

Chapter 6 Review Quiz, page 348

1. true

3. true

5. true

7. true

9. false

11. false

13. true

15. true

17. true

19. false

21. $-3 + 6i$

23. $\frac{125}{26} - \frac{25}{26}i$

25. $|z - 2 - i| = \frac{1}{13}$

27. $1/R$

29. $\frac{1}{5}(z+1) - \frac{1}{5^2}(z+1)^2 + \frac{1}{5^3}(z+1)^3 - \cdots$; $R = 5$

31. $\dfrac{7}{(z+1)^2} + \dfrac{4}{z+1} - 2 + (z+1)$

33. 1

35. $\frac{1}{6}$

37. 1, $\dfrac{1}{z-\pi}$, π

39. **(a)** $\pi + \pi i$

(b) 0 for $n = 0$; 0 for $n = 1$; $2\pi i(1/1!)$ for $n = 2$; 0 for $n = 3$; $2\pi i(-1/3!)$ for $n = 4$; 0 for $n = 5$; $2\pi i(1/5!)$ for $n = 6$; and so on.

Chapter 7

Exercises 7.1, page 357

1. f is not conformal at $z = \pm 1$.

3. f is not conformal at $z = (2n+1)\pi i$, $n = 0, \pm 1, \pm 2, \dots$.

5. f is not conformal at $z = \frac{1}{2}(2n+1)\pi$, $n = 0, \pm 1, \pm 2, \dots$.

11. $w = \cos\dfrac{\pi z}{2}$ by entry H-4 of Appendix III.

13. $w = \left(\dfrac{1+z}{1-z}\right)^{1/2}$ by entries H-5 and E-4 of Appendix III.

15. $w = \left(\dfrac{e^{\pi/z}+e^{-\pi/z}}{e^{\pi/z}-e^{-\pi/z}}\right)^{1/2}$ by entries H-6 and E-4 of Appendix III.

Exercises 7.2, page 368

1. $T(0) = \infty$, $T(1) = i$, $T(i) = 1$, $T(\infty) = 0$

3. $T(0) = -1$, $T(1) = i$, $T(i) = \infty$, $T(\infty) = 1$

5. $|w| \geq 1$ and $u \geq \frac{1}{2}$ 7. $u \leq 0$ and $|w - 1| \geq 2$

9. $v \geq 0$ and $\left|w - \frac{1}{2}\right| \geq \frac{1}{2}$ 11. $v \geq 0$ and $u \leq 1$

13. The image consists of a set of all points $w = u + iv$ such that $\left|w + \frac{1}{3}\right| \geq \frac{2}{3}$ and $v \leq 0$.

15. The image consists of a set of all points $w = u + iv$ such that $\left|w + \frac{1}{20}\right| \geq \frac{9}{20}$ and $u \geq -\frac{1}{2}$.

17. (a) $S^{-1}(z) = \dfrac{z + 1}{z - i}$ (b) $S^{-1}(T(z)) = \dfrac{(1 + i)z - 1}{2z + i}$

19. (a) $S^{-1}(z) = \dfrac{z - 2}{z - 1}$ (b) $S^{-1}(T(z)) = \dfrac{3}{z}$

21. $T(z) = \dfrac{2z + 2}{-z + 2}$ 23. $T(z) = \dfrac{2z}{z + i}$

25. $T(z) = \dfrac{3z - 3i}{(1 + 4i)z - (4 + i)}$

Exercises 7.3, page 376

1. the first quadrant $u \geq 0$, $v \geq 0$

3. the region bounded by the ray $u = 0$, $0 \leq v < \infty$, the line segment $v = 0$, $0 \leq u \leq 1$, and the ray $u = 1$, $-\infty < v \leq 0$, and containing the point $1 + i$

5. the region bounded by the ray $v = 1$, $-\infty < u \leq 0$, the line segment $u = 0$, $0 \leq v \leq 1$, the line segment $v = 0$, $0 \leq u \leq 1$, and the ray $\arg(z - 1) = \pi/4$, and containing the point $1 + i$

7. $f'(z) = A(z + 1)^{-1/2} z^{-1/2} (z - 1)^{-1/2}$ 9. $f'(z) = A(z + 1)^{-1/3} z^{-1/3}$

Exercises 7.4, page 384

1. $\phi(x, y) = \dfrac{1}{\pi}[\text{Arg}(z + 1) - 2\text{Arg}(z) + \text{Arg}(z - 1)]$

3. $\phi(x, y) = 5 + \dfrac{1}{\pi}[\text{Arg}(z + 2) - 2\text{Arg}(z + 1) + \text{Arg}(z) - 5\text{Arg}(z - 1)]$

5. $\phi(x, y) = \dfrac{2x - 1}{\pi}\left[\tan^{-1}\left(\dfrac{x}{y}\right) - \tan^{-1}\left(\dfrac{x - 2}{y}\right)\right] + \dfrac{y}{\pi}\log_e\left[\dfrac{(x - 2)^2 + y^2}{x^2 + y^2}\right]$

7. $\phi(x, y) = \dfrac{y}{\pi} + \dfrac{x^2 - y^2}{\pi}\left[\tan^{-1}\left(\dfrac{x - 1}{y}\right) - \tan^{-1}\left(\dfrac{x}{y}\right)\right]$

    $+ \dfrac{xy}{\pi}\log_e\left[\dfrac{(x - 1)^2 + y^2}{x^2 + y^2}\right]$

9. (b) $\phi(x, y) = e^{-y}\cos x$

Exercises 7.5, page 398

1. (a) $w = z^2$ (b) $\phi(x, y) = \dfrac{1}{\pi}\left[-\text{Arg}\left(z^2 + 1\right) - \text{Arg}\left(z^2\right) + 2\text{Arg}\left(z^2 - 1\right)\right]$

3. (a) $w = \left(\dfrac{1+z}{1-z}\right)^2$

(b) $\phi(x,\,y) = 1 + \dfrac{1}{\pi}\left\{2\text{Arg}\left[\left(\dfrac{1+z}{1-z}\right)^2 + 1\right] + \text{Arg}\left[\left(\dfrac{1+z}{1-z}\right)^2\right]\right.$

$\left. -2\text{Arg}\left[\left(\dfrac{1+z}{1-z}\right)^2 - 1\right]\right\}$

5. (a) $w = \sin\left(\dfrac{\pi}{4}z\right)$

(b) $\phi(x,\,y) = 1 + \dfrac{1}{\pi}\left\{-3\text{Arg}\left[\sin\left(\dfrac{\pi}{4}z\right) + 1\right] + 3\text{Arg}\left[\sin\left(\dfrac{\pi}{4}z\right)\right]\right.$

$\left. -\text{Arg}\left[\sin\left(\dfrac{\pi}{4}z\right) - 1\right]\right\}$

7. (a) $w = \dfrac{1}{z}$ **(b)** $\phi(x,y) = \dfrac{-2x}{x^2 + y^2} + 2$

9. (a) $w = \dfrac{2z - 1 - \sqrt{3}}{\left(4 + 2\sqrt{3}\right)z + 1 + \sqrt{3}}$

(b) $\phi(x,y) = \dfrac{10}{\log_e\left(7 - 4\sqrt{3}\right)}\log_e\left|\dfrac{2z - 1 - \sqrt{3}}{\left(4 + 2\sqrt{3}\right)z + 1 + \sqrt{3}}\right|$

11. (a) $w = \sin^{-1} z$ **(b)** $\phi(x,\,y) = 5 + \dfrac{10}{\pi}\text{Re}\left[\sin^{-1} z\right]$

13. $\Omega(z) = z^4$ **15.** $\Omega(z) = \cosh z$

17. (a) $z = \pi i - \frac{1}{2}\left[\text{Ln}(w + 1) + \text{Ln}(w - 1)\right]$

(b) $z(t) = \pi i - \frac{1}{2}\left[\text{Ln}(t + 1 + ic_2) + \text{Ln}(t - 1 + ic_2)\right]$

19. (a) $z = \dfrac{1}{\pi}\left[\left(w^2 - 1\right)^{1/2} + \cosh^{-1} w\right]$

(b) $z(t) = \dfrac{1}{\pi}\left\{\left[(t + ic_2)^2 - 1\right]^{1/2} + \cosh^{-1}(t + ic_2)\right\}$

21. $\Omega(z) = \text{Ln}\left(z^4 + 4\right) - \text{Ln}\left(z^4 - 16\right)$

Chapter 7 Review Quiz, page 403

1. false   **3.** false

5. true   **7.** true

9. false   **11.** true

13. true   **15.** true

17. sense   **19.** triples

21. circle   **23.** $\dfrac{z - z_1}{z - z_3}\dfrac{z_2 - z_3}{z_2 - z_1}$

25. $\pi/2,\ 3\pi/4$ **27.** $\pi/5$

29. $\left(\dfrac{1 + t + ic_2}{1 - t - ic_2}\right)^2$

Indexes

Symbol Index

Word Index

Unity:
   definition of, 4
   roots of, 24

V
Value, absolute, 9
Value of a complex function, 46
Vector, 9
Vector field:
   and analyticity, 257
   conservative, 150
   complex representation of, 120
   definition of, 120
   gradient, 150
   irrotational, 151, 256
   normalized, 121
   solenoidal, 257
   two-dimensional, 120
   velocity, 256
Velocity:
   complex, 261
   field, 122
   of a fluid, 122, 256
   potential, 151
Vertices of a polygonal region, 370
Vortex, 267

W
w-plane, 53

X
x-axis, 9

Y
y-axis, 9

Z
Zero(s):
   of an analytic function, 304
   of the complex number system, 4
   definition of, 304
   of hyperbolic functions, 188, 190, 192
   isolated, 305
   location of, 329-330
   of multiplicity n, 304
   number of, 251, 329
   of order n, 304
   of polynomial functions, 251
   simple, 304
   of trigonometric functions, 185
z-plane, 9